Hugo Campos • Oscar Ortiz
Editors

The Potato Crop

Its Agricultural, Nutritional and Social Contribution to Humankind

Editors
Hugo Campos
International Potato Center
Lima, Peru

Oscar Ortiz
International Potato Center
Lima, Peru

https://doi.org/10.1007/978-3-030-28683-5

This book is an open access publication.

Foreword

As the wise and old saying goes, "it takes a village to raise a child"; it also takes a large number of authors to develop a book like this one. Over sixty to be more precise, coincidentally close to the nearly 50 years of the International Potato Center (CIP), which has been working with potato and sweetpotato since 1971 and 1985, respectively. Along with maize, the potato is one of the most priceless and significant botanical presents among all the crops that the American continent has ever given to the entire world. And because of potato's plasticity and ability to be planted in the vast majority of geographies, latitudes, and altitudes occupied by mankind, it has become one of our main staple crops.

This crop has achieved not only a global agronomic and economic footprint, but also attained a significant social and artistic footprint, which is reflected in pre-Inca and Inca ceramics, "The Potato Eaters" painting of Vincent van Gogh, and the "Ode to Potato" composed by the Literature Nobel Prize winner Pablo Neruda.

Potato is one of the crops becoming more and more relevant to address food and nutrition security, and climate change challenges globally, which is reflected in its production in developing countries having already surpassed that from developed countries several years ago. Indeed, the two main potato producers on a global basis are China and India, respectively. Going forward, this trend sheds light on the main research and development drivers to be addressed by scientists in industrial, academic, and international development domains, as well as by policy makers.

No single volume could attempt to describe, even succinctly, the tremendous number of scientific discoveries in the vast world of potato research, nor does this book attempt this. Instead, its aim is to give readers a solid, current, and comprehensive understanding of the major contributions of diverse science disciplines related to this fascinating crop. This book does masterfully weave together very diverse domains of scientific prowess related to potato into one cohesive body of work. Furthermore, it not only addresses agronomic or biological aspects of potato, it also delves into pivotal social scientific aspects which affect not only its production and utilization, but also covers the adoption of new technologies by farmers. The fact that the ultimate adoption of agricultural innovations is, by and large, a behavioral change process warrants the blend of biological/agronomic aspects with social science-related ones in a book of this nature.

We commend Drs. Campos and Ortiz for organizing a talented group of top experts, not only from the International Potato Center, but equally importantly from many academic and private organizations who share the passion for this crop, in the diverse scientific and economic aspects addressed in this book. Furthermore, we welcome their zeal in making this updated and well-organized body of work widely available to readers as an open access book, therefore maximizing the readership and impact of the book. This should be particularly welcomed by producers, processors, scientists, graduate students, and policy makers based in developing countries who enjoy less opportunities to access updated, relevant, and comprehensive information than their peers from developed countries.

In closing, we invite you to read this book so you can understand why the editors and we share a strong passion and commitment to deliver the benefits of potato to as many people in the world as feasible. We promise you will not regret the reading journey ahead of you.

Barbara Wells
International Potato Center, Lima, Peru

Rodney Cooke
Rome, Italy

Preface

Why preparing a textbook providing the current status of several areas of knowledge and research in potato? Is there a need for such a textbook? The framework to address in a positive manner such valid questions can be articulated as follows:

The potato is the third largest crop in terms of human consumption after wheat and rice, and it stands out over many crops since it exhibits remarkable phenotypic plasticity and the ability to adapt to a wide array of production environments. Indeed, it can be successfully planted and able to sustain high productivities from the Peruvian coastline to highlands situated 5,000 m above the sea level in the Andes and in addition across vast differences in day length such as those existing between Ecuador and Finland. It is also cultivated in tropical regions, such as the African highlands, the volcanic mountains of West Africa, and Southeast Asia, and in subtropics, such as North India and Southern China. Moreover, potato is a main crop in temperate environments such as the European continent, Central Asia, and several regions of Canada and the United States.

The global pattern of potato production is rapidly changing and will remain doing so: whereas the main production of potato has been historically associated with developed countries, it has shifted toward developing countries, with a strong growth in production in Asia and Africa. Indeed, the current first and second largest producers of potato are China and India, respectively, and the production of potato in developing countries has already exceeded that of the developed world since the year 2005. Among the reasons explaining a lesser demand for fresh potatoes in developed countries, particularly in Europe, the replacement of potato as a feedstuff by cereals and the shift of diets toward low-calorie food stand out. Although European countries such as France, Denmark and Belgium have increased potato production over the last decade to serve the needs of the processing industry, such increase in potato production does not offset the global reduction of potato production in developed countries.

Unlike other staple crops such as wheat, corn, rice, and soybeans, which are commercialized on a global basis as commodities and therefore can be subject to a myriad of financial, commercial, and political factors which affect their price, availability, and relative scarcity, the global commerce of potato is negligible. The fact that potatoes are mostly produced and consumed locally and regionally is very relevant, particularly in developing countries, since it means that in relative terms, potato is therefore more resilient to price volatility at global scale and it can be used to smooth disruptions in global food supply and demand.

Finally, the potato represents not only a source of income to producers in developing countries, since it is an additional and important contributor in terms of food and nutrient security. Potato can contribute in supplementing diets with its vitamins, particularly vitamin C, mineral content, and high-quality protein. Current breeding efforts to develop so-called biofortified potato varieties able to deliver high contents of Fe and Zn will further increase its nutritive value to fight hidden hunger and diseases such as anemia.

Progress in crop science is happening in a very rapid path, and there is the need to look at the evolution of different areas of knowledge from those related to genes and genomes to that related to value chain, market development, and utilization. These different levels are increasingly interconnected and starting to converge, which justifies a potato book with this multilevel perspective.

While designing this book and its content, our main focus was to develop an updated reference for its users, built upon recent scientific progress. Because of the speed at which new knowledge is created, and former becomes obsolete, no book can claim to comprehensively address the current entire knowledge available in any given crop, and particularly in potato. We aimed to strike a balanced perspective across different topics and levels in the agri-food systems, and across the myriad geographies where potato is an important crop, the main aspects related to the crop, and also the way forward through which the potato can increase its contribution to humankind, both from an income perspective and also as a provider of well-being and food security to the millions of potato producers and consumers and their families. Though we are cognizant about, and acknowledge, the fact that the current main production of potato takes place in developing countries, the majority of the insight and information provided across the diverse chapters of this book pertains to both groups of countries. To achieve this, we were fortunate to assemble a selected, committed group of 61 authors who represent 25 countries and 23 research and/or academic organizations. Furthermore, 45% of the authors are not affiliated with CIP to provide a wide, balanced, and comprehensive perspective on the many aspects this book addresses.

The book is organized into four main themes. The first part, the global and dietary relevance of potato, provides updated perspectives on the value this crop represents globally, both from a production and nutrition/diet perspectives, respectively. It also covers the increasing role that potato represents in value chains, particularly in developing countries.

The second theme of the book provides a wide and updated perspective about the genetic improvement of potato and related fields such as genetics, cytogenetics, its genes and genomes, and the role of the ex situ conservation of its genetic resources.

The third theme of the book addresses several of the many agronomic aspects needed for an efficient and effective production of potato.

Finally, its fourth theme focuses on two aspects which are more relevant to developing than to developed countries, such as gender considerations and the role of participatory research approaches in potato.

We trust this book will encourage more researchers, particularly young ones, both males and females, to consider devoting their scientific pursuits to such a fascinating crop as the potato and to inform decision-makers about the increasing importance of the potato in the food security of developing countries. If such accomplishments take place, our work and that of the chapter contributors would be more than fully justified.

Lima, Peru

Hugo Campos
Oscar Ortiz

Acknowledgments

First of all, the editors would like to unreservedly thank all the coauthors who generously devoted part of their most priceless asset, time, to contribute several versions and revisions of the draft chapters that became the final chapters assembled in this volume. We are equally grateful to the organizations that they are affiliated with for allowing them to take on this writing task. Regardless, all remaining errors are solely our responsibility.

We would also like to sincerely thank the Springer team, namely, Joao Pildervasser, Rahul Sharma, Susan Westendorf, Lavanya Venkatesan, and Anthony Dunlap, for their expert and patient editorial and production support throughout this writing project, for gently keeping us in line with our declared deadlines, and for putting up with unplanned delays. Special thanks must also go to Jacco Flipsen from Springer for trusting us with this book.

Finally, we are very grateful to all the donors and supporting organizations for enabling the potato research carried out at CIP and globally. Much of this research was undertaken as part of the CGIAR Research Program on Roots, Tubers and Bananas (RTB) and supported by the CGIAR Trust Fund contributors: https://www.cgiar.org/funders/.

Hugo Campos would like to express his gratitude to many former mentors and colleagues, to his current colleagues for their generosity with their expertise and insight, and to CIP for the support and encouragement to finish this writing project. He also wishes to wholeheartedly thank his wife, Orietta, and their children, Noelia and Ignacio, for their constant love, support, understanding, encouragement, and for putting up with the arrival of yet another writing assignment. This book is dedicated to them.

Oscar Ortiz would like to thank the number of colleagues who have worked at CIP and its partners to develop potato knowledge, technologies, and innovations over more than 20 years of his work experience at CIP; and also a special gratitude to his wife, Rosa Amelia, and his daughters, Tamara and Mariana, for their support over the years during the long working hours and extensive traveling in missions to develop potato work globally for the benefit of resource-poor farmers and consumers.

Contents

Contributors

Birgit Adolf Technische Universität München, Freising, Germany

Andrei Alyokhin University of Maine, Orono, ME, USA

Walter R. Amoros International Potato Center, Lima, Peru

Jorge Andrade-Piedra International Potato Center, Lima, Peru

Christelle Andre The New Zealand Institute for Plant and Food Research Limited/ Luxembourg Institute of Science and Technology, Auckland, New Zealand

Noelle Anglin International Potato Center, Lima, Peru

Jeffery W. Bentley Independent Consultant, Cochabamba, Bolivia

Francisco Bittara Molina J. R. Simplot Company, Boise, Idaho, USA

Meredith W. Bonierbale International Potato Center, Lima, Peru
Duquesa Business Centre, Manilva, Malaga, Spain

Gabriela Burgos International Potato Center, Lima, Peru

Hugo Campos International Potato Center, Lima, Peru

Amy O. Charkowski Colorado State University, Fort Collins, CO, USA

Oswaldo Chavez International Potato Center, Lima, Peru

Walter de Jong Cornell University, Ithaca, NY, USA

André Devaux International Potato Center, Quito, Ecuador

David S. Douches Michigan State University, East Lansing, MI, USA

David Ellis International Potato Center, Lima, Peru

John Elphinstone Food and Environment Research Agency (Fera) Ltd., Sand Hutton, York, UK

A. R. Figueira Department of Plant Pathology, Lavras Federal University, Lavras, Minas Gerais, Brazil

Gregory A. Forbes Independent Consultant, Servas, Gard, France

Marcel Gatto International Potato Center, Hanoi, Vietnam

Elmar Schulte-Geldermann International Potato Center, Nairobi, Kenya
TH Bingen, University of Applied Sciences, Bingen, Germany

Marc Ghislain International Potato Center, Nairobi, Kenya

Jean-Pierre Goffart Walloon Agricultural Research Centre, Gembloux, Belgium

Rene Gomez International Potato Center, Lima, Peru

Guy Hareau International Potato Center, Lima, Peru

Hans Hausladen Technische Universität München, Freising, Germany

A. Jeevalatha ICAR-Central Potato Research Institute, Shimla, Himachal Pradesh, India

R. A. C. Jones Institute of Agriculture, University of Western Australia, Crawley, Perth, Australia

J. F. Kreuze International Potato Center, Lima, Peru

Peter Kromann International Potato Center, Nairobi, Kenya

Jürgen Kroschel Independent Consultant for International Agricultural Research & Development, Filderstadt, Germany

Stan Kubow McGill University, Montreal, Quebec, Canada

Alison Lees The James Hutton Institute, Invergowrie, Dundee, UK

Hannele Lindqvist-Kreuze International Potato Center, Lima, Peru

Elisa Mihovilovich Independent Consultant, Lima, Peru

Netsayi Noris Mudege International Potato Center, Nairobi, Kenia

Norma Mujica Independent Consultant, Lima, Peru

Diego Naziri International Potato Center, Hanoi, Vietnam, and Natural Resources Institute, University of Greenwich, Chatham, UK

Rebecca Nelson Cornell University, Ithaca, NY, USA

Julius Okello International Potato Center, Kampala, Uganda

Joshua Okonya International Potato Center, Kampala, Uganda

Miguel Ordinola International Potato Center, Lima, Peru

Oscar Ortiz International Potato Center, Lima, Peru

Rodomiro Ortiz Swedish University of Agricultural Sciences (SLU), Alnarp, Sweden

Monica L. Parker International Potato Center, Nairobi, Kenya

Willmer Perez International Potato Center, Lima, Peru

Athanasios Petsakos Formerly CIP, Sevilla, Spain

Vivian Polar CGIAR Research Program on Roots, Tubers and Bananas (RTB), Lima, Peru

Jaroslaw Przetakiewicz Plant Breeding and Acclimatization Institute, Radzikow, Blonie, Poland

Alberto Salas International Potato Center, Lima, Peru

Elisa Salas International Potato Center, Lima, Peru

Silvia Sarapura Escobar Royal Tropical Institute, Amsterdam, The Netherlands

Gary A. Secor Department of Plant Pathology, North Dakota State University, Fargo, ND, USA

Kalpana Sharma International Potato Center, Nairobi, Kenya

J. A. C. Souza-Dias Instituto Agronômico de Campinas (IAC)/APTA/SAA-SP, Campinas, São Paulo, Brazil

Victor Suarez International Potato Center, Lima, Peru

Graham Thiele CGIAR Research Program on Roots, Tubers and Bananas, Lima, Peru

J. P. T. Valkonen Department of Agricultural Sciences, University of Helsinki, Helsinki, Finland

Claudio Velasco International Potato Center, Quito, Ecuador

Thomas Zum Felde Genetics Genomics and Crop Improvement, International Potato Center, Lima, Peru

Part I
Food Security, Diets and Health

Chapter 1
Global Food Security, Contributions from Sustainable Potato Agri-Food Systems

André Devaux, Jean-Pierre Goffart, Athanasios Petsakos, Peter Kromann, Marcel Gatto, Julius Okello, Victor Suarez, and Guy Hareau

Abstract In the coming decades, feeding the expanded global population nutritiously and sustainably will require substantial improvements to the global food system worldwide. The main challenge will be to produce more food with the same or fewer resources. Food security has four dimensions: food availability, food access, food use and quality, and food stability. Among several other food sources, the potato crop is one that can help match all these requirements worldwide due to its highly diverse distribution pattern, and its current cultivation and demand, particularly in developing countries with high levels of poverty, hunger, and malnutrition. After an overview of the current situation of global hunger, food security, and agricultural growth, followed by a review of the importance of the potato in the

A. Devaux (✉)
International Potato Center, Quito, Ecuador
e-mail: a.devaux@cgiar.org

J.-P. Goffart
Walloon Agricultural Research Center, Gembloux, Belgium
e-mail: j.goffart@cra.wallonie.be

A. Petsakos
Formerly CIP, Seville, Spain

P. Kromann
International Potato Center, Nairobi, Kenya
e-mail: p.kromann@cgiar.org

M. Gatto
International Potato Center, Hanoi, Vietnam
e-mail: m.gatto@cgiar.org

J. Okello
International Potato Center, Kampala, Uganda
e-mail: j.okello@cgiar.org

V. Suarez · G. Hareau
International Potato Center, Lima, Peru
e-mail: v.suarez@cgiar.org; g.hareau@cgiar.org

H. Campos, O. Ortiz (eds.), *The Potato Crop*,
https://doi.org/10.1007/978-3-030-28683-5_1

current global food system and its role played as a food security crop, this chapter analyzes and discusses how potato research and innovation can contribute to sustainable agri-food systems with reference to food security indicators. It concludes with a discussion about the challenges for sustainable potato cropping considering the needs to increase productivity in developing countries while promoting better resource management and optimization.

1.1 Introduction: The Current Situation of Global Hunger, Food Security, and Agricultural Growth

A growing earth population and the increasing demand for food is placing unprecedented pressure on agriculture and natural resources. Today's food systems do not provide sufficient nutritious food in an environmentally sustainable way to the world's population (Wu et al. 2018). Around 821 million are undernourished while 1.2 billion are overweight or obese. At the same time, food production, processing, and waste are putting unsustainable pressure on environmental resources. By 2050, a global population of 9.7 billion people will demand 70% more food than is consumed today (FAO et al. 2018). Feeding this expanded population nutritiously and sustainably will require substantial improvements to the global food system—one that provides livelihoods for farmers as well as nutritious products to consumers while minimizing today's environmental footprint (Foley et al. 2011). A critical challenge is to produce more food with the same or fewer resources.

According to the Global Hunger Index (GHI), substantial progress has been made in terms of hunger reduction for the developing world (Von Grebmer et al. 2017). The GHI ranks countries on a 100-point scale with 0 being the best score (no hunger) and 100 being the worst. Whereas the 2000 GHI score for the developing world was 29.9, the 2017 GHI score is 21.8, showing a reduction of 27%. Yet, there are great disparities in hunger at the regional, national, and subnational levels, and progress has been uneven.

Sub-Saharan Africa (SSA) and South Asia (SA) have the highest 2017 GHI scores, at 29.4 and 30.9, respectively. These scores are still on the upper end of the serious category (20.0–34.9), and closer to the alarming category (35.0–49.9) than to the moderate one (10.0–19.9). These data show that persistent and widespread hunger and malnutrition remain a huge challenge in these two regions. In other parts of the developing world within the low range, are also countries with serious or alarming GHI scores, including Tajikistan in Central Asia (CA); Guatemala and Haiti in Latin America and the Caribbean (LAC); and Iraq and Yemen in the Near East and North Africa (NENA) regions. Black et al. (2013) estimate that undernutrition causes almost half of all child deaths globally.

The current rate of progress in food supply will not be enough to eradicate hunger by 2030, and not even by 2050. Despite years of progress, food security is still a serious threat. Conflicts, migration, and climate change are hitting the poorest people the hardest and effectively maintaining parts of the world in continuous crisis.

The 2017 GHI report emphasizes that hunger and inequality are inextricably linked. Most closely tied to hunger, perhaps, is poverty, the clearest manifestation of societal inequality. Both are rooted in uneven power relations that often are perpetuated and exacerbated by laws, policies, attitudes, and practices.

According to FAO (2002) "*Food security exists when all people, at all times, have physical, social and economic access to sufficient, safe and nutritious food to meet their dietary needs and food preferences for an active and healthy life.*" Food security has four key dimensions: (a) food availability, (b) food access, (c) food quality and use, and (d) food stability.

Food availability refers to the supply of food at the national or regional level which ultimately determines the price of food. Improved availability of food is necessary to reduce food insecurity and hunger but is insufficient to completely end malnutrition, particularly because access to other services such as potable water, sanitation, and health services is also required.

Food access refers to the ability to produce one's own food or buy it, which implies having the purchasing power to do so. Given that a large portion of the poor worldwide are farmers, there remains considerable attention to promoting agriculture to enhance food access. The emphasis on an agricultural pathway to increase food access is twofold, since increased agricultural production provides income to purchase food as well as direct access to food for consumption obtained from own production.

Food use and quality refers to the level of nutrition obtained through food consumption from a nutritional, sanitary, sensory, and sociocultural point of view.

Food stability incorporates the idea of having food access at all times thus incorporating issues such as price stability and securing incomes for vulnerable populations (FAO 2006a).

This widely accepted FAO definition reinforces the multidimensional nature of food security that requires multisector approaches. Such approaches should combine the promotion of broad-based agricultural growth and rural development with programs that directly target the food insecurity as well as social protection programs focused on nutrition including a gender approach (Salazar et al. 2016). Agricultural growth results in rural development and prosperity through a series of multiplier effects, that is, through backward and forward linkages, due to increased incomes. These effects typically stimulate enhanced investment in both farm and non-farm sectors (Hazell and Haggblade 1989; Pandey 2015). Growth in rural farm sector increases demand for goods and services produced by the non-farm sector, further increasing purchasing power and effective demand, thus deepening growth in non-farm sector. Further, Haggblade et al. (2007) argue that the increased income earned in rural non-farm sector can kick off a series of reverse linkages in which such income is invested in agriculture to further strengthen its growth and improve livelihoods of farm households.

During the 2014 World Economic Forum, in a debate on "Rethinking Global Food Security," Shenggen Fan, Director of the IFPRI, argued that tackling hunger and malnutrition is not only a moral issue but also one that makes economic sense. The world loses 2–3% Gross Domestic Product (GDP) per year because of hunger, while investing US$1 in tackling hunger yields a return of US$30. Ajay Vir Jakhar, Chairman of Bharat Krishak Samaj (Farmers' Forum) in India, added that farmers do not think in terms of food security at the global level, but in their own households. While policy makers tend to think in terms of global and national issues and solutions, localized solutions and help from the public and private sectors are also needed to support the bulk of farmers who are farming small plots of land and which have a critical role as engines of food productivity growth and social development. By declaring 2014 the International Year of Family Farming, the United Nations acknowledged the importance of family farming in reducing poverty and improving global food security. Localized, technical, and commercial solutions with the support of both public and private sectors are needed in combination with global food security policies.

Therefore, enhancing food security requires policies that improve households' ability to obtain food through production and better income. Growth in agricultural productivity is key to reducing rural poverty since most of the poor depend on agriculture and related activities for their livelihoods. Because the potato is one of the global crops with a most diverse distribution pattern (Haverkort et al. 2014) and is grown in areas with high levels of poverty, hunger, and malnutrition, it can be particularly effective crop for enabling smallholder families to attain food security and climb out of poverty. Hence, innovations based on potato science can be a significant vehicle for targeting the poor and hungry as part of a broader set of research and development activities.

This chapter first presents the importance of the potato in the current global food system and its value as a food security crop. It then discusses the role of agriculture and the potato for their contribution to food security in its different dimensions: analyzing opportunities and challenges on how potato research and innovation can enhance productivity and how potato agri-food systems can contribute to food security at a global scale using natural resources in a sustainable way. A list of key research and technology options that can contribute to sustainable agri-food systems intensification approaches is suggested. The chapter concludes with a discussion about the challenges for sustainable potato cropping combining the needs to increase productivity in developing countries while promoting better input management and optimization. These conclusions emphasize also the need to integrate better agriculture sustainable intensification and food security indicators.

1.2 The Potato in the Global Food System

Potato is currently grown on an estimated 19 million hectares of farmland globally, and the potato production worldwide stands at 378 million tons (Table 1.1). The highest concentrations are found in the temperate zone of the northern hemisphere where the crop is grown in summer during the frost-free period. In these regions, potato is mainly grown as a cash crop and is therefore an important source of income. In tropical regions, the crop is significant in the highlands of the Andes, the

Table 1.1 Potato production indicators

Region	2014–2016			Average annual growth rate								
	Production (000 tons)	Area (000 ha)	Yield (t/ha)	Production (%)	Area (%)	Yield (%)	Production (%)	Area (%)	Yield (%)	Production (%)	Area (%)	Yield (%)
				1961–1963 vs. 1988–1990			1988–1990 vs. 2014–2016			1961–1963 vs. 2014–2016		
Africa	25,270	1756	14.4	4.9	3.7	1.1	4.5	3.2	1.1	4.7	3.6	1.1
Asia[a]	190,617	9975	19.1	3.7	2.4	1.3	4.3	2.8	1.4	4.0	2.6	1.3
Europe	119,551	5547	21.6	−1.0	−1.8	0.8	−1.3	−2.4	1.3	−1.1	−2.2	1.1
North America	24,430	763	32.0	1.2	−0.2	1.4	0.8	0.6	0.1	1.0	0.2	0.8
Latin America and Caribbean	18,334	1023	17.9	2.2	0.0	2.2	1.5	0.1	1.4	1.8	0.0	1.8
World	378,202	19,063	19.8	0.1	−0.8	0.9	1.3	0.2	1.1	0.7	−0.3	1.0

Source: FAOSTAT (2017), accessed Oct 2018

n.a. = not available, a = U.S. Department of Agriculture, 2016

[a]Asia + Oceania

African highlands, and the Rift valley, and the volcanic mountains of West Africa and Southeast Asia, where production is both for food and cash (Muthoni et al. 2010). In the subtropics, the crop is grown as a winter crop during the heat-free period such as in the Mediterranean region, North India, and southern China. It is only in the tropical lowlands that potato is not a main staple, largely because the temperatures in these areas are too high for tuber development and growth in traditional potato cultivars (Haverkort et al. 2014). Figure 1.1 illustrates the current pattern of the potato distribution worldwide (You et al. 2014; FAO 2016).

Potato is now the world's third most important food crop in terms of human consumption, after wheat and rice (FAOSTAT 2013) despite the large proportion of potato produce used for seed and as animal feed (Fig. 1.2). Consumption of fresh potatoes accounts for approximately two-thirds of the harvest, and around 1.3 billion people eat potatoes as a staple food (more than 50 kg per person per year) including regions of India and China.

1.3 Potato Production and Demand Trends by Region

Across global landscapes, the versatility of the potato crop coupled with notable increases in production in many countries over the last two decades is unparalleled, although this increase has been mainly driven by area expansion and secondarily by yield improvements. Global statistics also indicate that potato production is shifting towards developing countries especially with strong increase in production in Asia and Africa, especially in East Africa (Fig. 1.3). In fact, the developing world's potato production exceeded that of the developed world for the first time in 2005

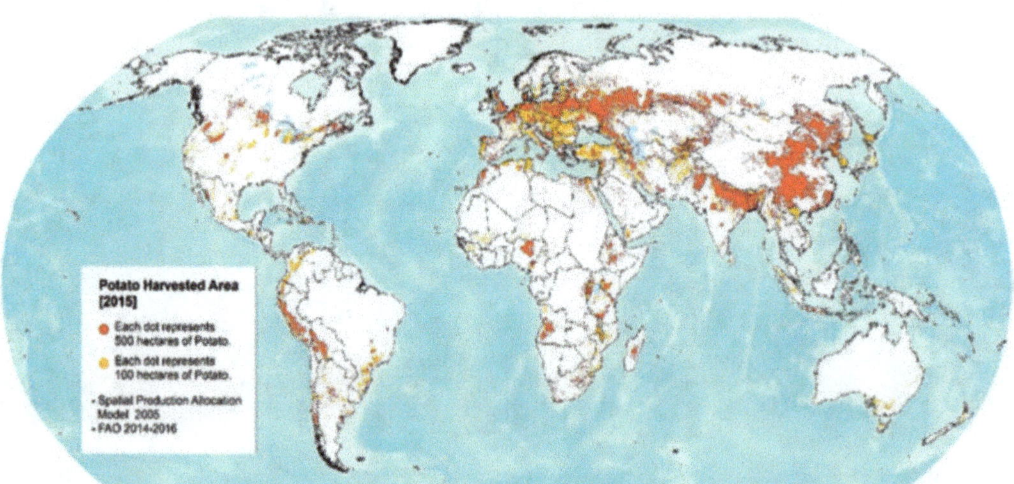

Fig. 1.1 Potato distribution worldwide, harvested area (You et al. 2014; FAOSTAT 2016)

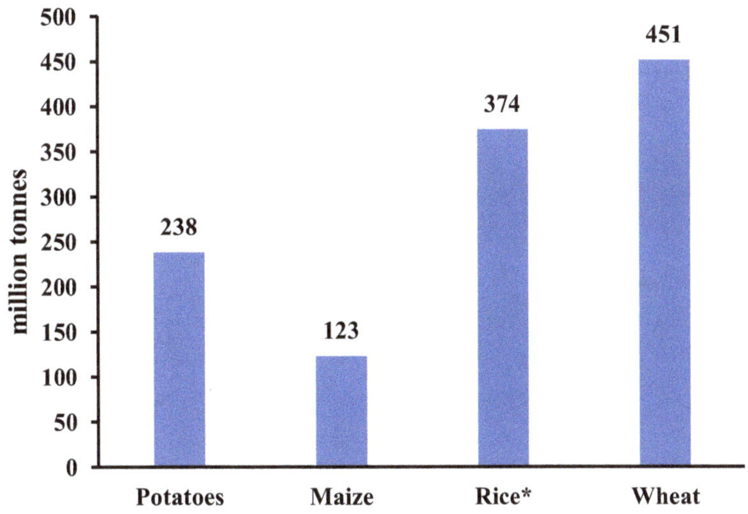

Fig. 1.2 World food consumption per year for the four main nutritious crops. (Source: FAOSTAT 2013)

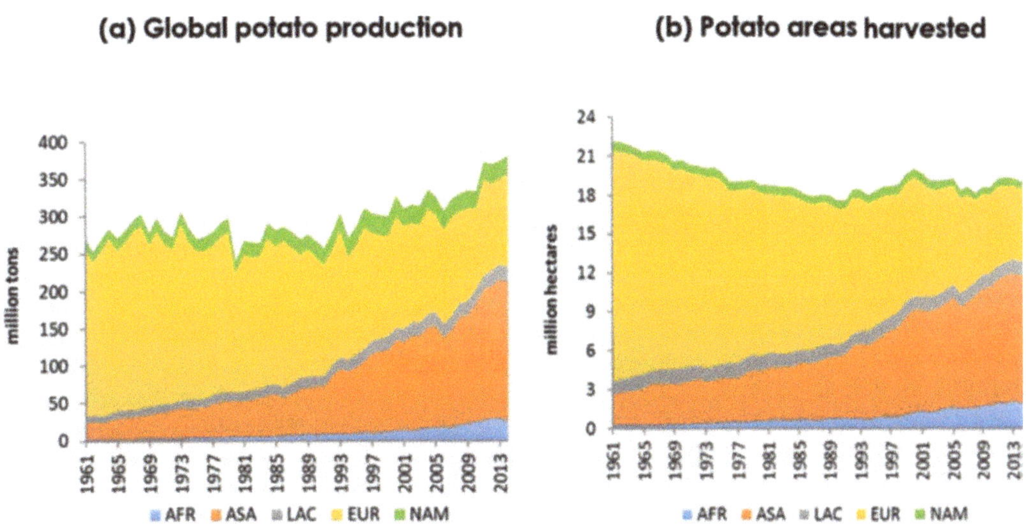

Fig. 1.3 Potato supply (**a**) and (**b**) 1961–2013. *AFR* Africa, *ASA* Asia, *LAC* Latin America and Caribbean, *EUR* Europe, *NAM* North America. (Source: FAOSTAT 2017, accessed Nov 2017). (**a**) Global potato production, (**b**) potato areas harvested

(FAO 2014). It reaffirms the increasing importance of potatoes as a source of food, employment, and income in Asia, Africa, and Latin America.

As shown in Fig. 1.4a, Africa has registered large increases of harvested area over the last 20 years, but despite the impressive growth, total production and harvested areas are still much smaller compared to Europe and Asia (Fig. 1.3a). In Africa, the increase in potato production has largely been through increase of area under production, which more than doubled since 1994 and now exceeds that of the Latin

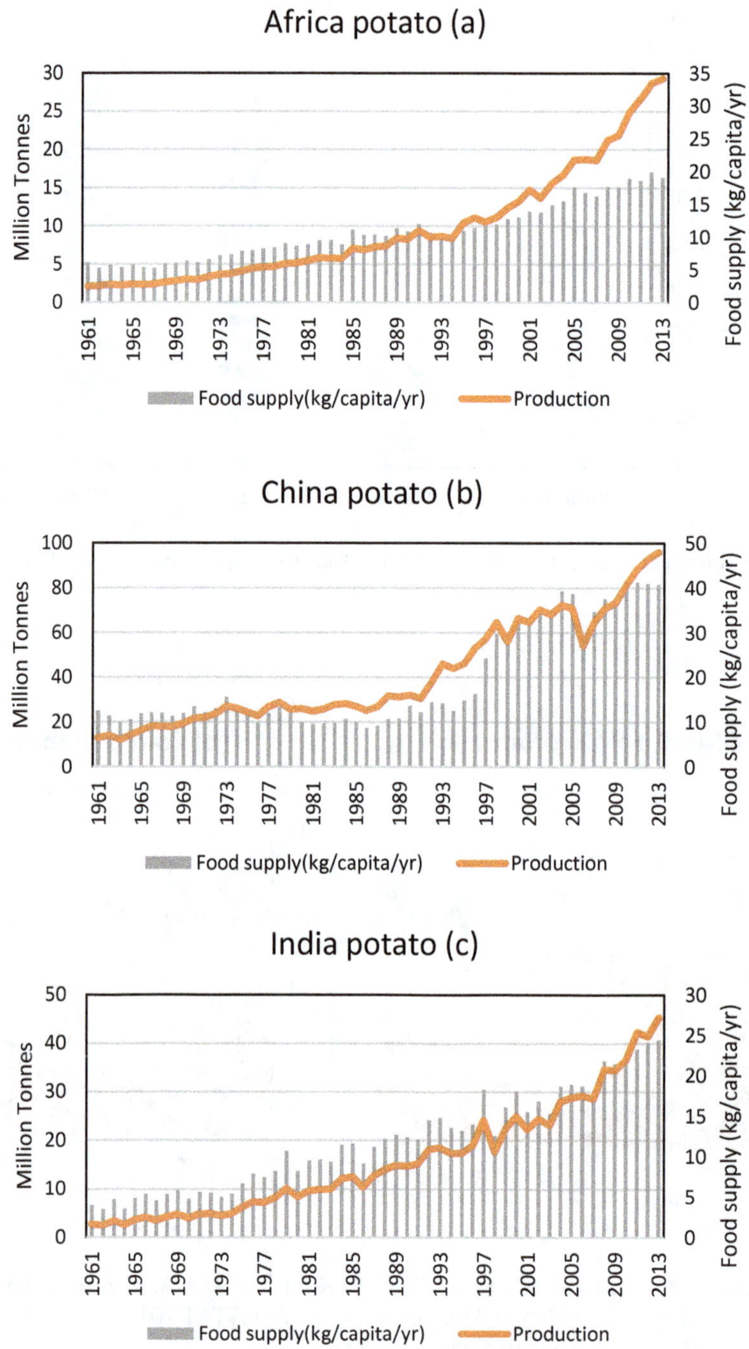

Fig. 1.4 Relative development of potato production and food supply (kg capita^{-1} year^{-1}) in Africa (**a**) China (**b**), and India (**c**). FAOSTAT (2013)

America and Caribbean (LAC) region. In the tropical highlands of East Africa, farmers grow potato both for food and cash (Muthoni et al. 2010). The increase of potato production in East African countries over the last years has been impressive, suggesting a higher contribution of the crop to local food systems. In Tanzania, for instance, potato supply has almost tripled between 2000 and 2014 (FAOSTAT 2017), while in Rwanda potato is included in the national priority list of crops due to its role in national food security (approximately 125 kg per capita consummed per year; FAO 2009). As world population levels are predicted to show the greatest rise in Africa in the coming decades, increased contribution of potato to local food systems in this region is of considerable importance (Birch et al. 2012).

In Asia, China and India have experienced nearly a half century of steady growth in potato production (Fig. 1.4b, c). Both countries also have ambitious growth targets for future years. For some decades now, the Chinese state has been working to increase national potato consumption, also launching a campaign since 2014 to promote both the cultivation and the consumption of this tuber (The Wall Street Journal 2015). China became the world's largest potato producer in 1993 and currently accounts for almost one quarter of global potato production and about 28% of total cultivated areas (FAO 2015a, b). Potato in China is mainly used for food, both as a vegetable and in processed forms, while a smaller part is also consumed as animal feed (Scott and Suarez 2012a). Potato in India is mainly grown in the Indo-Gangetic plain, either as monoculture or in rotation with maize, wheat, and/or rice and it is regarded as both an important staple and a cash crop. Following the growth in production volumes, potato yields in India have also increased significantly, at an average of 2% per year, because of successful breeding programs, quality seed systems, and storage infrastructure that have reduced post-harvest losses (Scott and Suarez 2011).

Looking at other Asian countries, potato is the principal vegetable in Bangladesh and the second most important crop behind rice. Its cultivation is widely distributed across the country where it is grown mainly as a cash crop (Scott and Suarez 2012b). Potato production in Bangladesh has greatly expanded during the last decades, especially after 2000 when output surged from about 1.5 million tons to more than 8 million tons in 2013 (FAO 2015a, b). This impressive growth, besides the rising domestic demand because of population growth and the "westernization" of dietary preferences in urban areas (Pingali 2006), can also be attributed to the introduction of several improved high-yielding varieties and the development of cold storage facilities which facilitated near year-round availability of potato. At the same time, producers also gained significant price advantages (Reardon et al. 2012). In Nepal, potato is the second most important staple food crop after rice. The potato has also become a significant source of rural income in Pakistan where production is concentrated in Punjab, with spring and autumn crops accounting for 85% of the harvest. Expansion of irrigated Pakistani land has resulted in substantial increases in potato production (up 254% from 1990 to 2009) and area under cultivation.

Regarding potato production in LAC over the past 60 years, the annual average potato domestic supply has increased from 7.2 million tons in the 1961–1963 period to 19.6 million tons in 2011–2013, which represents an average annual growth rate of 2%. By way of comparison, growth rates for potato production in ASA and AFR averaged over 4% for a similar period, i.e. more than double those of LAC (Scott 2011). Most of the production is oriented towards human consumption (74%, maintaining this trend throughout the period) and it highlights a relatively low processing level of 1% (FAOSTAT 2017).

The role that potato plays in the diets in LAC vary—from basic staple, producer/consumers in the Andean highlands to complementary vegetable for urban households in most of South America, to a relatively expensive complimentary vegetable in much of Central America and the Caribbean, and to a popular fast food in the form of French fries in urban markets throughout the region (Scott 2011). Per capita consumption of potatoes in Latin America increased slightly from 22 kg/person on average between 1961 and 1963 to 25 kg/person between 2011 and 2013. But these regional trends do not reflect the important differences in trends at the subregional and country levels. Peru, is one of the countries where potato consumption has grown significantly, reaching in 2015 a figure of 85 kg/person. This is due to various public–private policies, rural infrastructure, expansion of supermarket trade focused on potatoes and a strong relationship with the gastronomy sector promoting Andean food including the native potato and its products. Brazil and Mexico have increased their consumption, although their absolute values, 18.5 and 14.8 kg/ person respectively still remain low compared to other countries in LAC. The cases of Argentina and Colombia are showing a downward trend.

The United States is the fifth largest potato producer in the world with more than 420,000 ha harvested in 2013 and a total output of almost 20 million tons (FAO 2015a, b). Although in the United States potato is no longer the traditional staple of the past, it is nevertheless gaining increased appreciation by nutritionists because of its nutrient density and its contribution to a more balanced diet (Bohl and Johnson 2010). There is also a large demand by the processing industry for producing commodities like frozen French fries and chips for both the local and foreign markets. Potato yields in the United States have more than doubled over the last 50 years, rising from 22 tons ha^{-1} in 1961 to 49 tons ha^{-1} in 2016. This increase in yields has been suggested to be primarily driven by improvements in management rather than genetic improvements, since most breeding programs have traditionally focused on quality traits such as dry matter content and storage longevity to meet the demands of the processing industry and the consumer (Douches et al. 1996).

In Europe, Germany, France, Netherlands, the United Kingdom, and Belgium are together the strongest potato producers in the European Union (EU), due to potato yields higher than 40 tons ha^{-1} in this area of northwestern Europe (Fig. 1.5) and to the strong links of production with the dynamic European potato processing industry. Potato is also prevalent in Eastern European countries, particularly in Russia, Ukraine, and Poland where per capita consumption has traditionally exceeded 100 kg annually. Although Eastern Europe constitutes the region with the highest use of potato as animal feed globally, feed use of potato has been steadily declining over the last 20 years and being replaced by cereals, most notably in Poland. This decline in feed use, together with the shift of diets towards low-calorie food and a trend to spend less time on cooking observed in Western European countries, has led to a significant decrease in demand for fresh potatoes, and therefore potato production in the continent is falling (European Commission 2007). Despite the aforementioned decline, some European countries like France, Denmark, and Belgium have increased production over the last decade, due to growth of the processing industry (French fries, crisps) and starch production (Eurostat 2017).

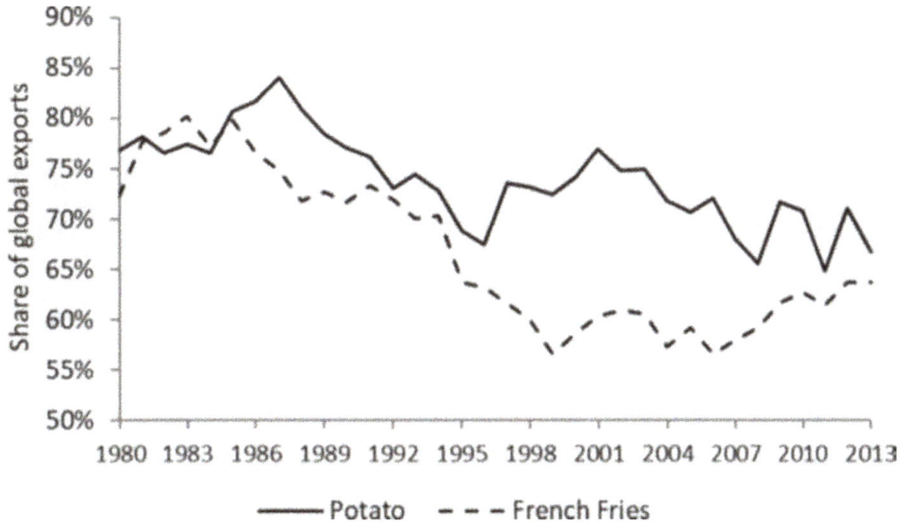

Fig. 1.5 European share of global exports for potato (fresh and seed) and French fries 1980–2013. (Source: FAOSTAT 2017)

Table 1.2 Potato utilization, consumption and trade indicators

	Utilization				Consumption	Trade	
Region	Food (%)	Feed (%)	Seed (%)	Other (%)	Per capita (kg year^{-1}) 2011–2013	Import Quantity (000 tons) 2011–2013	Export Quantity (000 tons)
Africa	69	4	8	19	19.2	967	830
Asia[a]	67	12	5	11	29.1	6,042	2,683
Europe	52	19	15	11	83.4	17,890	19,477
North America	84	1	6	9	55.3	3,615	5,354
Latin America and Caribbean	73	2	8	16	23.8	2,040	439
World	64	12	9	12	34.4	30,554	28,784

Source: FAOSTAT (2017), accessed Oct 2018
n.a. = not available, a = U.S. Department of Agriculture, 2016
[a]Asia + Oceania

Moreover, the competitiveness of the potato industry has established Europe as the world's biggest net potato exporter, amounting for more than 60% of all exports of fresh potato and a similar percentage of global exports of French fries (Fig. 1.5). These statistics concern mainly intra-EU trade, and also export of seed potato to non-EU countries, primarily to Mediterranean African countries like Egypt and Algeria (FAOSTAT 2017).

Tables 1.1 and 1.2 give a synthesis of the potato indicators by region confirming the expansion of potato in Asia, which is now the major potato producer continent. In Africa, the potato growth rate has also been strong with Egypt, Malawi, South Africa, Algeria, and Morocco producing more than two-thirds of all the potatoes in the region. In many countries of Latin America and the Caribbean, potato areas have

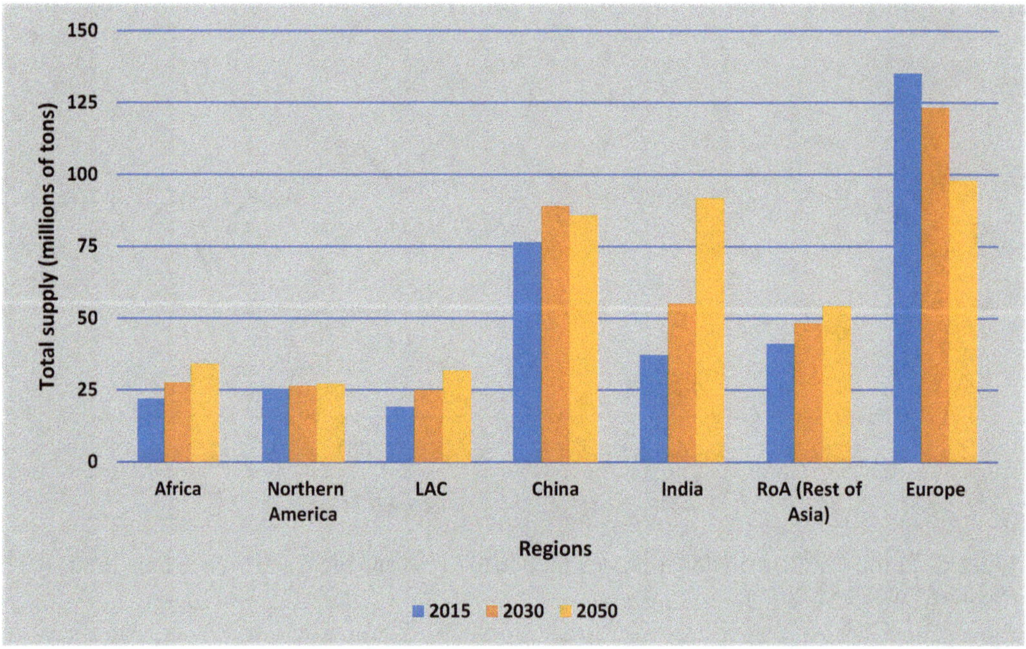

Fig. 1.6 The future of potato production. Adapted from Rosegrant et al. (2017)

declined although output has risen due to improvements in productivity. Potato production growth rate in Europe and North America has declined due to a significant decrease in demand for fresh potatoes that was partially compensated by a large demand from the processing industry as described before. In North America, yields increased rapidly between 1961 and 1990 and then somehow stagnated, suggesting yields are near their potential in the region and there may exist genetic limits. In Europe, on the contrary, the relative larger yields growth rates occurred after 1990.

Future trends by region indicate a major production increase in Asia and Africa as compared to other regions (Fig. 1.6). Considering some underlying assumptions such as population growth, climate change, and economic growth pathways, the UN projects a population decline in China and growth of per capita GDP which will affect diet composition. Therefore, the future supply of potato in China will not continue to grow faster than in the past. According to Rosegrant et al. (2017), it is in India where potato supply will almost triple because of the very high population growth, especially under certain socioeconomic scenarios.

1.4 The Potato Remains a Food Security Crop in the Developing World's "Nutrition Transition"

In many developing countries, and especially in urban areas, the globalization, the emergence of fast food outlets and supermarkets and the rising levels of income are driving a "nutrition transition" led by major shifts in the availability, affordability, and acceptability of different types of food, especially toward more energy-dense foods and prepared food products. It is translated into major and rapid shifts in dietary patterns. As an example of this nutrition transition process, demand for potato is increasing in many developing countries in Africa and Asia (Fig. 1.3). In South Africa, potato consumption has been growing in urban areas as part of the staple food consumption of the middle class, although maize remains the primary staple in rural areas. In China, higher income and increased urbanization have led to increased demand for processed potatoes. In this context, potato plays a role in diet diversification in many countries where family agriculture and smallholders continue to supply local markets with fresh and affordable agricultural produce. Potato is still an important staple in rural food systems where it is produced, and emerging food systems which are urbanized but where consumers still rely on staples such as potatoes. In Industrial food systems, highly urbanized in Northern countries in Europe and North America with the development of the processing industry there is a lower dependence on traditional staples (Gillespie and Van den Bold 2017).

As mentioned by Haverkort and Struik (2015) potato used to be a "local for local" crop and it still is in many countries because of the bulkiness and the limited storability of the seed and ware tubers. Compared to other staples, and except for processed potato products, fresh potato is a thinly traded commodity in global markets and is absent in major international commodity exchanges. It is therefore subject to less price volatility at global scale. Thus, potato can be relied upon to smooth the disruptions in global food supply and demand that have an impact on other commodity prices, such as witnessed during the 2007–2008 and subsequent food price spikes (FAO 2009).

Potato also improves food security because it is a source of employment and income, both of which have direct links to household food access and nutrition (Pinstrup-Andersen 2014; Kanter et al. 2015). The comparatively short maturity period, nutritious characteristics, employment, and income opportunity that characterize potato make it a resilient crop that can secure vulnerable livelihoods under the effects of climate change and changing market environments. Moreover, potato yields more food more rapidly on less land than any other major crop.

Where other staple crops are available to meet energy requirements, potato should not replace them but rather supplement the diet with its vitamins and mineral content and high-quality protein. Potato can be promoted as a healthy and versatile component of a nutritious and balanced diet including other vegetables and whole grain foods. Likewise, it contributes to combat micronutrient deficiency, also referred to as hidden hunger, which is a major global public health problem, affecting an estimated two billion people globally (Bailey et al. 2015). Potato contains interesting amount of health promoting components such as vitamin C, phenolic compounds, and iron and has protein content comparable to that of cereal grains (Burlingame et al. 2009). When eaten with its skin, a single medium-sized potato of 150 g provides nearly half the daily adult requirement (100 mg) of vitamin C. It is also a good source of vitamins B_1, B_3, and B_6; minerals such as iron, potassium, phosphorus, and magnesium; and contains folate, pantothenic acid, and riboflavin. Vitamin B deficiency has both short- and long-term impacts, including poor cognitive and pregnancy outcomes and poor child development and life outcomes. Vitamin C, on the other hand, is important for body metabolism and iron absorption (FAO 2009).

As an example, potato plays an important role in the food basket of most Peruvians; it continues to make up a relatively high proportion of daily calorie availability, reflecting its importance as traditional source of energy. To enhance the nutritional contribution of the potato in rural Andean highlands, the International Potato Center (CIP) coordinated activities to promote innovation in potato-based systems with the objective of contributing to food and nutrition security of rural highland populations enhancing native potato production and promoting dietary diversity. In rural areas of Peruvian highlands, there is a high prevalence of chronic malnutrition amongst children (42% in children under 2 years according to INEI Peru 2017). Native potato varieties are commonly grown in the Andes and are a significant part of the local diets. Some of them have higher contents of micronutrients (Zn and Fe) and are rich in antioxidants compared to commercial improved varieties. Creed-Kanashiro et al. (2015) explored the relationships between agricultural production characteristics and nutritional status of young children of families in rural areas of Peru whose livelihood is based on potato production systems combining native and commercial varieties. The results showed a positive relationship between percentage of recommended dietary intake (RDI) for both Fe and Zn intakes by children and production of native potatoes for home consumption, raising small animals for consumption and sale (e.g., guinea pigs and poultry) and the area of production of commercial potatoes that allowed the families to improve their incomes and diversifying the family diet.

Finally, it should be mentioned that potatoes have been related to increased risks of obesity mainly because of their high glycemic index. Recent reviews of clinical intervention and observational studies centered on the potato concluded that these studies did not provide convincing evidence to suggest an association between intake of potato and risks of obesity, Type II diabetes (T2D), or cardiovascular disease (CVD) (Borch et al. 2016). However, as part of the trend towards urbanization and associated lifestyles, raising incomes, and greater consumption of "convenience foods", demand for fried potatoes is increasing. Overconsumption of these high-energy products, along with reduced physical activity, can lead to overweight and obesity. Therefore, the role of fried potato products in the diet must be taken into consideration in efforts to prevent overweight and diet related noncommunicable diseases, including heart disease and diabetes. For these reasons, even though the potato is a nutritious staple crop, it has often been associated to a meal component with no specific attributes in many societies, even in developing countries. This image is reinforced by some negative myths such that potato is fattening, requires intense use of fertilizers and pesticides for its cultivation and can contribute to soil erosion. Negative images can be mitigated with better information about the nutritional value of the potato and its importance in the diet, while promoting sustainable and environmentally friendly production systems, thus addressing the question of natural resource management.

1.5 Policies and Strategies for the Development of the Potato as a Food Security Crop

The role potato can play as a food security crop at national scale has been addressed in some developing countries with different policies, either sectoral and crop-specific or at the macro level. In its quest to improve food security for a rising population, the Government of China is developing a national plan to increase production and consumption of potato and promoting the crop as a staple instead of a vegetable. This status can give access to important complementary policies and resources at national and regional level and to subsidies from the central government. It also recognizes the double role of potato in current China. Potato is still a major staple for poor rural areas where local governments continue to provide subsidized inputs (e.g., clean seeds of selected varieties), while at the same time being at the forefront of an increasing private sector-led processing industry, accompanying rising incomes in urban populations and diversification of diets (Scott and Suarez 2012a, c).

In Peru, the major center of origin of the potato, a large effort began in early 2000s to develop a competitive and inclusive native potato value chain for domestic markets. Initially led by the International Potato Center, the initiative gathered several private and not-for-profit actors to add value to the native potato grown by small farmers while developing a niche market. Several new products were developed in the process, for example selected native potato varieties for fresh consumption sold as gourmet potatoes in innovative packages in large supermarket chains, snacks such as colored native potato chips, and culinary innovations in the gastronomy sector featuring native potato as a central component of sophisticated dishes. The innovations in the value chain continued and a second round of new products emerged, including for example frozen native potato fries, native potato-based liquors, and even cosmetics made from potato. Although no specific sectoral policies were behind this initiative, the development of the native potato value chain took advantage of Government policies at the macro level promoting private sector and market-led developments and the fast growth of Peru's economy and of the purchasing power of the population since the beginning of the twenty-first century. While the Government of Peru focused on public investments to promote export-oriented agricultural growth, the experience with the native potato value chain has proven successful to link small potato farmers to domestic markets and to develop a more inclusive growth strategy of the highly diverse agricultural sector of the country (World Bank 2017).

A final example of policies that have been adopted to promote the potato sector in developing countries is through seed laws and regulations. Seeds are an important input of production of the potato crop and can affect yields since they are vehicle for important diseases. Seed degeneration due to viruses is one of the most common constraints affecting potato productivity, and therefore large research and extension efforts are made to improve availability and access to quality seeds for small farmers. One particular aspect of potato seed is that it is vegetatively propagated. In most developing countries, however, seed systems were first established following developed country standards for cereals and grains, promoting the use of certified seed under a formal seed system. This has led to very low use (less than 10%) of certified potato seed in most developing countries. To increase access to quality seed by small farmers, the Food and Agriculture Organization (FAO) of the United Nations promoted the definition of a new seed category, the Quality Declared Seed (QDS), that relaxes some of the standards required for Certified Seeds and recognizes the importance of seed producers in providing seed of enough quality through the informal seed system (FAO 2006b). Ethiopia adopted the QDS definition in a new seed law passed in 2016 without distinction of crops. Peru (in 2018) and Ecuador (in 2013) have modified the seed regulation for potato to accept the use of QDS. However, differences still exist on how countries are beginning to adopt this category. While in Peru, the new regulation defines the category as QDS similar to the FAO definition, in Ecuador the regulation defines the new category as "common seed" and introduces aspects of the FAO definition for QDS.

Other countries are updating the regulations regarding seed quality assurance systems for potato to increase availability and access of quality seed by farmers. A broad range of changes are proposed, from relaxing some of the standards required for certified seed to allowing the use of private inspection services to increase the number of seed producers that can be inspected each season (e.g., Kenya). One of the motivations of these changes is the increasing recognition of the role of the potato crop for national food security. Some concerns have been raised, however, on the potential consequences of relaxing seed quality standards on the incidence of seed-borne diseases (e.g., *Ralstonia solanacearum*).

1.6 Food Security Challenges and Perspectives for Potato Research and Development

1.6.1 Potato in a Global Food Security Context

The potato, because of its adaptability, its yielding capacity and its nutrition contribution, and as an important component of diversified cropping systems, has a long history of helping relieve food insecurities, and contributing to improve household incomes in times of crisis and today's population expansion. Among important issues and challenges at global level, the European Association for Potato Research (EAPR) Conference in 2017 identified three broad concerns: (1) food security and food safety for a growing population considering consumer's needs; (2) sustainable and environmentally friendly production addressing the question of natural resource management taking advantage of new technologies available such as breeding techniques, biocontrol and big data management; (3) innovation in practice turning scientific results into products and processes to improve the performance of agri-food systems (Andrivon 2017).

Potato's support to food availability can be achieved through improved productivity, either by increasing yields or expanding production areas, combined with technologies that reduce post-harvest losses. Newly developed potato technologies and production concepts can potentially help create solutions in areas where there is a huge need to increase food production. Figure 1.7 shows the global distribution of yields and the low yield levels in most of the developing world, where observed actual yield is usually much lower than the attainable yield. The actual yield is the expression of a potato cultivar in a specific agro-ecological environment and depends on availability of inputs, the economically optimal use of available inputs given the farmers' conditions, and externalities such as local meteorological variations and climate change as well as the technology level and the quality of crop management. Actual yields range from below 5 tons of fresh tubers ha^{-1}(median yield in Uganda: Gildemacher et al. 2009) to well above 100 tons fresh tubers ha^{-1} (in Columbian Basin, USA: Kunkel and Campbell 1987).

The yield gap, expressed as the difference between actual yield in farmers' fields and the attainable yield—using best agricultural practices—leaves a great potential for improvement considering that, in developing countries, the full expression of the crop's yielding capacity has not yet been achieved. Much improvement is needed in agronomic practices, quality seed production, and varieties tolerant/resistant to abiotic and biotic threats (Birch et al. 2012). The high nutritional value of potato mentioned above reinforces the potential of potato to respond to food security challenges.

To reach impact on food availability, access and better use in diets, proper selection of target areas for potato research and identification of the most important constraints to potato production are crucial for defining priority interventions. As a step towards achieving this and to prioritize options for potato research, CIP led an expert survey to assess priorities for potato research across the developing world (Kleinwechter et al. 2014), and an ex-ante evaluation of the economic relevance of

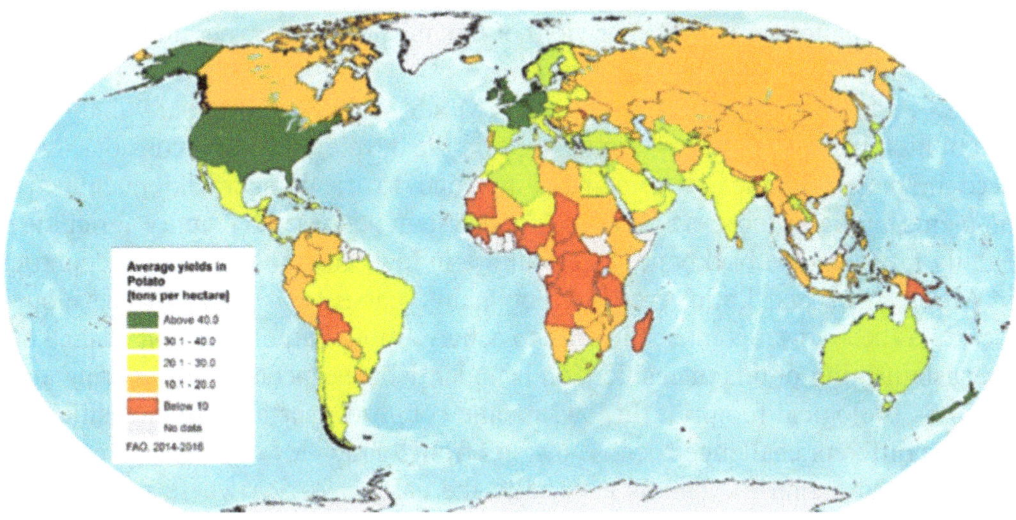

Fig. 1.7 Global distribution of potato yields (tons ha^{-1}). FAOSTAT, 2014–2016

these options (Hareau et al. 2014). Most of these research options for potato have been defined to be relevant in numerous countries of Asia, Africa, and LAC. They include research in potato late blight, drought tolerance/water use efficiency, seed systems and development of farmer organizations and farmers' links to markets (e.g., Harahagazwe et al. 2018 for research priorities in Africa). However, the current and upcoming contextual changes, especially considering the climate change requires to revisit some of these issues with a new perspective. Many simulation exercises based on IPCC (Inter-governmental Panel on Climate Change) scenarios and biology models are underway and suggest that future potato cropping systems could differ from those we know today with the implication that new cultivars will be required to respond these new conditions (Andrivon 2017; Quiroz et al. 2018).

1.6.2 Research and Innovation for Sustainable Potato Cropping

As suggested by Haverkort and Struik (2015), the future prospects of food security challenges for agricultural production can be expressed by the formula: $P = G \times E \times M \times S$ where Performance (P) is determined by Genotype or varieties (G), the Environment or agro-ecological conditions where the crop is grown (E)—which as mentioned above is evolving and will further change in the future—its Management and adaptation to local socioeconomic conditions (M) and the Societal requirements (S) driven by society's demands for food and the need to make agriculture more environmentally and consumer friendly with a focus on food safety. The societal or consumer requirements will vary according to the context. In high-income countries, consumers are looking principally for healthy and easy to prepare foods while in the developing world, the consumers' needs are driven by the food and nutrition challenges and demand concerns principally food availability in both quantity and quality. The actual performance of potato can be expressed as yield of fresh or dry matter per unit area or yield of the finished product per unit area recovered from the raw material after processing.

This Performance analysis ($P = G \times E \times M \times S$), suggested originally in the context of high-income countries, can be used for the developing world considering the need in most countries to principally respond to the hunger and malnutrition problems through a more efficient agri-food system but based on family agriculture. Family farms are the backbone of agriculture in low- and middle-income countries in Africa, Asia, and Latin America. For many years, the trend in the developed world has been towards intensification to achieve more outputs per unit of land but the sustainability of this intensification is under debate especially considering agriculture's ecological footprint. In low-income countries, sustainable intensification (SI) is a different challenge because it starts from a much lower level of inputs than in developed countries. This is especially the case in Africa where potential for increasing production through area expansion is diminishing, partly due to high population growth (Headey and Jayne 2014). For instance, Wu et al. (2018) and Jayne et al. (2014) argue that even though Africa has a high cropping intensity gap[1], closing this gap sustainably must focus on input intensification rather than area expansion. The relevant question is how to promote technology options that allow for increased output quantity and quality (especially from the nutrition point of view), while considering agriculture's environmental impact, preserving land and other resources in both developed and developing countries. In this context, sustainable intensification of potato cropping goes beyond production aspects and considers strong socioeconomic, demographic, and environmental trade-offs to optimize performance. Institutional incentives to support innovation involving diverse stakeholders, with emphasis on research partnerships, are also required to respond to hard-to-find compromises that will vary between different cropping systems (Hall et al. 2001). Multidisciplinary approaches contribute to recognize and solve practical problems at the level of the crop, the cropping system and the agri-food system to achieve sustainable food security in its four dimensions.

The issues mentioned above indicate that sustainable potato production and efficient use of resources will require future adjustments and redesigns of the cropping and processing systems. Two main options can be considered to increase food security, which remains a main goal for future potato research: (1) produce more with less through better input management and optimization; (2) to produce just as much but waste less both before and after harvest through better value chain management, better storage, processing and marketing operations, and responding to increased involvement and awareness of consumers (Andrivon 2017).

As an attempt to analyze how to combine and score different research and technology options according to their effect on sustainable agri-food system indicators in developing countries and their relation to the four dimensions of food security, Table 1.3 attached suggests a list of priority research and technology options in the spheres of G and M. The assessment is based on the literature and the authors' expert opinions. We scored key options using some critical indicators of sustainable agri-food system intensification related to productivity, agriculture income, human well-being and environmental sustainability according to Smith et al. (2017). We scored the research and development options using a simple scale of high, medium, and low effects according to their relation to sustainable intensification indicators and their main contribution to one of the four dimensions of food security.

The performance of any of these technologies or development approaches will, as expressed by the Performance equation ($P = G \times E \times M \times S$), depend on external factors both socio economic and agro-ecological. Their potential contribution to food security will be strongly influenced by the environment (E) and the societal (S) requirements in the local context where they are implemented—e.g., enabling environment, policies for financial and nonfinancial services are also key components for achieving efficient food systems.

[1] Wu et al. (2018, p. 2) define cropping intensity gap as "the amount of incremental cropping intensity that is possibly available if all croplands in a given region are fully intensively used."

Table 1.3 Qualitative evaluation of key research options to enhance food security dimensions through the performance of potato production systems (based on the authors appreciations and the literature related to indicators of contribution to sustainable agri-food systems intensification)

	Food security dimensions						
	Availability				Access	Use/quality	Stability
	Indicators of contribution to sustainable agri-food systems intensification						
Key research options to enhance the performance of potato production systems	Water use efficiency	Land use efficiency	Nitrogen and phosphorus use efficiency	Pesticide use efficiency	Farmer incomes	Increase in calorie and nutrient production efficiency	Reduction of environmental footprint (soil, water, air)
Breeding and variety development (Genotype)							
High yield		***	Neutral/negative. High productive crops might need additional N	Neutral/negative. High productive crops might need additional pesticides	***	**	
Disease resistance (e.g., late blight, viruses)		***		***	**	**	***
Tolerance to drought/heat/salinity	***	***	*		**	**	*
Biofortification (e.g., Fe and Zn)					*	***	
Earliness	***	***	***	**	***	***	***
Potato seed production (Management)							
High quality seed production and distribution		**	*	*	***	*	Neutral/negative high-quality seed use might motivate increased chemical input

Farmer-based seed production		*	*	*	**	*	**
Decision support tools and farming practices (Management)							
Decision support and diagnostic tools for pest and disease control		*		***	**		***
SMART agriculture approaches (sensor use for precision in NPK and water use)	***	**	***		**	*	***
Intensification of cereal-based systems with introduction of early potato varieties	*** No irrigation needed, higher productivity per unit of water	** Uses current fallow land already under agriculture	*	**	**	**	**
Sustainable resource management							
Ecosystem management and biodiversity use	*		*	**	* Policy dependent	*** Use of potato biodiversity can increase nutrient production efficiency	***
Soil and water management	***	**	**		**	***	**
Efficient and inclusive value-chains (Management and Society)							
Value-chain innovation					***	**	***
Post-harvest management (Management and Society)							
Post-harvest losses assessment and reduction					**	***	***

*** = High effect; ** = Medium effect; * = Low effect

In the sections below, some key research and technology options identified in Table 1.3 are briefly described.

1.6.3 Potato Breeding, a Driving Force Towards More Efficient Potato Production

For genotype development (G), priority should be given to achieve a combination of traits to enhance stress tolerance and nutritional aspects to better respond to contextual changes, especially climate and local needs. The recent development of participatory breeding helps to best define the crucial trait combinations required and to facilitate acceptance of new genotypes by growers (Schulte-Geldermann et al. 2012). With the recent findings on the potato genome sequence (PGSC 2011) and the possibilities occurring with new breeding technologies (NBTs), potato breeding appears as the number one opportunity to improve potato production for global food security (Birch et al. 2012).

In many developing and in-transition countries governments have substantially invested in breeding improved varieties. A total of about 840 improved varieties[2] with various combination of traits have been released, most of which in Asian countries like China and India (Table 1.4). CIP has considerably contributed to global crop improvement through supporting NARS and providing access to advanced breeding material. About 43% of total releases are CIP-related (i.e., NARS-bred varieties distributed/facilitated by CIP, NARS selection from CIP crosses, NARS crosses from CIP progenitors). At the regional level, CIP-related varieties are most prominent in Africa where 70% of total releases are CIP-related. This points to the importance of CIP in the regions and the support many national breeding programs require.

Table 1.4 Total and CIP-related number of releases by region

Region	Year	Total	CIP-related[a]	
Asia[b]	2015	518	180	35%
Africa[c]	2010	178	124	70%
Latin America[d]	2007	141	60	43%
Total		**837**	**364**	**43%**

[a]NARS-bred varieties distributed/facilitated by CIP, NARS selection from CIP crosses, NARS crosses from CIP progenitors; calculations for Asia based on Gatto et al. (2018), for Africa on Labarta (2015), for Latin America on Thiele et al. (2008)
[b]Bhutan, Bangladesh, China, India, Indonesia, Nepal, Pakistan, Philippines, Sri Lanka, Vietnam
[c]Burundi, D.R. Congo, Ethiopia, Kenya, Madagascar, Malawi, Rwanda, Tanzania, Uganda
[d]Bolivia, Colombia, Ecuador, Peru, Venezuela

[2] The current true total number of releases is likely higher given that the most recent data available dates back more than a decade.

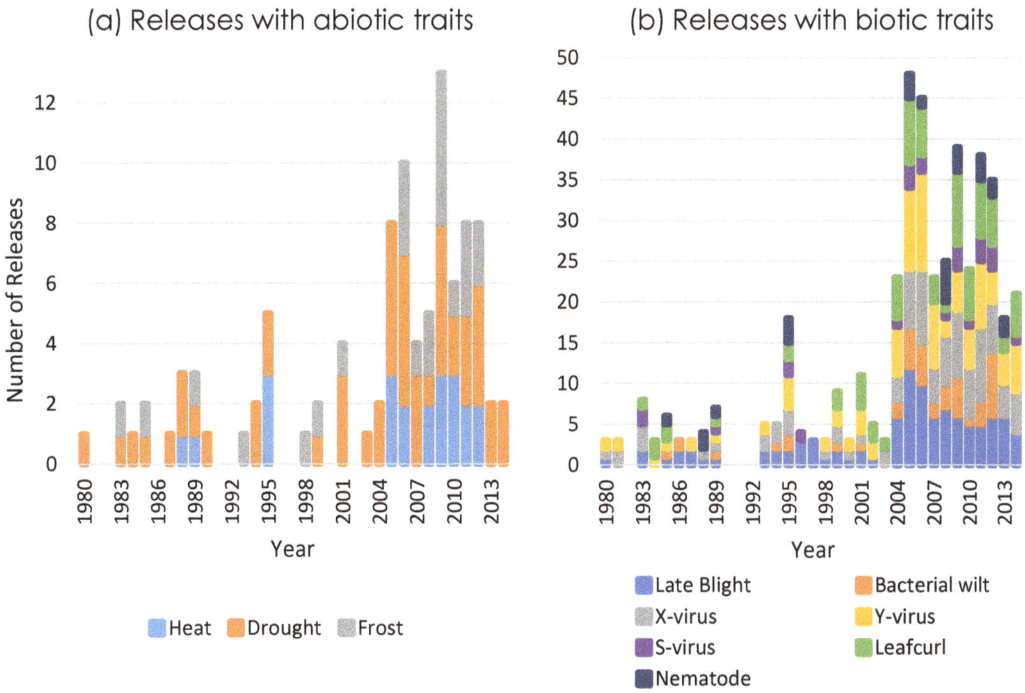

Fig. 1.8 Total releases with abiotic and biotic traits in Asia between 1980 and 2014. Notes: release is for *high* resistant category only. *Medium* and *low* resistant, and *susceptible* categories are not shown. (Source: Gatto et al. 2018)

The development of early and high-yielding varieties with resistance to *P. infestans* has been a longstanding potato breeding objective. Genotypes with resistance to viruses (PVY, PLRV, PVS, PVX), nematodes (mainly *Globodera* and *Meloidogyne* species), bacterial wilt, and a broader spectrum of cultivars tolerant to abiotic stresses like heat, drought, and saline conditions, and focus on beneficial root traits, can increase productivity and expand potato production to new areas.

The importance to develop and release varieties with high tolerance to abiotic stresses and high resistances to biotic stresses has increased, as Fig. 1.8 depicts for Asia. Especially starting in the early 2000, major traits have been bred into released varieties likely as a result of adjusting breeding objectives aiming increasingly at mitigating the adverse effects associated with climate change and variability.

New resilient varieties will potentially expand potato production to new areas and produce more nutritious food under current and future stress factors. Genetic biofortification through conventional and new breeding techniques can help to overcome micronutrient malnutrition and support the consumption of better-quality tubers. This crop improvement approach aims to positively influence human health, as a complement to diet supplementation and food fortification. In recent years, CIP has initiated the development of Fe and Zn biofortified potatoes, under the umbrella of the HarvestPlus Program (http://www.harvestplus.org/), a global interdisciplinary alliance for developing biofortified varieties of staple crops. Food security programs working to deploy biofortified crops will strongly benefit from nutritional education efforts and awareness programs considering gender roles in the beneficiary communities.

1.6.4 Seed Quality and Availability, the Key to Harvest Success

As sustainable potato production depends on a constantly renewed supply of disease-free planting material, improving quality seed production and seed distribution is another strong avenue of research opportunity related to crop management (M) in potato development. The conceptual framework underpinning the concept of seed security contemplates different types of seed insecurity: poor seed quality, lack of availability, limited access to high quality seed, lack of access to preferred and adapted varieties, inefficient seed systems (FAO 2016; CGIAR/RTB 2016). As a vegetative propagated crop, the growth, development, yield and quality of the potato is strongly influenced by the quality of the seed tubers planted affected by their physical, physiological and health status. More than 90% of seed potatoes in developing economies is produced in the farmer-based category and is considered to be of poor quality (Thomas-Sharma et al. 2015). Potato seed production systems should support the access to high quality seed potato tubers of improved varieties by combining rapid multiplication technologies (e.g., aeroponics or sand hydroponics) with decentralized seed multiplication, e.g., promotion of quality declared seed systems (FAO 2006b; Fajardo et al. 2010). It should be complemented with on-farm seed maintenance (e.g., positive selection, small seed-plot technique and improved storage) in an integrated approach (Gildemacher et al. 2011; Schulte-Geldermann et al. 2012; Thomas-Sharma et al. 2015; Obura et al. 2016). Improving technologies for farmer-based seed production and distribution of high quality planting material of existing and new varieties have the potential to reach high numbers of beneficiaries with strong impacts on poverty reduction and food availability.

1.6.5 Potato Crop Management and Farming Practices to Increase Productivity and Sustainability

The highly adaptable potato can fit to many types of environments (E) from sea level to high mountain conditions where small-scale farmers predominate. Beside temperature regime and solar radiation, there are many factors that affect its productivity as for example, soil characteristics, water use efficiency, nutrient availability and hazards such as night frost or heat waves that may drastically impact resource use, and thereby sustainability. The adaptation of the crop depends on the genotype but also on the crop management practices (M) that need to evolve according to the specific agro-ecological conditions, the socioeconomic context and the local production systems. Crop management is context specific and should consider local knowledge that can be improved promoting new tools and approaches.

Smart agriculture is a novel avenue for resource use optimization based on new monitoring and decision support tools. Remote sensing and global information system (GIS) tools coupled with decision support systems (DSS) and precision agriculture technologies may contribute to increased productivity while interaction among biophysical and social disciplines for sustainable food production intensification can at the same time contribute to resource use optimization. Fertilizer (N, P and K) recommendation systems are now using field scale models as well as tractor, drone and satellite embedded spectral sensors to monitor crop nutrient status to supplement fertilization according to inter and within field variability (Goffart et al. 2008, 2017). These more sophisticated technologies are still mainly used in the high-income countries, but massive and varied data management could foster new models to be developed and contribute to decision support systems under developing country conditions. CIP is adapting such system to the Andean condition developing a strategy to manage late blight that combines host resistance and a decision support system to optimize the use of fungicides. The vulnerability of the crop to many pests and diseases, which the current global climate change can worsen, remains one of the most severe threats for a wider potato diffusion and its sustainable cropping. To improve crop health, portable molecular diagnostic tools and decision support systems for early warning and control of pests and diseases (e.g., for efficient fungicide use to control late blight) will contribute to better crop production monitoring and input use efficiency. Research to develop biocontrol is very active and considered to grow substantially in the coming decade, but there are still few confirmed successes from the field, and specific management tools (Decision Support Tools) are still missing (Velivelli et al. 2014). CIP and EAPR are coordinating actions in Europe and Latin America to promote biocontrol and compare the efficiency of biocontrol agents using defined protocols (Devaux et al. 2017).

Specific intensification practices can be developed under specific cropping systems such as in the cereal-based systems in India through "Double-Transplanting (DT)" of rice and planting early maturing potato between the two rice crops as a valid alternative to the traditional potato-boro rice and kharif (monsoon) rice-boro rice. This cropping pattern contributes to enhance system productivity without sacrificing area or productivity of either of the two crops, thus creating new opportunities for potato cultivation for small-scale producers (Arya et al. 2015).

To enhance ecological sustainability, the objective is to implement management practices that increase the level of provision of ecosystem services such as natural soil fertility and biological control. Natural regulation of pests and diseases is an important element in potato agro-ecosystems. In organic farming, it has been demonstrated that the development of natural antagonist associations of the Colorado beetle, such as auxiliary insects and useful pathogens, can significantly improve the control of such a pest for the potato crop (Crowder et al. 2010). Another example of ecosystem management is the delivery of nitrogen through natural fixation and mineralization, which can be enhanced by cropping practices such as cover crops, legume-based intercropping systems and application of organic soil amendments before the potato crop. Biodiversity based agricultural approaches that rely on the design and management of on-farm agrobiodiversity to generate ecosystem services is another avenue to reduce potato's ecological footprint and increase farmer's resilience to cope with frost risks as it is the case in the Andes. The management of ecosystems at field and landscape level can provide a series of production benefits to reduce the need for off-farm inputs. The analysis of beneficial microbial communities and their impact on potato plant phenotypes expressions still needs to be developed as discussed at a EAPR-CIP workshop on biostimulant and biocontrol agents (Devaux et al. 2017).

1.6.6 Integrating Food Security and Value Chain Development

Although potato remains a staple food in rural areas in developing countries, it is also increasingly becoming a cash food for farmers in Asia, Africa and LAC (DeFauw Sherri et al. 2012). The majority of potato producers are smallholders who depend strongly on agriculture, including the potato crop, for income, food security and employment. Potato production reaches consumers via multilevel marketing systems, not directly from the farmer's field. Thus, the challenge to achieve food and nutrition security as well as prosperity for these smallholders will be obtained or lost by the way agricultural value chains are coordinated. Value chain development and organized markets through farmer associations, storage facilities, and better links with traders and consumers are then required to allow potato producers to access better value markets to get higher and steady incomes from their production. In the recent years, research activities to improve the efficiency of the value chain and coordination among its actors have evolved to achieve more inclusiveness in the value chain development approaches (Devaux et al. 2018). Several factors are contributing to this evolution: changes in consumer demands, new or emerging markets with strict standards, including food safety, processing technologies, and better access to market information. To respond to these changes and the need to make agriculture more environmentally and consumer friendly responding to the Society's requirements (S), research should be characterized by an interaction between natural and social sciences and should be market-driven considering the needs and challenges of all value chain stakeholders. In a compendium about perspectives on the status of innovation for Value Chain Development, Devaux et al. (2016) analyze the opportunities emerging from new markets for agricultural produce and identify challenges to smallholder participation in these markets, approaches for increasing access to markets through strengthening value chain stakeholders' relationships, enhancing innovation and improving an enabling environment. Linked to the value chain efficiency, the assessment of food losses across the value chain and the quality of marketed potatoes also require further research efforts to optimize food availability and consumer access to quality potato products.

1.6.7 Post-harvest Management: Reducing Food Losses

As indicated above, another way to face the food security challenge is to produce just as much, but waste less through better post-harvest management. Post-harvest management in potato, including storage, processing and value chain efficiency, is a much larger problem than cereals and deserves special attention. Reduction of food losses appears as a key opportunity. The basics of storage management have not changed, but the implementation and application of the basics are evolving worldwide, according to diversity in location, climate and market criteria, that will influence storage management structures and management decisions (Olsen 2014). In developing countries, recent studies have analyzed food loss across the potato value chain, as for example in Ecuador and Peru, by collecting qualitative and quantitative data to provide a comprehensive identification and characterization of losses. The results show that the most important losses occur in the production node, ranging from 90 to 95% of the total losses in the chain. On average farmers suffer this highest loss across the value chain ranging between 8 and 20% of their production at or before harvest before moving on to the next node of the chain. The main causes of losses are poor crop and harvest management, infested tubers by pest and diseases, high percentage of small tubers and weather conditions: frost and heavy rains (Delgado et al. 2017).

1.7 Concluding Remarks: Towards Future Potato Research for Global Food and Nutrition Security

The analysis of leveraging potato agri-food systems for global food security issues and challenges in this chapter emphasizes the need for making agricultural research programs and food system interventions more responsive to food security dimensions. The multidimensional nature of food security requires multisector innovation in approaches that allow to use the knowledge available and transform scientific results into products and processes to improve the performance of agri-food systems, considering the challenge to produce more food with the same or fewer resources.

In both developed and developing countries, innovations resulting from potato research should be incremental through a step by step improvement of an existing structure promoting technologies adapted to the local context. This is particularly true for smallholder family agriculture in developing countries where there is a great need to increase potato production in a sustainable way. While this approach has the advantage of not destabilizing an existing system, it may also suffer of a systemic lock-in or a lack of enabling environment that keeps agriculture and food systems on less efficient pathways as developed by Baret (2017). An example of lock-in is the use of pesticides and their promotion by agro-chemical companies and technical support services that influence farmers' decision making, restraining the use of more environmental friendly options such as decision support tools for efficient pest control with a more rational pesticide use. The valuable use of varieties tolerant or resistant to pests and diseases can also be limited by processing companies that promote varieties for their processing characteristics regardless of their environmental footprint. In developing countries, low infrastructure quality, weak institutions and policies create also huge limitations to the adoption of new and more sustainable technologies. To reach food security goals, a stronger emphasis must be put towards promoting evidence-based policies for communicating information and influence decision makers. It is also important to favor affordable and better-adapted technologies that can respond to the needs of small-scale farmers and, significantly limit negative impacts towards the environment. The research and technology options proposed in this chapter will require policy support, financial and nonfinancial services to have a chance to be adopted and used by local farmers. They will also necessitate a better access to discovery and creative ideas through better services from the potato research community at national and international levels.

Local calibration/validation and demonstration are two essential phases towards local end-user uptake, either involving farmers or extension services representatives. Public or private investments are also required to support such actions to enable farmers to have access to new technologies, and to be trained in their use. With the upcoming of new communication technologies such as smartphones, expansion of mobile broadband and access to local online platforms integrating large amounts of local data and links to Decision Support Systems, we have yet to fully exploit the potential of information technologies especially in developing countries. Local farmers, especially the younger ones, are expected to be able to have increasing access to such new adapted tools, i.e. for Late Blight management in the Andes. But this will only be possible globally if technological innovation is accompanied by capacity building and institutional innovation (associativity, access to credit, communication network) in rural areas.

There is a great dichotomy between research activities in developed versus developing countries that highlights the need for more exchange, knowledge sharing, and collaboration. Since 2014, the interaction between EAPR-linked research organizations and CIP has been looking at mechanisms to enhance partnership between European partners and CIP involving research partners in the Southern hemisphere to promote collaborative research activities, links between research networks such as Euro and Latin Blight (Acuña et al. 2017) as well as facilitating short and long-term training with universities in Europe (Durroux-Malpartida 2014).

To reach the strongest impact on food security, potato research and development efforts need to move towards food systems engineering, rather than focus explicitly on technology/solution development. In this paper, we are analyzing the different components that contribute to the performance of the potato using the key relation $P = G \times E \times M \times S$, enabling a list of key research and technology options to guide agriculture research and technology development toward sustainable intensification approaches responding to farmers' needs both for food security and better income. The argument is that agricultural programs need to integrate better agriculture sustainable intensification and food security indicators considering also other dimensions such as quality, diversity of products, health impacts and climate change effects. Multidisciplinary approaches and a better understanding of the evolving food systems are required to recognize and solve practical problems of the whole potato value chain to achieve sustainable food security. Policies, investments and services that support agricultural productivity, sustainability and expand risk management capacity are also required to give potato farmers the best chance to meet future needs, while increasing their adaptability and resilience to foster food security.

Acknowledgements The authors are grateful and sincerely appreciate the valuable contributions of Mr. Henry Juarez and Dr. Jorge Andrade-Piedra. Their inputs, opinions, and suggestions have greatly contributed to improve this chapter.

References

Acuña I, Restrepo S, Lucca F, Andrade-Piedra J (2017) Recent developments: late blight in Latin America. In: Schepers HTAM (ed) Proceedings of the sixteenth EuroBlight Workshop, Aarhus, Denmark, 14–17 May 2017. PAGV special report no. 18

Andrivon D (2017) Potato facing global challenges: how, how much, how well? Potato Res 60:389. https://doi.org/10.1007/s11540-018-9386-z

Arya S, Ahmed M, Bardhan Roy SK, Kadian MS, Quiroz R (2015) Sustainable intensification of potato in rice-based system for increased productivity and income of resource poor farmers in West Bengal, India. Int J Trop Agric. ISSN 0254-8755 33(2):203–208

Bailey RL, West KP Jr, Black RE (2015) The epidemiology of global micronutrient deficiencies. Ann Nutr Metab 66(Suppl 2):22–33

Baret PV (2017) Acceptance of innovation and pathways to transition towards more sustainable food systems. Potato Res 60:383. https://doi.org/10.1007/s11540-018-9384-1

Birch PRJ, Bryan GJ, Fenton B, Gilroy EM, Hein I, Jones JT, Prashar A, Taylor MA, Torrance L, Toth IK (2012) Crops that feed the world 8: potato: are the trends of increased global production sustainable? Food Secur 4:477–508. https://doi.org/10.1007/s12571-012-0220-1

Black RE, Victora CG, Walker SP, Bhutta ZA, Christian P, de Onis M, Ezzati M, Grantham-McGregor S, Katz J, Martorell R, Uauy R (2013) Maternal and child undernutrition and overweight in low-income and middle-income countries. Lancet 832(9890):427–451

Bohl WH, Johnson SB (2010) Commercial potato production in North America. The Potato Association of America Handbook. http://potatoassociation.org/wp-content/uploads/2014/04/A_ProductionHandbook_Final_000.pdf

Borch D, Juul-Hindsgaul N, Veller M, Astrup A, Jaskolowski J, Raben A (2016) Potatoes and risk of obesity, type 2 diabetes, and cardiovascular disease in apparently healthy adults: a systematic review of clinical intervention and observational studies. Am J Clin Nutr 104(2):489–498

Burlingame B, Mouillé B, Charrondiére UR (2009) Review: nutrients, bioactive non-nutrients and anti-nutrients in potatoes. J Food Compos Anal 22:494–502

CGIAR Research Program on Roots, Tubers and Bananas (2016) Multi-stakeholder framework for intervening in RTB seed systems: user's guide. Lima (Peru), 13 p. RTB Working Paper. ISSN 2309-6586, no. 2016-1

Creed-Kanashiro H, Hareau G, Devaux A, Maldonado L, Ordinola M, Fonseca C, Suarez V, Astete L, Marin M, Penny M (2015) Agriculture-nutrition linkages: analyzing nutritional outcomes of interventions in potato-based production systems of Peru. In: 2nd international conference on global food security, October 11–14, 2015. Cornell University, Ithaca

Crowder DW et al (2010) Organic agriculture promotes evenness and natural pest control. Nature 466:109–112

DeFauw Sherri L, He Z, Larkin RP, Mansour SA (2012) Sustainable potato production and global food security. In: He Z, Larkin R, Honeycutt W (eds) Sustainable potato production: global case studies. Springer, Amsterdam, 531 pp

Delgado L, Schuster M, Torero M (2017) The reality of food losses: a new measurement methodology. IFPRI discussion paper 1686. International Food Policy Research Institute (IFPRI), Washington, DC. http://ebrary.ifpri.org/cdm/ref/collection/p15738coll2/id/131530

Devaux A, Torero M, Donovan J, Horton DE (eds) (2016) Innovation for inclusive value-chain development: successes and challenges. International Food Policy Research Institute (IFPRI), Washington, DC. https://doi.org/10.2499/9780896292130

Devaux A, Goffart JP, Kromann P, Toth I, Braguard C, Declerck S (2017) Report on CIP-EAPR Workshop 2017 on biocontrol and biostimulant agents for the potato crop. Potato Res 60:291. https://doi.org/10.1007/s11540-018-9385-0

Devaux A, Torero M, Donovan J, Horton D (2018) Agricultural innovation and inclusive value-chain development: a review. J Agribus Dev Emerg Econ 8(1):99–123. https://doi.org/10.1108/JADEE-06-2017-0065

Douches DS, Maas D, Jastrzebski K, Chase RW (1996) Assessment of potato breeding progress in the USA over the last century. Crop Sci 36(6):1544–1552

Durroux-Malpartida V (2014) EAPR and CIP strengthen collaboration to tap potatoes' potential. Potato Res 57:367. https://doi.org/10.1007/s11540-014-9276-y

European Commission (2007). The potato sector in the European Union. Commission staff working document. SEC(2007) 533

Eurostat (2017) The EU potato sector—statistics on production, prices and trade. http://ec.europa.eu/eurostat/statistics-explained/index.php/The_EU_potato_sector_-_statistics_on_production,_prices_and_trade#Publications

Fajardo J, Lutaladio N, Larinde M, Rosell C, Barker I, Roca W et al (2010) Quality declared planting material: protocols and standards for vegetatively propagated crops: expert consultation, Lima, 27–29 November 2007. Food and Agriculture Organization of the United Nations, Rome

FAO (2002) The state of food insecurity in the world 2001, Rome, pp 4–7

FAO (2006a) Food security. Policy brief. FAO, Rome

FAO (2006b) Quality declared seed system. FAO plant protection and protection paper # 185, Rome

FAO (2009) International year of the potato 2008: new light on a hidden treasure. http://www.fao.org/potato-2008/en/events/book.html

FAO (2014) FAO statistical databases FAOSTAT. http://faostat3.fao.org/

FAO (2016) FAOSTAT Database. http://www.fao.org/faostat/en/#data/QC

FAO (2015a) Household seed security concepts and indicators. Discussion paper, 10 pp

FAO (2015b) FAOSTAT database. http://www.fao.org/faostat/en/#data/FBS

FAO (2016) Seed security assessment: a practitioner's guide, Rome

FAO, IFAD, UNICEF, WFP and WHO (2018) The state of food security and nutrition in the world 2018. Building climate resilience for food security and nutrition. FAO, Rome

FAOSTAT (2013) Food balance sheet. http://www.fao.org/faostat/en/#data/FBS

FAOSTAT (2017) Food and agriculture data. http://www.fao.org/faostat/en/#data/QCinfo

Foley JA et al (2011) Solution for a cultivated planet. Nature 478:337–342

Gatto M, Hareau G, Pradel W, Suarez V, Qin J (2018) Release and adoption of improved potato varieties in Southeast and South Asia. International Potato Center (CIP), Lima, Peru, 42 p. Social sciences working paper no. 2018-2. ISBN 978-92-9060-501-0

Gildemacher PR, Kaguongo W, Ortiz O, Tesfaye A, Woldegiorgis G, Wagoire WW, Kakuhenzire R, Kinyae P, Nyongesa M, Struik PC, Leeuwis C (2009) Improving potato production in Kenya, Uganda and Ethiopia: a system diagnosis. Potato Res 52(2):173–205

Gildemacher PR, Schulte-Geldermann E, Borus D, Demo P, Kinyae P, Mundia P et al (2011) Seed potato quality improvement through positive selection by smallholder farmers in Kenya. Potato Res 54:253–266

Gillespie S, Van den Bold M (2017) Agriculture, food systems, and nutrition: meeting the challenge. Global Chall 1:1600002. https://doi.org/10.1002/gch2.201600002

Goffart JP, Olivier M, Frankinet M (2008) Potato crop nitrogen status assessment to improve N fertilization management and efficiency: past–present–future. Potato Res 51(3):355–383. https://doi.org/10.1007/s11540-008-9118-x

Goffart JP, Gobin A, Delloye C, Curnel Y (2017) Crop spectral reflectance to support decision making on crop nutrition. Paper presented to the International Fertiliser Society at a conference in Cambridge, United Kingdom, on 7th December 2017. Proceedings 812, pp. 29. www.fertiliser-society.org © 2017 International Fertiliser Society—ISBN 978-0-85310-449-0

Haggblade S, Hazell PBR, Reardon T (2007) Research perspectives and prospectives on the rural nonfarm economy. In: Haggblade S, Hazell PBR, Reardon T (eds) Transforming the rural nonfarm sector: opportunities and threats in the developing world. The Johns Hopkins University Press, Baltimore

Hall A, Bockett G, Taylor S, Sivamohan MVK, Clark N (2001) Why research partnerships really matter: innovation theory, institutional arrangements and implications for developing new technology for the poor. World Dev 29(5):783–797

Harahagazwe D, Condori B, Barreda C, Bararyenya A, Byarugaba AA, Kude DA, Lung'aho C, Martinho C, Mbiri D, Nasona B, Ochieng B, Onditi J, Randrianaivoarivony JM, Tankou CM, Worku A, Schulte-Geldermann E, Mares V (CIP), de Mendiburu F, Quiroz R (CIP) (2018) How big is the potato (*Solanum tuberosum* L.) yield gap in Sub-Saharan Africa and why? A participatory approach. Open Agric 3(2):180-189. ISSN 2391-9531 .

Hareau G, Kleinwechter U, Pradel W, Suarez V, Okello J, Vikraman S (2014) Strategic assessment of research priorities for Potato. Lima (Peru). CGIAR Research Program on Roots, Tubers and Bananas (RTB). RTB working paper 2014-8. www.rtb.cgiar.org

Haverkort AJ, Struik PC (2015) Yield levels of potato crops: recent achievements and future prospects. Field Corp Res 182:76–85

Haverkort A, de Ruijter FJ, van Evert FK, Conijn JG, Rutgers B (2014) Worldwide sustainability hotspots in potato cultivation. 1. Identification and mapping. Potato Res 56:343–353

Hazell P, Haggblade S (1989) Agricultural technology and farm–nonfarm growth linkages. Agric Econ 3:345–364

Headey DD, Jayne TS (2014) Adaptation to land constraints: is Africa different? Food Policy 48:18–33

INEI Peru (2017) IV Censo Nacional Agropecuario (CENAGRO). Lima. INEI Perú (2017). Encuesta demográfica y de salud familiar 2016. Lima, Peru

Jayne TS, Chamberlin J, Headey DD (2014) Land pressures, the evolution of farming systems, and development strategies in Africa: a synthesis. Food Policy *48*:1–17

Kanter R, Walls HL, Tak M, Roberts F, Waage J (2015) A conceptual framework for understanding the impacts of agriculture and food system policies on nutrition and health. Food Secur 7(4):767–777

Kleinwechter U, Hareau G, Suarez V (2014) Prioritization of options for potato research for development—results from a global expert survey. Lima (Peru). CGIAR Research Program on Roots, Tubers and Bananas (RTB). RTB working paper 2014-7. www.rtb.cgiar.org

Kunkel R, Campbell GS (1987) Maximum potential potato yield in the Columbia Basin, USA: model and measured values. Am Potato J 64(7):355

Labarta R (2015) The effectiveness of potato and sweetpotato improvement programmes from the perspectives of varietal output and adoption in Sub-Saharan Africa. In: Walker T, Alwang J (eds) Crop improvement, adoption, and impact of improved varieties in food crops in Sub-Saharan Africa. CABI, Wallingford

Muthoni J, Mbiyu MW, Nyamongo DO (2010) A review of potato seed systems and germplasm conservation in Kenya. J Agric Food Inform 11(2):157–167

Obura B, et al (2016) Cost benefit analysis of seed potato replacement strategies among smallholder farmers in Kenya. Tropentag, September 18–21, 2016, Vienna, Austria. http://www.tropentag.de/2016/abstracts/links/Obura_eAhVYqsS.pdf. Accessed 7 Sept 2018

Olsen N (2014) Potato storage management: a global perspective. Potato Res 57:331–333. https://doi.org/10.1007/s11540-015-9283-7

Pandey A (2015) Rural farm and non-farm linkages in Uttar Pradesh. J Land Rural Stud *3*(2):203–218

PGSC (The Potato Genome Sequencing Consortium) (2011) Genome sequence and analysis of the tuber crop potato. Nature 475:189–195

Pingali P (2006) Westernization of Asian diets and the transformation of food systems: implications for research and policy. Food Policy 32(3):281–298

Pinstrup-Andersen P (2014) Food systems and human nutrition: relationships and policy interventions. In: Thompson B, Amoroso L (eds) Improving diets and nutrition: food-based approaches (Chapter 2). CABI, Wallingford, pp 8–20

Quiroz R, Ramírez DA, Kroschel J, Andrade-Piedra J, Barreda C, Condori B et al (2018) Impact of climate change on the potato crop and biodiversity in its center of origin. Open Agric 3:273–283

Reardon T, Chen K, Minten B, Adriano L (2012) The quiet revolution in staple food value chains: enter the dragon, the elephant and the tiger. Asian Development Bank (ADB) and International Food Policy Research Institute (IFPRI), Mandaluyong City

Rosegrant MW, Sulser TB, Mason-D'Croz D, Cenacchi N, Nin-Pratt A, Dunston S, Zhu T, Ringler C, Wiebe K, Robinson S, Willenbockel D, Xie H, Kwon H-Y, Thomas TS, Wimmer F, Schaldach R, Nelson GC, Willaarts B (2017) Quantitative foresight modeling to inform the CGIAR research portfolio. Project report for USAID. International Food Policy Research Institute (IFPRI), Washington, DC

Salazar L, Aramburu J, González M, Winters P (2016) Food security and productivity: impacts of technology adoption in small subsistence farmers in Bolivia. Food Policy 65:32–52

Schulte-Geldermann E, Gildemacher PR, Struik PC (2012) Improving seed health and seed performance by positive selection in three Kenyan potato varieties. Am J Potato Res 89:429–437. https://link.springer.com/content/pdf/10.1007%2Fs12230-012-9264-1.pdf. Accessed 13 Sept 2018

Scott G (2011) Growth rates for potatoes in Latin America in comparative perspective: 1961–07. Am J Potato Res 88:143–152

Scott GJ, Suarez V (2011) Growth rates for potato in India and their implications for industry. Potato J 38(2):100–112

Scott GJ, Suarez V (2012a) From Mao to McDonald's: emerging markets for potatoes and potato products in China 1961-2007. Am J Potato Res 89(3):216–231

Scott GJ, Suarez V (2012b) The rise of Asia as the center of global potato production and some implications for industry. Potato J 39(1):1–22

Scott GJ, Suarez V (2012c) Limits to growth or growth to the limits? Trends and projections for potatoes in China and their implications for industry. Potato Res 55(2):135–156

Smith A, Snapp S, Chikowoa R, Thorne P, Bekunda B, Gloverd J (2017) Measuring sustainable intensification in smallholder agroecosystems: a review. Glob Food Sec 12:127–138

The Wall Street Journal (2015) Pushing the potato: China wants people to eat more 'Earth Beans'. https://www.wsj.com/amp/articles/BL-CJB-25584?responsive=y

Thiele G, Hareau G, Suarez V, Chujoy E, Bonierbale M, Maldonado L (2008) Varietal change in potatoes in developing countries and the contribution of the International Potato Center: 1972–2007. International Potato Center (CIP), Lima, Peru. Working paper 2008-6, 46 p

Thomas-Sharma S, Abdurahman A, Ali S, Andrade-Piedra JL, Bao S, Charkowski AO et al (2015) Seed degeneration in potato: the need for an integrated seed health strategy to mitigate the problem in developing countries. Plant Pathol 65:3–16. https://www.researchgate.net/publication/280913576_Seed_degeneration_in_potato_The_need_for_an_integrated_seed_health_strategy_to_mitigate_the_problem_in_developing_countries. Accessed 13 Sept 2018

Velivelli SLS, Sessitsch A, Prestwich BD (2014) The role of microbial inoculants in integrated crop management systems. Potato Res 57(3–4):291–309

Von Grebmer K et al (2017) Global hunger index: getting to zero hunger. Welthungerhilfe/International Food Policy Research Institute/Concern Worldwide, Bonn/Washington, DC/Dublin. https://doi.org/10.2499/9780896292710

World Bank (2017) Gaining momentum in Peruvian agriculture: opportunities to increase productivity and enhance competitiveness. World Bank, Washington, DC

Wu W, Yu Q, You L, Chen K, Tang H, Liu J (2018) Global cropping intensity gaps: increasing food production without cropland expansion. Land Use Policy *76*:515–525

You L, Wood-Sichra U, Fritz S, Guo Z, See L, Koo J (2014) Spatial Production Allocation Model (SPAM) 2005 v3.2. http://mapspam.info

Chapter 2
The Potato and Its Contribution to the Human Diet and Health

Gabriela Burgos, Thomas Zum Felde, Christelle Andre, and Stan Kubow

Abstract Potato has contributed to human diet for thousands of years, first in the Andes of South America and then in the rest of the world. Its contribution to the human diet is affected by cooking, potato intake levels, and the bioavailability of potato nutrients. Generally, the key nutrients found in potatoes including minerals, proteins, and dietary fiber are well retained after cooking. Vitamins C and B_6 are significantly reduced after cooking while carotenoids and anthocyanins show high recoveries after cooking due to an improved release of these antioxidants.

In many developed countries potatoes are consumed as a vegetable with intakes that vary from 50 to 150 g per day for adults. On the other hand, in some rural areas of Africa and in the highlands of Latin American countries, potato is considered a staple crop and is consumed in large quantities with intakes that vary from 300 to 800 g per day for adults. These marked differences in the potato intake affect significantly the contribution of potato nutrients to the human dietary requirements.

In recent years, information about nutrient bioaccessibility and bioavailability from potatoes has become available indicating higher bioaccessibility of minerals and vitamins in potato as compared with other staple crops such as beans or wheat. Bioavailability refers to the fraction of an ingested nutrient that is available for utilization in normal physiological functions and/or for body storage while bioaccessibility refers to the amount that is potentially absorbable from the gut lumen.

In addition, potatoes have shown promising health-promoting properties in human cell culture, experimental animal and human clinical studies, including anticancer, hypocholesterolemic, anti-inflammatory, anti-obesity, and antidiabetic

G. Burgos (✉) · T. Zum Felde
International Potato Center, Lima, Peru
e-mail: g.burgos@cgiar.org; t.zumfelde@cgiar.org

C. Andre
The New Zealand Institute for Plant and Food Research Limited/Luxembourg Institute of Science and Technology, Auckland, New Zealand
e-mail: christelle.andre@plantandfood.co.nz

S. Kubow
McGill University, Quebec, Canada
e-mail: stan.kubow@mcgill.ca

H. Campos, O. Ortiz (eds.), *The Potato Crop*,
https://doi.org/10.1007/978-3-030-28683-5_2

properties with phenolics, anthocyanins, fiber, resistant starch, carotenoids as well as glycoalkaloids contributing to the health benefits of potatoes.

2.1 Introduction

Diverse studies have demonstrated that potato is an important source of carbohydrates, resistant starch, quality proteins, vitamins C and B_6 as well as potassium (Camire et al. 2009). Potato is also a source of antioxidants that can contribute to prevent both degenerative and age-related diseases with lutein and zeaxanthin being present in high levels in yellow-fleshed potatoes (Burgos et al. 2009) and anthocyanins being present in purple and red-fleshed potato landraces (Burgos et al. 2013b) commonly grown and eaten in the Andean highlands of Peru, Bolivia, Ecuador, and Colombia. Potatoes also contain glycoalkaloids, which in high concentrations can be toxic to humans but in low concentrations can have beneficial effects such as inhibition of the growth of cancer cells (Friedman 2015). The nutritional composition of potatoes is summarized in Fig. 2.1. The concentration of energy, starch, protein, lipids, dietary fiber, potassium, phosphorus, magnesium, iron, zinc, vitamin C, vitamin B_6, chlorogenic acid, and glycoalkaloids has a range of variation independent from the flesh color. Yellow-fleshed potatoes have a carotenoid concentration higher than white-fleshed potatoes while purple potatoes have a higher anthocyanin concentration than red- or white-fleshed potatoes.

Like other plant foods, the nutritional composition of potatoes is affected by different pre-harvest (environment, cultural practices, maturity at harvest, biotic and abiotic stresses, etc.) and post-harvest (processing, storage, transport, etc.) conditions.

Potato has contributed to the human diet for thousands of years, first in the Andean region and then in the rest of the world. Its contribution is affected by cooking, the amount of potato intake, and the bioavailability of the nutrients. Generally, the key phytonutrients found in potatoes including minerals, proteins, and dietary fibers are well retained after cooking. Vitamins C and B_6 are significantly reduced after cooking while carotenoids and anthocyanins show high recovery after cooking due to an improved release of these antioxidants from the food matrix after cooking (Tian et al. 2016). In this chapter, the range of nutrient concentrations is expressed on a fresh weight (FW) basis and ranges refer to both raw and cooked potatoes. However, for calculating their contribution to the diet, only values of cooked potatoes are considered.

The worldwide mean potato intake is equivalent to 93 g per day (FAO 2013). However, this value has a large range of variation. In many developed countries potatoes are consumed as a vegetable and served as a part of a larger meal with intakes that vary from 50 to 150 g per day for adults. On the other hand, in some rural areas of Africa and in the highlands of Latin American countries, potato is considered a staple crop and consumed alone in large quantities as a complete meal with intakes that vary from 300 to 800 g per day for adults (De Haan et al. 2019).

Energy:	96 to 123 Kcal	Magnesium:	16 to 40 mg
Starch:	16 to 20 g	Iron:	0.29 to 0.69 mg
Protein:	1.76 to 2.95 g	Zinc:	0.29 to 0.48 mg
Lipids:	0.1 to 0.5 g	Vitamin C:	7.8 to 20.6 mg
Dietary fiber:	1.8 g to 2.1 g	Vitamin B6:	0.299 mg
Potassium:	150 to 1386 mg	Chlorogenic acid:	19 to 399 mg
Phosphorus:	42 to 120 mg	Glycoalkaloids:	0.7 to 18.7 mg

Fig. 2.1 Nutritional composition of potatoes per 100 g FW

Implications of the contribution of potatoes to the human diet as related to the magnitude of potato intake are also described in this chapter.

Bioavailability refers to the fraction of an ingested nutrient that is available for utilization in normal physiological functions and/or for its contribution towards body stores (La Frano et al. 2014). Many factors affect the bioavailability of a compound; these may be divided into exogenous factors such as the complexity of the food matrix, the chemical form of the compound of interest, structure and amount of co-ingested compounds as well as endogenous factors including mucosal mass, intestinal transit time, rate of gastric emptying, intestinal and hepatic metabolism, and the extent of conjugation and protein-binding in blood and tissues (Holst and

Williamson 2008). A prerequisite for bioavailability of any compound is its bioaccessibility in the gut, defined as the amount that is potentially absorbable from the lumen (Fernández-Garcia et al. 2009). Bioavailability can be limited by a low bioaccessibility, which can be affected by the nature of the food matrix, location within the plant, food processing, gastric and luminal digestion, in addition to the physico-chemical properties of the compound itself. In this chapter, the bioaccessibility and bioavailability of phytonutrients in potato will be reported and discussed.

The contribution of potato to human health will be described in terms of the evidence concerning the anticancer, hypocholesterolemic, anti-inflammatory, anti-obesity, and anti-diabetic role of potatoes.

2.2 Contribution to Diet

2.2.1 *Energy*

The energy provided by 100 g of boiled tubers of potatoes varies from 96.33 to 123.17 kcal (De Haan et al. 2019), which is similar to the energy provided by 100 g of cooked rice (130 kcal) but lower than the energy provided by 100 g of wheat (361 kcal), 100 g of cooked cassava (160 kcal) and soybeans (173 kcal) (King and Slavin 2013). Potato has a low energy density with 100 g of boiled potatoes contributing between 4 and 6% of the requirement of energy of an adult of between 50 and 90 kg of weight (considering 1.90 as basal metabolic rate factor, FAO/OMS/UNU 2004). However, preparing and serving potatoes with ingredients with a high fat content raises greatly the caloric value of the dish. One hundred grams of potato chips and French fries provide 529 and 564 kcal, respectively.

In areas where potato is considered as a staple food, the amount of potato intake is high and consequently the contribution of potatoes towards meeting dietary requirements is much higher. In Huancavelica, a location in the Peruvian central highlands, women have an average daily consumption of 840 and 645 g during abundance and scarcity period of potato, respectively. In those regions, potatoes are mainly eaten as boiled and provide between 28 and 38% of the recommended total energy requirements for women (De Haan et al. 2019).

2.2.2 *Carbohydrates*

2.2.2.1 Starch

Starch is the predominating carbohydrate in potato ranging from 16.5 to 20.0 g/100 g FW (Liu et al. 2007). Biochemically, potato starch is composed of amylose and amylopectin with the latter molecule typically making up 70–80% of the available starch in the tuber and the remaining portion being composed of amylose (Zeeman et al. 2010). Starch can also be classified by levels of digestibility within the human intestinal tract, i.e. rapidly digested (RDS), slowly digested (SDS), or resistant (RS) starch (Englyst et al. 1992). RDS and SDS represent the portion of starch digested within the first 20 and 21–120 min post-ingestion, respectively. The remaining resistant starch (RS) is undigested and fermented when it reaches the large intestine with the production of short-chain fatty acids (Raigond et al. 2014). Because of the resistance of the amylose structure to digestion, more of the RS component is expected to be composed of amylose rather than amylopectin. The rapid breakdown of amylopectin to digestion is the reason that it is more prevalent in RDS and SDS fractions (Bach et al. 2013). Potential health benefits attributed to SDS include satiety, improved physical performance, glucose tolerance enhancement and blood lipid level reduction in healthy individuals and in those with hyperlipidemia (Miao et al. 2015). Possible health benefits of RS include prevention of colon cancer, hypoglycemic effects, substrate provision for growth of gut probiotic microorganisms, reduction of gall stone formation, hypocholesterolemic effects, inhibition of fat accumulation, and increased absorption of minerals (Sajilata et al. 2006).

Monro et al. (2009) determined the RDS, SDS, and RS concentration of freshly cooked potatoes from nine potato varieties and found concentrations ranging from 9 to 15 g/100 g FW, from 0 to 1.72 g/100 g FW, and from 0.58 to 1.05 g/100 g FW, respectively. These authors also found that cooking and then cooling potatoes significantly increased SDS and RS (up to 7.7 g and 1.96 g/100 g FW, respectively), while RDS was significantly reduced to 7.3 g/100 g FW. This latter phenomenon is referred to as starch retrogradation, which is based upon rearrangement of the molecules of amylose and amylopectin causing decreased starch digestion (Leeman et al. 2005).

Glycemic index (GI) is a measure of the extent of the change in blood glucose content (glycemic response) following consumption of digestible carbohydrate, relative to a standard such as glucose (Venn and Green 2007). A higher GI value represents a more rapid entry of a larger quantity of glucose from a test food into the bloodstream. Based on in vivo postprandial GI, high RDS content in foods has been significantly correlated with a high glycemic index response (Champ 2004), whereas low RDS levels are associated with low to medium GI values (Lynch et al. 2007). A wide variability in GI values of potatoes has been noted ranging from high to medium to low values based on cultivar differences (Ek et al. 2012). Such variations could partly be related to differences in the amylopectin to amylose ratio as amylose-rich starches are digested more slowly due to their difficulty to gelatinize and swell as opposed to starches with a high amylopectin content (Brennan 2005). Bach et al. (2013) defined low RDS and high SDS as the optimal profile for potatoes that leads to low GI values, and identified two genotypes with this profile. Tuber cooking followed by cooling (forming retrograded starch) is also another way for the consumer to obtain lower postprandial glucose levels, and thereby benefit from reduced GI following potato intake (Fernandes et al. 2005). Lowering the dietary GI load has been associated with body weight loss, improved blood pressure, and decreased risk of cardiovascular diseases, whereas habitual intake of high GI foods has been linked to type 2 diabetes and other chronic heart issues (McGill et al. 2013). As the GI does not take into account the typical portion size, the GI value and the quantity of carbohydrates are combined to generate the glycemic load (GL) value, which can better quantify the glycemic impact of a food (Salmeron et al. 1997). Initial studies involving potatoes were limited by the sole use of GI for their glycemic evaluation (Crapo et al. 1977; Soh and Brand-Miller 1999), which categorized them with a high GI. In contrast, potatoes have been generally noted to have a medium to low glycemic impact based on the GL estimation (Lynch et al. 2007).

2.2.2.2 Sugars

Potato tubers also contain significant quantities of free sugars with glucose and fructose as the principal monosaccharides and sucrose as the major disaccharide. Glucose, fructose, and sucrose concentrations in raw tubers of tetraploid potatoes range from 3.25 to 255 mg/100 g FW, from 2.5 to 153.7 mg/100 g FW, and from 43 to 159.7 mg/100 g FW, respectively (Amrein et al. 2003; Rodríguez et al. 2010). Higher levels of glucose, fructose, and sucrose have been recently reported for diploid potato group *Phureja* with concentrations ranging from 11.5 to 701 mg/100 g FW for glucose, from 7.25 to 605 mg/100 g FW for fructose and from 159 to 737 mg/100 g FW for sucrose (Duarte-Delgado et al. 2016). The reducing sugars glucose and fructose as well as free asparagine are acrylamide precursors. Acrylamide is formed through the Maillard reaction during high temperature cooking such as frying, roasting, or baking (Muttucumaru et al. 2008). Acrylamide has been classified as "probably carcinogenic to humans" by the WHO and the International Agency for Research on Cancer. Therefore, the reducing sugar content in potatoes has been recommended not to be greater than 100 mg/100 g FW in order to keep acrylamide formation on a low level (Kumar et al. 2004). Importantly, cold storage (2–4 °C) may induce an accumulation of reducing sugars in tuber tissue leading to undesirable browning, production of bitter flavors, and increased levels of acrylamide with cooking (Neilson et al. 2017).

2.2.3 Protein

According to Camire et al. (2009), the protein content of potatoes generally ranges from 1 to 1.5 g/100 g FW depending on the cultivar. De Haan et al. (2019) reported higher levels of protein in cooked tubers of Peruvian floury landraces (1.76–2.95 g/100 g FW). Potato protein content is generally low compared with other major staples like maize and beans although potato yields more protein per unit growing area than do cereals (Bamberg and Del Rio 2005). Also, the quality of the potato protein, which reflects its digestibility and indispensable amino acid content, is very good. The biological value of potato protein—the proportion retained for growth or maintenance divided by the amount absorbed—is high. Depending on the cultivar, the biological value of potato protein is between 90 and 100 and is very similar to the biological value of whole egg protein (100) and is higher than that of soybeans (84) and legumes (73) (Camire et al. 2009).

The levels of lysine, methionine, threonine, and tryptophan are likely to limit the protein quality of mixed diets consumed by humans. Potatoes exceed the recommended levels of these indispensable amino acids, demonstrating that potato protein is of high quality. Compared with pasta, white rice, and whole grain cornmeal, potatoes are the only staple food meeting the recommended lysine level. However, sulfur-containing amino acids (methionine + cysteine) are lower in potatoes than in the other common plant staple foods (King and Slavin 2013).

2.2.4 *Lipids*

Total lipids of potatoes are low and range from 0.1 to 0.5 g/100 g FW and consist mainly of phospholipids (47%), glycol and galactolipids (22%), which are structural elements of biological membranes as well as neutral lipids (21%) such as acylglycerols and free fatty acids (Ramadan and Oraby 2016). More than 94% of the tuber lipids contain esterified fatty acids. The essential polyunsaturated fatty acids with one to three double bounds consist of mainly linoleic acid (C18:2 cis-9,12, an *n*-6 fatty acid) and linolenic acid (C18:3 cis-9,12,15, an *n*-3 fatty acid) (70–75%), precursors of a wide range of bioactive compounds generated endogenously (Galliard 1973). The composition of the fatty acids of the potato lipids is nutritionally advantageous. For example, potato consumption in the United Kingdom was estimated to provide 10 and 13% of the dietary *n*-6 and *n*-3 polyunsaturated fatty acid intake, respectively (Gibson and Kurilich 2013). In contrast, potato intake provided only 4% of saturated fatty acid and 6% trans fatty acid intake, which was largely attributed to the addition of fats and oils such as butter and margarine to potato dishes.

2.2.5 *Fiber*

Dietary fiber represents the undigested and unabsorbed carbohydrate part in the diet. These resistant carbohydrates may be fermented in the large intestine. Soluble fibers lower serum lipids, whereas insoluble fibers increase stool weight (Slavin 2008). Potatoes contain dietary fiber in their cell walls, especially in the thickened cell walls of the peel (Camire et al. 2009). Cooked potatoes without the skin provide 1.8 g fiber/100 g, FW, whereas cooked potatoes with the skin provide 2.1 g fiber/100 g FW. Potatoes contain less fiber than whole-grain cornmeal (7.3 g/100 g), but more fiber than white rice (0.3 g/100 g) or whole-wheat cereal (1.6 g/100 g). Although potatoes cannot therefore be considered a high-fiber food, they can be a significant source of fiber for individuals regularly eating potatoes, particularly in developed countries where fiber intake is generally far below recommended levels (Auestad et al. 2015). In that regard, potatoes have been indicated to contribute 14.4–26.2% of daily fiber intake in men and women living in the USA based on the National Health and Nutrition Examination Survey (NHANES) data (2009–2010).

2.2.6 Minerals

Potassium is the most abundant mineral in potato with concentrations varying from 150 to 1386 mg/100 g FW (Nassar et al. 2012). Potassium functions as an important electrolyte in the nervous system. High intake levels of potassium can help control high blood pressure and may decrease the risk of stroke (Bethke and Jansky 2018). One hundred grams of boiled potatoes can contribute up to 16% of the Adequate Intake (AI) of potassium recommended for adults (4700 mg per day).

Phosphorus and magnesium are also present in potato in moderate quantities ranging from 42 to 120 mg/100 g FW and from 16 to 40 mg/100 g FW, respectively (Bonierbale et al. 2010). One hundred grams of boiled potatoes can contribute up to 11% of the Estimated Average Requirement (EAR) of phosphorus and magnesium for adults (42–120 and 265–340 mg per day, respectively). Calcium is present in minor quantities in potato ranging from 2 to 20 mg/100 g FW; contributing no more of 2% of the EAR of calcium for adults (800–1100 mg per day).

Iron and zinc concentrations from raw potatoes range from 0.25 to 0.83 mg/100 g FW and from 0.23 to 0.39 mg/100 g FW, respectively (Burgos et al. 2007). Iron and zinc concentrations are significantly affected by the growing environment. Interestingly, Lombardo et al. (2013) reported that soil composition affects the mineral concentration of crops, with a sandy texture of the soil favoring the iron oxidation processes to insoluble polymers and consequently reducing iron availability to the plant.

Burgos et al. (2007) reported iron and zinc concentration in cooked potatoes ranging from 0.29 to 0.69 mg/100 g FW and from 0.29 to 0.48 mg/100 g FW, respectively. These values are lower than iron and zinc concentrations reported for cereals and legumes but bioavailability of iron and zinc from potatoes may be higher due to the presence of high levels of ascorbic acid—which facilitates iron absorption in the human body—and low levels of phytic acid, an inhibitor of iron and zinc absorption. It has been recently demonstrated that the bioaccessibility of iron in potato is higher than that reported in crops such as wheat and beans. Approximately 63–79% of the potato iron is released from the food matrix after in vitro gastrointestinal digestion, and therefore available for intestinal absorption (Andre et al. 2015).

In the Andean highlands, where there is little access to meat and the levels of anemia and malnutrition are high, potatoes are an important dietary source of iron due to their high consumption. For example, in Huancavelica, in the Peruvian highlands, women and children consume on average 840 and 200 g of potato per day, respectively (De Haan et al. 2019). Similarly, in parts of Rwanda and other African countries, women consume an average of 400 g of potatoes per day. Therefore, improving the iron and zinc concentrations of potato and their bioavailability would have a real impact to contribute to reduce malnutrition and improve life quality in these and other areas where anemia and/or stunted growth are still pervasive.

The International Potato Center (CIP) has been working for the past 15 years on potato mineral biofortification to increase the concentration of iron and zinc in this crop. The CIP Biofortification Potato Program started from a baseline of 0.48 mg/100 g FW for iron and 0.35 mg/100 g FW for zinc. After three cycles of breeding and recurrent selection, concentrations of the first biofortified potatoes reach 0.73 mg iron and 0.63 mg zinc/100 g FW. Considering 400 g of potato consumption for women of the Andes, the consumption of biofortified potatoes would cover 41 and 37% of the EAR of iron and zinc in women.

Presently, CIP is combining the first products of its biofortification breeding program with advanced breeding lines to release new varieties that will be able to withstand major potato pests and diseases, tolerate heat and drought, providing high yields, and respond to preferences of farmers and consumers.

2.2.7 *Vitamins*

Potatoes are a good source of ascorbic acid (vitamin C) and pyridoxine (vitamin B_6). Vitamin C as an antioxidant plays an important role in protection against oxidative stress. Vitamin C is an important free radical scavenger of reactive oxygen species such as hydroxyl radicals, superoxide anions, singlet oxygen, and hydrogen peroxide that can cause tissue damage resulting from lipid peroxidation, DNA breakage or base alterations, which may contribute to degenerative diseases such as heart disease or cancer (Bates 1997). In addition, due to its participation in the oxidation of transition metal ions, vitamin C also plays an important role in enhancing the bioavailability of non-haem iron (Teucher et al. 2004) and serves as a cofactor in the synthesis of collagen needed to support cardiovascular function, maintenance of cartilage, bones, and teeth, as well as wound healing (Naidu 2003).

Fresh potatoes have varying concentrations of vitamin C, which can reach 50 mg/100 g FW (Han et al. 2004) when they are freshly harvested. Significant variation in vitamin C concentrations of potatoes occur due to genotype and environment and genotype by environment interactions (Andre et al. 2007; Burgos et al. 2009).

Cooking and storage reduce the concentration of vitamin C in potato tubers. In addition, there are differences in the degree of reduction of vitamin C content depending on the cooking types. Retention levels of vitamin C after boiling in 20 native landraces varied between 50 and 90%. The losses may be caused by: (1) leaching into cooking water, (2) destruction by heat treatment, and (3) oxidation. It is interesting to note that the peel forms a barrier preventing loss of nutrients during cooking. As a consequence, boiling potato when it is peeled results in 10% more loss of vitamin C or phenolic compounds than if it is cooked with the peel (Woolfe and Poats 1987). Retention levels after storing under farming conditions has been shown to vary between 22 and 62%, depending on the variety (Burgos et al. 2009). Retention levels of vitamin C in 12 genotypes grown in Colorado state in the USA after 7 months of cold storage was less than 50% (Külen et al. 2012).

One hundred grams of cooked potatoes with vitamin C levels around 20 mg/100 g FW can provide between 27 and 33% of the EAR of vitamin C for an adult (75 for males and 60 for females, according to FAO/WHO 2001). One hundred grams of cooked potatoes contains lower concentrations of vitamin C than 100 g of cooked broccoli (68–108 mg/100 g FW; depending on the way of cooking; Yuan et al. 2009), 100 g of cooked spinach (44–79 mg/100 g FW; depending on the way of cooking; Zeng 2013) and 100 g of raw red pepper (up to 200 mg/100 g; Wahyuni et al. 2011). However, it is noteworthy that the final contribution of a particular food to the total intake of vitamin C depends on the total amount consumed in the diet and so potatoes may therefore contribute to a significant extent to the total dietary intake of vitamin C (Love and Pavek 2008). In that respect, potatoes have been estimated to provide over 50% of the daily vitamin C requirement in the USA and approximately 20% of the dietary vitamin C intake in Europe (Love and Pavek 2008).

Vitamin B_6, also called pyridoxine, is a versatile cofactor for key metabolic processes (Hellmann and Mooney 2010) that plays a major role in various cellular reactions and also confers several health benefits for humans, which may be partly attributed to its antioxidant capabilities (Fitzpatrick et al. 2012). It helps in maintaining normal nerve function and plays a crucial role in the synthesis of neurotransmitters such as dopamine and serotonin. Vitamin B_6 also assists normal nerve cell communication and acts as a coenzyme in the breakdown and utilization of carbohydrates, fats and proteins. In plant, it is a potent antioxidant, critical for plant pathogen resistance (Spinneker et al. 2007).

Potatoes are considered to be a good dietary source of vitamin B_6, with concentrations ranging from 0.450 mg/100 g FW to 0.675 mg/100 g FW (Moonney et al. 2013). Physical and chemical factors such as heat, light exposure, and pH also influence vitamin B_6 content, but this vitamin is relatively stable during storage (Fitzpatrick et al. 2012).

The mean concentration of vitamin B_6 in cooked potatoes (0.299 mg/100 g FW) is higher than the mean concentration of other staple crops such as maize (0.139 mg/100 g FW), rice (0.050 mg/100 g FW), cassava (0.051 mg/100 g FW), and wheat (0.034 mg/100 g FW) (Fudge et al. 2017). One hundred grams of cooked potatoes can provide between 17 and 23% of the Recommended Dietary Allowance (RDA) of B_6 from an adult (1.3–1.7 mg per day).

Potato tuber contains also moderate amount of vitamin E (Chitchumroonchokchai et al. 2017). Vitamin E is the collective name for a set of eight related tocopherols and tocotrienols, characterized by a hydrophobic isoprenoid tail and a more hydrophilic chromanol head (Bramley et al. 2000). In potato, significant amount of α-tocopherol has been found in raw tubers of Andean genotypes, ranging from 68 to 517.5 μg/100 g FW (recalculated from Andre et al. 2007), whereas amounts in commercial varieties varied between 15 and 75 μg/100 g FW (recalculated from Andre et al. 2007 and Chun et al. 2006).

In humans, as in plants, vitamin E is located primarily within the phospholipid bilayer of cell membranes. It reacts with and quenches free radicals in cell membranes, preventing polyunsaturated fatty acids from damage by lipid oxidation. Vitamin E deficiency has been associated with an elevated risk of artherosclerosis and other degenerative diseases. It is generally assumed that increases of α-tocopherol in the diet may contribute to a decreased risk of chronic diseases (Andre et al. 2010). The EAR for vitamin E is of 15 mg for women and men (Otten et al. 2010). The consumption of high vitamin E containing potato tubers, such as the Andean varieties, could therefore significantly increase the dietary vitamin E intake.

2.2.8 Antioxidants

Potato is one of the most important sources of antioxidants in the human diet (Lachman and Hamouz 2005). As such, it supports the antioxidant defense that reduces cellular and tissue toxicities that result from free radical-induced protein, lipid, carbohydrate, and DNA damage (Andre et al. 2010). In this way, potato antioxidants may reduce the risk for cancers, cardiovascular diseases, and type 2 diabetes.

Based on metabolic relationships and structural composition, there are three major groups of antioxidants present in potato, as in most plants. The first group consists in the aromatic phenolic compounds, which encompasses flavonoids including anthocyanins and flavonols produced by the flavonoid pathway, hydroxycinnamic acids and their derivatives produced by the phenylpropanoid pathway, and the amino acids tyrosine, phenylalanine, and tryptophan produced by the shikimate pathway. The second group encompasses the isoprenoid antioxidants such as the carotenoids and tocopherols; and the third group includes antioxidants related to ascorbate and glutathione functions in a redox system of compound-recycling that include ascorbic acid (Lovat et al. 2016).

2.2.9 Phenolics

Phenolic compounds, also known as polyphenols, constitute one of the most widely distributed group of dietary antioxidants in the plant kingdom, presenting more than 10,000 different structures, ranging from relatively simple phenols to complex polymers such as lignans and suberins. Phenolic compounds are produced in the cytoplasm and are subsequently transported in the vacuole or deposited in the cell wall. Routes to the major classes of phenolic compounds involve: (1) the core phenylpropanoid pathway from phenylalanine to an activated (hydroxy)cinnamic acid derivative, as well as specific branch pathways for the formation of (2) simple phenolic acids, lignins and lignans, (3) flavonoids, (4) tannins, and (5) stilbenes (Andre et al. 2009). Their aromatic cycles can be further modified through hydroxylations, methylations, glycosylations, acylations, or prenylations, extending their variability and complexity (Winkel-Shirley 2001). Phenolic acids include chlorogenic, caffeic, ferulic, and sinapic acids. Among flavonoids, anthocyanins are natural pigments, responsible for the red-blue color of many fruits and vegetables. Anthocyanins can also impact the organoleptic characteristics of foods, which may influence their technological behavior during food processing and also have implications in the field of human health (Pascual-Teresa and Sanchez-Ballesta 2008). Flavonols represent one of the most widespread flavonoid classes in plant and include compounds like quercetin and kaempferol that are most commonly found in their glycosylated form, i.e., linked with glucose or rutinose. As compared to other phenolic compounds, flavonol concentrations are known to be largely influenced by the environmental conditions during plant growth (Lancaster et al. 2000).

Phenolic compounds are considered to be health-promoting phytochemicals as they have shown in vitro antioxidant activity and have been reported to exhibit beneficial anti-bacterial, hypoglycemic, anti-viral, anti-carcinogenic, anti-inflammatory and vasodilatory properties (Duthie et al. 2000; Mattila and Hellstrom 2006).

2.2.9.1 Chlorogenic Acid

Chlorogenic acid has been reported as the predominant phenolic acid in raw and boiled potato tubers (Burgos et al. 2013b). The main function of chlorogenic acid in the plant appears to defend against pathogens. Concentrations of chlorogenic acid as well as other hydroxycinnamic acids are significantly induced following pathogen invasion, and deposited to enforce the cell walls to arrest pathogen development (Yogendra et al. 2015). In humans, these compounds consumed through diet are increasingly considered as effective protective agents against reactive oxygen species (ROS), which are known to be involved in aging and many degenerative diseases (Liang and Kitts 2015).

The isomers of chlorogenic acid, neo-chlorogenic acid, and crypto-chlorogenic acid, as well as caffeic acid are also found in potato tubers (Andre et al. 2007). Potatoes contain three isomers of chlorogenic acid depending on whether the hydroxycinnamate is attached to 3-, 4-, or 5-position of the quinic acid moiety with 5-*O*-caffeoylquinic acid as the principal chlorogenic acid. The 5-*O*-caffeoylquinic acid (CQA) isomer is also the principal chlorogenic acid component of coffee and apples (Stalmach et al. 2010; Clifford 1999). In vitro and ex vivo studies have demonstrated a reduction in oxidation of human LDL following the consumption of coffee suggesting that 5-*O*-CQA protects against in vitro oxidation of human LDL, a key step in the formation of atherosclerotic plaques (Natella et al. 2007; Richelle et al. 2001). 5-*O*-CQA has also been shown to exert anti-carcinogenic effects in animal models (Stalmach et al. 2010).

Lachman et al. (2013) have reported chlorogenic acid concentrations of raw potatoes ranging from 7.87 to 60.07 mg/100 g FW in nonpeeled potatoes and from 5.11 to 46.13 mg/100 g FW in their peeled counterparts, while Burgos et al. (2013b) report a wider range of variation in the chlorogenic acid concentration of raw purple potatoes (ranging from 63 to 329.75 mg/100 g FW). Boiling, baking, and microwaving reduce the chlorogenic acid concentration of potatoes with boiled tubers having a higher retention of chlorogenic acid than baked and microwaved ones (Lachman et al. 2013). In a recent study conducted by Piñeros-Niño et al. (2017), the chlorogenic acid concentration of cooked tubers from 193 potato varieties ranged from 19.25 to 399 mg/100 g FW. In a previous study by Burgos (2014), the chlorogenic acid concentration in cooked tubers of purple-fleshed cultivars ranged from 36.17 to 395.73 mg/100 g FW and in red-fleshed cultivars from 14.45 to 48.60 mg/100 g FW.

The highest concentration of chlorogenic acid reported in 100 g of cooked potato tubers is similar to the maximum amount provided by a single cup of coffee (350 mg chlorogenic acid; Clifford 1999) and is tenfold higher than the maximum amount provided by whole apples (38.5 mg/100 g FW, Spanos and Wrolstad 1992).

Chlorogenic acid is only partially bioavailable and its bioactivity may be modulated by the gut microbiota that can generate bioactive secondary microbial phenolic metabolites such as caffeic acid that have much greater bioavailability (Tomas-Barberan et al. 2014; Olthof et al. 2003). Chlorogenic acid may also promote a healthy gut microbiome. In a batch culture fermentation model of the colon, chlorogenic acid was found to promote growth of Bifidobacterium bacterial species that could be beneficial for gut health (Mills et al. 2015).

2.2.9.2 Anthocyanins

Anthocyanins are a class of water-soluble flavonoids, which show a range of pharmacological effects, such as prevention of cardiovascular disease, obesity control, and anti-tumor activity. Their potential anti-tumor effects are reported to be based on a wide variety of biological activities including antioxidant, anti-inflammation, anti-mutagenesis, induction of differentiation, inhibiting proliferation by modulating signal transduction pathways, inducing cell cycle arrest, and stimulating apoptosis or autophagy of cancer cells; anti-invasion; anti-metastasis; reversing drug resistance of cancer cells and increasing their sensitivity to chemotherapy (Lin et al. 2017).

Anthocyanins are present in the flesh and skin of several purple- and red-fleshed potatoes such as those landraces found in the Andes, which show a wide range of anthocyanin structures and concentrations that are largely cultivar-dependent (Brown et al. 2003) and location-dependent (Ieri et al. 2011). Increased height above sea level, higher annual sum of precipitation, and lower annual average temperatures cause higher anthocyanin concentrations (Lachman et al. 2009).

The total anthocyanin concentration of raw and cooked purple-fleshed potatoes ranges from 63 to 588 mg/100 g FW and from 71 to 453 mg/100 g FW, respectively (Burgos et al. 2013a, b). Total anthocyanin concentration of cooked red-fleshed potatoes ranges from 8.2 to 55.3 mg/100 g FW (Burgos 2014).

Giusti et al. (2014) identified five major anthocyanidins (cyanidin, petunidin, pelargonidin, peonidin, and malvidin) in extract from purple potato and three major anthocyanidins (cyanidin, pelargonidin, and peonidin) in extracts of red potatoes. The extract of purple potatoes contained four major anthocyanins: cyanidin-3-rutinoside-5-glucoside, petunidin-3-rutinoside-5-glucoside, pelargonidin-3-rutinoside-5-glucoside, and peonidin-3-rutinoside-5-glucoside, with petunidin and peonidin glycosides being the most predominant. The extract of red-fleshed potatoes contained four major anthocyanins: cyanidin-3-rutinoside-5-glucoside, pelargonidin-3-rutinoside-5-glucoside, peonidin-3-rutinoside-5-glucoside, and pelargonidin-3-rutinoside, with pelargonidin-3-rutinoside-5-glucoside being the most predominant.

Burgos (2014) characterized the anthocyanin profile of 12 purple-fleshed accessions and 6 red-fleshed accession from CIP's genebank and found that in purple-fleshed accessions the predominant anthocyanin is petunidin-3-(coumaroyl) rutinoside-5-glucoside (petanin), representing from 37 to 78% of the total anthocyanins. It is followed by peonidin-3-(coumaroyl) rutinoside-5-glucoside, cyanidin-3-(coumaroyl) rutinoside-5-glucoside, and minor proportions of malvidin 3-(coumaroyl) rutinoside-5-glucoside and pelargonidin-3-(coumaroyl) rutinoside-5-glucoside. In red-fleshed accessions, the predominant anthocyanin is pelargonidin-3-(coumaryl) rutinoside-5-glucoside, representing 41–75% of the total anthocyanins. It is followed by peonidin-3-(coumaroyl) rutinoside-5-glucoside, pelargonidin-3-rutinoside, and cyanidin-3-(coumaroyl) rutinoside-5-glucoside in various proportions and then by pelargonidin-3-(coumaryl) rutinoside in minor proportions. Figure 2.2 shows as an example the anthocyanin profile for two purple-fleshed accessions (CIP 705534 and CIP 702363) and two red-fleshed accessions (CIP 703625 and CIP 702453). Pt3(c)R5G represented by the purple bar is dominant in the purple-fleshed accessions while PI3R(c)R5G represented by the pink bar is dominant in the red-fleshed potatoes.

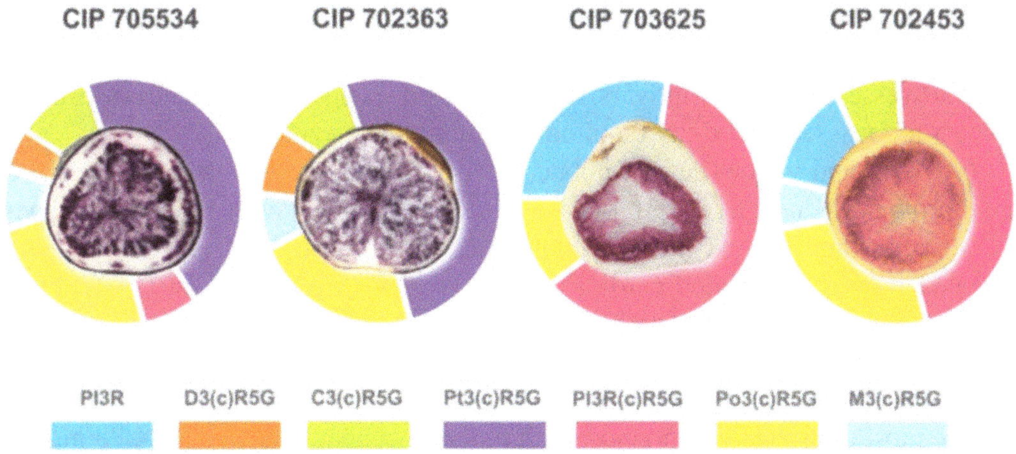

Fig. 2.2 Anthocyanin profile in purple-fleshed and red-fleshed potatoes. (Pl3R: pelargonidin-3-rutinoside, Pl3R(c)R5G: pelargonidin-3-(coumaroyl) rutinoside-5-glucoside, Pt3(c)R5G: petunidin-3-(coumaroyl) rutinoside-5-glucoside, Po3(c)R5G: peonidin-3-(coumaroyl) rutinoside-5-glucoside, C3(c)R5G: cyanidin-3-(coumaroyl) rutinoside-5-glucoside, M3(c)R5G: malvidin 3-(coumaroyl) rutinoside-5-glucoside)

The most prominent anthocyanins present in the red- and purple-fleshed accessions are acylated with hydroxycinnamic acid (Fossen and Andersen 2000). Three different cinnamic acids were found acylating the anthocyanins in the extract of purple and red potatoes: caffeic, *p*-coumaric and ferulic acid (Giusti et al. 2014). Acylated anthocyanins are known to be stable and hence can be considered as promising natural colorants for the food industry.

The highest anthocyanin concentration reported in a dark purple-fleshed potato (above 400 mg/100 g FW; Andre et al. 2007) is lower than in blueberries (558 mg/100 g, FW; Hosseinian and Bea 2007), cranberries (589 mg/100 g FW; Wada and Ou 2002), eggplant (750 mg/100 g FW; Wu et al. 2006), and purple corn (1642 mg/100 g FW; Cevallos-Casals and Cisneros Zevallos 2003). However, the contribution of purple-fleshed potatoes to the diet can be considerably higher considering the high mean intake of potatoes in some areas like the Andean highlands where consumption may reach 500 g per day, as compared to the mean intake of blueberries, cranberries, and eggplant (1 g per day in the United States; Wu et al. 2006).

Kubow et al. (2017) studied the biotransformation of anthocyanins from cooked purple-fleshed potatoes using a dynamic human gastrointestinal (GI) model that includes stomach, small intestine, and colonic vessels. After 24 h digestion, liquid chromatography-mass spectrometry identified 15–36 anthocyanin species throughout the GI vessels. Genetic background of the purple potato cultivars led to major variances in the pattern of anthocyanin breakdown and release during digestion composition. Diminished concentrations of several anthocyanin species in the colonic vessels indicated microbial biotransformation which is, in turn, associated to increased bioaccessibility.

The cytotoxicity and cell viability of colonic Caco-2 cancer cells and nontumorigenic colonic CCD-112CoN cells after 24 h exposure to colonic fecal water of purple-fleshed potato digests has been also tested by Kubow et al. (2017). The cultivar Leona showed a significant potency to induce cytotoxicity and decrease viability of Caco-2 cells. The differing microbial anthocyanin metabolite profiles in colonic vessels between cultivars were indicated to play a significant role in the impact of fecal water toxicity on tumor and nontumorigenic cells.

In white- and yellow-fleshed potato tubers, flavonols predominate in the flavonoid profile (Andre et al. 2007). Flavonols have been extensively studied in the past 10 years as they present a range of putative health-promoting effects, including reduced risk of cancer and cardiovascular diseases (Wang et al. 2016). Rutin in particular has shown strong antioxidative and anti-inflammatory effects at the cellular level (Habtemariam and Lentini 2015).

In potato tubers, rutin (quercetin-3-*O*-rutinoside) and kaempferol-3-*O*-rutinoside are the most important compounds, with reported concentrations of 0–4.78 mg and 0–5.68 mg/100 g FW, respectively, in Andean raw potato tubers (Andre et al. 2007). The influence of various cooking methods on potato flavonols has been investigated, which revealed the stability of the concentrations through treatment (Navarre et al. 2010). The bioaccessibility of these compounds was also high (close to 100% on average) when evaluated in a collection of 12 Andean potato genotypes (Andre et al. 2015).

2.2.10 Carotenoids

Potatoes contain lipophilic phytonutrients in the form of carotenoids that have numerous health-promoting properties including decreasing risk of several chronic diseases (Gammone et al. 2015; Wu et al. 2015). Carotenoids have been reported to exhibit chemoprevention by a variety of mechanisms including immune system activation, protection against oxidative stress, promotion of gap junction communication, inhibition of DNA damage, enhanced metabolic detoxification, and tumor suppressor action and inhibition of oncogene expression (Khachik et al. 1999; Fiedor and Burda 2014).

Potato carotenoid concentrations and profiles are related to the flesh color with dark yellow cultivars showing approximately tenfold higher concentrations of total carotenoids than white-fleshed varieties (Brown et al. 2005). Significant and predominant amounts of zeaxanthin and antheraxanthin are found in deep yellow-fleshed potatoes while the carotenoid profile of yellow potatoes is composed of violaxanthin, antheraxanthin, lutein, and zeaxanthin and that of cream-fleshed potatoes of violaxanthin, lutein, and β-carotene (Burgos et al. 2009).

The violaxanthin, antheraxanthin, lutein, and β-carotene concentration of raw tubers of potatoes from the *Tuberosum* group ranged from 1.5 to 87.8 μg/100 g FW, 0.6 to 15.8 μg/100 g FW; 1.6 to 35.1 μg/100 g FW; and 0.1 to 2.1 μg/100 g FW, respectively (Fernandez-Orozco et al. 2013), while the concentration of these carotenoids in tubers from the *Phureja* group ranged from 20.0 to 410 μg/100 g FW; 9.3 to 503 μg/100 g FW; 55 to 211 μg/100 g FW and 4.8 to 27 μg/100 g FW, respectively (Burgos et al. 2009), and in tubers from the Andigenum group from 14.3 to 173 μg/100 g FW, 7 to 16 μg/100 g FW; 43.3 to 442 μg/100 g FW and 10.5 to 54.8 μg/100 g FW, respectively (Andre et al. 2007).

Boiling does not affect the lutein and zeaxanthin concentration of potato; however, violaxanthin and antheraxanthin concentrations of potatoes are significantly reduced after boiling. Lutein and zeaxanthin concentrations of cooked yellow-fleshed potatoes ranged from 73 to 253 μg/100 g FW and from 0 to 1048 μg/100 g FW, respectively (Burgos et al. 2012) with deep yellow-fleshed potatoes being a significant source of zeaxanthin (above 500 μg/100 g FW).

Lutein and zeaxanthin are important dietary carotenoids that are selectively taken up into the macula of the eye, where they protect against development of age-related macular degeneration and cataracts (Wu et al. 2015). Moreover, these compounds have been reported to have other health-promoting effects, including immune-enhancement and reduction of the risk of developing degenerative diseases such as cancer and cardiovascular diseases (Krinsky and Johnson 2005). There is no recommended daily intake for lutein and zeaxanthin, but many studies show a health benefit for lutein supplementation at 10 mg per day and zeaxanthin at 2 mg per day (American Optometric Association 2009).

The highest values of lutein and zeaxanthin reported in 100 g of yellow-fleshed potatoes are lower compared to the amount of lutein provided by 100 g of lettuce (540 μg; Kimura and Rodriguez-Amaya 2003), broccoli (3250 μg; Khachick et al.

1992), parsley (5800 μg; Hart and Scott 1995), or spinach (4180 μg; Tee and Lim 1991); and lower than the amount of zeaxanthin provided by 100 g of maize at its highest zeaxanthin concentration (3800 μg/100 g) (Brenna and Berardo 2004) and of red paprika (2200 μg/100 g) (Müller 1997; Mínguez Mosquera and Hornero-Méndez 1994). However, it is important to consider that potato consumption can be as high as 500 g per day whereas the mean intake of the above-mentioned vegetables is less than 50 g per day; hence, the overall contribution of potato-based carotenoids to the dietary intake can be higher. Furthermore, the contribution of a food source to lutein and zeaxanthin intake depends on their digestive stability, bioaccessibility, and bioavailability in the respective food matrix. Bioaccessibility refers to the proportion of ingested carotenoid that is released from the food matrix and incorporated into micelles in the gastrointestinal tract, and thus available for intestinal absorption (Rodriguez-Amaya 2015). Bioavailability refers to the portion of the carotenoid that is absorbed in the body, enters in systemic circulation and becomes available for utilization in normal physiological functions or for storage in the human body (van Het Hof et al. 2000). The bioavailability of carotenoids from plant foods is influenced by the species and structure of carotenoids present in the food, composition, and release of carotenoids from the food matrix, absorption in the intestinal tract, transportation within the lipoprotein fractions, biochemical conversions, and tissue-specific depositions, as well as by the nutritional status of the ingesting consumer (Bohn 2017). Research on bioavailability of potato carotenoids is required to have more useful and complete information regarding their nutritional and health benefits.

Burgos et al. (2013a) evaluated the in vitro digestive stability and the efficiency of micellarization or bioaccessibility of lutein and zeaxanthin in yellow-fleshed potatoes. The gastric and duodenal digestive stability of lutein and zeaxanthin in boiled tubers ranged from 70 to 95% while the bioaccessibility ranged from 33 to 71% for lutein and from 51 to 71% for zeaxanthin. A more recent study has reported that bioaccessibility of lutein and zeaxanthin in yellow-fleshed clones range from 76 to 82% for lutein and from 24 to 55% for zeaxanthin (Andre et al. 2015).

The maximum bioaccessible lutein concentration reported in yellow-fleshed potatoes is around 300 μg/100 g FW while the maximum bioaccessible zeaxanthin concentration is around 600 μg/100 g FW. Considering the mean potato intake in the Andes of Peru, Ecuador, and Bolivia (500 g per day), potato tubers from the variety with the highest bioaccessible lutein could provide 14% of the suggested level of lutein intake for having health benefit (10 mg per day). Likewise, potato tubers of the variety with the highest bioaccessible zeaxanthin concentration could provide 50% more than the suggested level of zeaxanthin intake (2 mg per day). In Spain, potato was shown to contribute 13–20% towards the total dietary intake of zeaxanthin and was ranked as the third main contributor after citrus fruits and green vegetables (Garcia-Closas et al. 2004). More population studies are needed, however, regarding the nutritional and health contributions of carotenoids as provided by potatoes.

2.3 Antioxidant Activity

Antioxidant activity (AA) describes the capacity of redox molecules in foods and biological systems to scavenge free radicals considering the additive and synergistic effects of all antioxidants rather than the effect of single compounds, and may, therefore, be useful to study the potential health benefits of antioxidants on oxidative stress-mediated diseases (Puchau et al. 2010). In that context, mixtures of phytochemicals found in plant foods are more effective in improving antioxidant status than isolated phytochemicals (DeGraft-Johnson et al. 2007). The antioxidant activity of foods as assessed by indices such as ferric reducing ability of plasma (FRAP) has been indicated to be a valid and reproducible determinant of human plasma AA measurements (Rautiainen et al. 2008). Antioxidant capacities of food staples such as potatoes could thus potentially affect the antioxidant status of that population.

The antioxidants in potato are mainly hydrophilic (polyphenols, ascorbic acid, anthocyanins, and flavanols) (Fig. 2.3) (Reyes et al. 2005). In white- or yellow-fleshed potatoes, prevalent contributors of AA are chlorogenic acid, gallic acid, caffeic acid, and catechin (Reddivari et al. 2007a), while in purple- and red-fleshed potatoes the major contributors to AA are anthocyanins and chlorogenic acid (Lachman et al. 2009). Potatoes also contain lipophilic antioxidants (carotenoids and vitamin E) (Fig. 2.4).

Because antioxidant activity of potato anthocyanins results from the synergistic effect of each anthocyanin pigment (Hayashi et al. 2003), it is important to assess different pigmented potato cultivars for individual anthocyanidin content, as well as the contribution of the anthocyanidin composition to their antioxidant activity. A high degree of hydroxylation and/or methoxylation of individual anthocyanidins could contribute in conjunction with other phenolics to high AA (Lachman et al. 2009).

Burgos et al. (2013b) reported that boiled potatoes of purple-fleshed potato varieties have an AA ranging from 4017 to 17,304 mg Trolox equivalents (TEq)/g, FW as determined by the 2,2-azino-bis-3-ethylbenzthiazoline-6-sulfonic acid (ABTS) antioxidant capacity measure and from 2369 to 9754 mg TEq/g, FW as determined by 1,1-diphenyl-2-picryl-hydrazyl (DPPH) antioxidant capacity assay. Compared to other sources of antioxidants, potato has lower AA than strawberry, blackberry, and blueberry as determined by the ABTS assay (around 25,030–50,000 mg TEq/g, FW, Garcia-Alonso et al. 2004). However, as indicated above the overall contribution of potato to the antioxidant intake of a population will finally depend in the amount of potatoes typically consumed.

Ombra et al. (2015) reported after simulated gastrointestinal digestion, the extracts from purple potato have high in vitro antioxidant, antimicrobial, and anti-proliferative activities against the colon cancer cells Caco-2 and SW48 and the breast cancer cells MCF-7 and MDA-MB-231.

After digestion of cooked tubers from purple-fleshed potatoes using a dynamic human gastrointestinal model, Kubow et al. (2017) found an increased FRAP anti-

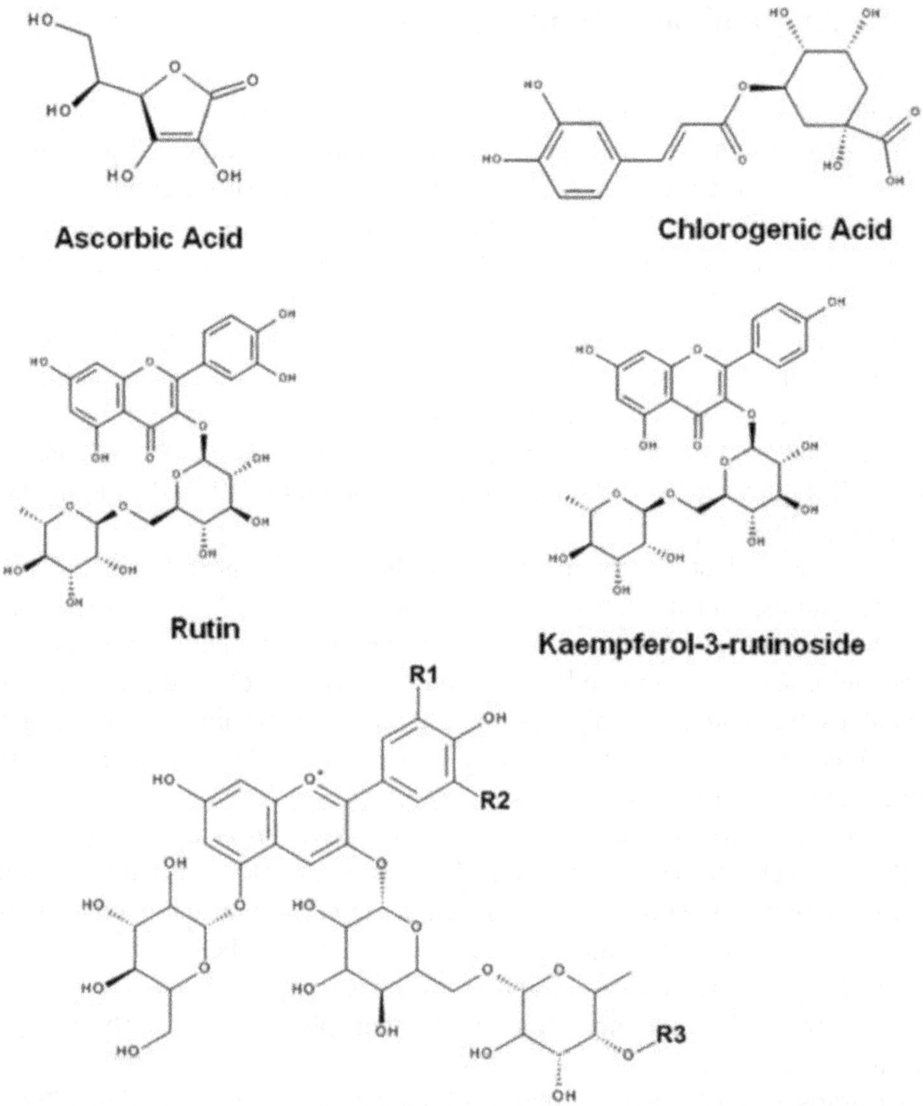

Anthocyanins	R1	R2	R3
pet-3-coum-rut-5-glc	OMe	OH	p-coumaric acid
peo-3-coum-rut-5-glc	OMe	H	p-coumaric acid
cyan-3-coum-rut-5-glc	OH	H	p-coumaric acid
mal-3-coum-rut-5-glc	OMe	OMe	p-coumaric acid
pel-3-coum-rut-5-glc	H	H	p-coumaric acid

Fig. 2.3 Chemical structure of hydrophilic antioxidants in potato

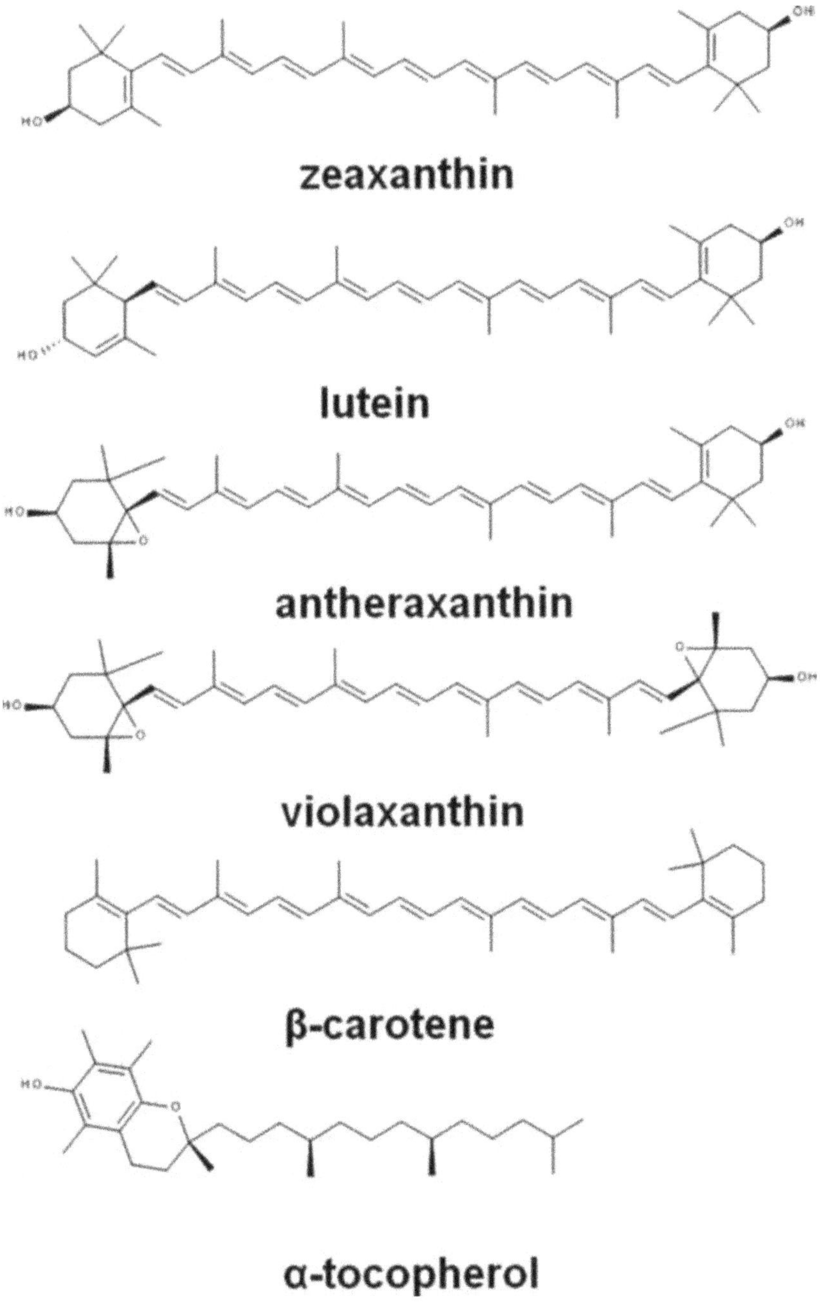

Fig. 2.4 Chemical structure of lipophilic antioxidants in potato

oxidant activity in the colonic reactors. Metabolic microbial breakdown of anthocyanins over a 24 h period appeared to generate sufficient amounts of microbial metabolites to produce an improvement in antioxidant capacity. Anthocyanins and their metabolites can, via antioxidant activity, provide protection for intestinal cells against oxidative stress in the gut, and hence alleviate gut inflammation, protect against colorectal cancer, and generally enhance colorectal health.

2.4 Glycoalkaloids

Glycoalkaloids are secondary plant metabolites that serve as natural defenses against bacteria, fungi, viruses, and insects (Friedman 2004). They can be toxic for humans when present in high concentrations, and can impart a bitter taste to potatoes. However, glycoalkaloids and hydrolysis products without the carbohydrate side chain (aglycones) also have beneficial effects that include: lowering of cholesterol (Friedman et al. 2003) and inhibition of the growing of cancer cells in culture as well as tumor growth in vivo (Friedman 2015).

Although there are many glycoalkaloids, α-chaconine and α-solanine make up 95% of the total glycoalkaloids present (Friedman et al. 1997); α-solanine is found in greater concentrations than α-chaconine, and α-solanine has only half as much specific toxic activity as a α-chaconine (Lachman et al. 2001).

Experiments with human taste panels revealed potato varieties with glycoalkaloid levels exceeding 14 mg/100 g FW tasted bitter (Friedman 2006). Those in excess of 22 mg/100 g FW also induced mild to severe burning sensations in the mouths and throats of panel members.

Glycoalkaloid levels vary greatly in different potato varieties and may be influenced by factors such as light, mechanical injury, and storage. They are also influenced by stress such as heat and drought during production. This raises concern for maintaining the quality of potatoes under climate change (Andre et al. 2009), and suggests increased attention may be needed to glycoalkaloid concentrations of potato varieties bred for or grown in warm environments.

Glycoalkaloid concentration of raw potatoes ranges from 0.7 to 18.7 mg/100 g FW (Friedman et al. 2003). Peeling significantly reduced the glycoalkaloid levels in the tubers: solanine to 43.6% and chaconine to 31% (Lachman et al. 2013). Cooking also significantly reduced the levels of glycoalkaloids (Tajner-Czopek et al. 2008), with boiling reducing the levels of glycoalkaloids more than baking and microwaving (Lachman et al. 2013).

Glycoalkaloid content in potato tubers should not exceed 20 mg/100 g FW, because this level is dangerous for human health (Ruprich et al. 2009). The toxicity of glycoalkaloids at appropriate high levels may be due to adverse effects such as anticholinesterase activity on the central nervous system and to disruption of cell membranes adversely affecting the digestive system and general body metabolism (Friedman et al. 2003). The toxicity of glycoalkaloids is associated with the synergistic interaction between two main components of glycoalkaloids: α-solanine and α-chaconine.

However, glycoalkaloids also have anti-carcinogenic properties. Exposure of cancer cells to glycoalkaloids produced potatoes (α-chaconine and α-solanine) or their hydrolysis products (mono-, di-, and trisaccharide derivatives and the aglycones solasodine, solanidine, and tomatidine) inhibits the growth of the tumor cells in culture as well as in vivo tumor growth (Friedman 2015). On the basis of the anti-carcinogenic properties of these potato components, it is conceivable that the levels typically noted in commercial potatoes might help to protect against multiple cancers. Epidemiological studies, however, are needed to substantiate this possibility.

2.5 Contribution to Health

Population-based epidemiological studies have emphasized the importance of nutrition to combat metabolic disorders emerging worldwide that have been associated with diet, such as diabetes, cancer, and cardiovascular diseases. In that regard, higher intakes of fruits and vegetables have been consistently indicated to exert protective effects against such chronic diseases (Dragsted et al. 2006). Potato has been underappreciated relative to other vegetables as it has been subject to controversy such as being labeled as a contributor to development of obesity and diabetes (Burlingame et al. 2009). On the other hand, potatoes contain relatively high concentrations of key phytonutrients that have shown bioactivities that could counteract chronic disease development (Ezekiel et al. 2013). Potatoes have shown promising health-promoting effects in human cell culture, experimental animals, and human clinical studies, including anti-cancer, hypocholesterolemic, anti-inflammatory, anti-obesity, and anti-diabetic properties. Nutritional compounds of potatoes such as phenolics, anthocyanins, fiber, starch as well as compounds considered anti-nutritional such as glycoalkaloids, lectins, and proteinase inhibitors are believed to contribute to the health benefits of potatoes (Fig. 2.5). As there are many biological activities attributed to the compounds present in potato, some of which could be beneficial or detrimental depending on specific circumstances, long-term studies investigating the association between potato consumption and diabetes, obesity, cardiovascular disease, and cancer while controlling for fat intake are needed (Visvanathan et al. 2016).

2.5.1 Anticancer Effect

Several studies have shown a reduction in proliferation of cancer cells when treated with potato extracts. Potato antioxidants such as phenolic acids and anthocyanins, glycoalkaloids, fiber, and proteinase inhibitors identified in potatoes have been implicated in the suppression of cancer cell proliferation in vitro and in vivo.

2.5.1.1 Role of Potato Antioxidants

Phenolic acids and anthocyanins are potato antioxidants that have reported anti-carcinogenic activity. Hayashi et al. (2006) reported that anthocyanins in steamed purple and red potatoes suppressed the growth of benzopyrene-induced stomach cancer in mice. Reddivari et al. (2007b) found that the anthocyanin fractions from potato extracts were cytotoxic to prostate cancer cells through activation of caspase-dependent and caspase-independent pathways. Madiwale et al. (2011) reported that purple flesh potatoes rich in anthocyanins suppressed proliferation and elevated

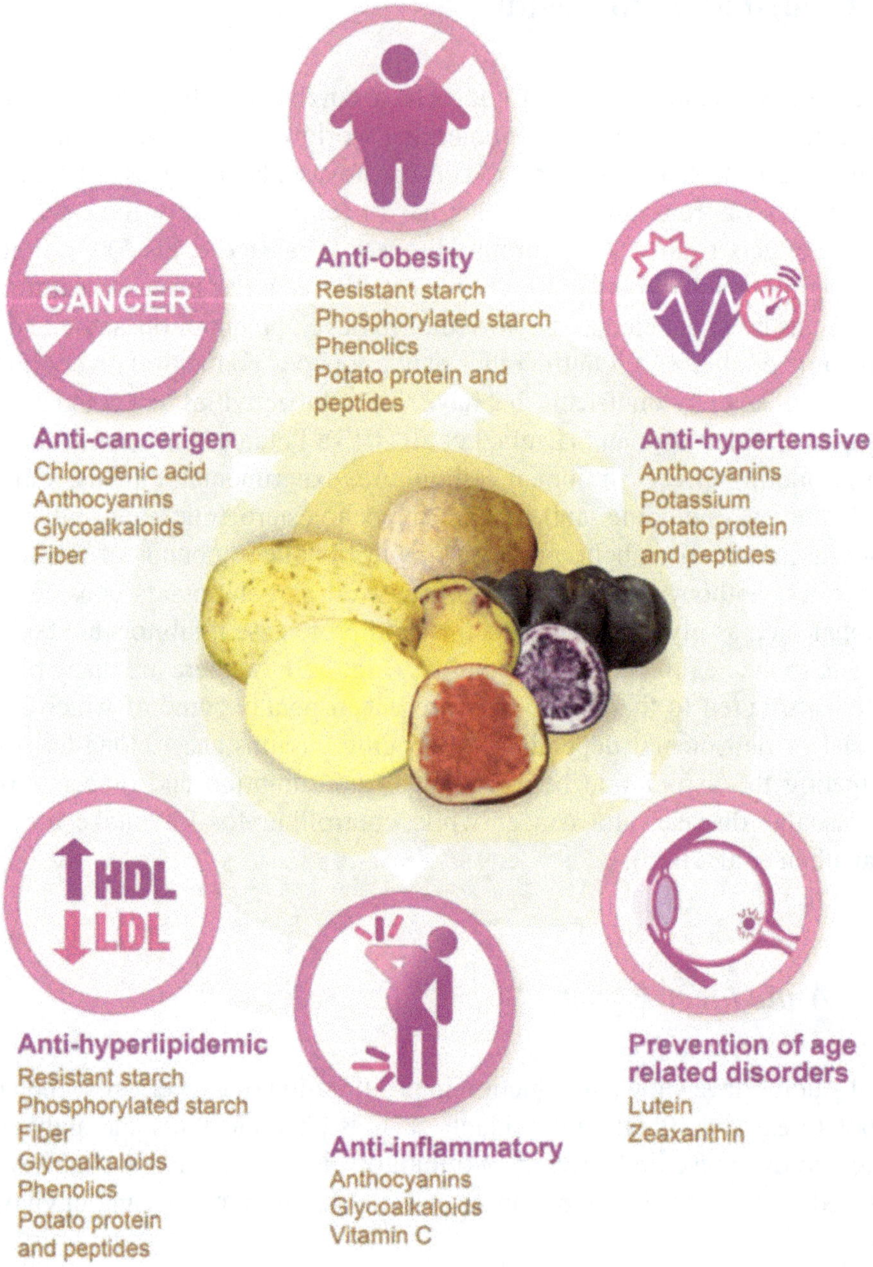

Fig. 2.5 Health benefits of potatoes

apoptosis of colon cancer cells compared with white and yellow flesh potatoes. In a more recent study, Charepalli et al. (2015) found that extracts of purple-fleshed potatoes suppress colon tumorigenesis via elimination of colon cancer stem cells. Chlorogenic acid, the main phenolic acid of potato, is effective against human liver, colon, and prostate cancer cells (Wang et al. 2011) and inhibits significantly the proliferation of colon cancer and prostate cancer cells.

2.5.1.2 Role of Potato Glycoalkaloids

α-Solanine and α-chaconine, the main steroidal glycoalkaloids in potatoes, are well studied for their antitumor properties (Friedman 2015). Lee et al. (2004) found that α-solanine exhibited growth inhibition and apoptosis induction in multiple cancer cells such human colon (HT29) and liver (HepG2) cancer cells. Friedman et al. (2005) evaluated the anti-carcinogenic effect of α-solanine and α-chaconine extracted from five fresh potato varieties (Dejima, Jowon, Sumi, Toya, and Vora Valley) and found that glycoalkaloids exerted anti-proliferative effects of the following human tumor cell lines: cervical (HeLa), liver (HepG2), lymphoma (U937), stomach (AGS and KATO III) cells, and on normal liver (Chang) cells. Friedman et al. (2005) also reported that the anti-proliferative effects of the glycoalkaloids were concentration dependent and that α-chaconine was more bioactive than α-solanine. Yang et al. (2006) found that α-chaconine induced the apoptosis of HT-29 human colon cancer cells through caspase-3 activation and inhibition of extracellular signal-regulated kinase phosphorylation.

Reddivari et al. (2010) showed that α-chaconine exhibited potent anti-proliferative properties and increased cyclin-dependent kinase inhibitor p27 levels in two prostate cancer cell lines, LNCaP and PC3. More recently, it has been reported that α-solanine, has a positive effect on the inhibition of pancreatic cancer cell growth in vitro and in vivo. Sun et al. (2014) demonstrated that α-solanine inhibited cancer cell growth through caspase 3-dependent mitochondrial apoptosis and that the expression of tumor metastasis-related proteins, MMP-2 and MMP-9, was also decreased in the cells treated with α-solanine. Lv et al. (2014) reported that α-solanine inhibited proliferation of PANC-1, sw1990, MIA PaCa-2 cells in a dose-dependent manner, as well as cell migration and invasion with a toxic dose and that the administration of α-solanine during 2 weeks in a xenograft model reduces the tumor volume and weight by 43–61%. These studies showed beneficial effects on pancreatic cancer in vitro and in vivo, which may be mediated via suppressing pathways involving proliferation, angiogenesis, and metastasis.

2.5.1.3 Role of Potato Fiber

Langner et al. (2009) reported that commercially available potato fiber extract (Potex) exhibited anti-proliferative effects in several tumor cell cultures. The fiber extract decreased cancer cell motility, induced apoptosis, and also caused morphological changes in tumor cells.

2.6 Anti-diabetic and Anti-obesity Effects

Potato consumption has often been associated in cohort studies with elevated risk of type 2 diabetes (Muraki et al. 2016) and obesity (Borch et al. 2016), which has been attributed to a relatively high glycemic index in some potato varieties and processed potato products containing added saturated and trans fats. A major confounding factor in such studies is typical Western dietary patterns associated with increased disease risk typically include potato consumption along with high intake of red and processed meat, refined grains, high-fat dairy products, fried foods and sugar (Pastorino et al. 2016). More research is needed to adjust association of food items such as potatoes in such dietary patterns (Hu 2002). Moreover, RDS present in cooked potatoes (especially amylose) tends to retrograde upon cooling generating appreciable amounts of slowly digestible starch (SDS) or resistant starch (RS) that contribute to dietary fiber content (Sajilata et al. 2006) and potentially positively impact health by slowing postprandial glucose release from cooked potatoes (King and Slavin 2013).

GI values below 56 are considered as low glycemic index while values above 74 are considered to indicate a high glycemic index. GI values in potato ranged from 56 to 94 for eight British cultivars (Henry et al. 2005) and from 53 to 103 for seven Australian cultivars (Wang et al. 2014). When boiled red potatoes were served hot to volunteers, a GI of 89.4 was found (Fernandes et al. 2005). When cooking is followed by cooling, amylose retrogrades to produce resistant starch. The GI response was only 56.2 when cooking was followed by refrigeration of 12–24 h.

Potato chips and French fries have been implicated by some nutrition researchers as major contributors to obesity risk as these products contain a high fat and caloric content. Potato servings, however, are not likely in themselves to promote obesity as potatoes are considered to have a low energy density as they are a low-fat food with a high-water content (Anderson et al. 2013). Potato-based foods with high calorie fat additions have been considered as a major culprit towards obesity risk (Camire et al. 2009).

Conversely, potatoes may have a role in controlling appetite and therefore weight gain, by contributing to satiety. Satiety is the feeling of fullness and the loss of hunger that occur after eating. Many factors influence satiety, including the rate of gastric emptying and the proportion of macronutrients in the food. Foods that increase satiety are thought to promote weight control by delaying subsequent meals and total calories consumed (Camire et al. 2009). Compared to rice and pasta, adult feeding studies have shown that satiating amounts of potatoes co-ingested with meat resulted in lower energy intake and postprandial insulin concentrations; and higher levels of ghrelin, which is a gastric orexigenic appetite-stimulating hormone that contributes to feeding regulation (Erdmann et al. 2007). Likewise, studies involving children showed that meat co-ingestion with boiled mashed potato resulted in an approximate 40% lower energy intake as compared to meat consumed together with either pasta or rice (Akilen et al. 2016). The stronger satiety of boiled mashed potato for the calories consumed was related to similar suppression of ghrelin postprandially relative to the other carbohydrate-rich foods despite the lower potato meal intake. Short-term intervention studies have generally indicated that high GI meals decrease satiety, and an increase in the return of hunger and energy intake at a later meal as opposed to low glycemic index meals containing potatoes (Roberts 2000).

2.7 Anti-hyperlipidemic, Anti-hypertensive and Anti-inflammatory Effects

A variety of animal feeding studies have shown cholesterol-lowering properties from potato intake that have been related to its content of protein, resistant and phosphorylated starch, fiber, glycoalkaloids (Friedman 2006), and phenolic compounds (Friedman 1997). Robert et al. (2006) found that consumption of cooked potatoes (consumed with skin) improved lipid metabolism in cholesterol-fed rats. Rats fed a potato-enriched diet for 3 weeks had lower concentrations of plasma cholesterol and triglycerides and reduced liver cholesterol content. Hashimoto et al. (2006) showed that retrograded starch from two varieties of potato pulp lowered serum total cholesterol and triglyceride concentrations. The authors indicated that the retrograded starch promoted the excretion of bile acids resulting in a low concentration of serum cholesterol; and that retrograded starch inhibited the synthesis of fatty acids at the mRNA levels of fatty acid synthase (FAS) and SREBP-1c, which might be related to the observed reduction of the serum triglyceride concentrations. Kanazawa et al. (2008) reported that gelatinized potato starch containing a high level of phosphate reduced concentrations of serum-free fatty acids and triglycerides and liver triglycerides.

Liyanage et al. (2008) have demonstrated the hypocholesterolemic effect of potato peptides. Rats fed a cholesterol-free diet containing 20% (w/w) potato peptides showed greater concentrations of serum high-density lipoprotein (HDL) cholesterol and increased fecal steroid output and lesser non-HDL cholesterol concentrations than rat fed diets containing 20% casein peptides. The results were attributed to inhibition of cholesterol absorption, possibly via suppression of micellar solubility of cholesterol. In a follow-up study, Liyanage et al. (2009) found that potato peptides reduced the serum non-HDL cholesterol concentrations by stimulating fecal steroid excretion, accelerated by cecal short-chain fatty acids in a hypercholesterolemic rat model. There is a lack of data, however, from randomized controlled trials to demonstrate a relationship between potato consumption and blood lipid parameters in humans.

Vinson et al. (2012) showed a significant lowering of systolic blood pressure in humans after supplementation to hypertensive subjects in a 4-week cross-over trial involving consumption of six to eight small purple potatoes twice daily versus no potato intake. The blood pressure lowering effect was related to high intake of polyphenols associated with the pigmented potatoes. This latter intervention trial is contrasted by an analysis from three large prospective cohort studies indicating increased hypertension risk in association with potato intake of four or more servings per month as opposed to one serving per month (Borgi et al. 2016). A major limitation of such trials is that co-ingestion of salt, high-salt foods, saturated or trans fats with potatoes could have contributed to the hypertension risk as opposed to potato per se, particularly since potatoes are typically eaten in a meal context (Miller and Stanner 2016). In support of this contention, a 3-year longitudinal study of Japanese people showed that adherence to a traditional Japanese dietary pattern exerted favorable effects on blood pressure that was partly associated with potato intake (Niu et al. 2016).

Relatively high intake of potassium is needed to counteract the blood pressure raising effects of a high sodium diet and so protect against hypertension (Camire et al. 2009). An increase in consumption of potassium-rich foods has been promoted to combat hypertension and cardiovascular disease (WHO 2012). In that regard, intake of potassium-rich foods has been indicated to protect against stroke risk (Adebamowo et al. 2015). As potatoes are rich in potassium and are naturally very low in sodium content, this food could counter development of hypertension-associated diseases. Additionally, Makinen et al. (2008) reported that a protein isolated from vascular bundle and inner tuber tissues of potato enhanced the inhibition of the angiotensin converting enzyme I, a biochemical factor affecting blood pressure that contributes to hypertension.

Kaspar et al. (2011) found anti-inflammatory effects in healthy men consuming white and pigmented potatoes with greater effects from pigmented potatoes. Potato phenolics and glycoalkaloids have shown evidence of anti-inflammatory activities (Kenny et al. 2013). Indigestible carbohydrates including resistant starch and fiber have demonstrated the ability to modulate inflammatory markers in both animal models (Vaziri et al. 2014) and human clinical trials (Jiao et al. 2015). Hence, the contribution of resistant starch or fiber from select potato products may have also direct impact on inflammatory stress in humans.

2.8 Potato and Its Relationship with Cardiovascular Diseases

As a key dietary source of potassium, vitamin C, and dietary fiber, potatoes contribute significantly to nutrients with defined roles in promoting cardiovascular health (McGill et al. 2013). Boiled potatoes have been shown to have favorable impact on several measures of cardiometabolic health in animals and humans, including lowering blood pressure, improving lipid profiles, and decreasing markers of inflammation (McGill et al. 2013). When eaten as a regular food item and consumed with skin, potato intake can significantly enhance cardioprotective fiber intake that is generally lacking in Western-type diets (Lockyer et al. 2016). Large prospective studies in Sweden involving a 13-year follow-up showed no adverse relationship of higher potato intake with cardiovascular risk for either morbidity or mortality (Larsson and Wolk 2016). Similarly, a systematic review of five observational studies carried out by Borch et al. (2016) showed no convincing evidence to support an adverse association between unprocessed potato intake and the risk of developing metabolic disorders including obesity, type 2 diabetes, and cardiovascular disease. On the other hand, processed potato products like French fries and potato crisps, with high lipid and trans fats content and added sodium can have adverse effect of the heart health and so should be minimized in the diet. In that regard, despite the lack of a relationship between chronic disease risk and potato intake in the above comprehensive review by Borch et al. (2016), they showed that French fries and fried potato were associated with an increased risk for obesity and type 2 diabetes. Likewise, a longitudinal study involving 4440 subjects with an 8-year follow-up showed no association between higher potato intake and mortality risk, whereas participants who consumed fried potatoes two to three times/week had an increased mortality risk (Veronese et al. 2017). Camire et al. (2009) have recommended preparing potatoes with minimum lipid addition and consume potatoes with peels to conserve their cardiovascular health promoting properties.

2.9 Concluding Remarks

Potato is an important source of carbohydrates, resistant starch, quality protein, vitamins C and B6 as well as potassium. Potatoes are also a source of antioxidants. Chlorogenic acid and glycoalkaloids are present in all potatoes independently of the flesh color while deep yellow-fleshed potatoes contain high amounts of lutein and zeaxanthin; and purple-fleshed potatoes contain high amounts of anthocyanins. Potatoes glycoalkaloids in high concentrations can be toxic to humans but in low concentrations can have beneficial effects such as inhibition of the growth of cancer cells.

The contribution of potato to the diet is affected by cooking, potato intake, and the bioavailability of potato nutrients. Potato vitamins are significantly reduced after cooking. However, 100 g of cooked potatoes provide around 30% of the requirement of vitamin C and 20% od the requirement of vitamin B6. Potato carotenoids and anthocyanins show high recoveries after cooking due to an improved release of these antioxidants. In vitro studies demonstrate that potato lutein and zeaxanthin have a high bioaccessibility and that potato phenolics undergo microbial transformation in the intestinal tract producing metabolites that may also promote a healthy gut microbiome. Further research in humans is needed to confirm the beneficial effect of potato phenolics in the gut.

In areas where potato is consumed in large quantities like in the highlands of Latin American countries, the potato contribution to the energy, protein, iron, and zinc intake is significant. In those areas, iron and zinc biofortified potatoes are expected to contribute to reduce malnutrition and anemia. Nevertheless, to assess the full potential of the biofortified potatoes, human studies are required to gain insight on how much of the iron from biofortified potatoes are absorbed by the human body.

Regarding its contribution to human health, potatoes have shown promising health-promoting effects in human cell culture, experimental animals, and human clinical studies. Potato compounds such as phenolic acids and anthocyanins, glycoalkaloids, fiber, and proteinase inhibitors have been implicated in the suppression of cancer cell proliferation in vitro and in vivo and are believed to contribute to the hypocholesterolemic, anti-inflammatory, anti-obesity and anti-diabetic properties of potato.

References

Adebamowo SN, Spiegelman D, Willett WC (2015) Association between intakes of magnesium, potassium, and calcium and risk of stroke: 2 cohorts of US women and updated metaanalyses. Am J Clin Nutr 101:1269–1277

Akilen R, Deljoomanesh N, Hunschede S, Smith CE, Arshad MU, Kubant R, Anderson GH (2016) The effects of potatoes and other carbohydrate side dishes consumed with meat on food intake, glycemia and satiety response in children. Nutr Diabetes 6:e195. https://doi.org/10.1038/nutd.2016.1

American Optometric Association (2009) Diet, nutrition and eye health. http://www.aoa.org/documents/nutrition/Diet_Nutrition_Eye_Health_booklet.pdf. Accessed 28 Sept 2017

Amrein T, Bachmann S, Noti A, Biedermann M, Barbosa M, Biedermann-Brem S, Grob K, Keiser A, Realini P, Escher F, Amadó R (2003) Potential of acrylamide formation, sugars, and free asparagine in potatoes: a comparison of cultivars and farming systems. Agric Food Chem 51(18):5556–5560

Anderson GH, Soeandy CD, Smith CE (2013) White vegetables: glycemia and satiety. Adv Nutr 4(Suppl):356S–367S

Andre CM, Oufir M, Guignard C, Hoffmann L, Hausman JF, Evers D, Larondelle Y (2007) Antioxidant profiling of native Andean potato tubers (*Solanum tuberosum* L.) reveals cultivars with high levels of β-carotene, alpha-tocopherol, chlorogenic acid, and petanin. Agric Food Chem 55:10839–10849

Andre CM, Schafleitner R, Legay S, Lefèvre I, Aliaga CA, Nomberto G, Hoffmann L, Hausman JF, Larondelle Y, Evers D (2009) Gene expression changes related to the production of phenolic compounds in potato tubers grown under drought stress. Phytochemistry 70:1107–1116

Andre CM, Larondelle Y, Evers D (2010) Dietary antioxidants and oxidative stress from a human and plant perspective: a review. Curr Nutr Food Sci 6(1):2–12

Andre CM, Evers D, Ziebel J, Guignard C, Hausman JF, Bonierbale M, zum Felde T, Burgos G (2015) In vitro bioaccessibility and bioavailability of iron from potatoes with varying vitamin C, carotenoid, and phenolic concentrations. J Agric Food Chem 63(41):9012–9021. https://doi.org/10.1021/acs.jafc.5b02904

Auestad N, Hurley J, Fulgoni V, Schweitzer CM (2015) Contribution of food groups to energy and nutrient intakes in five developed countries. Nutrients 7:4593–4618

Bach S, Yada RY, Bizimungu B, Fan M, Sullivan JA (2013) Genotype by environment interaction effects on starch content and digestibility in potato (*Solanum tuberosum* L.). J Agric Food Chem 61(16):3941–3948. https://doi.org/10.1021/jf3030216. Epub 2013 Apr 12

Bamberg JB, del Rio A (2005) Conservation of potato genetic resources. In: Razdan MK, Mattoo AK (eds) Genetic improvement of solanaceous crops. Volume I: Potato. Science Publishers, Inc., Plymouth, p 476

Bates C (1997) Vitamin Analysis. Ann Clin Biochem 34(6):599–626. https://doi.org/10.1177/000456329703400604

Bethke P, Jansky S (2018) The Effects of Boiling and Leaching on the Content of Potassium and Other Minerals in Potatoes. J Food Sci 73(5):H80–H85

Bohn T (2017) Bioactivity of carotenoids–chasms of knowledge. Int J Vitam Nutr Res 10:1–5. https://doi.org/10.1024/0300-9831/a000400

Bonierbale M, Burgos G, zum Felde T, Sosa P (2010) Composition nutritionnelle des pommes de terre. Cahiers de nutrition et dietétique 45:S28–S36

Borch D, Juul-Hindsgaul N, Veller M, Astrup A, Jaskolowski J, Raben A (2016) Potatoes and risk of obesity, type 2 diabetes, and cardiovascular disease in apparently healthy adults: a systematic review of clinical intervention and observational studies. Am J Clin Nutr 104:489–498

Borgi L, Rimm EB, Willett WC, Forman JP (2016) Potato intake and incidence of hypertension: results from three prospective US cohort studies. BMJ 353:i2351. https://doi.org/10.1136/bmj.i2351

Bramley PM, Elmadfa I, Kafatos A, Kelly FJ, Manios Y, Roxborough HE, Schuch W, Sheehy PJ, Wagner JH (2000) Vitamin E. J Sci Food Agric 80:913–938

Brenna O, Berardo N (2004) Applications of near-infrared reflectance spectroscopy (NIRS) to the evaluation of carotenoids in maize. J Agri Food Chem 52:5577–5582

Brennan CS (2005) Dietary fibre, glycaemic response, and diabetes. Mol Nutr Food Res 49:560–570

Brown CR, Wrolstad R, Durst R, Yang P, Clevidence B (2003) Breeding studies in potatoes containing high concentrations of anthocyanins. Am J Pot Res 80:241. https://doi.org/10.1007/BF02855360

Brown CR, Culley D, Yang CP, Durst R, Wrolstad R (2005) Variation of anthocyanin and carotenoid contents and associated antioxidant values in potato breeding lines. J Am Soc Hortic Sc 130:174–180

Burgos G (2014) Concentración y bioaccesibilidad de carotenoides y compuestos fenólicos en papas cocidas. Tesis doctoral. Universidad de la Laguna. Departamento de Química Analítica, Nutrición y Bromatología

Burgos G, Amoros W, Morote M, Stangoulis J, Bonierbale M (2007) Iron and zinc concentration of native Andean potato varieties from a human nutrition perspective. J Sci Food Agric 87(4):668–675

Burgos G, Salas E, Amoros W, Auqui M, Munoa L, Kimura M, Bonierbale M (2009) Total and individual carotenoid profiles in the Phureja group of cultivated potatoes: I. concentrations and relationships as determined by spectrophotometry and high performance liquid chromatography (HPLC). J Food Compos Anal 22:503–508

Burgos G, Amoros W, Salas E, Muñoa L, Sosa P, Díaz C, Bonierbale M (2012) Carotenoid concentrations of native Andean potatoes as affected by cooking. Food Chem 133:1131–1137

Burgos G, Muñoa L, Sosa P, Bonierbale M, Zum Felde T, Díaz C (2013a) In vitro bioaccessibility of lutein and zeaxanthin of yellow fleshed boiled potatoes. Plant Foods Hum Nutr 68(4):385–390

Burgos G, Amoros W, Muñoa L, Sosa P, Cayhualla E, Sanchez C, Diaz C, Bonierbale M (2013b) Total phenolic, total anthocyanin and phenolic acid concentrations and antioxidant activity of purple-fleshed potatoes as affected by boiling. J Food Compos Anal 30:6–12

Burlingame B, Mouille B, Charrondiere R (2009) Nutrients, bioactive non-nutrients and anti-nutrients in potatoes. J Food Compos Anal 22:494–502

Camire ME, Kubow S, Donnelly DJ (2009) Potatoes and human health. Crit Rev Food Sci Nutr 49:823–840

Cevallos-Casals B, Cisneros Zevallos L (2003) Stoichiometric and kinetic studies of phenolic antioxidants from Andean purple corn and red-fleshed sweet potato. J Agric Food Chem 51:3313–3319

Champ MMJ (2004) Physiological aspects of resistant starch and in vivo measurements. J AOAC Int 87:749–755

Charepalli V, Reddivari L, Radhakrishnan S, Vadde R, Agarwal R, Vanamala JK (2015) Anthocyanin-containing purple-fleshed potatoes suppress colon tumorigenesis via elimination of colon cancer stem cells. J Nutr Biochem 26(12):1641–1649. https://doi.org/10.1016/j.jnutbio.2015.08.005

Chitchumroonchokchai C, Diretto G, Parisi B, Giuliano G, Failla ML (2017) Potential of golden potatoes to improve vitamin A and vitamin E status in developing countries. PLoS One 12(11):e0187102

Chun J, Lee J, Ye L, Exler J, Eitenmiller RR (2006) Tocopherol and tocotrienol contents of raw and processed fruits and vegetables in the United States diet. J Food Compos Anal 19:196–204

Clifford M (1999) Chlorogenic acids and other cinnamates – nature, occurrence and dietary burden. J Sci Food Agric 79:362–372

Crapo PA, Reaven G, Olefsky J (1977) Postprandial plasma-glucose and -insulin responses to different complex carbohydrates. Diabetes 26:1178–1183

DeGraft-Johnson J, Kolodziejczyk K, Krol M, Nowak P, Krol B, Nowak D (2007) Ferric-reducing ability power of selected plant polyphenols and their metabolites: implications for clinical studies on the antioxidant effects of fruits and vegetable consumption. Basic Clin Pharmacol Toxicol 100:345–352

De Haan S, Burgos G, Liria R, Rodriguez F, Creed-Kanashiro H, Bonierbale M (2019) The Nutritional Contribution of Potato Varietal Diversity in Andean Food Systems: a Case Study. Am J Potato Res 96:151. https://doi.org/10.1007/s12230-018-09707-2

Dragsted LO, Krath B, Ravn-Haren G, Vogel UB, Vinggaard AM, Jensen PB, Loft S, Rasmussen SE, Sandstrom B, Pedersen A (2006) Biological effects of fruit and vegetables. Proc Nutr Soc 65:61–67

Duarte-Delgado D, Ñústez-López C, Narváez-Cuenca C, Restrepo-Sánchez L, Melo S, Sarmiento F, Kushalappa A, Mosquera-Vásquez T (2016) Natural variation of sucrose, glucose and fructose contents in Colombian genotypes of *Solanum tuberosum* Group Phureja at harvest. J Sci Food Agric 96(12):4288–4294. https://doi.org/10.1002/jsfa.7783. Epub 2016 Jun 21

Duthie G, Duthie S, Kyle J (2000) Plant polyphenols in cancer and heart disease: implications as nutritional antioxidants. Nutr Res Rev 13:79–106

Ek KL, Brand-Miller J, Copeland L (2012) Glycaemic effect of potatoes. Food Chem 133:1230–1240

Englyst HN, Kingsman SM, Cummings JH (1992) Classification and measurement of nutritionally important starch fractions. Eur J Clin Nutr 46:S33–S50

Erdmann J, Hebeisen Y, Lippl F, Wagenpfeil S, Schusdziarra V (2007) Food intake and plasma ghrelin response during potato-, rice- and pasta-rich test meals. Eur J Clin Nutr 46:196–203

Ezekiel R, Singh N, Sharma S, Kaur A (2013) Beneficial phytochemicals in potato—a review. Food Res Int 50:487–496

FAO (2013) FAO statistical databases FAOSTAT. http://www.fao.org/faostat/en/#data/FBS. Accessed October 12th 2017

FAO/OMS/UNU (2004) Human energy requirements. Report of joint FAO/WHO/UNU expert consultation. FAO, Rome

FAO/WHO (2001) Human vitamin and mineral requirements. Report of a joint FAO/WHO expert consultation. Bangkok, Thailand

Fernandes G, Velangi A, Wolever T (2005) Glycemic index of potatoes commonly consumed in North America. J Am Diet Assoc 105:557–562

Fernández-Garcia E, Carvajal-Lérida I, Pérez-Gálvez A (2009) In vitro bioaccessibility assessment as a prediction tool of nutritional efficiency. Nutr Res 29:751–760

Fernandez-Orozco R, Gallardo-Guerrero L, Hornero-Méndez D (2013) Carotenoid profiling in tubers of different potato (*Solanum* sp.) cultivars: accumulation of carotenoids mediated by xanthophyll esterification. Food Chem 141:2864–2872

Fiedor J, Burda K (2014) Potential role of carotenoids as antioxidants in human health and disease. Nutrients 6(2):466–488

Fitzpatrick TB, Basset GJ, Borel P, Carrari F, Della Penna D, Fraser PD, Hellmann H, Osorio S, Rothan C, Valpuesta V (2012) Vitamin deficiencies in humans: can plant science help? Plant Cell 24:395–414

La Frano M, de Moura F, Boy E, Lönnerdal B, Burri B (2014) Bioavailability of iron, zinc, and provitamin A carotenoids in biofortified staple crops. Nutr Rev 72(5):289–307. https://doi.org/10.1111/nure.12108.

Fossen T, Andersen Ø (2000) Anthocyanins from tubers and shoots of the purple potato, Solanum tuberosum. J Hortic Sci Biotechnol 75(3):360–363. https://doi.org/10.1080/14620316.2000.11511251

Friedman M (1997) Chemistry, biochemistry, and dietary role of potato polyphenols. A review. J Agric Food Chem 45:1523–1540

Friedman M (2004) Analysis of biologically active compounds in potatoes (Solanum tuberosum), tomatoes (Lycopersicon esculentum), and jimson weed (Datura stramonium) seeds. J Chromatogr A 1054(1–2):143–155

Friedman M (2006) Potato glycoalkaloids and metabolites: roles in the plant and in the diet. J Agric Food Chem 54:8655–8681

Friedman M (2015) Chemistry and anticarcinogenic mechanisms of glycoalkaloids produced by eggplants, potatoes, and tomatoes. J Agric Food Chem 63:3323–3337

Friedman M, McDonald G, Filadelfi-Keszi M. (1997) Potato glycoalkaloids: chemistry, analysis, safety, and plant physiology. Critical Reviews in Plant Sciences 16(1):55–132. https://doi.org/10.1080/07352689709701946

Friedman M, Roitman JN, Kozukue N (2003) Glycoalkaloid and calystegine contents of eight potato cultivars. J Agric Food Chem 51(10):2964–2973

Friedman M, Lee KR, Kim HJ, Lee IS, Kozukue N (2005) Anticarcinogenic effects of glycoalkaloids from potatoes against human cervical, liver, lymphoma, and stomach cancer cells. J Agric Food Chem 53:6162–6169

Fudge J, Mangel N, Gruissem W, Vanderschuren H, Fitzpatrick T (2017) Rationalising vitamin B_6 biofortification in crop plants. Curr Opin Biotechnol 44:130–137. https://doi.org/10.1016/j.copbio.2016.12.004

Galliard T (1973) Lipids of potato tubers. I. Lipid and fatty acid composition of tubers from different varieties of potato. J Sci Food Agric 24(5):617–622

Gammone MA, Riccioni G, D'Orazio N (2015) Carotenoids: potential allies of cardiovascular health? Food Nutr Res 59:26762. https://doi.org/10.3402/fnr.v59.26762

Garcia-Alonso M, Pascual-Teresa S, Santos-Buelga C, Rivas-Gonzalo J (2004) Evaluation of the antioxidant properties of fruits. Food Chem 84:13–18

Garcia-Closas R, Berenguer A, Tormo MJ, Sanchez MJ, Quiros JR, Navarro C, Arnaud R, Dorronsoro M, Chirlaque MD, Barricarte A, Ardanaz E, Amian P, Martinez C, Agudo A, Gonzalez CA (2004) Dietary sources of vitamin C, vitamin E and specific carotenoids in Spain. Br J Nutr 91:1005–1011

Gibson S, Kurilich A (2013) The nutritional value of potatoes and potato products in the UK diet. Nutr Bull 38:389–399

Giusti MM, Polit MF, Ayvaz H, Tay D, Manrique I (2014) Characterization and quantitation of anthocyanins and other phenolics in native Andean potatoes. J Agric Food Chem 62:4408–4416

Habtemariam S, Lentini G (2015) The therapeutic potential of rutin for diabetes: an update. Mini Rev Med Chem 15(7):524–528

Han J, Kosukue N, Young K, Lee K, Friedman M (2004) Distribution of ascorbic acid in potato tubers and in home-processed and commercial potato foods. J Agric Food Chem 52(21):6516–6521

Hart D, Scott KJ (1995) Development and evaluation of an HPLC method for the analysis of carotenoids in foods and measurement of the carotenoid content of vegetables and fruits commonly consumed in the UK. Food Chem 54:101–111

Hashimoto N, Ito Y, Han KH, Shimada K, Sekkikawa M, Topping DL (2006) Potato pulps lowered the serum cholesterol and triglyceride levels in rats. J Nutr Sci Vitaminol 52:445–450

Hayashi K, Mori M, Matsunani Y, Suzutan T, Ogasawara M, Yoshida I, Hosokawa TA, Azuma M (2003) Anti Influenza Virus Activity of a Red-Fleshed Potato Anthocyanin. Food Sci Technol Res 9(3):242–244

Hayashi K, Hibasami H, Murakami T, Terahara N, Mori M, Tsukui A (2006) Induction of apoptosis in cultured human stomach cancer cells by potato anthocyanins and its inhibitory effects on growth of stomach cancer in mice. Food Sci Technol Res 12:22–26

Hellmann H, Mooney S (2010) Vitamin B6: A Molecule for Human Health? Molecules 15:442–459

Henry C, Lightowler H, Strik C, Storey M (2005) Glycaemic index values for commercially available potatoes in Great Britain. Br J Nutr 94(6):917–921

Holst B, Williamson G (2008) Nutrients and phytochemicals: from bioavailability to bioefficacy beyond antioxidants. Curr Opin Biotechnol 19:73–82

Hosseinian FS, Bea T (2007) Saskatoon and wild blueberries have higher antho-cyanin content than other Manitoba berries. J Agric Food Chem 55(26):10832–10838

Hu FB (2002) Dietary pattern analysis: a new direction in nutritional epidemiology. Curr Opin Lipidol 13(1):3–9

Ieri F, Innocenti M, Andrenelli L, Vecchio V, Mulinacci N (2011) Rapid HPLC/DAD/MS method to determine phenolic acids, glycoalkaloids and anthocyanins in pigmented potatoes (*Solanum tuberosum* L.) and correlations with variety and geographical origin. Food Chem 125:750–759

Jiao J, Xu J-Y, Zhang W, Han S, Qin L-Q (2015) Effect of dietary fiber on circulating C-reactive protein in overweight and obese adults: a meta-analysis of randomized controlled trials. Int J Food Sci Nutr 66:114–119

Kanazawa T, Atsumi M, Mineo H, Fukushima M, Nishimura N, Noda T (2008) Ingestion of gelatinized potato starch containing a high level of phosphorus decreases serum and liver lipids in rats. J Oleo Sci 57:335–343

Kaspar KL, Park JS, Brown CR, Mathison BD, Navarre DA, Chew BP (2011) Pigmented potato consumption alters oxidative stress and inflammatory damage in men. J Nutr 141:108–111

Kenny OM, McCarthy CM, Brunton NP, Hossain MB, Rai DK, Collins SG, Jones PW, Maguire AR, O'Brien NM (2013) Anti-inflammatory properties of potato glycoalkaloids in stimulated Jurkat and Raw 264.7 mouse macrophages. Life Sci 92:775–782

Khachick F, Goli M, Beecher G, Holden J, Lusby W, Tenorio M, Barrera M (1992) Effect of food preparation on qualitative and quantitative distribution of major carotenoid constituents of tomatoes and several green vegetables. J Agric Food Chem 40:390–398

Khachik F, Cohen L, Zhao Z (1999) Metabolism of dietary carotenoidsand their possible role in prevention of cancer and macular de-generation. In: Shibamoto T, Terao J, Osawa T (eds) Functional foods for disease prevention I, American Chemical Society symposium series. Oxford University Press, New York, pp 71–85

Kimura M, Rodriguez-Amaya DB (2003) Carotenoid Composition of Hydroponic Leafy Vegetables. J Agric Food Chem 51:2603–2607. http://dx.doi.org/10.1021/jf020539b

King J, Slavin J (2013) White potatoes, human health, and dietary guidance. Adv Nutr 4:393S–401S. https://doi.org/10.3945/an.112.003525

Krinsky NI, Johnson EJ (2005) Carotenoid actions and their relation to health and disease. Mol Aspects Med 26(6):459–516. Review

Kubow S, Iskandar MM, Melgar-Bermudez E, Sleno L, Sabally K, Azadi B, How E, Prakash S, Burgos G, Felde TZ (2017) Effects of simulated human gastrointestinal digestion of two purple-fleshed potato cultivars on anthocyanin composition and cytotoxicity in colonic cancer and non-tumorigenic cells. Nutrients 9(9):pii: E953. https://doi.org/10.3390/nu9090953

Külen O, Stushnoff C, Holm DG (2012) Effect of cold storage on total phenolics content, antioxidant activity and vitamin C level of selected potato clones. J Sci Food Agric 93:2437–2444

Kumar D, Singh B, Parveen K (2004) An overview of the factors affecting sugar content of potatoes. Ann Appl Biol 145:247–256

Lachman J, Hamouz K (2005) Red and purple coloured potatoes as a significant antioxidant source in human nutrition – a review. Plant Soil Environ 51(11):477–482

Lachman J, Hamouz K, Orsak M, Pivec V (2001) Potato glycoalkaloids and their significance in plant protection and human nutrition – review. Rosstlinna Vyrova 47(4):181–191

Lachman J, Hamouz K, Sulc M, Orsak M, Pivec V, Hejtmankova A (2009) Cultivar differences of total anthocyanins and anthocyanidins in red and purple-fleshed potatoes and their relation to antioxidant activity. Food Chem 114:836–843

Lachman J, Hamouz K, Musilová J, Hejtmánková K, Kotíková Z, Pazderu K, Domkárová J, Pivec V, Cimr J (2013) Effect of peeling and three cooking methods on the content of selected phytochemicals in potato tubers with various colour of flesh. Food Chem 138:1189–1197

Lancaster J, Reay P, Norris J, Butler R (2000) Induction of flavonoids and phenolic acids in apple by UV-B and temperature. J Hortic Sci Biotechnol 75(2):142–148. https://doi.org/10.1080/14620316.2000.11511213

Langner E, Rzeski W, Kaczor J, Kandefer-Szersze M, Pierzynowski SG (2009) Tumour cell growth-inhibiting properties of water extract isolated from heated potato fibre (Potex). J Pre-Clin Clin Res 3:36–41

Larsson SC, Wolk A (2016) Potato consumption and risk of cardiovascular disease: 2 prospective cohort studies. Am J Clin Nutr 104:1245–1252

Lee KR, Kozukue N, Han JS, Park JH, Chang EY, Baek EJ, Chang JS, Friedman M (2004) Glycoalkaloids and metabolites inhibit the growth of human colon (HT29) and liver (HepG2) cancer cells. J Agric Food Chem 52:2832–2839

Leeman M, Östman E, Björck I (2005) Vinegar dressing and cold storage of potatoes lowers postprandial glycaemic and insulinaemic responses in healthy subjects. Eur J Clin Nutr 59:1266–1271

Liang N, Kitts D (2015) Role of chlorogenic acids in controlling oxidative and inflammatory stress conditions. Nutrients 8(1). https://doi.org/10.3390/nu8010016

Lin BW, Gong CC, Song HF, Cui YY (2017) Effects of anthocyanins on the prevention and treatment of cancer. Br J Pharmacol 174:1226–1243

Liu Q, Tarn R, Lynch D, Skjodt N (2007) Physicochemical properties of dry matter and starch from potatoes grown in Canada. Food Chem 105:897–907

Liyanage R, Han KH, Watanabe S, Shimada K, Sekikawa M, Ohba K (2008) Potato and soy peptide diets modulate lipid metabolism in rats. Biosci Biotechnol Biochem 72:943–950

Liyanage R, Han KH, Shimada KI, Sekikawa M, Tokuji Y, Ohba K (2009) Potato and soy peptides alter caecal fermentation and reduce serum non-HDL cholesterol in rats fed cholesterol. Eur J Lipid Sci Technol 111:884–892

Lockyer S, Spiro A, Stanner S (2016) Dietary fibre and the prevention of chronic disease—should health professionals be doing more to raise awareness? Nutr Bull 41:214–231

Lombardo S, Pandino G, Mauromicale G (2013) The influence of growing environment on the antioxidant and mineral content of "early" crop potato. J Food Compos Anal 32(1):28–35

Lovat C, Nassar A, Kubow S, Li X, Donnelly D (2016) Metabolic biosynthesis of potato (*Solanum tuberosum* L.) antioxidants and implications for human health. Crit Rev Food Sci Nutr 56(14):2278–2303

Love SL, Pavek JJ (2008) Positioning the potato as a primary food source of vitamin C. Am J Potato Res 85:277–285

Lv C, Kong H, Dong G, Liu L, Tong K, Sun H, Chen B, Zhang C, Zhou M (2014) Antitumor efficacy of α-solanine against pancreatic cancer in vitro and in vivo. PLoS One 9(2):e87868

Lynch DR, Liu Q, Tarn TR, Bizimungu B, Chen Q, Harris P, Chik CL, Skjodt NM (2007) Glycemic index: a review and implications for the potato industry. Am J Potato Res 84:179–190

Madiwale GP, Reddivari L, Holm DG, Vanamala J (2011) Storage elevates phenolic content and antioxidant activity but suppresses antiproliferative and pro-apoptotic properties of colored-flesh potatoes against human colon cancer cell lines. J Agric Food Chem 59:8155–8166

Makinen S, Kelloniemi J, Pihlanto A, Makinen K, Korhonen H, Hopia A (2008) Inhibition of angiotensin converting enzyme I caused by autolysis of potato proteins by enzymatic activities confined to different parts of the potato tuber. J Agric Food Chem 56:9875–9883

Mattila P, Hellstrom J (2006) Phenolic acids in potatoes, vegetables, and some of their products. J Food Compos Anal 20:152–160

McGill C, Kurilich A, Davignon J (2013) The role of potatoes and potato components in cardiometabolic health: a review. J Ann Med 45(7):467–473

Miao M, Jiang B, Cui S, Zhang T, Jin Z (2015) Slowly digestible starch–a review. Crit Rev Food Sci Nutr 55(12):1642–1657. https://doi.org/10.1080/10408398.2012.704434

Miller R, Stanner S (2016) Baked, mashed, boiled or fried—can potatoes increase the risk of hypertension? Nutr Bull 41:252–256

Mills CE, Tzounis X, Oruna-Concha MJ, Mottram DS, Gibson GR, Spencer JP (2015) In vitro colonic metabolism of coffee and chlorogenic acid results in selective changes in human faecal microbiota growth. Br J Nutr 113:1220–1227

Mínguez Mosquera M, Hornero-Méndez D (1994) Comparative Study of the Effect of Paprika Processing on the Carotenoids in Peppers (Capsicum annuum) of the Bola and Agridulce Varieties. J Agric Food Chem 42(7):1555–1560

Monro J, Mishra S, Blandford E, Anderson J, Genet R (2009) Potato genotype differences in nutritionally distinct starch fractions after cooking, and cooking plus storing cool. J Food Compos Anal 22(6):539–545

Moonney S, Chen L, Kuhn C, Navarre R, Knowles NR, Hellman H (2013) Genotype-specific changes in vitamin B_6 content and the PDX family in potato. Bio Med Res Int. Article ID 389723, 7 pages. https://doi.org/10.1155/2013/389723

Müller HZ (1997) Determination of the carotenoid content in selected vegetables and fruit by HPLC and photodiode array detection. Lebensm Unters Forsch 204:88. https://doi.org/10.1007/s002170050042

Muraki I, Rimm EB, Willett WC, Manson JE, Hu FB, Sun Q (2016) Potato consumption and risk of type 2 diabetes: results from three prospective cohort studies. Diabetes Care 39(3):376–384. https://doi.org/10.2337/dc15-0547

Muttucumaru N, Elmore J, Curtis T, Mottram D, Parry M, Halford N (2008) Reducing acrylamide precursors in raw materials derived from wheat and potato. J Agric Food Chem 56(15):6167–6172. https://doi.org/10.1021/jf800279d. Epub 2008 Jul 15

Naidu K (2003) Vitamin C in human health and disease is still a mystery? An overview. Nutr J 2:7

Nassar AM, Sabally K, Kubow S, Leclerc YN, Donnelly DJ (2012) Some Canadian-grown potato cultivars contribute to a substantial content of essential dietary minerals. J Agric Food Chem 60(18):4688–4696. https://doi.org/10.1021/jf204940t

Natella F, Nardini M, Belelli F, Scaccini C (2007) Coffee drinking induces incorporation of phenolic acids into LDL and increases the resistance of LDL to ex vivo oxidation in humans. Am J Clin Nutr 86:604–609

Navarre DA, Shakya R, Holden J, Kumar S (2010) The effect of different cooking methods on phenolics and vitamin C in developmentally young potato tubers. Am J Potato Res 87(4):350–359

Neilson J, Lagüe M, Thomson S, Aurousseau F, Murphy AM, Bizimungu B, Deveaux V, Bègue Y, JME J, Tai HH (2017) Gene expression profiles predictive of cold-induced sweetening in potato. Funct Integr Genomics 17(4):459–476

Niu K, Momma H, Kobayashi Y, Guan L, Chujo M, Otomo A, Ouchi E, Nagatomi R (2016) The traditional Japanese dietary pattern and longitudinal changes in cardiovascular disease risk factors in apparently healthy Japanese adults. Eur J Nutr 55:267–279

Ombra M, Fratianni F, Granese T, Cardinale F, Cozzolino A, Nazzaro F (2015) In vitro antioxidant, antimicrobial and anti-proliferative activities of purple potato extracts (Solanum tuberosum cv Vitelotte noire) following simulated gastro-intestinal digestion. Nat Prod Res 29(11):1087–1091. https://doi.org/10.1080/14786419.2014.981183

Olthof MR, Hollman PC, Buijsman MN, van Amelsvoort JM, Katan MB (2003) Chlorogenic acid, quercetin-3-rutinoside and black tea phenols are extensively metabolized in humans. J Nutr 133:1806–1814

Otten J, Hellwig P, Meyers L (2010) Dietary reference intakes: the essential guide to nutrient requirements. The National Academies Press, Washington, DC

Pascual-Teresa S, Sanchez-Ballesta MT (2008) Anthocyanins: from plant to health. Phytochem Rev 7:281–299

Pastorino S, Richards M, Pierce M, Ambrosini GL (2016) A high-fat, high-glycaemic index, low-fibre dietary pattern is prospectively associated with type 2 diabetes in a British birth cohort. Br J Nutr 115:1632–1642

Piñeros-Niño C, Narváez-Cuenca C, Kushalappa A, Mosquera T (2017) Hydroxycinnamic acids in cooked potato tubers from Solanum tuberosum group Phureja. Food Sci Nutr 5(3):380–389

Puchau B, Zulet MA, Gonzalez de Echavarri A, Hermsdorff HH, Martinez JA (2010) Dietary total antioxidant capacity is negatively associated with some metabolic syndrome features in healthy young adults. Nutrition 26(5):534–541

Raigond P, Ezekiel R, Raigond B (2014) Resistant starch in food. J Sci Food Agric 95(10):1968–1978. https://doi.org/10.1002/jsfa.6966. Review

Ramadan M, Oraby H (2016) Fatty acids and bioactive lipids of potato cultivars: an overview. J Oleo Sci 65(6):459–470. https://doi.org/10.5650/jos.ess16015

Rautiainen S, Serafini M, Morgenstern R, Prior RL, Wolk A (2008) The validity and reproducibility of food-frequency questionnaire–based total antioxidant capacity estimates in Swedish women. Am J Clin Nutr 87:1247–1253

Reddivari L, Hale AL, Miller JC (2007a) Determination of phenolic content, composition and their contribution to antioxidant activity in specialty potato selections. Am J Potato Res 84:275–282

Reddivari L, Vanamala J, Chintharlapalli S, Safe SH, Miller J (2007b) Anthocyanin fraction from potato extracts is cytotoxic to prostate cancer cells through activation of caspase-dependent and caspase-independent pathways. Carcinogenesis 28(10):2227–2235

Reddivari L, Vanamala J, Safe SH, Miller JC (2010) The bioactive compounds α-chaconine and gallic acid in potato extracts decrease survival and induce apoptosis in LNCaP and PC3 prostate cancer cells. Nutr Cancer 62:601–610

Reyes LF, Miller JC, Cisneros-Zevallos L (2005) Antioxidant capacity, anthocyanins and total phenolics in purple- and red-fleshed potato (*Solanum tuberosum* L.) genotypes. Am J Potato Res 82:271–277

Richelle M, Tavazzi I, Offord E (2001) Comparison of antioxidant activity of commonly consumed polyphenolic beverages (coffee, cocoa, tea) prepared per cup serving. J Agric Food Chem 49:3438–3442

Robert L, Narcy A, Rock E, Demigne C, Mazur A, Rémésy C (2006) Entire potato consumption improves lipid metabolism and antioxidant status in cholesterol-fed rat. Eur J Nutr 45:267–274

Roberts SB (2000) High-glycemic index foods, hunger, and obesity: is there a connection? Nutr Rev 58:163–169

Rodríguez B, Ríos D, Rodríguez E, Díaz C (2010) Influence of the cultivar on the organic acid and sugar composition of potatoes. J Sci Food Agric 90(13):2301–2309. https://doi.org/10.1002/jsfa.4086

Rodriguez-Amaya DB (2015) Status of carotenoid analytical methods and in vitro assays for the assessment of food quality and health effects. Curr Opin Food Sci 1:56–63. https://doi.org/10.1016/j.cofs.2014.11.005

Ruprich J, Rehurkova I, Boon PE, Svensson K, Moussavian S, Van der Voet H, Bosgra S, Van Klaveren JD, Busk L (2009) Probabilistic modeling of exposure doses and implications for health risk characterization: glycoalkaloids from potatoes. Food Chem Toxicol 47:2899–2905

Sajilata MG, Singhal RS, Kulkarni PR (2006) Resistant starch—a review. Compr Rev Food Sci Food Saf 5:1–17

Salmeron J, Manson JE, Stampfer MJ, Colditz GA, Wing AL, Willett WC (1997) Dietary fiber, glycemic load, and risk of non-insulin-dependent diabetes mellitus in women. J Am Med Assoc 277:472–477

Slavin JL (2008) Position of the American Dietetic Association: health implications of dietary fiber. J Am Diet Assoc 108:1716–1731

Soh NL, Brand-Miller J (1999) The glycaemic index of potatoes: the effect of variety, cooking method and maturity. Eur J Clin Nutr 53:249–254

Spanos GA, Wrolstad RE (1992) Phenolics of apple, pear, and white grape juices and their changes with processing and storage. A review. J Agric Food Chem 40:1478–1487

Spinneker A, Sola R, Lemmen V, Castillo MJ, Pietrzik K, González-Gross M (2007) Vitamin B_6 status, deficiency and its consequences—an overview. Nutr Hosp 22(1):7–24

Stalmach A, Steiling H, Williamson G, Crozier A (2010) Bioavailability of chlorogenic acids following acute ingestion of coffee by humans with an ileostomy. Arch Biochem Biophys 501:98–105

Sun H, Lv C, Yang L, Wang Y, Zhang Q, Yu S, Kong H, Wang M, Xie J, Zhang C, Zhou M (2014) Solanine induces mitochondria-mediated apoptosis in human pancreatic cancer cells. BioMed Res Int 2014:805926

Tajner-Czopek M, Jarych-Szyszka M, Lisińska G (2008) Changes in glycoalkaloids content of potatoes destined for consumption. Food Chem 106(2):706–711

Tee E, Lim C (1991) Carotenoid composition and content of Malaysian vegetables and fruits by the AOAC and HPLC methods. Food Chem 41:309–339

Teucher B, Olivares M, Cori H (2004) Enhancers of iron absorption: ascorbic acid and other organic acids. Int J Vitam Nutr Res. 74(6):403–419

Tian J, Chen C, Ye X, Chen S (2016) Health benefits of the potato affected by domestic cooking: a review. Food Chem 202:165–175

Tomas-Barberan F, García-Villalba R, Quartieri A, Raimondi S, Amaretti A, Leonardi A, Rossi M (2014) In vitro transformation of chlorogenic acid by human gut microbiota. Mol Nutr Food Res 58(5):1122–1131. https://doi.org/10.1002/mnfr.201300441. Epub 2013 Dec 23

van Het Hof KH, West CE, Weststrate JA, Hautvast JG (2000) Dietary factors that affect the bioavailability of carotenoids. J Nutr 130(3):503–506

Vaziri ND, Liu SM, Lau WL, Khazaeli M, Nazertehrani S, Farzaneh SH, Kieffer DA, Adams SH, Martin RJ (2014) High amylose resistant starch diet ameliorates oxidative stress, inflammation, and progression of chronic kidney disease. PloS One 9:e114881

Venn B, Green T (2007) Glycemic index and glycemic load: measurement issues and their effect on diet-disease relationships. Eur J Clin Nutr 61(Suppl 1):S122–S131

Veronese N, Stubbs B, Noale M, Solmi M, Vaona A, Demurtas J, Nicetto D, Crepaldi G, Schofield P, Koyanagi A, Maggi S, Fontana L (2017) Fried potato consumption is associated with elevated mortality: an 8-y longitudinal cohort study. Am J Clin Nutr 106:162–167

Vinson J, Demkosky C, Navarre D, Smyda M (2012) High-antioxidant potatoes: acute in vivo antioxidant source and hypotensive agent in humans after supplementation to hypertensive subjects. J Agric Food Chem 60(27):6749–6754. https://doi.org/10.1021/jf2045262

Visvanathan R, Jayathilake C, Chaminda Jayawardana B, Liyanage R (2016) Health-beneficial properties of potato and compounds of interest. J Sci Food Agric 96(15):4850–4860. https://doi.org/10.1002/jsfa.7848. Epub 2016 Jul 7. Review

Wada L, Ou B (2002) Antioxidant activity and phenolic content of Oregon cane-berries. J Agric Food Chem 50(12):3495–3500

Wahyuni Y, Ballester A, Sudarmonowati E, Bino RJ, Bovy AG (2011) Metabolite biodiversity in pepper (Capsicum) fruits of thirty-two diverse accessions: variation in health-related compounds and implications for breeding. Phytochemistry 72(11–12):1358–1370

Wang Q, Chen Q, He M, Mir P, Su J, Yang Q (2011) Inhibitory effect of antioxidant extracts from various potatoes on the proliferation of human colon and liver cancer cells. Nutr Cancer 63:1044–1052

Wang S, Copeland L, Brand-Miller J (2014) Discovery of a low-glycaemic index potato and relationship with starch digestion in vitro. Br J Nutr 111(4):699–705. https://doi.org/10.1017/S0007114513003048

Wang W, Sun C, Mao L, Ma P, Liu F, Yang J, Gao Y (2016) The biological activities, chemical stability, metabolism and delivery systems of quercetin: a review. Trends Food Sci Technol 56:21–38

WHO (World Health Organization) (2012) Guideline: potassium intake for adults and children. World Health Organization, Geneva

Winkel-Shirley B (2001) Flavonoid biosynthesis. A colorful model for genetics, biochemistry, cell biology, and biotechnology. Plant Physiol 126:485–493

Woolfe J, Poats S (1987) The Potato in the Human Diet. Cambridge University Press. 0521326699, 9780521326698

Wu X, Beecher GR, Holden JM, Haytowitz DB, Gebhardt SE, Prior RL (2006) Concentration of anthocyanins in common foods in the United States and estimation of normal consumption. J Agric Food Chem 54(11):4069–4075

Wu J, Cho E, Willett WC, Sastry SM, Schaumberg DA (2015) Intakes of lutein, zeaxanthin, and other carotenoids and age-related macular degeneration during 2 decades of prospective follow-up. JAMA Ophthalmol 133(12):1415–1424. https://doi.org/10.1001/jamaophthalmol.2015.3590

Yang SA, Paek SH, Kozukue N, Lee KR, Kim JA (2006) Alpha-chaconine, a potato glycoalkaloid, induces apoptosis of HT-29 human colon cancer cells through caspase-3 activation and inhibition of ERK 1/2 phosphorylation. Food Chem Toxicol 44:839–846

Yogendra KN, Kushalappa AC, Sarmiento F, Rodriguez E, Mosquera T (2015) Metabolomics deciphers quantitative resistance mechanisms in diploid potato clones against late blight. Funct Plant Biol 42:284–298

Yuan G, Sun B, Yuan J, Wang Q (2009) Effects of different cooking methods on health-promoting compounds of broccoli. J Zhejiang Univ Sci B 10(8):580–588. https://doi.org/10.1631/jzus.B0920051

Zeeman SC, Kossmann J, Smith AM (2010) Starch: its metabolism, evolution, and biotechnological modification in plants. Annu Rev Plant Biol 61:209–234

Zeng C (2013) Effects of different cooking methods on the vitamin C content of selected vegetables. Nutr Food Sci 43(5):438–443. https://doi.org/10.1108/NFS-11-2012-0123

Chapter 3
Enhancing Value Chain Innovation Through Collective Action: Lessons from the Andes, Africa, and Asia

André Devaux, Claudio Velasco, Miguel Ordinola, and Diego Naziri

Abstract The development community has shown increasing interest in the potential of innovation systems and value chain development approaches for reducing poverty and stimulating greater gender equity in rural areas. Nevertheless, there is a shortage of systematic knowledge on how such approaches have been implemented in different contexts, the main challenges in their application, and how they can be scaled to enable large numbers of poor people to benefit from participation in value chains. This chapter provides an overview of value chain development and focuses on the International Potato Center's experiences with the Participatory Market Chain Approach (PMCA), a flexible approach that brings together smallholder farmers, traders, processors, researchers, and other service providers in a collective process to explore potential business opportunities and develop innovations to exploit them. The PMCA is an exemplary case of South–South knowledge exchange: it was first developed and implemented in the Andes, but has since been introduced, adapted, and applied to different market chains in Africa and Asia, where it has contributed to improved rural livelihoods. The experiences of adjusting and implementing the approach in these different contexts and the outcomes of those interventions, and complementary approaches, are examined in this chapter. Lessons learned from these experiences are shared with a goal of informing the promotion, improvement, and scaling of value chain approaches in the future.

A. Devaux (✉) · C. Velasco
International Potato Center, Quito, Ecuador
e-mail: a.devaux@cgiar.org; c.velasco@cgiar.org

M. Ordinola
International Potato Center, Lima, Peru
e-mail: cip-incopa@cgiar.org

D. Naziri
International Potato Center, Hanoi, Vietnam, and Natural Resources Institute, University of Greenwich, Chatham, UK
e-mail: d.naziri@cgiar.org

H. Campos, O. Ortiz (eds.), *The Potato Crop*,
https://doi.org/10.1007/978-3-030-28683-5_3

3.1 Introduction

Agriculture is rapidly changing in developing countries in response to a variety of factors that include enhanced access to technologies, increased urbanization, shifting in diets, improvements in farmers' education and health, institutional and policy reforms, and investments in rural infrastructure. At the same time, the growing number of supermarkets—referred to as "supermarket revolution" (Reardon and Hopkins 2006)—and the smallholder agriculture integration in changing food markets are helping to create new income opportunities and a more dynamic environment; one in which new technologies and agricultural practices can contribute to helping smallholder farmers respond to these changes and associated challenges. Facilitating sustainable access to high-value markets can enable poor farmers to increase their incomes, making it an effective strategy for reducing poverty (Wiggins and Keats 2013). But those farmers are often at a disadvantage when it comes to producing for and doing business in high-quality food chains and ensuring compliance to both public and private standards (Henson and Humphrey 2010), given their limited access to financial and other services, and their largely poor organizational capacity for collective marketing. Furthermore, their market connections are mostly informal and characterized by high levels of distrust, uncertainty, and transaction costs. This is particularly true for perishable crops, such as potatoes, which are grown on small farms in mountainous areas. For agricultural research to benefit such farmers and make value chains more inclusive, it must be complemented by other efforts to improve the regulatory environment, alleviate resource constraints, and build local capacity to respond to evolving technological and economic challenges and opportunities. It may also be necessary to act to influence the incentives and constraints faced by market actors, so that they can communicate more effectively with farmers and establish mutually beneficial and enduring business relationships.

There are clear signs that agro-industries are having a significant impact on economic development and poverty reduction globally, both in urban and rural communities (FAO 2013). Despite the risks associated with high-value markets, changes in and around the agri-food sector can contribute to the development of better support services for farmers, such as technology, extension, and financial products. There is evidence that small producers with access to technical support services are more willing to adopt new technologies and make investments to take advantage of emerging market opportunities (Royer et al. 2016).

Traditionally, different organizations have designed and implemented different types of interventions in agriculture and associated markets. While public agricultural research and extension programs have focused mainly on increasing agricultural production and productivity, nongovernmental organizations (NGOs) and other entities have focused on commercialization and the development of inclusive value chains. The impact of interventions in these areas has primarily been constrained by the lack of holistic approaches that address the challenges and opportunities along the value chain while taking into consideration the needs and capacities of different value chain actors, from input suppliers to farmers, agro-industries, and consumers. Accordingly, development practitioners have progressively introduced the so-called innovation system approaches which bring together different value chain stakeholders in recognition of the fact that systematic change occurs through the interactions of multiple actors—both individual and institutional—and sources of agricultural knowledge and innovation (Biggs 2008; Schut et al. 2016). However, the practical application of agricultural innovation systems and inclusive value chain approaches—and in particular the integration of these two approaches—is a challenge. This has been documented by studies, such as those presenting the results of recent work by the CGIAR[1] consortium and partners in Africa, Asia, and Latin America, in which the opportunities arising from new and expanding markets for agricultural products are analyzed and the challenges for smallholder participation in these markets, and benefits derived from participation, are identified (Devaux et al. 2016). In Latin America, Devaux et al. (2009) present the case of the Papa Andina Initiative in the Andes, which used collective action to promote innovation in the market chain through two approaches: The Participatory Market Chain Approach (PMCA) (Devaux et al. 2009) and Innovation Platforms (Thiele et al. 2011b). Both approaches sought to promote the interaction of small-scale potato producers with market actors and agricultural service providers, with a goal of establishing alliances and contractual agreements in response to new market opportunities.

This chapter offers some perspectives on value chain development and presents the experiences of the International Potato Center (CIP) with the PMCA. This approach, originally developed to increase competitiveness and improve the livelihoods of small-scale potato producers in the Andes, has proved useful in other market chains and in other parts of the world, such as East Africa and Asia. The chapter unfolds as follows. We first present the experiences of implementing the approach in the Andes, and the adjustments that have been made to it, while analyzing factors that have influenced its implementation. We then describe and analyze the experiences of replicating and validating the approach in different contexts and regions. Finally, we discuss the lessons learned in order to inform the design of interventions that use the PMCA approach as a research for development tool and provide insights for its replication and adaptation elsewhere.

3.2 General Concepts of Value Chain Development

The term "value chain" is used in different ways in literature. In this chapter, by "value chain" we refer to a set of actors that interact to transform inputs and services into products with attributes that consumers are willing to buy. The debate surrounding value chains in recent years has been about understanding the changes that

[1] CGIAR is a global research partnership for a food-secure future (http://www.cgiar.org/)

take place in some rapidly evolving markets for agricultural products, the consequences of these changes for the poorest actors in the market (specifically small producers and small-to-medium-sized companies), and effective alternatives that governments, development organizations, and the private sector can provide to support those value chain actors. Millions of people with low-incomes—a large portion of them women—participate in agricultural value chains as producers, traders, processors, and retailers (Fig. 3.1). Many millions more, including the majority of the poor in the developing world, participate in agricultural value chains as workers or consumers. Improving the performance of agricultural value chains can thus benefit a large number of people (Reardon and Timmer 2012; Reardon et al. 2012).

We refer to "value chain development" (which we abbreviate as VCD hereafter) as a type of intervention aimed at reducing poverty through improved links between companies, urban dwellers and rural producers. It has been defined as "A positive or desirable change in a value chain to extend or improve productive operations and generate social benefits: poverty reduction, income and employment generation, economic growth, environmental performance, gender equity and other sustainable development goals" (UNIDO 2011). From this perspective, many development agencies, donors, and governments have adopted the value chain approach as a key element of their strategies to reduce rural poverty (Humphrey and Navas-Alemán 2010). In contrast to traditional agricultural research and development (R&D) approaches, which focus on improving the capacities of small producers to increase their productivity or better manage natural resources, the VCD approach challenges R&D organizations to work with diverse actors to understand how a value chain functions and identify mutually beneficial options to improve its efficiency.

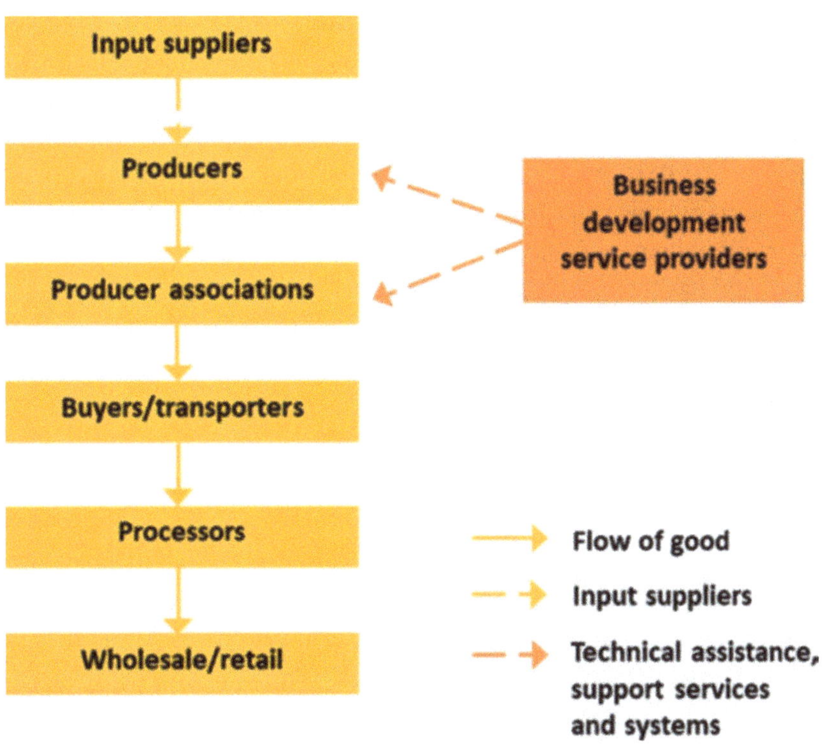

Fig. 3.1 Simple value chain (Devaux et al. 2016)

While the globalization of markets offers opportunities to market higher value products that simply did not exist before, these markets generally demand much more in terms of business acumen, efficiency and attention to quality and food safety standards than traditional product markets (Reardon et al. 2009). Participation in value chains for more demanding markets requires small-scale producers to deliver a regular supply of products of consistent quality in sufficient quantity at the right time and price. The fulfillment of these conditions requires access to land, inputs, technology, knowledge, organizational capacity, ability, and infrastructure, which may be lacking in some communities or groups of resource-poor farmers. Public policies are also needed to adapt government strategies to different situations and support the participation of smallholder farmers in more dynamic value chains. Smallholders are at a disadvantage compared to medium- and large-scale producers when it comes to taking advantage of these transformations, since they are often located in areas with less private and public infrastructure, are further away from markets, produce comparatively little amounts of products than need consolidation, and have less favorable conditions for high-yielding agriculture.

In this context, CIP, within the framework of its activities in the Andean region, developed a more integrated, participatory approach that combines agricultural, institutional, and value chain innovations while seeking synergies. Called the Participatory Market Chain Approach (PMCA), it is a flexible approach that tries to involve small-scale farmers, market agents (traders, companies, and processors among others), researchers, and service providers in a collective process to identify and explore potential business opportunities that can equitably benefit the diverse actors of a selected value chain (Bernet et al. 2006). It was developed and applied for the first time in Peru, to increase competitiveness in potato market chains, which are important components of local agri-food systems, and to contribute to improving the livelihoods of small farmers. Subsequently, thanks to CIP's global presence, the approach was introduced and adapted together with local organizations that applied it to different market chains in other countries in the Andes, Africa, and Asia, in an example of South–South knowledge exchange. The PMCA is most effective when it is implemented as part of a comprehensive strategy that includes support to farmers' organizations, business development, policy change, and public advocacy. CIP spearheaded the development of the PMCA and has supported the development of local capacities needed to facilitate successful innovation processes (Devaux et al. 2013).

3.3 The Participatory Market Chain Approach: Origin and Characteristics

The regional initiative known as *Papa Andina* (Andean Potato) was implemented from 1999 to 2010 to strengthen the capacity of R&D organizations in Bolivia, Ecuador and Peru with the overall objective of increasing competitiveness and improving the livelihoods of small-scale potato producers (Devaux et al. 2011).

In 2002, CIP social scientists working in the Papa Andina program and the Potato Innovation and Competitiveness Project in Peru (INCOPA)[2] began to experiment with a participatory approach known as Rapid Appraisal of Agricultural Knowledge Systems (RAAKS) (Engel and Salomon 1997). This approach brings together diverse stakeholders in a participatory process to stimulate collective learning, foster trust, and promote innovation. Papa Andina used RAAKS to identify market opportunities involving small-scale farmers, together with other value chain actors, researchers, and service providers.

The participation of traders, supermarkets, food processors, and chefs in a process of research for development was a radical break from previous participatory R&D efforts, which had been limited to researchers and farmers. Many researchers felt—and some still do—that working with market agents could distract research for development scientists from their focus, which was solving farmers' production problems. When the interventions were implemented, additional steps were added to RAAKS for the development of new products, resulting in a new approach, the PMCA, which was implemented in Peru and then validated in Bolivia in 2003. In subsequent years, this approach was applied in different contexts and thoroughly documented (Ordinola et al. 2009; Devaux et al. 2011).

The PMCA involves the individuals participating in a market chain and service providers, public and private, that support the chain—such as NGOs, financial services providers, development professionals, and researchers—in a facilitated process for identifying and developing innovations to exploit market opportunities. Those innovations can be technological, commercial, or institutional. The PMCA was conceived to be implemented in three phases that last between 9 and 18 months, according to the public and private context and the actors involved (Bernet et al. 2006; Antezana et al. 2008). However, when applied in different contexts, it became apparent that various activities might be required to consolidate progress in strengthening the relationships between the market chains' actors and follow up on technical and commercial innovations after the conclusion of the three phases.

The PMCA's phases (Fig. 3.2):

- **Phase 1. Getting to know market chain actors and their activities through an assessment**. The PMCA is initiated by an R&D organization that takes the lead in selecting the market chain in which it will work, identifying potential R&D partners and conducting exploratory and participatory diagnostic research on the chain. This phase, which can last from 2 to 4 months, concludes with a public event to discuss the assessment's results, generate ideas for possible innovations and motivate the actors of the market chain and service providers to participate in Phase 2.

[2] The INCOPA project, managed by CIP with support from the Swiss Agency for Development and Cooperation, was created to improve potato value chain competitiveness in Peru, with an emphasis on native potatoes grown by smallholder farmers in Andean highlands.

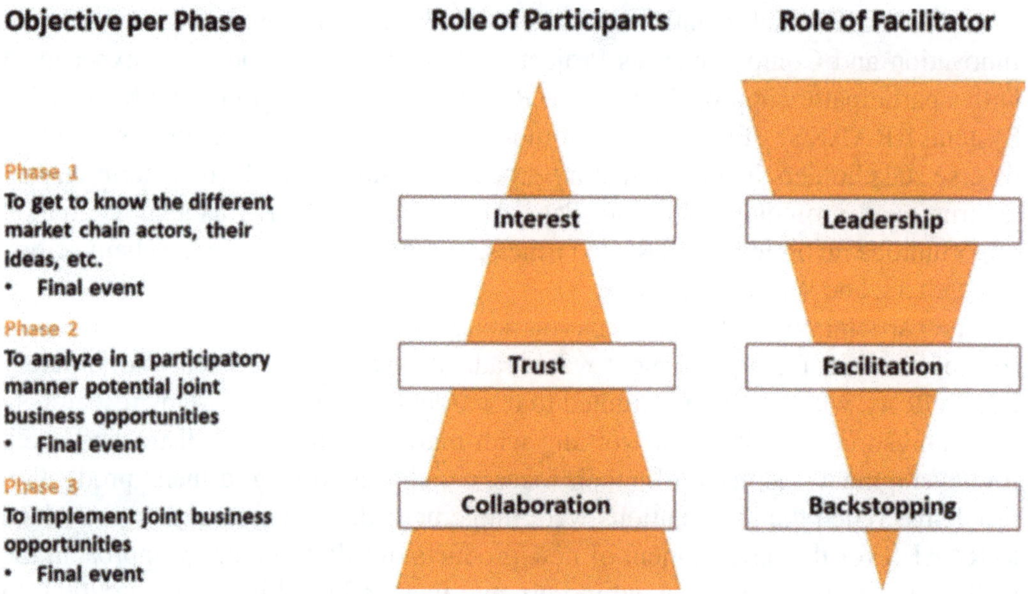

Fig. 3.2 The structure of the methodology: the three phases of the PMCA (Source: Bernet et al. 2006)

Phase 2. Joint analysis of potential market opportunities and innovations needed to take advantage of them. Representatives of participating R&D organizations facilitate the exploration and analysis of potential market opportunities. Actors of the value chain and service providers are organized in thematic groups that meet between 6 and 10 times to analyze potential opportunities, including thorough technical or market studies. A second public event takes place at the end of Phase 2 to discuss prioritized business opportunities and to encourage new stakeholders, with the appropriate knowledge and complementary experience, to join Phase 3.

Phase 3. Joint development of innovations. During this final phase, the group focuses on the development, market testing, and launch of specific innovations. This may require work in specialized areas such as processing, packaging, labeling or branding tests. Companies increasingly play a leading role in this phase. The PMCA exercise closes with a third public event in which the market innovations developed are presented to a wide selection of stakeholders such as public officials, potential donors, policy makers, decision makers from the private sector (supermarkets, processors, etc.), the public and representatives of the media.

Experiences show that the implementation of the PMCA can deviate from the orderly sequence of phases described above. While some groups have dissolved before producing any useful innovation, others have advanced and launched cost-effective innovations during Phase 2. Some groups have also continued to interact and generate innovations years after the conclusion of the PMCA exercise.

3.4 Examples of Value Chain Intervention Implemented in Several Geographies

Since the PMCA was developed, it has been adapted and applied to foster innovation in varied value chains in different countries, including the following:

- Potatoes in Bolivia, Peru, Ecuador, and Indonesia
- Dairy products in Bolivia
- Organic coffee in Peru
- Yams in Colombia
- Sweetpotatoes, tomatoes, and peppers in Uganda
- Potatoes, sweetpotatoes, bananas, and cassava in Uganda
- Aquaculture in Nepal and Bangladesh
- Amazonian fruits in Bolivia
- Fruits and vegetables in Nepal
- Organic and traditional products (e.g., home-made pasta, goat cheese, nuts, honey, mushrooms, tea, and dried fruits and vegetables) in Albania

Only a few of these applications have been well documented and evaluated. In the following sections, we analyze concrete examples of applications of the PMCA in interventions for which there is sufficient published information (Table 3.1).

3.4.1 *Peru's Native Potato Revolution*

From 2001 to 2010, more than 20 public organizations, NGOs and companies collaborated on the INCOPA project, within the framework of the Papa Andina regional initiative. The project focused on both commercial potato varieties and the

Table 3.1 Experiences with the PMCA around the globe

Experience	Product/s	Country/ies	Principal sources
(1) Peru's native potato revolution	Potato	Peru	Ordinola et al. (2009, 2013, 2014), Horton and Samanamud (2013)
(2) PMCA experiences in different value chains in the Andes	Coffee, milk, yam, and potato	Peru, Bolivia, Colombia	Horton et al. (2013c)
(3) The revalorization of native potatoes in Ecuador	Potato	Ecuador	Montesdeoca et al. (2013), Montesdeoca (2016)
(4) Building capacity for innovation in Ugandan value chains	Potato, sweetpotato, tomato, and hot peppers	Uganda	Mayanja et al. (2012, 2013), Horton et al. (2010)
(5) The PMCA and Farmer Business Schools in Indonesia	Potato, sweetpotato, vegetables, coffee, and cattle	Indonesia, Philippines	Horton et al. (2013a, b), International Potato Center (2017a)

native potato varieties that indigenous farmers in the Andes have traditionally grown and consumed. Its objective was to promote innovations in native potato production and commercialization in order to improve the competitiveness of this value chain for the benefit of the small-scale farmers that produce them. To this end, CIP and partners developed and used the PMCA as an applied research and development approach. INCOPA worked to link small-scale producers of native potatoes with researchers, development professionals, and a range of actors in potato market chains (supermarkets, processors, etc.) to capitalize on the biodiversity of native potatoes and their culinary, nutritional, and cultural attributes. The implementation of the PMCA was complemented by efforts to promote alliances and coordination between value chain actors through innovation platforms, to increase public awareness of the virtues of native potatoes and to support the formulation of policies in support of these chains. The PMCA acted as a catalyst for innovations that—together with different interventions in the technical, institutional and political spheres—triggered commercial, productive, and institutional processes that continue to this day, and contributed to what has been dubbed "Peru's native potato revolution" (Horton and Samanamud 2013).

In Peru, two cycles of the PMCA were implemented. The first application focused on the commercialization of improved potato varieties, whereas the second cycle focused on native potatoes. The exercises included the participation of not only researchers, Ministry of Agriculture officials, and typical market chain actors but also chefs and supermarket managers. This was the first time that such a diverse group met to collaborate on options for promoting the potato in Peru. The "new actors" brought new energy, perspectives, and ideas to discussions that had previously been dominated by R&D professionals.

The first application of the PMCA, which started in 2002 in Peru, gave rise to the country's first brand of selected fresh potatoes, *Mi Papa* (My Potato), which was distributed through the wholesale market in Lima. Subsequently, a new organization, CAPAC Peru, was created with the participants to promote the commercialization of high-quality local agricultural products, including potatoes marketed under the *Mi Papa* brand. The second application of the PMCA resulted in two new products made with Peru's native potatoes. *T'ikapapa* was launched as the first brand of gourmet, fresh native potatoes sold in supermarkets in the Peruvian capital, Lima. Shortly after that, *Jalca Chips*, an innovative potato chip product made of naturally colored (blue, red, yellow) native potatoes, was launched. Jalca Chips, sold in duty-free shops in Lima's international airport, opened the way for native potatoes into high-value, processed products markets. Both products were sold in grocery stores but were eventually replaced in the market by other brands of better quality using native potatoes, thanks to a process of creative imitation and improvement. Since then, more than 20 comparable native potato products have appeared on the market ranging from the ones produced by small provincial companies to the products of multinational companies, some of which are exported, as illustrated in Fig. 3.3 (Ordinola et al. 2009).

In addition to the new products, important ideas for policy initiatives, advocacy, and awareness campaigns to promote potato consumption emerged during the

Fig. 3.3 The PMCA as a trigger for innovation in the development of new potato products in Peru. (Source: Updated based on Ordinola et al. 2009)

second cycle of the PMCA. Perhaps the most remarkable one was the creation of a National Potato Day in Peru. It was established in 2005 and has been held annually since then, in both Lima and the provinces. The Minister of Agriculture (MINAGRI) established a special commission to organize the first event, which included organizations that participated in the PMCA. Today this commission continues to function as an institutionalized network and MINAGRI allocates an annual budget for the celebration of National Potato Day, which promotes potatoes to the Peruvian people. This commission supported the efforts of Peruvian government officials to get the United Nations to declare 2008 as the International Year of the Potato. After that proposal was accepted, the commission led numerous outreach activities to promote the celebration of the International Year of the Potato in Peru throughout 2008. The combination of innovations in the market chain, changes in supporting policies, and enhanced public awareness have contributed to improving the image and perceptions of native potatoes in Peru.

Encouraging the big players in the market During the application of the PMCA in Peru, some of the large players in the Peruvian market participated in the initial meetings, but then abandoned the process because they perceived the process time-consuming and saw few benefits in the short term. As a consequence, the first innovations involved mostly small processing companies and not all of their efforts were sustainable for different reasons, such as quality issues, lack of continuity and failure to position their products. The INCOPA project and its network of public-private collaborators backed the initiatives of the most innovative small entrepreneurs. As these products began to penetrate the markets, the large market players returned to learn about opportunities for developing new potato-based products. They supported the creation of institutional arrangements to organize an orderly supply of quality potatoes through commercial arrangements such as contracts to promote good practices that were backed by the project.

These efforts were supported by NGOs that worked in the regions where native potatoes are grown, and that helped producers to organize themselves so to be to supply a product that met the market's quality requirements. The CAPAC Peru platform,[3] which emerged as an institutional innovation through the PMCA, has supported the native potato value chain by promoting interactions between producers and market agents (Ordinola et al. 2009). With the revamped interest of bigger market players led to the creation of an array of new potato-based products and different brands of native potato chips of superior quality, which ranged from the products of multinationals such as Frito Lay and Grupo Gloria to those produced by small provincial companies, in response to market's diverse demands. The PMCA thus "triggered" a process whereby the second and third generation of innovations are often more important than the ones developed during the approach's initial application (Ordinola et al. 2013; Proexpansión 2011). Figure 3.3 illustrates this process.

Measuring impact In the province of Andahuaylas, in Peru's Apurímac Region, a study was implemented to evaluate the initial scope and impact of the INCOPA project's intervention through the CAPAC Peru platform, using the "impact pathway" methodological framework (Maldonado et al. 2011). In this region, INCOPA used the PMCA to promote the exploitation of market opportunities for native potato biodiversity and improvements in the competitiveness of the potato value chain. For this study, 80 producers in the intervention area were selected through stratified sampling (sampling units are grouped by geographical location), and additional 66 farmers in the same areas were identified as a counterfactual group.

The main conclusions of this case study include:

1. Potato producers in the studied area expanded and diversified their commercial relationships, mainly due to greater demand for native potatoes locally and from supermarkets and agribusinesses in Lima;
2. Small producers in the study area managed to develop business skills and improved their post-harvest management to the degree that they enjoyed increased demand from the new markets that they accessed;
3. Potato producers in the target group belong to organizations that maintained relationships and links with institutions which were engaged through the PMCA process and had continued to provide technical assistance and other services that allowed more efficient business operations;

[3] *Cadenas Productivas Agrícolas de Calidad en el Perú* (Quality Agricultural Production Chains), known as CAPAC Perú, was a second-level, social, economic, and technology promotion organization that provided specialized services for the development of production chains for potato and other tubers in Peru.

4. Farmers involved in the project attained higher average incomes through better prices (26% higher than the control group) and higher sales volumes of potatoes, especially native potatoes.

In a subsequent case study on the participation of potato growers in Peru's central highlands in native potato value chains, Tobin et al. (2016) studied the social differentiation between participants and nonparticipants within communities and the reasons for the inclusion or exclusion of households in the value chain. His findings indicate that participation in value chains is not necessarily beneficial for all small-scale farmers. For such programs to have broad social benefits, policies and other institutional arrangements are needed to minimize the risks associated with participation in such value chains and to provide support for participation in other types of less demanding markets. Since high-value markets often require more investment and assets, they are probably not the most appropriate option for lower-income households. Horton and Samanamud (2013) indicate that families with less land, less education, less access to credit, and less developed social networks have benefited less from new market opportunities. But despite these shortcomings, the same authors found that the innovation process generated by the PMCA helped to improve the image of native potatoes and link small-scale farmers with sufficient capacity to dynamic urban markets for potato-based products.

If we analyze the development of native potato value chains in Peru, it is clear that the country's economic and policy contexts have been favorable, including government support and beneficial policies, the participation of a committed private sector, a gastronomic sector willing to promote native products, and the support of international donors such as the Swiss Agency for Development and Cooperation, which strengthened the innovation process through training and coaching in the implementation of the PMCA and the development of innovations. This process made it possible to transform a situation in which native potatoes were nothing more than a subsistence food for poor farmers in the Andean highlands, with little prospect of market-oriented agriculture, to one in which they are now recognized as a noteworthy and nutritious Peruvian product that deserves a price premium in urban markets and gourmet restaurants.

More recently, CIP has built upon this achievement with support from the International Fund for Agricultural Development (IFAD) in a project to strengthen innovation to improve the incomes, food security, and resilience of potato producers in Bolivia, Ecuador, and Peru. This initiative complements IFAD's public investment projects in these three countries and has contributed to expanding the PMCA approach in the region.

3.4.2 Analysis of PMCA Experiences in Different Value Chains in the Andes

From 2007 to 2010, the Andean Change Alliance (*Alianza Cambio Andino*)[4] evaluated the processes and results of PMCA implementation in VCD interventions in the Andes. Eight applications of the PMCA were initiated under the leadership of professionals at agricultural R&D organizations in Bolivia, Colombia, Ecuador, and Peru. None of them had previously implemented the approach, but the Alliance provided them with training and backstopping during implementation. Out of the eight cases initiated, five were completed, four of which were analyzed based on the greatest learning potential with the resources available for the study (Horton et al. 2013c). Those cases are:

Case 1. Marketing high-quality coffee in San Martin, Peru
Case 2. Developing and marketing of new dairy products in Oruro, Bolivia
Case 3. Development of new markets for yams in Northern Colombia
Case 4. Conserving and marketing native potatoes in Northern Potosí, Bolivia

A summary of the results achieved and analysis of each case, according to Horton et al. (2013c), follows.

3.4.2.1 Case 1. Marketing of High Quality Coffee in San Martín, Peru

This case study focused on a women's group that took the lead in the development of a local market for locally produced coffee. Members of the group acquired knowledge and skills in coffee processing and marketing and established a new brand of coffee for the local market. Innovations included more careful selection of coffee beans and improvements in roasting, grinding, and packaging. The application of the PMCA motivated the creation of more networks and relationships among the different stakeholders. They participated in a public event to promote the region's coffee in 2010 that attracted local authorities, private sector players, media, and about 500 members of the public. Success in the commercialization of coffee helped to consolidate the women's group and raise its visibility in public and political circles, as well as in fairs and markets for organic products. They now play a more prominent role in the local agricultural system.

[4] Alianza Cambio Andino (Andean Change Alliance) was a regional program of cooperation among organizations and businesses in Bolivia, Colombia, Ecuador, and Peru that contributed to sustainable livelihoods in poor communities by improving the participation of small-scale farmers in innovation processes.

3.4.2.2 Case 2. Development and Marketing of New Dairy Products in Oruro, Bolivia

A local foundation, SEDERA, directed and facilitated implementation of the PMCA with the aim of diversifying the production of community dairy plants. SEDERA and a group of local farmers were successful in producing mozzarella cheese that met local quality requirements and was marketed under the brand *Vaquita Andina*. The PMCA motivated local dairy producers to diversify the types of cheese they produce and to improve the quality and sanitary standards of their product. While the original goal was to sell the new mozzarella cheese to local pizzerias, this did not materialize, mainly due to its relatively high price. Instead, the main consumers were high-income households willing to pay a premium for a locally produced cheese. The product has been sold in a store operated by SEDERA and in some high-end food markets, including a supermarket in Oruro. Although the economic benefits for small producers were limited, SEDERA gained experience in analyzing market chains and facilitating innovation processes and is now using a more market-oriented approach to its development work.

3.4.2.3 Case 3. Development of New Markets for Yams in Northern Colombia

In April 2008, the PBA Foundation facilitated the PMCA's implementation in 7 market chains over a period of 13 months. Some progress was made in improving the marketing of yams, but no new distinctive yam product was developed or marketed. To sell higher quality yams at better prices, small farmers increased plant density and improved the selection and cleaning of harvested tubers. Some shipments of fresh yams to the US were made, but the Colombian farmers faced strong competition from other Caribbean suppliers in that market. Commercial trials of high-quality yam fiber were hampered by a lack of resources for the construction of a pilot plant. In light of the small size of local farmers' organizations, the PBA Foundation worked to establish a regional network of local associations to improve their marketing performance. An unexpected result was the organization of suppliers within the local market to better coordinate the flow of products and stabilize price. The PBA Foundation has now incorporated elements of the PMCA into its portfolio of participatory methods.

3.4.2.4 Case 4. Conservation and Commercialization of Native Potatoes in Northern Potosí, Bolivia

PROINPA and the Center for Agricultural Development (CAD) have worked for years to conserve the biodiversity of potatoes and other Andean crops and to reduce rural poverty in northern Potosí. They facilitated the development of a new product called *Miskipapa*, which consisted of selected and washed native potatoes sold in meshes. To market Miskipapa, farmers had to improve the sorting and grading of their harvested potatoes. Although they expressed interest, there was little support from local government agencies. Miskipapa was sold for 3 years in supermarkets in La Paz and Cochabamba, in the shop of a mining union, two tourist hotels and farmers' markets. However, due to limitations in both supply and demand of native potatoes, the economic benefits for farmers were limited. Nevertheless, greater awareness of the crop's value contributed to renewed efforts to conserve the region's native potato biodiversity. The most significant result was the experience acquired by CAD, which changed its emphasis from enhancing production to promoting market chain innovation.

To analyze the factors that influenced PMCA implementation and results in these interventions, Horton et al. (2013c) used the "Institutional Analysis and Development (IAD) Framework" developed by Ostrom (2005) and modified Devaux et al. (2009) and Thiele et al. (2011a). This framework considers four main groups of factors that can influence the implementation and results of the PMCA:

- Macro context: which includes government policies, socio-economic conditions, and the agro-ecological characteristics of the region that can influence VCD.
- Market chain: biophysical and technological characteristics of the market chain in which the PMCA is being applied.
- Principal actors: attributes of the relevant market chain actors and service providers involved in the PMCA process.
- Rules in use: formal and (mainly) informal norms and customs that govern the behavior of participants.

The macro context The pro-market policies of Colombia and Peru provided a more favorable environment for the use of the PMCA than the policies of the Bolivian government, which emphasize the role of the state and "communitarian socialism." Agro-ecological environments can also have an effect on the implementation processes and results. In the Bolivian highlands, where poverty is very high and production conditions are adversely affected by climatic risks and the reduced use of inputs, there are severe limitations on the implementation of VCD approaches for the reduction of rural poverty.

The attributes of the chain Successful innovation is more likely in some market chains than in others. In the cases involving coffee in Peru and, to a lesser extent, dairy products in Bolivia, it was possible to bring in outside expertise to improve processing. On the contrary, in the cases of native potato and yam, the knowledge base for commercialization and processing was more restricted. One of the reasons was eventually because the participatory process did not involve the required diversity of participants because of the location and the opportunities linked to the value chain selected. Coffee and dairy products also presented more opportunities for processing, branding and product differentiation than potatoes and yams. As mentioned in the previous section, the processing of native potatoes into colorful potato chips has emerged as a viable enterprise that can respond to the demands of urban consumers in Peru and, to a lesser extent, in Bolivia. However, this type of industry is typically based in urban areas and was not considered to be a viable option for the potato farmers in northern Potosi, Bolivia.

The main actors The cases analyzed enabled the identification of three types of "champions" that can be crucial for the successful implementation of the PMCA, and for the approach's integration into the work of R&D organizations. The first type of champion is the facilitator who coordinates the group and supports the innovation processes; the second type is a manager/decision maker, who coordinates the VCD intervention and the implementation of the approach, including mobilization of resources; the third is a recognized leader within the market chain. In the first case study, for example, the facilitator, from the NGO Practical Action in Peru, played a key role in identifying and supporting local actors and facilitating the change processes. A senior manager of the NGO also provided strong institutional support. And the leader of the women's group led the development of the new coffee brand and the creation of networks with other actors in the local coffee sector. The leadership and investment capacity of the private sector are also crucial for the ultimate success of efforts to stimulate business innovations.

The rules in use This refers to social customs, norms and rules, both formal and informal, that guide human behavior on a day-to-day basis. It is an important component in the value chains that were analyzed, which were generally characterized by distrust and limited communication and interaction between the different actors in the chain (such as producers, intermediaries, processors and retailers), which of course limits coordination and collaboration. Ethnic and racial divisions and discrimination were notable in the value chains for potato in Bolivia and yam in Colombia, as compared to the coffee chain in Peru, where such divisions were not observed.

The rules in use (or "standard operating procedures") of the R&D organizations are also important. The PMCA is facilitated by individuals based at R&D organizations that have their own mandates, program structures, cultures, norms and external relationships. The mandate and culture at public agricultural research organizations can pose challenges for the successful implementation of the PMCA, as these organizations may be reluctant to work with the private entrepreneurs who process and market agricultural products. For example, the implementing organization in Peru, Practical Action, had a strong tradition of working with all sectors, so it easily incorporated the PMCA into its program to develop coffee markets in Peru. This is not always the case with public entities that are more focused on productive and technological issues.

Another important factor for the implementation of the PMCA is the process of training the people who will implement the approach, especially the group facilitators. If not done properly, deviations or failure to implement the approach may occur, as Horton et al. (2013c) documented. To support the training process, a guide for trainers was produced (Antezana et al. 2008).

3.4.3 The Revalorization of Native Potatoes in Ecuador

In November of 2008, through the Andean Change Alliance, CIP began collaborating with the Potato Program at Ecuador's National Agricultural and Livestock Research Institute (INIAP), a local NGO, Fundación Marco, and the Small-Scale

Potato Producers' Consortium (CONPAPA) on the implementation of PMCA in Ecuador, as part of an effort to create market opportunities based on the country's wealth of native potato varieties.

The qualitative assessment in the first phase of the PMCA involved 29 chefs and administrators of restaurants and hotels in different provinces of the country, in order to determine their knowledge and attitudes towards native potatoes and identify market niches. The results showed that these varieties were largely unknown in restaurant and hospitality sectors of the cities, where they could not be found in markets. Nevertheless, the participants agreed that they had an interesting business potential.

The final event of Phase 1 was thus dubbed as a "Meeting to seek for business opportunities in native potatoes," an activity attended by 35 participants that included chefs, farmers, processors, representatives of local NGOs and public officials. Two interest groups were formed in the event: one on culinary research with native potatoes and the other, considering the requirements of hotels and restaurants, focused on the supply of raw materials.

The participation of the chefs was central to the second phase of this PMCA experience to promote and give visibility to native potatoes in Ecuador. That phase culminated in a massive event organized by the Ecuador Gourmet Cooking School to present and share variations of some of the most traditional dishes in Ecuadorian cuisine made with native potatoes. This event raised the visibility of those varieties and subsequent efforts focused on taking advantage of opportunities in niche markets linked to gastronomy.

In the third phase, other restaurants joined the process in order to promote the consumption of native potatoes and visits by chefs to the farms of native potato growers were encouraged. Those who visited farms understood for the first time the reality of the cultivation of those potatoes, in mountainous areas with difficult access, and accepted that native potatoes should fetch a higher price in the market than commercial varieties. Farmers, for their part, had access for the first time to information concerning the quality and timeliness required for supplying potatoes to this type of market.

A year after initiating the PMCA, the *Papa Nativa* (Native Potato) brand was launched at the final event of the process with the slogan "Discovering the Andean flavor." The product consisted of selected and washed native potatoes in a mesh. The launch included press conferences and food fairs in the framework of the celebration of Ecuador's National Potato Day.

Developing the product as a commercial innovation triggered technological innovations at the farm level (fertilizer application, better seed management, pest management) and in the presentation of potatoes in the market (moving from an unsorted product to a selected one with a brand name). For the CONPAPA farmers' association, this meant raising members' capacities for postharvest handling and commercialization, since they began selling native potatoes in diverse presentations and linked with consumers such as restaurants and supermarkets in urban areas for the first time.

Although the "*Papa Nativa*—Discovering the Andean Flavor" brand has not prospered in the market due to supply problems, high transaction costs, and limited demand, the results of the PMCA were important because it showcased native potatoes in markets other than the local highland markets that they had traditionally been limited to. This result captured the attention of the company INALPROCES, which had been promoting high-quality products with social responsibility criteria for several years. Managers of INALPROCES were inspired by CIP's Papa Andina experience with native potatoes in Peru and foresaw interesting business opportunities in the production and sale of potato chips made with colorful native varieties in the Ecuadorian market and abroad.

INALPROCES partnered with INIAP, the Minga Foundation for Rural Action and Cooperation, CONPAPA and CIP's Ecuador office to identify native potatoes adequate for processing that could be produced and supplied by CONPAPA (Montesdeoca et al. 2013). Two colorful varieties obtained by INIAP by crossing selected native potatoes were chosen for the production of quality potato chips: Puca Shungo (Red Heart) and Yana Shungo[5] (Black Heart). Their main characteristic is the intense reddish or purple coloration in their flesh, which means they are rich in antioxidants, vitamins, and proteins, and differentiates them from traditional potatoes. In 2011, the brand of colorful native potato chips Kiwa was launched, promoted with support from international cooperation programs as a business model for corporate social responsibility. This business model included working in collaboration with a number of local actors to maintain consistent production levels and to ensure a supply of quality seed potatoes. The model also included training services for farmers based on the Farmers' Field Schools methodology. The business model promoted by INALPROCES was so successful from the beginnings that, in 2011, the company's work with Andean potato farmers was recognized as the best Corporate Social Responsibility project in Ecuador by the Ecuadorian-German Chamber of Industry and Commerce. It also won an international Award "Taste 11 Award for Top Innovation of Anuga—2011" in Germany for the native potato chips.

Since its launch on the market, the product has increased its sales in the domestic and export markets, including United States, Europe, and the Middle East (Fig. 3.4). This growth has promoted collaboration between CIP, INIAP, and CONPAPA farmers to carry out activities aimed at promoting the production of quality seed of the two selected varieties, increasing the volumes of their production and planning sales by CONPAPA to INALPROCES.

The development of Kiwa native potato chips is one of the most noteworthy examples of the evolution of the innovation process that, as we noted, continues and evolves long after the PMCA has formally ended. This commercial innovation has also catalyzed a range of technological innovations. For example, to satisfy the demand for seed potatoes of the varieties used in the production of the Kiwa product, these varieties were included in the seed multiplication program of the INIAP

[5] See technical data sheet: http://repositorio.iniap.gob.ec/handle/41000/3267

Fig. 3.4 Kiwa potato chips displayed at a supermarket in Saudi Arabia

high-quality seed production greenhouse. However, there are other challenges at the production level, mainly with changes in climate and frost in the Ecuadorian Andes that affected up to 80% of the production in some years. It is necessary to continue with the selection of varieties that are appropriate for quality processing and more tolerant to weather shocks. In the processing area, it is necessary to find more uses for these potatoes, such as mashed potatoes or frozen French-fry-cut native potatoes for export (Martin Acosta, Kiwa CEO, personal communication).

The Kiwa case is a good example of how market opportunities and desire of social responsible business can be driving forces in the establishment and operation of public–private partnerships for rural development. In particular, the partnership between INIAP, INALPROCES, and CONPAPA generated concrete research demands for the selection of varieties and specific requirements for the provision of high-quality seed and technology transfer services. Through this alliance, CONPAPA has strengthened the business management capacity of associated farmers and their ability to meet the demand of INALPROCES, access technical assistance services from public and private suppliers, and respond to other business opportunities. Currently CONPAPA, renamed "AGROPAPA" since 2014, produces certified seed under contract with the Ministry of Agriculture and has increased its customer base of fresh, quality ware potatoes in provincial markets.

Conclusions from This Case

If we apply the Institutional Analysis and Development (IAD) Framework as used by Horton et al. (2013c) to this case, we gain several insights. Regarding the macro context, it is clear that Ecuador's pro-market policies were not as favorable as Peru's in the case of the Native Potato Revolution. One of the great challenges has been to inspire entrepreneurship in face of a rather limited business mindset. Although Ecuador is one of the countries with the highest records of trademarks and inventions in the Andean region, these rarely translate into actual businesses and commercial uptake. As a result, fewer companies are initiated and fewer employment opportunities are created than in other countries (Wong and Padilla 2017). Furthermore, farmers have limited access to credit for both investment and working capital. As a matter of fact, none of the farmers involved in the PMCA had access to credit and, when necessary, they invested their limited family savings or got personal loans.

Regarding the value chain, native potatoes were unknown by urban consumers and it took a considerable effort to position them in that market as a fresh product and to connect producers with restaurants or potential buyers because there was no appropriate distribution system. Promoting native potatoes as colorful chips has been more successful. Given the smaller size of the Ecuadorian national market compared to Peru, the fact that the product is being placed in international markets has contributed to its success.

In terms of the principal actors, there was support from service providers and chefs in the initial implementation of the PMCA, but apparently, a recognized leader in the native potato chain was missing. When INALPROCESS launched the idea of producing colorful native potato chips, the leadership role was assumed by that company.

If we analyze the rules in use, the native potato value chain was not well organized; there was little communication and limited interaction between the different actors in the chain and service providers. The alliance between the private sector company INALPROCESS and the government institute INIAP was a new experience and it needed to be constructed giving consideration to the norms at INIAP, which has a more technological mandate than marketing and processing. Once the Kiwa native potato chips product was launched, the working model for providing training services and technical support by the company to help farmers respond to market requirements did not work systematically and was not connected with other service providers. For example, there was an unmet demand for a manual of procedures and best practices in crop management, post-harvest treatments, and potato selection, as well as clear and understandable rules and parameters to define the price paid to the farmer for his products.

Results: The implementation of the PMCA served to showcase native potatoes with the collaboration of the chefs and make them known to a broader public than the rural markets they have traditionally been sold in. The producers were able to meet the quality and supply schedule required by the processing company responding to demanding export potato markets, which has helped them access new market options for their potatoes.

Thanks to INALPROCESS, a colorful native potato chip product was developed which now has a presence in national and international markets. The product accounts today for 20% of the company's sales. They have been able to sell the Kiwa Native Andean Potato Chips (or Crisps) to retailers all over the world in presentations ranging from vending machine size to Club sizes, to retailers such as Costco in Canada and SnR in Philippines. The company is planning to expand mainly in the USA, the Middle East and Eastern Europe and to move into the organic-certified market (Martin Acosta, CEO Kiwa, personal communication).

3.4.4 *Building Capacity for Innovation in Ugandan Value Chains*

As part of CIP's efforts to facilitate South–South collaboration, the experiences of the PMCA in the Andes were shared with R&D professionals and actors in the value chains of various agricultural products in Uganda between 2005 and 2007.

The capacity development strategy implemented in Uganda included a series of complementary components. Two cross-learning visits were organized in the Andes for specialists from that country. This exchange generated enthusiasm and interest to apply the methodology in Uganda. Training workshops were organized at the beginning of each phase of the PMCA implementation in Uganda, which allowed members of the facilitating team to practice using tools in real-world situations. Participatory learning and joint decision-making by the facilitating team and thematic teams strengthened teamwork and empowered the participants.

The approach was subsequently validated in potato, sweetpotato, tomato, and hot pepper value chains in Uganda. Eight women from R&D organizations in the country led the application of the PMCA and facilitated the various meetings of the participants. Specialists from CIP and national entities from Peru and Bolivia provided capacity building and backstopping for the Ugandan team throughout the process (Mayanja et al. 2013; Horton et al. 2010).

Many organizations and individuals played roles in the introduction, validation and refinement of the PMCA in Uganda, including representatives of academic and research institutions, governmental and non-governmental organizations, the private sector and business organizations. Institutions in each of these categories had clear functions. The R&D institutions were responsible for introducing and facilitating the process of implementing the methodology. They also identified the actors, including service providers, in the market chain that ended up forming the thematic

Table 3.2 Participants in the first application of the PMCA in Uganda

R&D actors	Market chain actors
Potato work group	
Ssemwanga Center	Tom Cris (processor)
Africa 2000 Network—Uganda	Nyamarogo potato farmers
Agribusiness Initiative Trust	Uganda National Sweet Potato Producers Association (UNSPPA)
International Institute of Tropical Agriculture (IITA)/Foodnet	Merchants from Owino market

Adapted from Mayanja et al. (2012)

groups during the implementation of the approach. These different actors worked together to identify and exploit market opportunities. Through this process, innovations were generated.

The potato chain group, for example, was led by the Ssemwanga Center for Agriculture and Food Ltd., a consulting company with members who are also team leaders in the Uganda branch of the Africa 2000 network, the International Institute of Tropical Agriculture (a CGIAR center) and the Agribusiness Initiative Trust. The core team facilitated two thematic groups that were comprised of key actors, such as representatives of farmers' associations, processors and traders, who worked together in an attempt to address the challenges and opportunities identified in the market chain (Table 3.2). The diversity of organizations involved reflects the important role of partnerships in the promotion of innovations for market-driven development. Similarly, different organizations participated in the thematic groups that worked in sweetpotato and vegetable market chains.

The PMCA experience generated a series of results for participants that included the acquisition of new knowledge, skills, social networks and the ability to innovate. The market chain actors generated a series of viable commercial, technological and institutional innovations. The R&D actors (outside the core team members) and other service providers provided technical assistance, guidance and important links to the industry that contributed to the generation of innovations. Innovations that resulted from the PMCA included packaging and a better brand design for potato chips destined for high value markets, orange-fleshed sweetpotato flour for the local market, and sliced and dried hot pepper for export (Mayanja et al. 2012), some of these market innovations such as the potato chips continue to be produced.

It is important to mention that after the conclusion of the PMCA cycle, several activities were carried out to consolidate the progress in strengthening the relationships between the market chains' actors, follow up on commercial innovations, and promote and institutionalize use of the PMCA by R&D organizations in Uganda and other countries in the region. All these activities were carried out by the original facilitators, who voluntarily supported the effort after the initial funding for the PMCA had been exhausted. Advice and support was given to market chain actors to help them develop and market new products and present successful proposals for financing. Some of the original facilitators also served as trainers in workshops organized by other development programs in Uganda (Mayanja et al. 2013).

The PMCA was institutionalized by different entities through the implementation of projects that included the approach and promoted private sector participation. One example is the Mukono Zonal Institute for Agricultural Research and Development, whose director used the PMCA in other value chains, such as pineapple. The NGO VEDCO (Volunteer Efforts for Development Concerns) and local universities also used the approach for innovation in other market chains.

Development of Gender Guidelines for the PMCA in Uganda

During the different applications of the PMCA in Uganda, CIP researchers came to realize that the approach lacked ways to identify, analyze and address gender and generational differences among value chain actors, which reduced the approach's potential for having a positive impact in terms of ensuring that participants of both genders and all ages have equitable access to opportunities and benefits along value chains. To that end, CIP researchers and partners in Africa and the Andes developed a series of practical tools in recent years to integrate the gender approach into the PMCA's different phases of analysis and intervention. These tools are being tested and validated in projects both in Africa and the Andes Region, and there is an online guide available for their application (Mayanja et al. 2016). The guide seeks to create capacities among the facilitators of the PMCA to carry out a gender analysis to generate a better understanding of the different problems that men and women face in value chain interventions. The results of this analysis guide the development of gender strategies to promote equal opportunities for men and women to benefit from the PMCA.

Conclusions of PMCA Application in Uganda

The context in Uganda was favorable to the development of the market chain approach. Academic and research institutions, governmental and nongovernmental organizations and the private sector were actively involved in the concrete implementation of the PMCA.

From the beginning, the approach was applied to different chains, due to stakeholder interest, and its implementation was supported by different international cooperation projects. But despite the institutionalization of the approach by some organizations, it is important to recognize the need for adequate financing and institutional arrangements to accompany the work of the facilitators of the innovation process and to be able to follow up on it. After completing the PMCA, it is necessary to continue supporting the development process of emerging innovations and prototypes through business development services to entrepreneurs. Unfortunately, financing mechanisms for this type of services were not considered in the programs of the participating institutions.

The systematic process of training representatives of the R&D organizations that supported PMCA implementation was key to its development, and subsequent promotion in different value chains. The fact that women from different organizations led the process contributed to adjusting the approach to pay greater attention to gender issues along the chain. A recurring theme in the application of the PMCA in different contexts was the need to provide support to small-scale farmers to improve their productive and business capacity to respond to changing market demands. This is part of a development strategy that should be taken into account in national rural policies.

3.4.5 The PMCA and Farmer Business Schools in Indonesia

Between 2008 and 2009, CIP implemented the PMCA in fresh and processed potato value chains in West Java, Indonesia. Training and support were provided for a team of local implementers. The process resulted in 13 different, duly documented innovation processes, most of which generated new or improved processed products (mainly chips and sandwiches) rather than marketing fresh potatoes (Horton et al. 2013a, b).

During a project review, PMCA implementers identified limited business skills and inefficient farmer organizations as the main limitations that prevented the establishment of more effective links to the market. Farmers had little understanding of market opportunities and inadequate access to much needed market information as price trends and actual demand.

To address such shortcomings, CIP researchers integrated in the PMCA some elements of the Farmers Field School approach (FFS)—a group-based process that had primarily been used to promote integrated pest management by smallholder farmers (Orrego et al. 2009). This resulted in a new VCD approach that was named Farmer Business School (FBS). The FBS is a participatory action-learning process that aim to link groups of farmers to agricultural value chains (International Potato Center 2017a). It differs from other agricultural development approaches in that it focuses on the equitable and effective inclusion of small-scale farmers in value chains, instead of focusing only on farm production activities. The FBS builds farmers' capacity through a series of group learning activities based on real experiences during a farm production/marketing cycle while promoting interaction with other value chain actors. As a tangible outcome of FBS, participants are expected to have actual micro/small enterprises and business initiated or strengthened upon completing the FBS learning process. A FBS guide developed by (International Potato Center 2017b) covers five key issues:

1. Identification of market opportunities
2. Evaluation of market chains
3. Development of market-oriented innovations
4. Development of business plans
5. Provision of business support services

The FBS experience that CIP presented in Indonesia proved useful for strengthening farmers' organizations and building business skills and capacity. It was subsequently adapted by the FoodSTART project to the context of the highlands of the Cordillera region, in the Philippines, primarily to integrate it within the framework of a broader rural development investment project of the International Fund for Agricultural Development (IFAD) called CHARMP2. Between 2012 and 2013, the FBS approach was piloted with 6 groups of farmers (approximately 120 people) dedicated to commercial enterprises based on root and tuber crops, vegetables, coffee and livestock. Following the initial piloting, FBS was scaled to additional 66 groups (approximately 1600 people) between 2014 and 2015, through the collaboration between FoodSTART and CHARMP2. It is encouraging that, even after the formal partnership (and any financial and technical support from FoodSTART) ceased, CHARMP2 continued scaling the FBS approach with additional 25 farmers groups (approximately 1000 people). FBS is still the main VCD approach implemented by the Department of Agriculture through CHARMP2 and being FBS graduated has become a pre-requisite for accessing the livelihood assistance from the project (DA-CHARMP2 and CIP-FoodSTART+ 2018).

Based on the successful experience with CHARMP2, the FBS approach is currently being scaled in Asia by FoodSTART second phase (FoodSTART+) through partnerships with four IFAD investment projects in the Philippines, India and Indonesia. In the framework of FoodSTART+, FBS is being used as catalyst for introducing a number of farming and postharvest innovations to participant farmers groups: these include crop varieties, agronomic practices, processing technologies and marketing strategies needed to develop the new food products demanded by the market.

Comparison of the FFS, FBS and PMCA approaches Farmer field schools, farmer business schools and the participatory market chain approaches all employ the method of learning by doing to improve the well-being of smallholder farming families, but in different ways. The following is a comparative analysis of the three approaches that considers the scope and extent of implementation (Table 3.3).

This comparative analysis indicates that according to results achieved with the FBS in Asia, it could be an appropriate approach when the main challenge is for small-scale farmers' groups to exploit an existing well-defined and clear market

Table 3.3 Comparison of the FFS, FBS, and PMCA approaches

Criteria	FFS	PMCA	FBS
Types of actors involved	Smallholder farmers' groups	A diverse set of actors along the value chain	Focus on smallholder farmers' groups with interaction with other stakeholders in the chain
Themes covered	Agricultural production techniques	Participation to value chains and the promotion of interaction and collaboration among actors	Farmer organization and market development themes
System of focus	Crop system	Innovation with value chain actors to promote inclusive businesses	Agricultural business and value addition
Area of intervention	Strengthening farmer organization	Building "social capital bridges" to strengthen interaction and collaboration among multiple actors in different areas of the value chain	Strengthening farmer organization and innovation processes for enhanced participation in value chain
Implementation time frame	One crop cycle	Lasts as long as necessary to develop successful innovations in a value chain	Covers one farming and commercialization cycle

opportunity. PMCA, on the other hand, has demonstrated to be a preferred approach when the market opportunity is less defined or understood and where there are significant potential benefits for research and innovation in the processing or marketing stages (Devaux et al. 2011). PMCA needs commitment from research organizations while in the case of FBS it is less the case since the engagement with other chain actors and stakeholders is more sporadic. That said, the two approaches can be complementary when combined as the PMCA would contribute to conduct market research and to identify potential market opportunities to be developed. Then FBS can be run to help farmers group to exploit these market opportunities reinforcing farmers' business capacities and developing technological innovations required to respond to these opportunities.

The experience demonstrated the capacity of the actors who led the PMCA implementation process to adapt the approach to contexts and needs and adjust it. In Asia, the implementers developed the Farmer Business School approach to improve farmer's capacities to benefit from market opportunities. As indicated above, the FBS has been integrated into the framework of broader research for development projects, such as FoodSTART, as a tool for developing the farmers' capacity to access more dynamic markets, with also a gender focus, and considering the challenges of climate change. CIP and local partners are now validating the FBS in the Andes in combination with the application of PMCA. It is a process that is still in full development and needs to be further documented.

3.5 Lessons from Value Chain Approach Applications in Different Contexts

From the market chain experiences in different contexts described here we can draw lessons that can contribute to improving the design of future VCD interventions that use the PMCA as a collective research and innovation approach. It should be noted that in all of these cases, creative adaptations were made to the protocols set out in the implementation guide (Bernet et al. 2006). Following the analytical framework developed by Horton et al. (2013c), we can highlight the main factors that influenced these adaptations:

- the attributes of the context, which includes the policy environment and institutional framework;
- the attributes of the value chain;
- the characteristics of the participants and the facilitation role of the leading organization;
- the "rules in use" governing behavior and relationships between the value chain actors;
- and the importance of building the capacities of the professionals from R&D institutions responsible for implementing the approach.

The interaction of these variables and, especially the external environment of each context determined the implementation strategy and the forces that drove each implementation of the PMCA in the Andes, Asia, and Africa. Each of them was different, as were their scope and size.

Agricultural change creates opportunities for smallholders An important element for the promotion of the PMCA is the ever-growing integration of the agricultural sector into markets and value chains, which increases market opportunities for small-scale farmers. There are more opportunities for them to move beyond the practice of selling directly to rural consumers or traders in local markets to access modern value chains that can open doors to other buyers, including urban traders, processors and supermarket chains. When they have alternatives and the ability to participate in more competitive value chains, few smallholder farmers will limit themselves to local markets; the majority will try to sell to these new players, in order to reap greater benefits (Reardon et al. 2012).

By applying the PMCA, facilitating interactions between farmers and other value chain actors, and conducting market analyses, it is possible to identifying new commercial opportunities and market niches in which small-scale farmers can have a comparative advantage. In the PMCA user guide (Bernet et al. 2006), there is a tool called "impact filter" that provides a rapid, qualitative assessment of the expected impact of different market opportunities for the benefit of small-scale producers, while considering social and environmental factors. This tool allows R&D organizations to plan and guide interventions more effectively.

A holistic approach like the PMCA represents a new way of doing agricultural R&D Instead of undertaking research and then trying to transfer the results to farmers, this approach brings together a range of relevant actors—farmers, traders, processors, researchers and service providers—to set priorities and jointly develop innovations that respond better to the demands of the value chain. The approach has made it possible to achieve concrete results under different socioeconomic conditions, despite the fact that most development support for farmers continues to focus on agricultural production. This situation is reflected in the experiences analyzed in this chapter, which demonstrate a shortage of available or accessible services to accompany value chain innovation processes. The inclusive value chain development approach makes it clear that, in the debate on food security, more attention should be paid to post-harvest and value chain interventions, as opposed to focusing only on the farming sector. In the quest to improve food security, value chain efficiency, processing productivity and post-harvest management deserve almost as much attention as approaches to improve farm production.

Innovations take time and cannot always be programmed The PMCA triggers innovation processes that often continue long after the PMCA has formally ended. Second and third generation innovations tend to be even more important than the first ones, developed during the PMCA exercise, as is confirmed by the experiences in Peru, Ecuador and Uganda. However, this requires an investment in services and institutional mechanisms that allow monitoring and supporting innovations that arise beyond the life of a specific project. The lack of accompaniment of value chain innovations can be a limitation that reduces the scope of the innovation process. In the case of Peru, for example, the continued support of the Swiss Agency for Development and Cooperation allowed monitoring and accompaniment of innovation process beyond the formal conclusion of PMCA implementation.

Monitoring and evaluation challenges Most of the cases studied lacked a systematic process of monitoring and evaluating value chain innovations. The fact that these processes can take a long time, and that many innovations are developed after the PMCA implementation, with unexpected results, makes it more challenging to do impact analysis. In many cases, the optimal time for evaluating impacts is well after project financing has concluded, when it is not feasible to obtain specific financing for evaluation processes. An implementation guide and impact pathway can be useful guidelines for the implementation and evaluation of participatory approaches such as the PMCA and can serve as a basis for reflection and learning. In research and development projects with limited resources and tight deadlines, it is difficult to justify the time and resources necessary to develop actions and models of change. For this reason, when planning participatory interventions, specific resources should be allocated for the development of actions and models of change that can then be refined and tested by local implementers as the innovation process progresses.

Holistic and participatory approaches such as the PMCA are not easily "scaled" or "transferred" How can the initiatives described here be scaled or replicated better? The answer is not found in a particular or specific arrangement. The processes and approaches, and not so much the forms, are what lead to effective interactions, collaboration between actors and development of a market chain (Wiggins and Keats 2013). What needs to be scaled, in the sense of replication and adaptation, are the habilitation, facilitation and learning processes, supported by a necessary architecture that includes catalysts and leadership, forums and services to analyze and respond to specific problems and, especially, support for the farmers' groups that are organized and trained. The catalysts can be private companies and/or non-governmental organizations (NGOs). Both have their advantages and disadvantages. In the case of the private sector, the challenge is how to establish an equitable relationship with small-scale producers. In the case of NGOs, they will play the role of catalyst, based on their mission, only as long as they can be financed. To achieve the replication of successful approaches, it is important to promote innovation, learning and dissemination of experiences. To date, investment in learning and dissemination of these experiences has not reached a level comparable to that of the promotion of practical initiatives in the field. The experiences of collaboration with IFAD's investment projects, first in Asia, with FoodSTART, and more recently in South America, with the FIDA-Andes project, coordinated by CIP, has facilitated learning from and diffusion of these experience. In that sense the experience of FoodSTART in the Philippines is innovative and particularly encouraging as, following successful piloting and strengthening of the capacities of the partner's development project staff, the FBS approach has been institutionalized by the Department of Agriculture that is replicating and scaling it throughout the region. Similar outcomes are expected from the ongoing second phase of FoodSTART.

The PMCA has proven to be adaptable to a range of contexts Although the PMCA was developed to respond to a specific set of problems with a specific crop in a specific place, subsequent efforts of promotion, exchange of experiences and capacity building have contributed to its successful application to other crops that face different challenges and opportunities in other countries and regions. In the trajectory of its applications in the Andes, Uganda and Indonesia, the PMCA was not simply "transferred" from one place to another. In each case, it was necessary to adapt the approach to local circumstances and needs and to strengthen capacities to promote and accompany innovations.

There is no "one-size-fits-all" solution to the challenges of value chains There is no single factor of success. In each case presented here, a combination of interventions at different levels was needed to address the value chain development processes. A set of policy measures, development programs and technical capacity building for value chain actors are needed to facilitate innovation processes and strengthen effective links to promote the participation of small-scale producers in expanding markets. A comparative study of the variables that influence and determine the adaptation of participatory methodologies such as PMCA in different contexts would allow a better definition of the conditions needed for the replication, adaptation and scaling of these methodologies.

References

Antezana I, Bernet T, López G, Oros R (2008) Enfoque Participativo en Cadenas Productivas (EPCP): Guía para capacitadores. International Potato Center, Lima, 189 pp

Bernet T, Thiele G, Zschocke T (2006) Participatory market chain approach (PMCA)—user guide. International Potato Center, Lima. http://cipotato.org/publications/pdf/003296.pdf

Biggs S (2008) Learning from the positive to reduce rural poverty and increase social justice: institutional innovations in agricultural and natural resources research and development. Exp Agric 44(1):37–60

DA-CHARMP2 and CIP-FoodSTART+ (2018) Most significant change in the Cordilleras: a pilot evaluation. Department of Agriculture-Second Cordillera Highland Agricultural Resource Management Project and International Potato Center-Food Resilience Through Root and Tuber Crops in Upland and Coastal Communities of the Asia-Pacific, Manila, Philippines

Devaux A, Horton D, Velasco C, Thiele G, López G, Bernet T, Reinoso I, Ordinola M (2009) Collective action for market chain innovation in the Andes. Food Policy 34:31–38

Devaux A, Ordinola M, Horton D (eds) (2011) Innovation for development: the Papa Andina Experience. International Potato Center (CIP), Lima, Peru, 314 pp

Devaux, A., Ordinola, M., Mayanja, S., Campilan, D. and Horton, D. (2013) The participatory market chain approach: from the Andes to Africa and Asia. Papa Andina innovation brief 1. International Potato Center, Lima, Peru, 4 pp

Devaux A, Torero M, Horton D, Donovan J (eds) (2016) Innovation for inclusive value chain development: successes and challenges. International Food Policy Research Institute (IFPRI), Washington, DC, 529 pp

Engel P, Salomon M (1997) Facilitating innovation for development: a RAAKS resource box. Royal Tropical Institute, Amsterdam. https://www.kit.nl/sed/publications/?search=raaks&q_author=0&q_year=0&display=5

FAO (2013) Agroindustrias para el desarrollo. Roma. http://www.fao.org/3/a-i3125s.pdf

Henson S, Humphrey J (2010) Understanding the complexities of private standards in global agri-food chains as they impact developing countries. J Dev Stud 46(9):1628–1646. https://doi.org/10.1080/00220381003706494

Horton D, Samanamud K (2013) Peru's native potato revolution. Papa Andina innovation brief 2. International Potato Center, Lima, Peru, 6 pp

Horton D, Akello B, Aliguma L, Bernet T, Devaux A, Lemaga B, Magala D, Mayanja S, Sekitto I, Thiele G, Velasco C (2010) Developing capacity for agricultural market chain innovation: experience with the 'PMCA' in Uganda. J Int Dev 22:367–389

Horton D, Campilan D, Prasetya B, Gani H, Pakih M, Kusmana (2013a) The PMCA, business development services and farmer business schools in Indonesia. Papa Andina innovation brief 5. International Potato Center, Lima, Peru, 6 pp

Horton D, Campilan D, Prasetya B, Gani H, Pakih MR, Kusmana (2013b) Agricultural market chain development in Indonesia: experiences with the 'Participatory Market Chain Approach', 'Farmer Business Schools', and 'Business development services'. Social Science working document 2013-1. International Potato Center (CIP), Lima

Horton D, Rotondo E, Paz Ybarnegaray R, Hareau G, Devaux A, Thiele G (2013c) Lapses infidelities, and creative adaptations: lessons from evaluation of a participatory market development approach in the Andes. Eval Program Plann 39:28–41

Humphrey J, Navas-Alemán L (2010) Value chains, donor interventions and poverty reduction: a review of donor practice. IDS research report 63. Institute of Development Studies, Brighton

International Potato Center (2017a) Farmer Business School, a climate smart and gender responsive approach. Brief. The International Potato Center-food security through Asian Roots and Tubers (CIP-FoodSTART+), IFAD funded Project, The Philippines, 4 pp

International Potato Center (2017b) Farmer Business Schools in a changing world: a gender responsive and climate smart manual for strengthening farmer entrepreneurship. International Potato Center, Lima. 2 Volumes

Maldonado L, Ordinola M, Manrique K, Fonseca C, Sevilla M, Delgado O (2011) Estudio de caso: Evaluación de impacto de la intervención del proyecto INCOPA/CAPAC en Andahuaylas. CIP, Lima

Mayanja S, Akello B, Horton D, Kisauzi D, Magala D (2012) Value chain development in Uganda: lessons learned from the application of the participatory market chain approach. BANWA 9:64–96

Mayanja S, Akello B, Horton D, Kisauzi D, Magala D (2013) Building capacity for market-chain innovation in Uganda. Papa Andina innovation brief 4. International Potato Center, Lima, 4 pp

Mayanja S, Barone S, McEwan M, Thomas B, Amaya N, Terrillon J, Velasco C, Babini C, Thiele G, Prain G, Devaux A (2016) Prototype guide for integrating gender into participatory market chain approach. International Potato Center, Lima

Montesdeoca L (2016) Reporte trianual 2014–2016 de TRIAS—AGROPAPA-CONPAPA, 14 pp

Montesdeoca L, Acosta M, Quishpe C, Monteros C, Andrade-Piedra J, Pavez I (2013) Rescatando variedades ancestrales: Innovación de las papas nativas en Ecuador. En: "Innovaciones de Impacto, lecciones de la agricultura familiar en América Latina y el Caribe"/Editado por Priscila Enriquez, Hugo Li Pun, San José de Costa Rica. IICA, BID, 224 pp

Ordinola M, Devaux A, Manrique K, Fonseca C, Thomann A (2009) Generando innovaciones para el desarrollo competitivo de la papa en el Perú. Centro Internacional de la Papa (CIP), Lima

Ordinola M, Devaux A, Manrique K, Fonseca C (2013) Innovaciones en la cadena de la papa en el Perú: El valor de la biodiversidad. Leisa, Revista Agroecológica, Volumen 29, número 2, Nuevos mercados, nuevos valores

Ordinola M, Devaux A, Bernet T, Manrique K, Lopez G, Fonseca C, Horton D (2014) The PMCA and potato market chain innovation in Peru. Papa Andina innovation brief 3. International Potato Center, Lima, 8 pp

Orrego R, Ortiz O, Pradel W, Arévalo A, Barrantes C, Macedo O (2009) Sistematización de la implementación de las Escuelas de Campo de Agricultores (ECAs) en Andahuaylas. Lima (Perú). Centro Internacional de la Papa, Cooperación al Desarrollo—CESAL, Junta de Comunidades de Castilla—La Mancha, 42 pp. ISBN 978-92-9060-371-9

Ostrom E (2005) Understanding institutional diversity. Princeton University Press, Princeton

Proexpansión (2011) Cambios del sector papa en el Perú en la última década: Los aportes del proyecto Innovación y Competitividad de la Papa (INCOPA). Centro Internacional de la Papa, Lima

Reardon T, Hopkins R (2006) The supermarket revolution in developing countries: policies to address emerging tensions among supermarkets, suppliers and traditional retailers. Eur J Dev Res 18:522. https://doi.org/10.1080/09578810601070613

Reardon T, Timmer P (2012) The economics of the food system revolution. Ann Rev Resour Econ 4:225–264

Reardon T, Barrett CB, Berdegué JA, Swinnen JFM (2009) Agrifood industry transformation and small farmers in developing countries. World Dev 37(11):1717–1727

Reardon T, Chen K, Minten B, Adriano L (2012) The quiet revolution in staple food value chains: enter the dragon, the elephant, and the tiger. Asian Development Bank/International Food Policy Research Institute, Mandaluyong City/Washington, DC

Royer A, Bijman J, Bitzer V (2016) Linking smallholder farmers to high quality food chains: appraising institutional arrangements. In: Bijman J, Bitzer V (eds) Quality and innovation in food chains. Wageningen Academic Publishers, Wageningen, pp 33–62. Chapter 2

Schut M, Klerkx L, Sartas M, Lamers D, McCampbell M, Ogbonna I, Kaushik P, Atta-Krah K, Leeuwis C (2016) Innovation platforms: experiences with their institutional embedding in agricultural research for development. Exp Agric 52(4):537–561

Thiele G, Quirós CA, Ashby J, Hareau G, Rotondo E, López G, Paz Ybarnegaray R, Oros R, Arévalo D, Bentley J (eds) (2011a) Métodos participativos para la inclusión de los pequeños productores rurales en la innovación agropecuaria: Experiencias y alcances en la región andina 2007–2010. Programa Alianza Cambio Andino, Lima, 197 pp

Thiele G, Devaux A, Reinoso I, Pico H, Montesdeoca F, Pumisacho M, Andrade-Piedra J, Velasco C, Flores P, Esprella R, Thomann A, Manrique K, Horton D (2011b) Multi-stakeholder platforms for linking small farmers to value chains: evidence from the Andes. Int J Agric Sustain 9(3):1–11

Tobin D, Glenna L, Devaux A (2016) Pro-poor? Inclusion and exclusion in native potato value chains in the central highlands of Peru. J Rural Stud 46:71–80

UNIDO (2011) Pro-poor value chain development: 25 guiding questions for designing and implementing agroindustry projects. United Nations Industrial Development Organization, Vienna

Wiggins S, Keats S (2013) Leaping and learning: linking smallholders to markets in Africa. Agriculture for Impact, Imperial College and Overseas Development Institute, London

Wong N, Padilla M (2017) Emprendimiento social para la recuperación de la producción agrícola de patatas nativas de los Andes de Ecuador: Caso KIWA chips. En Experiencias de emprendimiento social en Iberoamérica. Editorial Universidad de Almería. Colección: Libros Electrónicos No 67, pp 77–81

Part II
Genetic Resources, Genetics and Genetic Improvement

Chapter 4
Ex Situ Conservation of Potato [*Solanum* Section *Petota* (Solanaceae)] Genetic Resources in Genebanks

David Ellis, Alberto Salas, Oswaldo Chavez, Rene Gomez, and Noelle Anglin

Abstract Conserving the genetic diversity of potato is critical for the long-term future of potato improvement programs. Further, it is the social and ethical responsibility of the present generation to ensure future generations have the same opportunities to use, exploit, and benefit from the genetic diversity that exists today. Genebanks and the ex situ conservation of potato genetic resources are the only way to ensure this happens; in situ conservation plays a complementary role, but it can never ensure that the vast diversity that exists on earth today is still there for use in the future. Material in ex situ genebanks not only serve as a reservoir of ready-to-use genetic material when needed but also provide invaluable tools for research now and in the future of cultivated potato and its wild relatives.

4.1 Ex Situ Conservation of Potato

The conservation of crop diversity, such as potato diversity, outside of its natural habitat, or ex situ, is held in botanical gardens and genebanks throughout the world. While botanical gardens conserve diversity and display this diversity for the public to enjoy and learn from, as well for scientific research, the overall mandate of botanical gardens is not generally to promote the use of and share the plant diversity they hold. In contrast, the mandate of genebanks *is* to provide access to the genetic resources they hold and to promote the use of these genetic resources for training, breeding, and research. Unfortunately, many national genebanks holding potato germplasm do not have the capacity, resources, nor a national mandate to widely distribute their holdings, and thus, access to potato germplasm is mostly left to the genebanks in developing countries, United States Department of Agriculture (USDA), and the potato genebank of the CGIAR, which holds potato germplasm at the International Potato Center, or CIP after its Spanish acronym located in Lima, Peru.

D. Ellis (✉) · A. Salas · O. Chavez · R. Gomez · N. Anglin
International Potato Center, Lima, Peru
e-mail: d.ellis@cgiar.org

H. Campos, O. Ortiz (eds.), *The Potato Crop*,
https://doi.org/10.1007/978-3-030-28683-5_4

Potato genetic resources held ex situ are divided into potato wild relatives (maintained as seed populations) and cultivated varieties which are maintained either in field plantings (where tubers are planted and harvested annually), seed, in vitro/clonal material, or cryopreserved material. The same genetic material could be held in two or more different forms in a single genebank (in vitro and cryopreserved) or between two genebanks (seed and in vitro). What is critical for a genebank, is that the material it holds is available in a form that can be readily used by all who need it, including breeders, large and small-holder farmers, industry, researchers from developing or developed countries, and for teaching and educational purposes. Thus, the wealth of potato diversity (Figs. 4.1, 4.2, 4.4, and 4.5) held and made available by genebanks is an asset for humanity, as it safeguards known genetic traits which are useful today, along with a myriad of known and unknown traits ready to be deployed to meet the needs for the threats and challenges potato farmers face in the future.

The more uniform the crop in general, the easier it is to manage as an ex situ collection. Unfortunately, cultivated potato is an incredibly diverse crop, and while virtually all the potato planted and used commercially worldwide are tetraploid potato varieties, diploid, triploid, tetraploid, and even pentaploid landrace potatoes are still grown throughout the South American Andes. This diversity has been collected, maintained and is available for use in breeding and research programs worldwide from numerous genebanks. The diversity of cultivated potato is reflected not only in its vast ploidy range but also at the species level where over 30 years ago cultivated

Fig. 4.1 Diversity of flower color and shape in cultivated potato from CIP germplasm collection

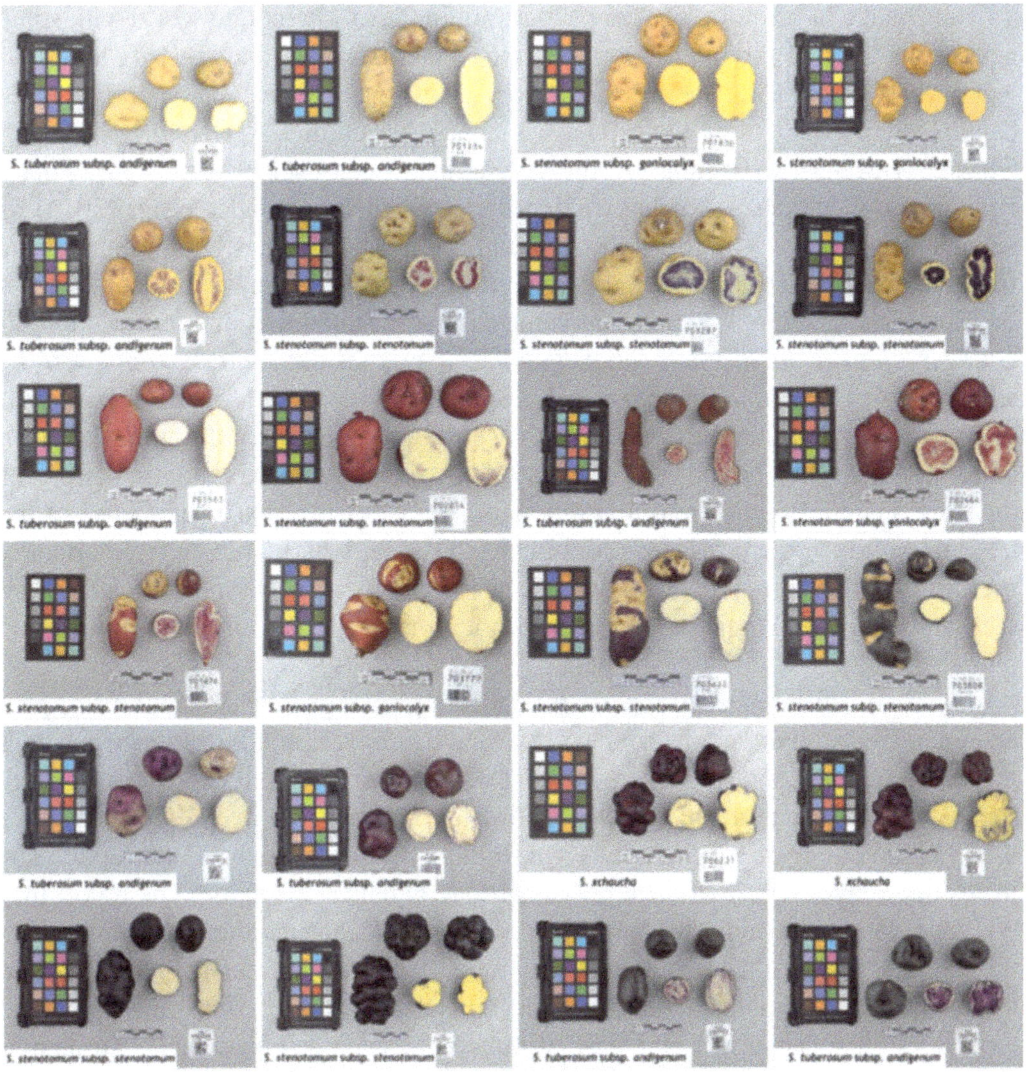

Fig. 4.2 Diversity of tuber shape and morphology and skin and flesh color in cultivated potato from the CIP germplasm collection. The color card in each photograph is used for standardizing measurement of colors of the tubers

potato was divided into seven species and nine taxa [*Solanum tuberosum* (C. Linneo) *subsp. andigenum* (Juz. & Bukasov), *S. tuberosum* (C. Linneo) subsp. *tuberosum, S. xchaucha* (Juz & Bukasov), *S. stenotomum* subsp. *goniocalyx* (Juz. & Bukasov), *S. stenotomum* subsp. *stenotomum* (Juz. & Bukasov), *S. xjuzepczukii* (Bukasov), *S. phureja* (Juz. & Bukasov), *S. xajanhuiri* (Juz. & Bukasov), and *S. xcurtilobum* (Juz. & Bukasov); (Hawkes 1990)] (Figs. 4.1 and 4.2). More recently the taxonomy of potato has been revised and cultivated potato has been regrouped into four species (*Solanum tuberosum, S. xjuzepczukii*, *S. xajanhuiri*, and *S. xcurtilobum*; Spooner et al. 2014). A similar taxonomic revision has occurred in the taxa of potato wild relatives with the previous taxonomic classification of Hawkes (1990) recognizing 228 wild species in 21 series. However, the recent taxonomic classification by Spooner et al. (2014) reduced the wild relatives of potato (*Solanum* section *Petota*) to 107 species.

The change in taxonomy has created a split in the ex situ potato world with some genebanks (such as the International Potato Center (CIP)] still organizing their collection based on the taxonomy of Hawkes (1990) while other major potato genebanks [such as the United States Department of Agriculture-Agricultural Research Service (USDA-ARS)] adopting the taxonomy of Spooner et al. (2014). Potato curators in all genebanks acknowledge the revision by Spooner et al. (2014) as needed and an advance in the field, yet the groupings by Hawkes (1990) make sense for managing an ex situ collection. Spooner et al. (2014) foresaw such challenges in their revision of the potato taxonomy and therefore stated that due to the "*complicating biological factors in section Petota*", they consider their "*taxonomy to be subject to critique and modification*" (Spooner et al. 2014, p 325). Two final comments are that there is consensus that ploidy in potato is not a good factor in delimitating species boundaries and that the use of different taxonomic names for the same material in different major potato genebanks is a detriment to use of the materials for the potato research community. That said, harmonizing the taxonomy used in global potato genebanks will need to be a focus in the next decade along with an extensive genetic comparison of potato collections to identify unique and redundant material between genebanks.

Taxonomic differences aside, ex situ collections and genebanks continue to play an ever increasingly dynamic role in providing the tools needed for food security in the future. The Second Report on The State of the World's Plant Genetic Resources for Food and Agriculture (FAO 2010) estimates that there are over 1750 genebanks worldwide holding approximately 7.4 million accessions; however, the report estimates that only 25–30% of these accessions are genetically unique. The report lists 98,285 potato accessions conserved in 174 genebanks around the world, and using the estimate above of the percentage of unique accessions, we can assume that there are at most an estimated 24,500–29,500 *unique* potato accessions are conserved worldwide in genebanks. In the case of potato, the accuracy of such an estimate of unique accessions is difficult to access until a thorough rationalization-comparison amongst the potato ex situ collections is carried out. According to this same report, 6 genebanks hold 41% of the global potato accessions: The French National Institute for Agricultural Research (INRA) in France (11%), Vavilov Institute in Russia (9%), The International Potato Center (CIP) in Peru (8%), The Leibniz Institute of Plant Genetics and Crop Plant Research (IPK) in Germany (5%), USDA-ARS in USA (5%), and The National Institute of Agrobiological Sciences (NIAS) in Japan (3%) and 20 global genebanks hold over 1000 potato accessions each. The listed global potato holdings collectively consist of 15% wild relatives of potato, 20% cultivated potato, 16% research and breeding materials, 14% advanced breeding lines, and 35% uncategorized accessions in these genebanks.

The last Global Strategy for Ex Situ Conservation of Potato (2006) analyzed 23 global potato collections which collectively maintain "*nearly 59,000*" potato accessions. This summary states that the genebanks in Latin America contain principally native cultivars while those in Europe and North America contain modern cultivars, breeding materials, and wild relatives. Such generalizations oversimplify the collections and indeed, although there is specialization among the potato genebanks, most major collections have a good representation of both wild and cultivated accessions.

Information about the particular attributes for each individual accession is critical for users to select which one or two of the thousands of accessions contain the trait(s) the user needs. Unfortunately, the information on accessions in global potato genebanks is generally incomplete and lacking the trait information users want and need most. Further, information on qualitative, complex traits (i.e., drought, frost tolerance, or yield) can be very specific to the physiological age at which the plant is subjected to the stress, the location, soil type, and degree of stress and, thus, is not always easy to list in a genebank database. Further, these complex traits require multiyear, multilocation evaluations in order to understand the complexities of the traits and genes associated with the trait. The USDA-ARS collection contains a large list of descriptors and traits recorded with 137 morphological, biochemical, nutritional, or physiological traits listed for potato (https://npgsweb.ars-grin.gov/gringlobal/cropdetail.aspx?type=descriptor&id=73), while CIP lists 72 descriptors or traits for their potato holdings (http://genebank.cipotato.org/gringlobal/search.aspx) and the European Union Genetic Resources of Potato lists 58 descriptors or traits (http://ecpgr.cgn.wur.nl/eupotato/). All these data are accessible in English on public internet-accessible databases, as is information from other global potato genebanks; although far too much of the information of the potato ex situ collections is not available publicly and outside of a few genebanks, much of the available listings from genebanks contain little information about the accessions. Fortunately, efforts such as Genesys, the global Gateway to Genetic Resources (https://www.genesys-pgr.org/welcome), provides descriptor and trait information from multiple genebanks in a single database, which facilitates use by facilitating the location and finding information on potato accessions worldwide.

4.2 Collection of New Potato Germplasm

There are estimations that 20% of plant species are in danger of extinction (Jansky et al. 2013) and such losses will be accelerated by climate change and habitat destruction. Further, for some plants, a large amount of their diversity may already only be able available from genebanks (Jarvis et al. 2011), which might be the future for potato where the wild relative species exist only in the fragile ecosystems of the Andes (Hijmans and Spooner 2001), where a changing climate is already having a substantial impact. Lingering questions for ex situ and in situ management of potato revolve around how to define diversity there is/was in cultivated and the large secondary genepools (wild potato relatives) of potato. For conservationists, the big question is how much of the diversity in potato is securely conserved, what diversity has been lost or is in imminent danger of being lost, and what is the economic value, as well as potential future value, of the potato diversity that is not securely conserved.

A benchmark study (Castañeda-Álvarez et al. 2015) evaluated the status of ex situ collections for 73 species (according to Spooner et al. 2014) of potato wild relatives and used environmental niche modelling (ENM) techniques to estimate potential geographic ranges of each species. Their data assigned a high priority status for collecting 32 species of wild potato, 43.8% of the species studied, due to severe gaps in the ex situ collections. Further, 20 more species were assigned medium priority for collection and only three species were determined to have good diversity representation in the ex situ collections. As part of the study, they also looked at in situ sites which were potentially threatened including that of *S. rhomboideilanceolatum* whose native habitat in Peru is increasingly threatened by road building and overgrazing. Potato wild relatives have made critical contributions to potato disease resistance, enhanced yield, and improved quality in past 50 years (Jansky et al. 2013). Therefore, it should be the obligation of present day potato researchers and breeders to do all we can to leave future potato scientists the same opportunities to explore, use, and reap benefits from the diversity of wild potato that we have today.

The geographic range of wild potato species is broadly spread out in 16 countries confined to the Americas (Mexico, United States, Costa Rica, Guatemala, Honduras, Panama, Argentina, Bolivia, Brazil, Chile, Colombia, Ecuador, Paraguay, Peru, Uruguay, and Venezuela (Fig. 4.3). For thousands of years, the ecological niches of the Andes have been the natural habitat and the center of the genetic diversity of wild potato species and innumerable native varieties. Long evolutionary processes have allowed the accumulation of genetic components that are valuable resources for the improvement of potato cultivation, such as the ability to survive adverse biotic conditions (diseases, pests) and abiotic (drought, frost, climate change), as well as, vast morphological differences (Figs. 4.4 and 4.5). However, this valuable genetic diversity is threatened by "genetic erosion." The main threats are urbanization, alternative land use, and the destruction of natural areas. Examples include overgrazing, the construction of new routes and villages, and the destruction of forests.

The domestication of potato is thought to have occurred ~10,000 years ago in the Lake Titicaca Basin on the border between Peru and Bolivia, most likely from the *S. brevicaule* complex with *S. candolleanum* as a potential ancestor (Spooner et al. 2014). The use of these wild potato species for food is likely, yet the high glycoalkaloid content needed to be dealt with in some fashion to avoid toxicity. Rumold and Aldenderfer (2016) studied early potato starch grains from tools found at a village site in the Titicaca basin (1700–3500-year-old site) which resembled freeze-dried samples rather than starch from fresh tubers. They hypothesize that a process such as drying and grinding may have played a role in the detoxification of the wild potato tubers, as in other root crops. This theory is consistent with the natural freeze drying of tubers in the Andes today known as "chuño." Recent supporting evidence for use of wild potatoes for human consumption comes from *S. jamesii* starch grains found among other 6000-year-old early human ruins in the Southwest US (Louderback and Pavlik 2017). It is interesting to note that *S. jamesii* was still used until recently as a food source for native communities in the area and that existing patterns of present day occurrence of this species suggests that early humans may

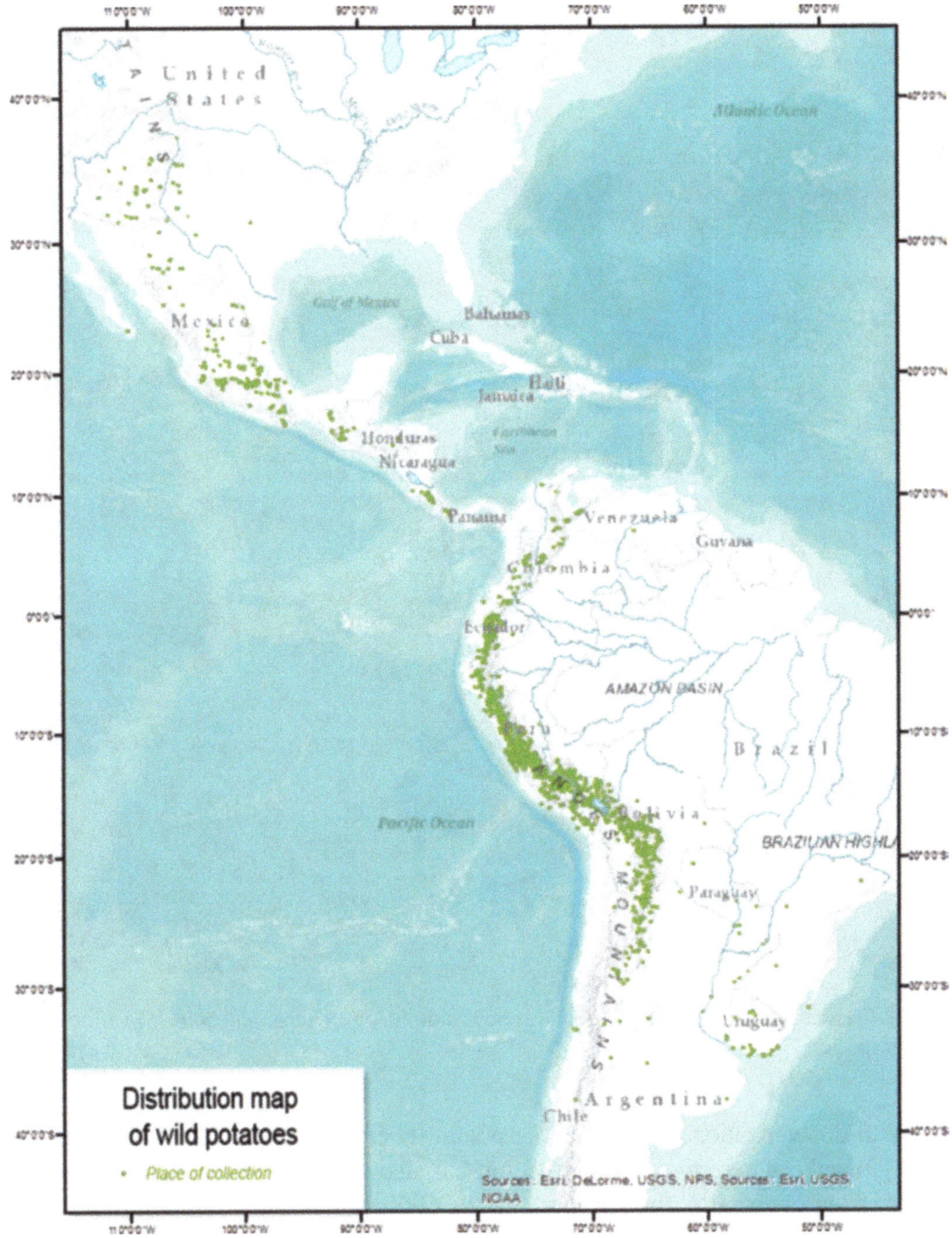

Fig. 4.3 A map of the distribution of wild potato accessions in the CIP genebank. Greens dot indicates origin of the accession based on passport data (https://cipotato.org/genebankcip/process/potato/potato-wild/)

have cultivated the tubers (Kinder et al. 2017). Domestication is believed to have given rise to increased tuber size through selection and cultivation which was associated with an increase in leaf carbon fixation and transport, reduction in glycoalkaloids, adaptation to long-day photoperiod, and reduced sexual fertility (Ryoko 2015).

Fig. 4.4 Examples of the diversity found in leaf morphology, shape, and color of various wild potato species

With domestication of early edible potato tubers, it is logical this also was the start not only of selection and cultivation but also of movement of selected potato tubers from one site to another. As early as 5000 years ago, maize, originating in Mexico, was already in S. America (Grobman et al. 2012), hence there is no reason to think that the north to south movement of food materials was not also reciprocated with a south to north movement of food such as cultivated potato. The movement of domesticated varieties most certainly also gave rise to the movement of new genetic material through the zone of native wild populations followed by intercrossing leading to increased diversity in the domesticated varieties and further selection for adaptive traits. These same adaptive traits and the diversity associated with them contributed to the adaptations which were selected for by early human populations and these same traits still exist in the wild populations today and will continue to be of critical value to breeders as long as potato improvement programs exist.

Fig. 4.5 Examples of the diversity found in fruit morphology, shape, and color in wild potato species

Thus collecting, movement, and exchange of potato and potato wild relatives has occurred for millennia, yet was mainly confined to the Americans until the Spanish conquest when potato was collected and shipped to Europe where it underwent further selection. Wild relatives were likely also collected before the last century, but reliable documentation of the collection of wild potato for genebanking is more recent. Darwin, in his famous voyage of the Beagle, collected potato from the Chilóe Islands in Chile in the 1830s which he named *S. maglia*, now known as S. *tuberosum* (Ristaino and Pfister 2016). At this time, Darwin erroneously assumed potatoes originated from these islands due to the diversity and wide spread presence. In his journals, he referenced that the crew "*stocked up on game and potatoes*." Fuentes (2014) references collection expeditions as yearly as the 1830–1833 by A. D'Orbigny from France in Bolivia. Perhaps the best documented collections are those done by the Vavilov Institute (Supreme Council of People's Economy) in Russia starting in 1925 by S.M. Bukasov and S.V. Juzepzuk (Loskutov 1999), after whom the wild potato species *S. bukasovii* and *S. xjuzepczukii* were named, respectively. Although detailed notes from these expeditions were never published, it is known that many species of wild potato were collected in Peru, Bolivia, and Chile (Ochoa 2004), and that some of this material still is maintained at the Vavilov Institute (Prof. N. Dzyubenko, personal communication).

Collections of wild potato continued by numerous collectors from Europe, Asia, North America, and Latin American institutions in an ad hoc manner until the early 1970s (Correll 1962; Hawkes 1990; Ochoa 1962) when CIP organized a meeting of experts to form and develop an International Germplasm Bank at CIP in Lima, Peru. Three Planning Conferences were held dealing with topics ranging from exploration, collection, conservation, and evaluation of potato genetic resources ending with the recommendation of collection strategies and to make the collected material available to researchers for their use for the benefit of humanity (International Potato Center 1973–1974, 1976, 1979). In the last 50 years, CIP has organized or been involved in ~300 systematic exploration missions for potato. These included the collection of wild and native potato germplasm varieties under the direction of C.M. Ochoa and collaborating with some 245 national and international scientists including A. David, J.B. Bamberg., J.G. Hawkes, A.M. Van Harten, J.P. Hjerting, W. Hondelmann, R. Hoopes, K.A. Okada, A. Salas, D. Spooner, and J.J.C. Van Soest. These collecting missions organized from Peru, where the greatest genetic diversity of the *Solanum* section Petota (Solanaceae) wild and cultivated exists, included exploring and collecting in the 16 countries of the American continent. Numerous accessions of each of the wild potato species were collected in the unique ecological niches along the geographical distribution of each species. Of interest was the rediscovery of wild species previously described, but known only as herbarium specimens in addition to the discovery of about 20 new species (including, *S. amayanum* (Ochoa 1989a, b, c), *S. bill-hookeri* (Ochoa 1988a, b), *S. bombycinum* (Ochoa 1983a, b), *S. chilliasense* (Ochoa 1981a, b, c, d), *S. incasicum* (Ochoa 1981a, b, c, d), *S. irosinum* (Ochoa 1981a, b, c, d), *S. longiusculus* (Ochoa 1987), *S. neovavilovii* (Ochoa 1983a, b), *S. orocense* (Ochoa 1980a, b, c), *S. ortegae* (Ochoa 1998), *S. peloquinianum* (Ochoa 1980a, b, c), *S. salasianum* (Ochoa 1989a, b, c), *S. sarasarae* (Ochoa 1988a, b), *S. simplicissimum* (Ochoa 1989a, b, c), *S. sucubunense* (Ochoa 1980a, b, c), *S. tapojense* (Ochoa 1980a, b, c), *S. taulisense* (Ochoa 1981a, b, c, d)) which were characterized and described by Ochoa (Hawkes 1989, 1990; Ochoa 1962, 1990, 1999; Spooner et al. 1999, 2001). These accessions are all documented with full passport data (country and location of collection) and many are available for use globally through the CIP website (https://cipotato.org/genebank-cip/process/potato/potato-wild/).

Most collection missions only have resources to focus collections from a single site and rarely can make multiple visits to the same site. Therefore, collectors are at the mercy of annual changes in rainfall and other abiotic and biotic factors that affect flowering or tuber germination. Exceptions are the collection trips by Bamberg and del Rio (2011) and Bamberg et al. (2003) where annual expeditions have been made to monitor in situ sites of *S. jamesii* and *S. stoloniferum* (formerly *S. fendleri*) since 1992 with over 200 visits. The expeditions were initiated to recollect from sites originally sampled in 1958 and 1978 to evaluate and characterize what changes could be measured over a 40+ year period and to maximize the ex situ collection of diversity found in this species (del Rio et al. 2001). Their research has yielded insight into the collection from wild potato, and perhaps, upset some long-held dogma. For example, as expected, they found more diversity using random amplified polymorphic DNAs (RAPDs) in seed collections, compared with the collection of tubers from individual plants in *S. jamesii*, but the opposite phenomena was observed for *S. stoloniferum*, where collection of tubers had greater diversity than seeds (Bamberg et al. 2009). In another report (del Rio et al. 1997), they found using RAPD markers, there were significant differences in allele frequencies between genebank-conserved material and recollected material from the same site in all seven comparisons of *S. jamesii* (diploid outcrosser), and 12 of 16 comparisons within *S. stoloniferum* (tetraploid inbreeder). While in situ biologists would conclude that this makes sense, since the argument is that evolutionary processes continue in situ, for the authors, this meant that recollecting from the same site was not only a good idea but also is desirable to capture new alleles (Bamberg et al. 2003). Other observations included a comparison of "easy" (road side collection sites) versus "remote" (sites one needed to hike to) where both collection methods yielded valuable alleles (Bamberg et al. 2010) and that patterns of geographic structure were not valuable predictors of genetic differences in these two species inferring that wide sampling of each population was needed to capture diversity.

4.3 International Fora for Plant Genetic Resources

The Convention of Biological Diversity (https://www.cbd.int/; CBD) is the most significant international treaty for the access to genetic resources globally; yet, it has had a negative impact on the collection of potato germplasm throughout the native range of wild potato. This is due to CBD reaffirming the sovereign rights of a nation to the genetic resources falling within the country boundaries, and hence, the country of origin has exclusive rights over terms for accessing genetic resources, including cultivated and wild relatives of potato, found within its boundaries. The treaty also reaffirmed the country of origin's right to "*fair and equitable sharing of benefits arising*" from the use of their genetic resources. CBD further asserts that nations are responsible for the long-term protection and *access* to genetic resources over which they have jurisdiction. CBD also acknowledged that access to, and sharing of, genetic resources, and *relevant technologies*, are critical for meeting food and health needs of a growing world population. Thus, it is a country's responsibility to not only protect but also make available for use and share in the benefits from such use of the genetic resources it holds. Unfortunately, most countries, and the international community generally, have been ill prepared for the responsibility of regulating access and for ensuring the sharing of benefits by third-party use of genetic resources, such as germplasm related to cultivated and wild potato relatives. Mechanisms to grant permission for the collection of plant genetic resources have taken time to implement, and in the meantime the collecting of plant genetic resources ceased. The challenge has been how to make genetic resources available for use on the one hand, while on the other, to ensure equitable sharing of benefits from the commercial utilization of the genetic resources, or derivatives thereof, by multiple third-party stakeholders' years later. Further, definition of the terms under which genetic resources could be accessed and benefits shared have been defined in the Nagoya Protocol on Access and Benefit Sharing (https://www.cbd.int/abs/, *Nagoya*). All potato species, except for *S. phureja*, are covered by the International Treaty for Plant Genetic Resources for Food and Agriculture (https://www.fao.org/plant-treaty/; ITPGRFA) which provides a legal framework to distribute the genetic material for research, training, and breeding with the acceptance of a Standard Material Transfer Agreement (SMTA) by the recipient.

Presently, there is a lack of a defined, functional mechanism for granting permission for collecting germplasm from the wild in most Latin American nations. The collection of potato germplasm for ex situ conservation in the countries where the clear majority of cultivated potato landraces and wild relatives originated and exist, has been virtually nonexistent since 2000. The exception is the United States of America (USA), which is not a party to CBD, and hence germplasm from potato wild relatives have continued to be collected in the USA (Bamberg et al. 2003), although only two species [*S. stoloniferum* (previously *S. fendleri*) and *S. jamesii*] are confirmed to exist within the boundaries of the USA. In 2016, a significant change occurred when the Peruvian National Institute for Innovative Agriculture (Instituto Nacional de Innovacion Agraria, INIA) was granted a permit for the collection of potato wild relatives within Peru, and in 2017 three collection missions were carried out collecting more than 70 populations of 14+ species (putatively *S. acaule, S. acroscopicum, S. aymaraesense, S. bukasovii, S. coelestispetalum, S. marinasense, S. pillahuatense S. raphanifolium, S. sandemani, S. sparsipilum, S. tacnaense, S. urubambae, S. velardei, S. yungasense*) under the auspices of the ITPGRFA. It is hoped that this will provide a model for further collections of wild and cultivated potato within its native range.

The collection of potato in its native range should be facilitated by the inclusion of cultivated potato (except for *S. phureja*) and all wild potato species in the list of Annex 1 crops under the ITPGRFA. The ITPGRFA established a global Multilateral System (MLS) that clearly defines the terms under which access for the use of genetic resources and their derivatives would be granted for use in research, training, and breeding under a Standard Material Transfer Agreement (SMTA) (http://www.fao.org/plant-treaty/areas-of-work/the-multilateral-system/the-smta/en/), which clearly outlines benefit sharing options. Of nations where wild potato species are native, only Colombia and Mexico are not parties to ITPGRFA, and thus, it is hoped that further facilitated access to collecting, conservation, and use of cultivated and wild potato species for research, training, and breeding will commence. For Colombia and Mexico, collecting, access, and use of native potato germplasm would be subject to bilateral agreements under the Nagoya protocol framework. To our knowledge, no other collections of wild potato species for inclusion in the MLS of the ITPGRFA have been carried out since the 1990s.

4.4 Methods of Conservation of Potato Germplasm

A major challenge for ex situ genebanks is to maintain the diversity held, often thousands of different accessions, in a form where every accession is readily available and accessible for use at a moment's notice. Collections of cultivated crops are often composed of various traditional varieties, along with improved cultivars, which often do not produce large quantities of seeds/tubers making them challenging to reproduce in field conditions. Additionally, in general, wild species are more challenging for ex situ genebanks to regenerate than cultivated due to the multitude of ecological niches they have become adapted to, often requiring unique environmental conditions (i.e., photoperiods) to thrive and reproduce, along with some genotypes/species, which genetically produce limited flowers, seeds, or tubers.

With the lifespan of most research grants being less than 5 years, delays in obtaining germplasm can mean missing critical planting windows which could result in a year's delay or failure to meet deliverables for a project. Thus, availability of material from the genebanks is critical. Most crop genebanks maintain their germplasm collections as botanical seed which for the majority of plants, including potato, stores very well and can be made available with a few weeks' notice if adequate seeds exist for that accession. Genebanks will only provide seed to requestors if they have adequate seed in cold storage with good viability to ensure that the material is safeguarded and there is enough seed to keep regenerating the accession. However, unlike most crops, cultivated potato is also often maintained as clones to preserve the unique allelic combinations in the selected material which poses additional challenges for maintenance and distribution.

Where desirable allelic combinations have not been fixed by selection, as in wild potato germplasm, the collections are maintained as botanical seed or true-potato-seed (TPS). A representative number of individuals (20–50) are typically collected from a wild population and seed is regenerated and combined to form a unique genebank accession of heterogeneous seed theoretically representative of the alleles found in that population. The accession is then distributed as a population of 50–100 seed where each seed represents a distinct genotype from that population. Thus, what is distributed is a heterogeneous mix of genotypes (seed) with each genotype representing a portion of the genetic make-up of the accession or population. Different collections of the same species (individual accessions) may or may not contain the same trait(s) desired by a breeder or even have this trait at the same frequency in different seed populations distributed. For example, when assessing late blight (*Phytophthora infestans*) resistance in wild potato accessions, Perez et al. (2001) found very high uniform resistance in three accessions of *S. orophilum* (100%, 98%, and 96% resistant individuals, respectively from 48 seedlings from each accession); whereas, in three accessions of *S. wittmachii*, resistance was high in one accession but very low in another (75%, 46% and 0.02% resistant individuals, respectively from 48 seedlings from each accession).

Unlike the wild species, cultivated potato is generally maintained as clones; however, in some cases seed populations of cultivated material may be just as useful to users. One of the main advantages of maintaining cultivated potato as seed is the overall cost and staffing required to maintain a seed collection is greatly reduced compared to in vitro collections which require subculturing every year or two. Further, sterile laboratory conditions with specialized equipment are not needed to reproduce potato if maintained as seed rather than in vitro clones. Seed populations can be produced and if adequate seed is obtained, a curator would not need to regenerate the accession for several decades unless the vigor of the seed decreases or the seeds are exhausted. Additionally, seed populations would have more diversity and new allele combinations that could lead to new trait discovery.

In contrast to maintenance of cultivated material as seed, maintenance of cultivated potato as a clone, preserving the unique allelic makeup of the landrace or variety, has advantages for either direct planting or facilitating the transfer of unique allelic combinations to breeding or research programs. Maintenance of clonal potato collections can be in the field, greenhouse, or in vitro as tissue culture. If maintained in the field or greenhouse, the collection is harvested, tubers stored and annually replanted as maintaining tuber viability for longer than a year is unreliable. Field or greenhouse maintenance of the collection also carries a risk of losing accessions due to biotic (insects and diseases) or abiotic (hail or wind) challenges. Therefore, many clonal potato collections are maintained in tissue culture as in vitro plantlets, which virtually eliminates the challenges of external biotic and abiotic factors. In vitro maintenance also facilitates international distribution of the collection as tissue culture plants are maintained in a sterile environment, and once cleaned of diseases, they can be certified as disease-free material for import purposes internationally.

The cultivated potato germplasm collection at CIP originated in the early 1970s with material collected or donated from around the world. In the early 1980s, research was initiated to place the collection into in vitro culture. Few in vitro collections have been maintained for 30 years or that have been used as extensively as the CIP cultivated potato collection. CIP currently maintains the largest in vitro potato collection with 8354 potato in vitro accessions (as of October 2018), the majority of which (89.8%) are landraces (native potatoes or "papa nativas" in Spanish) originating mainly from the Andean region, with the remaining accessions being improved varieties and breeding lines.

Since the vast majority of potato germplasm in genebanks around the world was collected over 20 years ago, and some, such as the collection at the Vavilov Institute in Russia, almost 90 years ago, most potato ex situ collections would be virtually impossible to replicate or replace today if lost. For this reason, Standards 4.9 of the Genebank Standards for Plant Genetic Resources for Food and Agriculture (FAO 2014) clearly state that "*A safety duplicate sample for every original accession should be stored in a geographically distant area, under the same or better conditions than those in the original genebank.*" CIP, along with other genebanks in the CGIAR, whose collections are held in trust under the ITPGRFA, takes this a step further with a guideline for backing up the collection in two distinct locations, one nationally and a second backup internationally. In the case of TPS accessions, such as the wild relatives of potato, a sealed package of seed equivalent to one to three times what is needed for a regeneration are backed up in a remote island archipelago in Norway at the Svalbard Global Seed Vault (https://www.nordgen.org/sgsv/, SGSV). The SGSV was established 10 years ago and funded by the Norwegian government with operations coordinated by the Nordic Gene Resource Centre and the Global Crop Diversity Trust. Shipment and coordination of all seed deposits are done by the Nordic Gene Resource Centre under a MTA which clearly states that the objective of the vault is "*...to provide a safety net for the international conservation system of plant genetic resources, and to contribute to securing the maximum amount of plant genetic diversity of importance to humanity for the long-term....*" (https://www.nordgen.org/sgsv/index.php?page=welcome). Under this agreement, the material is kept by the SGSV as a *black-box* safety back-up which denotes that the depositor retains full ownership and rights to the material, with the back-up facility holding the material only as a service of storage for the depositor. As of the end of 2017, SGSV contained almost 900,000 samples of seed from different crops including seed from 136 species of wild potato, five wild potato natural hybrids and six species of cultivated potato (taxonomy according to Hawkes 1990). Under conditions at the SGSV (airtight containers, properly dried seed, stored at −18 °C), good quality TPS can survive and maintain viability for 50–100 years.

Unfortunately, there are no analogous safety backup facilities to the SGSV for clonal material. In the case of clonal collections maintained in the field, the expense is often too high to allow distinct duplicate field planting sites, and hence, these collections are all too often not backed up and diminish over time. In the case of in vitro collections, the collections can be safely backed-up by shipping duplicate copies of in vitro cultures to a distant location for safety back-up. The facility backing-up the in vitro material generally needs to have lighted controlled temperature chambers (5–10 °C) for sustained slow growth of the material. In the case of the back-up for the USDA National Potato Genebank in vitro collection, cultures were shipped to the National Laboratory for Genetic Resources Preservation in Fort Collins, CO, where they are maintained as mini-tubers at 5 °C requiring subculture and reinitiating of the mini-tubers once every 2–3 years. For other in vitro potato collections, such as the one at CIP, the entire collection is shipped periodically (semi-annually) to two sites, one in a geographically distinct location in the high Andean Plateau at a CIP field station in Huancayo, Peru and the other facility outside of Peru at the *Brazilian Agricultural Research Corporation* (EMBRAPA) labs in Brasilia, Brazil. At both sites, the material is kept in slow growth conditions at 7 °C where the cultures can survive approximately 2 years without subculturing. Again, as with the seed deposits at the SGSV, these materials are stored in a black-box arrangement.

As genebanks are the stewards of the genetic resources that they maintain, with this stewardship comes great responsibility to ensure that no known pathogens are introduced into other countries with the distribution of germplasm creating a new disease problem. Vegetatively propagated crops such as potato, sugarcane, cassava, sweetpotato, yam, banana, citrus, and strawberry can be challenging because they must be multiplied continuously when maintained in the field to keep them in optimal growth. Although field maintenance of any crop is subject to abiotic challenges, as well as, biotic challenges such as diseases and pests, and with each multiplication or movement from one field to another, there are ample opportunities for the introduction of viruses and viroids (Sastry and Zitter 2014). Diseases caused by plant viruses limit sustainable production of vegetatively propagated staple food crops and in developing countries, this problem is confounded as agronomic practices to limit disease incidence and spread is marginal if done at all. As well, the use of new clean planting material to lessen virus titers is a luxury which most small holder farmers cannot afford. Additionally, low virus titers in planting materials generally display low symptom severity which makes infected plants not easily recognized as disease containing, and subsequently problematic if used for vegetative propagation (Bosch et al. 2007) because it can further perpetuate the diseases. Often human actions and climate change can be linked to the spread and outbreak of disease in conjunction with the movement and exchange of diseased plants and the international food supply chain are drivers of new disease threats (Wilkinson et al. 2011).

Actual examples of the human spread of pathogens with potato germplasm, other than by word of mouth, are difficult to find, yet published reports do exist to verify the potential challenge. The recent introduction of East Africa cassava mosaic-like viruses from African to the South West Indian Ocean islands follows human and insect transmission of the disease (De Bruyn et al. 2012), while the spread of the disease to Oman is likely directly related to human movement of plant material in the 1960s (Khan et al. 2012). Spread of viruses across international borders can also happen by insect vectors as in the case of the expansion and subsequent pandemic of the cassava mosaic disease (CMD) in the Great Lakes region of East and Central Africa which followed tightly the outbreaks of whitefly populations (Legg et al. 2011).

Therefore, for international distribution of germplasm from genebanks, plant material is tested, treated to eliminate diseases and certified to be clean and free of pathogens of phytosanitary importance as defined by import permits prior to transporting across any international border. For virus-free production of TPS, regeneration is carried out under strict sanitary conditions in a controlled screen house with limited access and maintained free of insects and thus vectors for viruses. Arracacha virus B–oca (arracacha and oca are two Andean Root and Tubers crops cultivated and eaten throughout Peru) strain (AVB-O), potato virus T (PVT), potato spindle tuber viroid (PSTVd), and alfalfa mosaic virus (AMV) have been shown to transmit virus particles to seed when pollen from an infected plant is used for pollinations (Jones 1982; Valkonen et al. 1992; FAO/IPGRI Technical Guidelines for the Safe Movement of Germplasm n.d.)]. Therefore, at CIP, for individual TPS production mother plants are tested pre-flowering for these seed transmitted viruses (AVB-O, PVT, PSTVd, and AMV) as well as for Andean potato latent virus (APLV), potato yellowing virus (PYV), and tobacco mosaic virus (TMV). If an individual mother plant tests positive for any of the viruses listed above, it is discarded and not used as a pollen donor or as a maternal parent for TPS production.

Pathogen elimination from clonal material is much more laborious, costly, and time-consuming. This is one rationale for in vitro culture as once the plantlets are clean, the in vitro culture methodology will maintain the investment of the phytosanitary cleaning. If endogenous bacteria are present in the cultures, these are usually eliminated with antibiotics prior to the elimination of viruses. A typical and successful virus elimination scheme starts with placing the in vitro cultures into thermotherapy (32–34 °C for 1 month), followed by isolation of 0.1–0.3 mm meristems and then plantlet regeneration. The regenerated plants are tested for the presence of viruses by serological, molecular, host range testing, and grafting onto indicator plants for final assurance of virus-free status. The entire process takes 12–24 months for one round and longer if not successful on the first run. At CIP, this method is highly efficient with 90% of all potato material going through virus elimination becoming virus-free after one run. The viruses of import/export importance in clonal material include potato virus T (PVT), potato spindle tuber viroid (PSTVd), potato virus X (PVX), potato virus Y (PVY), potato leaf roll virus (PLRV), potato virus S (PVS), Andean potato mottle virus (APMV), Andean potato latent virus (APLV), potato yellowing virus (PYV), and Arracacha virus B (AVB-O).

Advances in next-generation sequencing (NGS) hold great promise for reducing the labor, time, and cost of disease screening in germplasm collections. One such technology is the detection of viruses using small RNA sequencing and assembly (sRSA) technology. This technology is being tested on potato and it is hoped to be able to reduce the time necessary for phytosanitary cleaning of clonal potato germplasm from 12–24 months to less than 6 months (Kreuze 2014). Importantly, this technology could be invaluable in the future as new viruses or diseases of import importance develop so that entire germplasm collections can be rescreened quickly and efficiently ensuring the germplasm is readily available when needed.

The maintenance of in vitro potato clonal material has its advantages, including being readily available and the maintenance of disease-free status, but it is also undisputedly one of the most expensive long-term conservation methods available. This is due to the need for sterile plant tissue culture facilities, climate-controlled growth rooms, and highly trained personnel. Maintenance of clonal material as seed *is* one alternative, yet to maintain the allelic combinations present in an accession, maintenance of the clone is the only option. Cryopreservation, freezing tissue in liquid nitrogen at −196 °C, has been used for the long-term preservation of clonal plant genetic resources. Although research in the cryopreservation of plants has been ongoing for over 50 years (Reed 2008) and high confidence exists with the technology in animal and human systems (Di Santo et al. 2012), the application of long-term preservation of plant material, despite many successes, still has its skeptics. Limited long-term studies of plant material have shown largely positive and optimistic results (Volk et al. 2008), which support the theoretical long-term cryopreservation of plant meristems for centuries. Potato was one of the early crop plants used in cryopreservation research (Bajaj 1977; Grout and Henshaw 1978; Towill 1981a, b) and there are reports of over 20 potato species having been tested for cryopreservation using several different methods (Kaczmarczyk et al. 2011; Vollmer et al. 2017). To date the most successful methods for potato cryopreservation of genebank accessions included the IPK genebank using dimethyl sulfoxide (DMSO) droplet vitrification (Kaczmarczyk et al. 2011) with over 1000 cryopreserved accessions and the CIP genebank using a Plant Vitrification Solution (PVS2) droplet vitrification (Vollmer et al. 2017) method with over 2500 accessions cryopreserved as of the end of 2017.

Although the number of accessions or samples distributed from potato ex situ collections tells us little about *how* the germplasm was used or if it was useful at all, it does provide a measure of interest in the collections. In marketing, if a product is not available and there is little information on the attributes and benefits of the product, few will buy or order it. The same is true with germplasm collections, if the germplasm is not phytosanitary cleaned, in a form which can be distributed or due to some other factor(s), such as regulatory hurdles, is unavailable, it is not accessible for use and hence will not be used. Unfortunately, the physical accessibility of potato accessions from global collections is scattered with only 11/23 (~50%) of the collections surveyed having distributed material internationally in the period from 2004 to 2006 (Global Strategy for Ex situ Conservation of Potato 2006). While this is the only statistics found on distribution from multiple global potato global genebanks and it predates the International Treaty for Plant Genetic Resources for Food and Agriculture (Carputo et al. 2013), it serves as a good example and is likely the status today that less than 50% of potato ex situ material held in genebanks globally are available today beyond the walls of the genebank or the borders of the nation in which the genebank resides. This is a reflection of the state of the art of many factors and the international regulatory environment is just one factor (reviewed above). We have also discussed the need for phytosanitary cleaning and a propagule which can be distributed as disease-free. TPS could be an asset and a valuable tool for genebanks lacking the technology for long-term maintenance in tissue culture, yet is capable of virus screening of mother plants for TPS regeneration.

Another major limitation to the use/distribution of potato germplasm is the lack of information publicly accessible about potato germplasm in the various genebanks. While Genesys (https://www.genesys-pgr.org/welcome) offers a one site shop for information on the location of 19,066 potato accessions comprising over 175 species from 21 different countries, as of the end of 2017 it does not have information on 3 (INIA, France; Vavilov, Russia; and NIAS, Japan) out of 5 of the largest potato genebanks identified in the Second Report on The State of the World's Plant Genetic Resources for Food and Agriculture (FAO 2010). This highlights the difficulty for users to find information on the accessions conserved. Even the European Union Potato Project Database (http://ecpgr.cgn.wur.nl/eupotato/) does not contain information on the French or United Kingdom potato collections, yet it has information from both CIP and USA collections. If information on the holdings of the global potato germplasm collections is not readily available, or if available but not in a universal language, it will be impossible for most users to find information on what is held in ex situ collections.

As mentioned above, one measure of the value of genetic resources conservation is whether the collection is desired by the user community. Using the *in trust* potato collection at CIP as an example of the value of conservation of diversity of ex situ collections to the breeding and research community, since 2007, CIP has received 1,399 requests for potato germplasm. A total of 48,533 samples were distributed in the last 10 years (2007–2017), following the legal framework of the ITPGRFA. This represents over 1,200 unique accessions per year or almost 25% of the *in-trust* collection distributed annually. Material was sent to 96 countries with Peru, USA, Australia, and China requesting the most potato germplasm. All material is transferred using the Standard Material Transfer Agreement (SMTA) under the ITPGRFA. It could be expected that in the genomics era, DNA distributions would steadily increase; however, there is no trend in this as physical genetic material remains in high demand suggesting use in breeding programs still requires plants for crossing. Interestingly, in a recent (2015) survey (Shirey, unpublished data), late blight resistance and yield were the most cited traits of interest among users.

In 1997, a program was initiated to return traditional cultivars and landraces to the native, indigenous farmers in Peru (repatriation), whose ancestors preserved the potato diversity. For centuries, these potato farming communities have nurtured and planted multiple, 20–40, varieties per family as an insurance policy to ensure food availability. The rationale was that in one year one group of landraces would produce well, while in other years a different group might flourish. By conserving and planting diversity, they sustained their livelihood. In recent times, many communities have lost important cultivars they had planted for generations, due to terrorism, extreme weather, climate change, and/or disease/pest pressures. The CIP repatriation program is an example of the broader user base of ex situ genebanks where the *in-trust* collection served as a reservoir specifically for the landraces the current generation of potato guardians of the Andes (Papa Arariwa) needed to revive their traditional farming systems. In some cases, it allowed them to plant again the landraces they remembered as a child. Thus, repatriation is helping to restore the diversity and productivity in the traditional Andean potato farming region (Aguilar 2016). Further, the repatriation of genebank materials has benefits beyond restoring diversity and the traditions that come with this diversity, the added benefit is the distribution of pathogen-free stocks, which can increase productivity in a single season by as much as 40% thereby contributing to food security and poverty reduction for these communities. In total, 89 communities in Peru have received over 6,000 samples representing more than 1250 accessions of native landraces or over 50% of Peruvian landraces held in the in-trust potato collection.

4.5 Characterizing Potato Diversity in Genebanks

Molecular markers, are considered fixed landmarks in a genome, and thus, can reveal crucial genetic variability (Semagn et al. 2006). Markers have been used in many crop plants to assess genetic diversity, determine population structure, establishing trait-marker associations, discover and track quantitative trait loci (QTLs), produce genetic linkage maps, assist in selection for traits, understand the influence of genotypes on phenotypes, and more, all to improve or understand crop plants.

Loss of plant species in the past centuries has triggered genetic resource conservation with a need to accurately identify each accession. Molecular tools provide an easy less laborious method to assign plant taxa, as well as, characterize certain traits (Arif et al. 2010). To address the challenge of defining the biodiversity of a crop such as potato, biochemical and molecular markers have been used, yet the wide-scale use for looking at entire ex situ collections has only recently become more feasible due to advances in next generation sequencing (NGS) which has made genotyping projects more automated and affordable for large sample sizes.

Examples of where molecular markers have been employed to gain insight in potato are included below. Markers such as isozymes, simple sequence repeats (SSRs), and rapid amplified polymorphic DNA (RAPDs) have been employed in potato for varietal identification and genetic diversity assessments (Anoumaa et al. 2017; Carputo et al. 2013; Collares et al. 2004; Ghislain et al. 2009; Hoque et al. 2013; Rocha et al. 2010; Salimi et al. 2016; Xiaoyan et al. 2016). SSR markers have also been used to support a reevaluation of the taxonomic classification and structure of the gene pool of cultivated potato into four species (Spooner et al. 2007). Amplified fragment length polymorphisms (AFLPs) were utilized to hypothesize the domestication origin for cultivated potato from the northern species of the *S. brevicaule* group (Spooner et al. 2005) which was contrary to previous reports.

Even though different types of molecular markers exist and have been employed since the 1980s, single nucleotide polymorphisms (SNPs) are increasingly used predominantly due to recent advances in genome sequencing technology, the abundance of SNPs in most crop plants, reduced labor required to collect the data, and price per data point. The affordable cost and high-throughput nature of SNP markers have made them powerful tools for genetic analysis of plant species such as potato and highly useful in breeding (Bertioli et al. 2014). Discovery of SNPs in simple genomes is relatively easy requiring collection and evaluation of sequence data; however, in complex genomes such as potato, SNP detection is more challenging due to repetitive segments of the genome and multiple ploidy levels (Mammadov et al. 2012). Genome complexity reduction methods, such as genotyping by sequencing (GBS), diversity arrays technology (DArTseq), restriction site-associated DNA sequencing (RADseq), have been developed to aid in the discovery of novel SNPs; nevertheless, it is often challenging to identify SNP markers in polyploids such as potato, cotton, canola, and wheat (Bertioli et al. 2014; Logan-Young et al. 2015; Mammadov et al. 2012) due to separating allelic versus homoeologous SNPs or determining dosage in autopolyploids, both of which increase the rate of false positives (Clevenger and Ozias-Akin 2015).

SNP arrays have been developed which allow thousands to one million genome-wide SNP markers to be assessed simultaneously in an individual assay (LaFramboise 2009), thereby reducing the cost per marker data point. The Infinium 12K V2 Potato Array contains 12,720 SNPs, including the SNPs from the original SolCAP Infinium V1 8303 Potato Array with additional markers derived from the Infinium High Confidence SNPs (69K, Hamilton et al. 2011), which were selected for improved genome coverage, candidate genes, and regions with resistance genes. Both potato SNP arrays have been used in numerous studies as a genomic tool to improve cultivated potato or gain insight on genetic attributes. The SolCAP Infinium 8303 Potato Array were selected from 69,011 high-quality SNPs derived from six commercial potato cultivars "Atlantic," "Premier Russet," "Snowden," "Bintje," "Kennebec," and "Shepody" (Hamilton et al. 2011). These SNP markers were used to measure linkage disequilibrium for genome-wide association (GWA) mapping and population structure in European diploid and tetraploid germplasm (Stich et al. 2013). Genotyping a diversity panel of 250 lines of wild species, genetic stocks, and cultivated potato revealed that changes in heterozygosity and allele dosage has not occurred in over 150 years of breeding, but clear selection for alleles in biosynthetic pathways has occurred (Hirsch et al. 2013). The SolCAP Infinium 8303 Potato Array was used to develop linkage maps (Felcher et al. 2012), genotype populations for QTL analysis (Douches et al. 2014), and assess variation in glycoalkaloid biosynthesis (Manrique-Carpintero et al. 2013, 2014). It was also used to evaluate the genetic diversity and population structure of *S. tuberosum* sbsp. *andigenum* and *S. phureja* accessions from Colombia along with identifying 23 markers associated with nine morphological traits (Berdugo-Cely et al. 2017). In a wild species study, relationships deduced from the SNP markers were generally complementary to existing taxonomic classifications for 74 *Solanum* lines representing 25 wild taxa and were also effective in resolving complex taxa boundaries among germplasm with close genetic relationships (Hardigan et al. 2015).

Sequencing efforts including expressed sequence tags (ESTs) and the reference genome for potato (Potato Genome Sequencing Consortium 2011) allowed the development of the Infinium SOLCAP array that has provided the underpinning and a valuable tool for looking more closely at potato diversity. Additionally, the ease of data collection utilizing a SNP array has allowed genotyping of large sample numbers to finally be a reality. The CIP genebank has begun a project to genotype the entire landrace in vitro potato collection utilizing the Infinium V2 12K SNP array (http://solcap.msu.edu/potato_infinium.shtml) along with low density genotyping by sequencing (GBS) of 600 wild potato accessions and over 450 cultivated potatoes. This data is forthcoming; however, it will provide unique fingerprints of a large portion of the collection for users, diversity assessment in wild and cultivated material, identification of potential duplicated material, and genetic tools for genome-wide association studies (GWAS).

An initial assessment of the Infinium V2 SNP array was made by genotyping 250 potato accessions representing 7 taxa (Ellis et al. 2018). The genome-wide SNPs on the array were well distributed across the 12 potato chromosomes ranging from 798 to 1647 SNPs per chromosome. Of the 12,720 SNPs included in the Infinium 12K V2 Potato Array, the majority yielded good-quality signal intensities that were converted into genotypes. SNPs that did not produce a signal in $\geq$10% of the individuals or could not be clustered were filtered out of the dataset along with SNPs noted in previous studies to be poor or questionable (http://solcap.msu.edu/potato_infinium.shtml). The three-cluster diploid calling yielded 77% (9800) of the total SNPs on the array for use in subsequent analysis. In this data set only, 2.7% of the SNPs were monomorphic with only one of the three diploid genotypic classes scored among all the accessions of the diverse taxa used in this study. Aside from these few, most of the SNPs (97.3%) were polymorphic. A previous study with the 8303-array showed similar rates with a polymorphic rate of 75% (Stich et al. 2013); yet, this study included less taxa with only 44 genotypes. As expected, the five-cluster tetraploid calling had a lower rate of monomorphic markers with only 31 of 4859 (0.6%) of the SNPs being monomorphic; however only 39.5% of the total SNPs on the array could be utilized because of the difficulty in clustering/scoring the three heterozygous classes. Heterozygosity in this data set was high in many species/genotypes and ranged from 0.1 to 81.2% (Ellis et al. 2018).

A distance matrix was calculated from the SNP data and a phylogeny was constructed along with estimates on population structure (Ellis et al. 2018). Most of the taxa did not form distinct monophyletic clades and there were mixed taxa in each clade suggesting significant gene flow, low representation of certain taxa, and/or misclassified accessions. Further, branch lengths among individuals were short suggesting low diversity between taxa and the possibility of hybrids. *S. xjuzpeczukii* (2n = 3x = 36) was the only taxa to form a distinct monophyletic clade. The accessions of *S. xjuzpeczukii* were genetically redundant. This is interesting as these accessions originated from three countries (Bolivia, Peru, and Argentina); therefore, one would expect some genetic divergence. However, given that triploids are sterile and thus represent an evolutionary dead-end they were likely spread by human migration as clones. More of the genome needs to be evaluated (i.e., reduced representation sequencing) in these triploid accessions to elucidate if genetic differences among these accessions truly exists or if they are genetic duplicates before an archival process begins of putatively redundant material.

The phylogeny further demonstrated that *S.* x*chaucha*, *S. stenotomum* subsp. *goniocalyx*, and *S. stenotomum* subsp. *stenotomum* appear to be sister taxa with ancestry to *S. phureja*. Similar suggestions have emerged in previous studies using microsatellite data (Gomez, personal communication) and are supported by the revised taxonomy of Spooner et al. (2014) where these four taxa are lumped into a single species *S. tuberosum*. The phylogenetic tree also reveals putative discrepancies in species designation of accessions which will need to be verified in the future with phenotyping data. For example, multiple *S. stenotomum* subsp. *stenotomum* (17) and *S.* x*chaucha* (9) accessions grouped with the *S. tuberosum* subsp. *andigenum* accessions. In contrast, most of the *S. stenotomum* subsp. *goniocalyx* accessions clustered together or within the *S. stenotomum* subsp. *stenotomum* group (Ellis et al. 2018).

The analysis of the SNP data on 250 accessions from the genebank provided valuable information on intra- and interspecific relationships among taxa and provided additional support for targeting the collection of phenotypic data of suspected misclassified accessions. The data suggests that some of the accessions are hybrids and that gene flow has occurred between many of the taxa. The SNP array produced unique fingerprints with a few exceptions among this diverse panel of cultivated potato. These fingerprints provide a legacy for quality management systems (QMS) of the potato collection for years into the future to assure that as accessions are handled errors do not occur, and if they do occur, fingerprints can be used to validate genetic identity. Further, these data can be used to compare accessions between genebanks to determine where unique genetic material exists on a global scale and ensure it is safely preserved for the future. Overall, the genetic analysis facilitated a better understanding of the genetic diversity, population structure, and genetic relatedness of these potato taxa.

4.6 Climate Change and Genetic Resources Collections

The increased variability of the global climate, resulting in greater extremes of temperatures punctuated by increased dry, wet, cold, warm, and generally unpredictable climatic periods, will continue to have severe impacts on crop productivity and sustainability. Estimates of the extent of temperature increases range between 2 and 6 °C (Jarvis et al. 2011). The warming climates will be accompanied by increased atmospheric CO_2, and although both factors will favor plant productivity of C_3 plants, such as potatoes, this increased productivity will be accompanied by large increases in biotic (diseases and pests) and abiotic (drought, heat, adverse and unseasonable weather) challenges that could decrease plant yield and increase crop uncertainty. Previously adapted varieties may no longer be robust in the same geographic locality due to a lack of resistance/tolerance to these biotic and abiotic factors. These factors are interrelated, stressed resistance varieties may have attenuated resistance in different environments, resistance may be needed at different developmental stages than previously required, resistance could be overwhelmed by population sizes of pests and diseases, and new pests and diseases may appear. It must also be recognized that one cannot look at a single factor such as how climate affects insect vectors and thus the spread of diseases as an effect on insects will also affect insect pollinators reducing subsequent plant populations, insect predators, and insect populations by an extended growing season. It has been estimated that with a 2 °C rise in temperature, insects will go through one to five additional life cycles/year. In general, bacteria will respond better to moisture fluctuations regardless of temperature while fungi will respond to moisture fluctuations more in cooler weather (Beed 2011).

What is needed now and in the future is genes which can be introgressed into potato to help maintain productivity with the changing climate. Crop wild relatives conserved in genebanks offer "an enormous and unimaginable potential" for the discovery of valuable and desired traits (Machida-Hirano 2015). Wild relatives of potato have been a source of disease resistance for breeding programs for over 100 years (Hawkes 1958). An example is *Phytophthora infestans* (causal agent of late blight), which is the causal agent responsible for the devastation it inflicted during the Irish potato blight in 1845–1846. It is most severe in periods of high moisture with temperatures between 7.2 and 26.8 °C. It is predicted that for each degree of warming, late blight can occur 4–7 days earlier, extending the period of infections to between 10 and 20 days. With warmer temperatures starting earlier, there is a potential threat that late blight infestation will come earlier and earlier and last longer into the season resulting in increased fungicide applications leading to a greater pressure of the evolution of resistance to the fungicides (Beed 2011). Eleven hypersensitive-type resistance genes (R genes) have been characterized from *S. demissum*, a wild relative of potato of Mexican origin, thus *S. demissum* has been used extensively as a source for late blight resistance in European and North American cultivars (Asano and Tamiya 2016; Love 1999). Other sources of resistance in wild relatives of potato include *S. bulbocastrum, S. candolleanum, S. chacoense*, *S. pinnatisectum,* and *S. stoloniferon* (Machida-Hirano 2015).

Warmer temperatures can affect resistance genes rendering them ineffective. In tomato, increased temperatures caused a complete breakdown of resistance to *Ralsonia solanacearum* (bacterial wilt), another severe disease in potato (Kuun et al. 2001; Tung et al. 1992). Thus, increased temperatures affect not only the pathogenic organisms and vectors but also the resistance genes. Therefore, multiple forms of resistance for breeding programs may be needed and genebanks could provide such multiple forms of resistance. Examples of potato wild relatives where bacterial wilt resistance has been found are *S. chacoense, S. commersonii, S. phureja*, and *S. stonotonum* (Asano and Tamiya 2016; Machida-Hirano 2015).

Increased temperatures will also favor an increase in pests including aphids, weevils, and potato tuber moth (Asano and Tamiya 2016), and it has been stated that temperature is likely the single most important factor in cold blooded insect behavior, distribution, development, survival, and reproduction (Bale et al. 2002). Pest resistance in potato wild relatives is associated with high levels of glycoalkaloids, dense hairs, and trichomes found in species such as *S. chacoense, S. polyadenium,* and *S. tarijense* (Jansky et al. 2009). Nematode populations can also be influenced by climate as observed with the Southern root-knot nematode (*Meloidogyne incognita*) in coffee which has expanded its range in recent years associated with changing climatic conditions (Beed 2011). Nematode resistance in potato wild relatives can be found in *S. hougasii* for Columbia root-knot nematode (Brown et al. 1991) and *S. vernei* and *S. acuale* for cyst nematode (Hawkes 1994).

In addition to biotic challenges, climatic changes also can greatly affect potato survival and yield simply by the stresses imposed independent of effects on biotic factors. Increases in early and late season frosts, extreme heat, delay or unseasonable rains and other severe weather events. Tolerance to these abiotic factors is as important as resistance to pests and diseases. *S. acuale*, a potato wild relative can form a carpet of tiny plants at 4500 m.a.s.l displaying extreme frost and drought tolerance (personal observation). Other frost tolerant species include *S. demissum* and *S. xjuzcepczukii*. Drought tolerance can be found in *S. chillonanum*, *S. jamesii*, and *S. okadae* (Watanabe et al. 2011) and *S. xcurtilobum* (Cabello et al. 2012).

4.7 Concluding Comments

It is fortunate that the vast majority of the diversity of cultivated potato, as well as much of that of its wild relatives, is already safely conserved in numerous genebanks around the world. These resources represent an investment to ensure potato's role in food security well into the future. These conserved potato genetic resources will provide the traits and alleles needed to continue potato's dominance on plates around the world. Increased knowledge of the molecular and biochemical pathways is critical to maximize not only the discovery but also the use of the needed attributes in breeding programs and food systems. The benefits of these ex situ collections will continue to be delivered as the hidden characteristics, some with unknown value today, continue to be uncovered. One cannot however sit back complacently because as long as there is diversity which is threatened, and not securely conserved, there will be valuable alleles, individuals, and populations that could be critical in the future.

References

Aguilar AM (2016) Informe final de siembra de papas nativas repatriadas en la comunidad capesina de Santa Cruz de Pichiu, Chana, Huari, Ancash, 51p

Anoumaa M, Yao NK, Kouam EB, Kanmegne G, Machuka E, Osama SK, Nzuki I, Kamga YB, Fonkou T, Omokolo DN (2017) Genetic diversity and core collection for potato (*Solanum tuberosum* L.) cultivars from Cameroon as revealed by SSR markers. Am J Potato Res 94:449–463

Arif IA, Bakir MA, Khan HA, Al Farhan AH, Al Homaidan AA, Bahkali AH, Al Sadoon M, Shobrak M (2010) A brief review of molecular techniquest to assess plant diversity. Int J Mol Sci 11:2079–2096

Asano K, Tamiya S (2016) Breeding of pest and disease resistant potato cultivars in Japan using classical and molecular approaches. Jpn Agric Res Q 50(1):1–6

Bajaj YPS (1977) Initiation of shoots and callus from potato-tuber sprouts and axillary buds frozen at −196 °C. Crop Improv 4:48–53

Bale JS, Masters GJ, Hodkinson ID, Awmack C, bezemer TM, Brown VK, Bitterfield J, Buse A, Coulson JC, Farrer J, Good JEG, Harrington R, Hartley S, Jones TM, Lindroth RL, Press MC, Symrnooudis I, Watt AD, Whittaker JB (2002) Herbivory in global climate change research: direct effects of rising temperatures on insect herbivores. Glob Chang Biol 8:1–16

Bamberg J, del Rio A (2011) Use of native potatoes for research and breeding. Hortscience 46(11):1444–1445

Bamberg J, del Rio A, Moreyra R (2009) Genetic consequences of clonal versus seed sampling in model populations of two wild potato species indigenous to the USA. Am J Potato Res 86:367–372

Bamberg J, del Rio A, Fernandez C, Salas A, Vega S, Zorrilla C, Roca W, Tay D (2010) Comparison of "remote" versus "easy" in situ collection locations for USA wild solanum (potato) germplasm. Am J Potato Res 87:277–284. https://doi.org/10.1007/s12230-010-9133-8

Bamberg JB, del Rio A, Huamán Z, Vega S, Salas A, Pavek J, Fernandez C, Spooner DM (2003) A decade of collecting and research on wild potatoes of the Southwest USA. Am J Potato Res 80:159–172

Beed F (2011) The impact of climate change on interdependence for microbial genetic resources for agriculture. In: Fujisaka S, Williams D, Halewood M (eds) The impact of climate change on countries interdependence on genetic resources for food and agriculture. Background Study Paper No. 48. FAO Commission on Genetic Resources for Food and Agriculture, Italy, pp 38–47

Berdugo-Cely J, Valbuena RI, Sanchez-Betancourt E, Barrero LS, Yockteng R (2017) Genetic diversity and association mapping in the Colombian central collection of *Solanum tuberosum*

L. Andigenum group using SNP markers. PLoS One 12(3):e0173039. https://doi.org/10.1371/journal.pone.0173039

Bertioli DJ, Ozias-Akins P, Chu Y, Dantas KM, Santos SP, Gouvea E, Guimaraes PM, Leal-Bertioli SC, Knapp SJ, Moretzsohn MC (2014) The use of SNP markers for linkage mapping in diploid and tetraploid peanuts. G3 (Bethesda) 4:89–96

Bosch FVD, Jeger MJ, Gilligan CA (2007) Disease control and its selection for damaging plant virus strains in vegetatively propagated staple food crops; a theoretical assessment. Proc R Soc [Biol] 247:11–18

Brown CR, Mojtahedi H, Santo GS (1991) Resistance to Columbia root-knot nematode in *Solanum* spp. and in hybrids of *S. hougasii* with tetraploid cultivated potato. Am Potato J 68:445–452

Cabello R, De Mendiburu F, Bonierbale M, Monneveux P, Roca W, Chujoy E (2012) Large-scale evaluation of potato improved varieties, genetic stocks and landraces for drought tolerance. Am J Potato Res 89:400–410

Carputo D, Alioto D, Aversano R, Garramone R, Miraglia V, Villano C, Frusciante L (2013) Genetic diversity among potato species as revealed by phenotypic resistances and SSR markers. Plant Genetic Resources 11(2): 131–139. https://doi.org/10.1017/S1479262112000500

Castañeda-Álvarez NP, de Haan S, Juárez H, Khoury CK, Achicanoy HA, Sosa CC, Bernau V, Salas A, Heider B, Simon R, Maxted N, Spooner D (2015) Ex situ conservation priorities for the wild relatives of potato (*Solanum* L. Section Petota). PLos One 10(14):e0122599

Clevenger JP, Ozias-Akin P (2015) SWEEP: a tool for filtering high quality SNPs in polyploid crops. G3 (Bethesda) 5:1797–1803

Collares EAS, Choer E, Pereira AS (2004) Characterization of potato genotypes using molecular markers. Pesqui Agropecu Bras 39:871–878

Correll DS (1962) The potato and its wild relatives. Contributions from the Texas Research Foundation 4, p 606. Texas Research Foundation, Renner, Texas

De Bruyn A, Villemot J, Lefeuvre P, Villar E, Hoareau M, Harimalala M, Abdoul-Karime AL, Abdou-Chakour C, Reynaud B, Harkins GW, Varsani A, Martin DP, Lett J-M (2012) East African cassava mosaic-like viruses from Africa to Indian Ocean islands: molecular diversity, evolutionary history and geographical dissemination of a bipartite begomovirus. BMC Evol Biol 228(12):228–246

del Rio AH, Bamberg JB, Huaman Z, Salas A, Vega SE (1997) Assessing changes in the genetic diversity of potato gene banks. 2. In situ vs ex situ. Theor Appl Genet 95:199–204

del Rio AH, Bamberg JB, Huaman Z, Salas A, Vega SE (2001) Association of ecogeographical variables and RAPD marker variation in wild potato populations of the USA. Crop Sci 41:870–878

Douches D, Hirsch CN, Manrique-Carpintero NC, Massa AN, Coombs J, Hardigan M, Bisognin D, De Jong W, Buell CR (2014) The contribution of the Solanaceae coordinated agricultural project to potato breeding. Potato Res 57:215–224

Ellis D, Chavez O, Coombs J, Soto J, Gomez R, Douches D, Panta A, Silvestre R, Anglin NL (2018) Genetic identity in genebanks: application of the SolCap 12K SNP array in fingerprinting the global in trust potato collection. Genome 61:523–537

FAO (2010) The second report on the state of the world's plant genetic resources for food and agriculture. FAO, Rome, 368p

FAO (ed) (2014) Genebank standards for plant genetic resources for food and agriculture, rev edn. FAO, Rome

FAO/IPGRI Technical Guidelines for the Safe Movement of Germplasm. https://cropgenebank.sgrp.cgiar.org/images/file/learning_space/potato_tech_guid_safe_move_germplasm.pdf

Felcher KJ, Coombs JJ, Massa AN, Hansey CN, Hamilton JP, Veilleux RE, Buell CR, Douches DS (2012) Integration of two diploid potato linkage maps with the potato genome sequence. PLoS One 12:e36347

Fuentes XC (2014) Conserving the genetic diversity of Bolivian wild potatoes. PhD thesis, Wageningen University, Wageningen, NL, 229p, ISBN 978-94-6257-168-67

Ghislain M, Nunez J, Herrera MDR, Pignataro J, Guzman F, Bonierbale M, Spooner DM (2009) Robust and highly informative microsatellite based genetic identity kit for potato. Mol Breed 23:377–388

Global Strategy for Ex situ Conservation of Potato 2006. https://www.genebanks.org/wp-content/uploads/2017/01/Potato-Strategy-2007.pdf

Grobman A, Bonavia D, Dillehay TD, Pipernod DR, Iriartef J, Holst I (2012) Preceramic maize from Paredones and Huaca Prieta, Peru. Proc Natl Acad Sci U S A 109(5):1755–1759

Grout BWW, Henshaw GG (1978) Freeze preservation of potato shoot-tip cultures. Ann Bot 42:1227–1229

Hamilton JP, Hansey CN, Whitty BR, Stoffel K, Massa AN, Deynze AV, De Jong WS, Douches DS, Buell CR (2011) Single nucleotide polymorphism discovery in elite North American potato germplasm. BMC Genomics 12:302

Hardigan MA, Bamberg J, Buell CR, Douches DS (2015) Taxonomy and genetic differentiation among wild and cultivated germplasm of *Solanum* sect Petota. Plant Genome 1:16–270

Hawkes JG (1958) Significance of wild species and primitive forms for potato breeding. Euphytica 7:257–270

Hawkes JG (1989) Nomenclatural and taxonomic notes on the infrageneric taxa of the tuber-bearing solanums (Solanaceae). Taxon 39(3):489–492

Hawkes JG (1990) The potato: evolution, biodiversity and genetic resources. Belhaven Press, London, 259p

Hawkes JG (1994) Origins of cultivated potatoes and species relationships. In: Bradshaw JE, MacKay GR (eds) Potato genetics. CAB International, Wallingford, pp 3–42

Hijmans RJ, Spooner DM (2001) Geographic distribution of wild potato species. Am J Bot 88(11):2101–2112

Hirsch CN, Hirsch CD, Felcher K, Coombs J, Zarka D, Deynze AV, Jong WD, Veilleux RE, Jansky S, Bethke P, Douches DS, Buell CR (2013) Retrospective view of North American potato (*Solanum tuberosum* L.) breeding in the 20th and 21st centuries. G3 (Bethesda) 3:1003–1013. https://doi.org/10.1534/g3.113.005595.

Hoque ME, Huq H, Moon NJ (2013) Molecular diversity analysis in potato (*Solanum tuberosum* L.) through RAPD markers. SAARC J Agric 11:95–102

International Potato Center (1973–1974) Germplasm exploration and taxonomy of potatoes. In: Program planning conferences 1973–74, Lima, Perú, pp 5–48

International Potato Center (1976) Report of the planning conferences on the exploration and maintenance of germplasm resources, Lima, Peru, Marzo 15–19, 1976, 130p

International Potato Center (1979) Report of the planning conferences on the exploration, taxonomy and maintenance of potato germplasm III. CIP, Lima, Peru, 193p

Jansky SH, Simon R, Spooner DM (2009) A test of taxonomic predictability: resistance to Colorado potato beetle in wild relatives of potato. J Econ Entomol 102:442–431

Jansky SH, Dempewolf H, Camadro EL, Simon R, Zimnoch-Guzowska E, Bisognin DA, Bonierbale M (2013) A case for crop wild relative preservation and use in potato. Crop Sci 53:746–754

Jarvis A, Ramirez J, Hansen J, Leibing C (2011) Crop and forage genetic resources: international interdependence in the face of climate change. In: Fujisaka S, Williams D, Halewood M (eds) The impact of climate change on countries interdependence on genetic resources for food and agriculture. Background Study Paper No. 48. FAO Commission on Genetic Resources for Food and Agriculture, Italy, pp 7–83

Jones RAC (1982) Tests for transmission of four potato viruses through potato true seed. Ann Appl Biol 100:315–320

Kaczmarczyk A, Rokka V-M, Keller ERJ (2011) Potato shoot tip cryopreservation. A review. Potato Res 54:45–79. https://doi.org/10.1007/s11540-010-9169-7

Khan AJ, Akhtar S, Al-Matrushi AM, Briddon RW (2012) Introduction of East African cassava mosaic Zanzibar virus to Oman harks back to "Zanzibar, the capital of Oman". Virus Genes 46(1):195–198. https://doi.org/10.1007/s11262-012-0838-2

Kinder DH, Adams KR, Wilson HJ (2017) *Solanum Jamesii*: evidence for cultivation of wild potato tubers by ancestral Puebloan groups. J Ethnobiol 37(2):218–240

Kreuze J (2014) siRNA deep sequencing and assembly: piecing together viral infections. In: Gullino ML, Bonants PJM (eds) Detection and diagnostics of plant pathogens, plant pathology in the 21st century, vol 5. Springer, Dordrecht, 200p. https://doi.org/10.1007/978-94-017-9020-8_2

Kuun GK, Okole B, Bornmann L (2001) Protection of phenypropanoid metabolism by prior heat treatment in *Lycopersicon esculentum* exposed to *Ralstonia solanacearum*. Plant Physiol Biochem 39:871–880

LaFramboise T (2009) Single nucleotide polymorphism arrays: a decade of biological, computational and technological advances. Nucleic Acids Res 37:4181–4193

Legg JP, Jeremiah SC, Obiero HM, Maruthi MN, Ndyetabula I, Okao-Okuja G, Bouwmeester H, Bigirimana S, Tata-Hangy W, Gashaka G, Mkamilo G, Alicai T, Kumar LP (2011) Comparing the regional epidemiology of the cassava mosaic and cassava brown streak virus pandemics in Africa. Virus Res 159(2):161–170

Logan-Young CJ, Yu JZ, Verma SK, Percy RG, Pepper AE (2015) SNP discovery in complex allo-tetraploid genomes (*Gossypium spp.*, Malvaceae) using genotyping by sequencing. Appl Plant Sci 3:1400077. https://doi.org/10.1007/s12229-014-9146-y

Loskutov IG (1999) Vavilov and his institute. A history of the world collection of plant genetic resources in Russia. International Plant Genetic Resources Institute, Rome. ISBN 92-9043-412-0

Louderback LA, Pavlik BM (2017) Starch granule evidence for the earliest potato use in North America. Proc Natl Acad Sci U S A 114(29):7606–7610

Love SL (1999) Founding clones, major contributing ancestors, and exotic progenitors of prominent North American potato cultivars. Am J Potato Res 76:263–272

Machida-Hirano R (2015) Diversity in potato genetic resources. Breed Sci 65:26–40

Mammadov J, Aggarwal R, Buyyarapu R, Kumpatla S (2012) SNP markers and their impact on plant breeding. Int J Plant Genomics 2012:728398. https://doi.org/10.1155/2012/728398

Manrique-Carpintero NC, Tokuhisa JG, Ginzberg I, Holliday JA, Veilleux RE (2013) Sequence diversity in coding regions of candidate genes in the glycoalkaloid biosynthetic pathway of wild potato species. G3 (Bethesda) 3:1467–1479

Manrique-Carpintero NC, Tokuhisa JG, Ginzberg I, Veilleux RE (2014) Allelic variation in genes contributing to glycoalkaloid biosynthesis in a diploid interspecific population of potato. Theor Appl Genet 127:391–405

Ochoa CM (1962) Los Solanun Tuberiferos silvestres del Peru Secc. Tubernarium, sub-secc. Hyperbasarthrum), Lima, Peru

Ochoa CM (1980a) New taxa of Solanum from Peru and Bolivia. Phytologia 46(4):223–225

Ochoa CM (1980b) New tuber-bearing Solanum from Colombia. Phytologia 46(7):495–497. ISSN 0031-9430

Ochoa CM (1980c) Solanum peloquinianum: a new wild potato species from the Peruvian Andes. Am Potato J 57(1):33–35

Ochoa CM (1981a) [Solanum chilliasense, a new tuber bearing species of the Series Piurana]. Solanum chilliasense, nueva especie tuberifera de la serie Piurana. Lorentzia (Argentina) 4:9–11

Ochoa CM (1981b) [Solanum taulisense, a new Peruvian tuber bearing species]. Solanum taulisense, nueva especie tuberifera peruana. Lorentzia (Argentina). 4:13–15

Ochoa CM (1981c) Solanum irosinum, new Peruvian tuber bearing species resistant to Phytophthora infestans. Am Potato J 58(3):131–133. ISSN 0003-0589

Ochoa CM (1981d) Two new tuber-bearing Solanum from South America. Phytologia 48:229–232

Ochoa CM (1983a) Solanum bombycinum, a new tuber-bearing tetraploid species from Bolivia. Am Potato J 60(11):849–852. ISSN 0003-0589

Ochoa CM (1983b) Solanum neovavilovii: a new wild potato species from Bolivia. Am Potato J 60(12):919–923. ISSN 0003-0589

Ochoa CM (1987) Solanum longiusculus (Sect. petota), nova specie peruviana. Phytologia 63(5):329–330. ISSN 0031-9430

Ochoa CM (1988a) [Solanum sarasarae (Sect. Petota), a new species from Perú]. Solanum sarasarae (Sect. Petota), nova specie peruviana. Phytologia 64(4):245–246

Ochoa CM (1988b) Solanum bill-hookeri: new wild potato species from Peru. Am Potato J 65(12):737–740

Ochoa CM (1989a) [Solanum Ser. Simplicissima, new serie Tuberifera of Sect. Petota (Solanaceae)]. Solanum Ser. Simplicissima, nueva serie Tuberifera de la Sect. Petota (Solanaceae). Revista de la Academia Colombiana de Ciencias Exactas, Físicas y Naturales 17(65):321–323. ISSN 0370-3908

Ochoa CM (1989b) Solanum amayanum: a new wild Peruvian potato species. Am Potato J 66(1):1–4. ISSN 0003-0589

Ochoa CM (1989c) Solanum salasianum: new wild tuber-bearing species from Peru. Am Potato J 66(4):235–238. ISSN 0003-0589

Ochoa CM (1990) The potatoes of South America: Bolivia. Cambridge University Press, Cambridge, 512p. ISBN 0-521-38024-3

Ochoa CM (1998) Solanum ortegae: a new Peruvian species from Sect. petota. Phytologia 85(4):271–272. ISSN 0031-9430

Ochoa CM (1999) Las papas de Sudamérica: Perú (parte 1). Allen Press, Kansas, 1036p. ISBN 92-9060-197-3

Ochoa CM (2004) The potatoes of South America: Peru The wild species, part 1. International Potato Center, Peru, 1036p. ISBN 9290601981

Perez W, Salas A, Raymundo R, Huaman Z, Nelson R, Bonierbale M (2001) Evaluation of wild potato species for resistance to Late Blight. In: Scientist and farmer, partners in research for the 21st century, Program Report International Potato Center, Lima, Peru, 480p

Potato Genome Sequencing Consortium (2011) Genome sequence and analysis of the tuber crop potato. Nature 475:189–195

Reed BM (2008) Cryopreservation—practical considerations. In: Reed BM (ed) Plant cryopreservation a practical guide. Springer Science, New York, 513p

Ristaino JB, Pfister DH (2016) "What a painfully interesting Subject": Charles Darwin's studies of potato late blight. BioScience 66(12):1035–1045

Rocha EA, Paiva LV, de Carvalho HH, Guimaraes CT (2010) Molecular characterization and genetic diversity of potato cultivars using SSR and RAPD markers. Crop Breed Appl Biotechnol 10:204–210

Rumold CU, Aldenderfer MS (2016) Late archaic–early formative period microbotanical evidencefor potato at Jiskairumoko in the Titicaca Basin of southern Peru. Proc Natl Acad Sci U S A 13(48):13672–13677

Ryoko M-H (2015) Diversity in potato genetic resources. Breed Sci 65:26–40

Salimi H, bahar M, Mirlohi A, Talebi M (2016) Assessment of the genetic diversity among potato cultivars from different geographical áreas using the genomic and EST microsatellites. Iran J Biotechnol 14:270–277

Di Santo M, Tarozzi N, Nadalini N, Borini A. (2012) Human sperm cryopreservation: Update on techniques, effect on DNA integrity, and implications for ART. Adv Urol 2012:854837. https://doi.org/10.1155/2012/854837.

Sastry KS, Zitter TA (2014) Plant virus and viroid diseases in the tropics, volume 2: epidemiology and management. Springer, Dordrecht, pp 149–480. https://doi.org/10.1007/978-94-007-7820-7_2

Semagn K, Bjornstad A, Ndjiondjop MN (2006) An overview of molecular marker methods for plants. Afr J Biotechnol 5:2540–2568

Spooner DM, Salas Lopez A, Huaman Z, Hijmans RJ (1999) Wild potato collecting expedition in southern Peru (Department of Apurimac, Arequipa, Cusco, Moquegua, Puno, Tacna) in 1998: taxonomy and new genetic resources. Am J Potato Res 76(3):103–119. ISSN 1099-209X

Spooner DM, Salas AR, Huaman Z, Torres Maita RV, Hoekstra R, Schuler K, Hijmans RJ (2001) Taxonomy and new collections of wild potato species in central and southern Perú. Am J Potato Res 78(3):197–208. ISSN 1099-209X

Spooner DM, McLean K, Ramsay G, Waugh R, Bryan GJ (2005) A single domestication for potato based on multilocus amplified fragment length polymorphism genotyping. Proc Natl Acad Sci U S A 102:14694–14699

Spooner DM, Nunez J, Trujillo G, Herrera MDR, Guzman F, Ghislain M (2007) Extensive simple sequence repeat genotyping of potato landraces supports a major reevaluation of their gene pool structure and classification. Proc Natl Acad Sci U S A 104:19398–19403

Spooner DM, Ghislain M, Simon R, Jansky SH, Gavrilenko T (2014) Systematics, diversity, genetics, and evolution of wild and cultivated potatoes. Bot Rev 80:283–383. https://doi.org/10.1007/s12229-014-9146-y.

Stich B, Urbany C, Hoffmann P, Gebhardt C (2013) Population structure and linkage disequilibrium in diploid and tetraploid potato revealed by genome wide high density genotyping using the SolCap SNP array. Plant Breed 132:718–724

Towill LE (1981a) *Solanum etuberosum*—a model for studying the cryobiology of shoot-tips in the tuberbearing Solanum species. Plant Sci Lett 20:315–324

Towill LE (1981b) Survival at low temperatures of shoot-tips from cultivars of *Solanum tuberosum* group Tuberosum. CryoLett 2:373–382

Tung PX, Hermsen JG, Zaag P, Vander S, Schmiedche P (1992) Effects of heat tolerance on expression of resistance to *Pseudomonas solanacearum* in potato. Potato Res 35:321–328

Valkonen JPT, Pehu E, Watanabe K (1992) Symptom expression and seed transmission of alfalfa mosaic virus and potato yellowing virus (SB-22) in *Solanum brevidens* and *Setuberosum*. Potato Res 35:403–410

Volk GM, Wadell J, Bonnart R, Towill L, Ellis D, Lauffman M (2008) High viability of dormant *Malus* buds after 10 years of storage in liquid nitrogen vapour. CryoLetters 29:89–94

Vollmer R, Villagaray R, Egúsquiza V, Cardenas JE, Castro M, Chavez O, Anglin NA, Ellis D (2017) A large-scale viability assessment of the potato cryobank at the International Potato Center (CIP). In Vitro Cell Dev Biol Plant 53(4):309–317. https://doi.org/10.1007/s11627-017-9846-1

Watanabe KN, Kikuchi A, Shimazaki T, Asahima M (2011) Salt and drought stress tolerance in transgenic potatoes and wild species. Potato Res 54:319–324

Wilkinson K, Grant WP, Green LE, Hunter S, Jeger MJ, Lowe P, Medley GF, Mills P, Philipson J, Poppy GM, Waage J (2011) Infectious diseases of animals and plants: an interdisciplinary approach. Philos Trans R Soc B Biol Sci 366(1573):1933–1942

Xiaoyan S, Chunzhi Z, Ying L, Shuangshuang F, Qing Y, Sanwen H (2016) SSR analysis of genetic diversity among 192 diploid potato cultivars. Hortic Plant J 2:163–171

Chapter 5
The Genes and Genomes of the Potato

Marc Ghislain and David S. Douches

Abstract During the last decade, genomics research has generated new insights into potato genetics and made possible new strategies for varietal improvement. The most commonly grown and eaten potato is an autotetraploid, highly heterozygote crop suffering from rapid inbreeding depression. The genetic improvement of the potato presents numerous challenges using conventional tetraploid breeding techniques. However, novel breeding technologies are now available to increase precision and gains for varietal improvement. The public availability of the first potato genome sequence has created new ways to identify the genetic determinants of key traits of the potato as well as ways to use this knowledge for speeding up variety development. Genomic selection applied to tetraploid breeding promises to increase prediction of progeny performance by a more efficient selection of parents. Diploid hybrid breeding is finally making its way two decades after discovering a suppressor gene of the self-incompatibility locus of diploid potatoes. Direct gene transfer into existing varieties of major genes for key traits has been successful but biotech potato development has been constrained by public perception and issues related to the regulation of the technology. Although genome or gene editing is still in its primary stage in potato, it has already been successful in modifying gene expression in a controlled way, and it might face a lower regulatory burden and easier adoption than biotech, transgenic potatoes. Concluding on an optimistic note, we have many reasons, and evidence is starting to mount, that potato crop improvement is finally benefiting from decades of investment in molecular genetics and that the future hold the promises of faster releases of more robust varieties to pest, disease, and climatic extremes, as well as nutritionally enhanced varieties to feed an ever-growing world population.

M. Ghislain (✉)
International Potato Center, Nairobi, Kenya
e-mail: m.ghislain@cgiar.org

D. S. Douches
Michigan State University, East Lansing, MI, USA
e-mail: douchesd@msu.edu

H. Campos, O. Ortiz (eds.), *The Potato Crop*,
https://doi.org/10.1007/978-3-030-28683-5_5

5.1 At the Crossroad of Potato Improvement

5.1.1 The Numerous Challenges of Tetraploid Potato Breeding

Potato breeding has a long and successful history of improved variety released after breeding from mostly advanced breeding clones and landraces of tetraploid nature from essentially two genepools, the short-day adapted upland Andigenum and the long-day adapted lowland Chilotanum (Bonierbale et al., this volume; Spooner et al. 2007). The latter group has given rise to the modern cultivar well-adapted to the Northern hemisphere referred to as the Tuberosum group (Gavrilenko et al. 2013). In the US and Canada, variety replacement has been particularly disappointing in potato since turnover is in the range of several decades, unlike grain crops (Walker 1994). In developing countries, there has been marked change in variety adoption as evidenced by a study on CIP-related varieties which grew from nothing to beyond one million hectares in 35 years (Thiele et al. 2008). A more recent study by Gatto et al. (2018) provides further insight on the impact of CIP's breeding efforts, and documents that in China, the world's largest potato producer, there are over one million hectares planted with varieties that trace back to CIP pedigrees. Furthermore, CIP's genetic footprint in China reaches above 35% of all varieties currently in use, be it through the registration of CIP advanced breeding lines as varieties or through using CIP's elite breeding lines as parents in crosses initiated by Chinese breeding programs. The main bottleneck explaining the difference in varietal turnover is market-driven mostly by processing industry in the US and Canada, unwilling to adjust their manufacturing processes to the cooking and frying attributes of the new varieties. Regardless, potato breeding has shown the ability to deliver varieties with market-demanded processing qualities and new traits for higher resilience against climatic extremes, pest and diseases threats, and with enhanced nutritional qualities.

Phenotypic recurrent selection has been the method of choice to select for improved potato lines starting usually with about 100,000 seedlings from 200 to 300 crosses followed by clonal selection over many years (Bradshaw 2009, 2017). A recent review of the history of conventional potato breeding revealed many examples of important varieties been released after 30 years or more of crossing and clonal selection when an optimal timeline should be 13–14 years (Bradshaw 2009; Jansky and Spooner 2017). The main reason for such long cycle for variety development is the quantitative nature of most important traits, the rapid inbreeding depression, and the low intensity of selection in early generations. The propagation through tubers adds on delays due to low multiplication rate and ease of contamination with pathogens which delay bulking enough quality seed tubers for multilocation field selection.

The genetic base of potato varieties grown in large commercial area is relatively narrow compared to the accessible gene pool for conventional breeding of the potato. This is likely due to the narrow genetic base of the original tetraploid sources from Andigenum and Tuberosum which were used as starting material for breeding.

Ellis et al. (this volume) provide a detailed current status of the gene pool and the germplasm of potato. Many wild potato species can be crossed with cultivated potatoes directly or via another wild species used as a bridge (Plaisted and Hoopes 1989). However, only a fraction of useful genes from wild species have been introgressed successfully into modern potato varieties. About 40% of the wild species carry interesting genetic traits value for pests, diseases and abiotic stresses (Sood et al. 2017). Major genes for disease resistance from wild relatives of the potato were introgressed into breeding lines for late blight (*S. demissum, S. bulbocastanum*), viruses (*S. stoloniferum*), and nematodes (*S. spegazzini, S. vernei*) resistance (Bradshaw and Ramsay 2005; Bradshaw 2009; Finkers-Tomczak et al. 2011).

The real contribution of wild relatives to modern potato varieties is likely under-estimated due to uncertainty in pedigree information and quantitative nature of many important traits of the potato. *S. acaule,* which has been a source of disease resistance and abiotic stress tolerance but has been more used as a bridge between wild species and the cultivated potato (Watanabe et al. 1992). From the first cross between *S. bulbocastanum* bearing late blight-resistance genes and the bridging wild species *S. acaule*, 46 years of crossing and selection with cultivated potato, first diploid *S. Phureja* and then tetraploid *S. tuberosum*, were necessary to release the late blight-resistant varieties "Bionica" and "Toluca" (Haverkort et al. 2009). Introgression of wild species genomes into the cultivated groups has been facilitated by unreduced (*2n*) gametes in diploid potatoes and was shown to be highest in group Tuberosum because of intense breeding effort using a dozen of wild species (Plaisted and Hoopes 1989; Hardigan et al. 2017). Wild species carry genes for wild characteristics which are introduced as genetic drag with the disease-resistance genes lead to the notion that there could be a tradeoff between disease resistance and yield (Ning et al. 2017). This might have contributed to the limited use of wild species in potato breeding.

Assuming allelic combination has to consider one positive allele from a wild species and three neutral alleles from cultivated potato, the introgression of only ten positive alleles from wild species is only one in a million genotypes [$(1/4)^n$ where $n = 10$ genes]. This number becomes quickly without practical reach considering epistatic effects from the cultivated potato alleles, and that each cross redistributes the 20 or so quantitative trait loci which are priority traits of modern varieties (Bradshaw 2017). Potato being an auto-tetraploid clonally propagated crop has also accumulated rare mutations and epigenetic changes in alleles otherwise identical. This complicates further the straightforward exploitation of emerging molecular breeding approaches (Visser et al. 2014).

Increased selection intensity before clonal selection has been proposed by progeny tests and full-sib family selection (Bradshaw et al. 1995, 2000). Marker assisted selection can also screen at early stage major genes and Quantitative Trait Loci (QTL) with large effects (Gebhardt 2013; Sharma et al. 2014). In recent years, markers flanking major genes and QTL were developed for resistance to viruses (Mihovilovich et al. 2014; del Rosario et al. 2018), tuber starch and yield (Schönhals et al. 2016), and other important traits (Ramakrishnan et al. 2015). If applied at the early clonal generation stage and multiplexed, marker-assisted selection can be cost-effective (Slater et al. 2013). Estimated breeding value for traits which can be inferred from pedigree information was also proposed to accelerate the intensity of section and result in shorter breeding cycle (Slater et al. 2014a, b). Genomic selection is also proposed to improve combining unknown QTL at an early stage in the breeding process (Slater et al. 2016). New high density and high throughput polymorphic marker systems have been developed for potato (Hamilton et al. 2011; Uitdewilligen et al. 2013; Vos et al. 2015). Using a panel of 83 cultivars of mostly European origin, a high frequency of relatively rare variants and/or haplotypes, with 61% of the variants having a minor allele frequency below 5%, was found which can be explained by the limited number of meiosis separating these cultivars (Uitdewilligen et al. 2013). Recent estimates of linkage disequilibrium in modern potato cultivar populations confirmed the relatively limited number of meiosis separating modern cultivars and therefore limited power of Genome Wide Assisted Studies (GWAS) for allele/gene discovery (Vos et al. 2017; Sharma et al. 2018).

Hence, ways to improve conventional potato breeding exist and are under development but the fundamental inherent limitations of the narrow genetic base of advanced tetraploid potato germplasm used by breeders, the rapid inbreeding depression, and the low multiplication rate of seed tubers call for new methods and tools to improve the potato which will be complementary to conventional tetraploid breeding for some and an alternative for others.

5.1.2 New Potato Breeding Technologies

Potato genetic improvement has taken shortcuts many times to circumvent the limited genepool accessible by crossing and the tedious phenotypic recurrent selection.

Mutagenesis has long been used to improve yield, quality, biotic and abiotic stress resistance, and tolerance of many crops (Maluszynski et al. 1995). According to this review, more than 1,700 mutant varieties involving 154 plant species have been officially released. However, the tedious process of segregating out the rare positive mutation from the negative ones represent a bottleneck for potato crop improvement. Nevertheless, a novel form of doing mutagenesis is making a surprising come-back for potato crop improvement as mentioned below.

Somatic hybridization has been used in potato to bypass sexual incompatibilities between cultivated potato and wild species for about 40 years (Tiwari et al. 2018). These authors reported successful fusion products obtained from 23 *Solanum* species that were characterized for multiple traits. Numerous studies were generated from somatic hybrids to understand the genetic architecture of important traits including isolating important genes. However, no variety has apparently been released from breeding somatic hybrids with cultivated potato likely due to the limitation of tetraploid potato breeding to efficiently remove undesirable alleles.

Although direct gene transfer through transgenics has a much shorter history in potato crop improvement than conventional breeding approaches, it has been highly successful, though for a limited number of traits for which natural genetic variation was not readily available. Virus resistance was the first trait successfully engineered in potato in the late 1980s, and a commercial cultivar was first reported with the combined resistance to PVX and PVY (Lawson et al. 1990). Soon after, an insect resistance was engineered and led to the production of new commercial cultivars (Perlak et al. 1993). This first generation of biotech potatoes were commercialized under the name of NewLeaf™ from 1995 through 2001 in the United States and Canada, but potato processors and retailers realized soon that the NewLeaf potatoes were going to increase their costs without a share of their benefits which precipitated their decline (Thornton 2003). By 2004, none of them were commercialized anymore. For the next decade, no new biotech potato was released. Recently, a series of potato biotech varieties have been released with reduced bruising and browning first and then with late blight resistance, low acrylamide potential, reduced black spot, and lowered reducing sugars while others are near to be released (see below). The opportunities for engineering new traits that bring benefits to the producer and the consumer are numerous and applicable to potato (Halterman et al. 2016). However, the public acceptance of biotech crops remains volatile and unpredictable, and the current lack of science-based regulatory frameworks in many potato producing countries constrain the scope of genetic engineering to products with sizable benefits to both producers and consumers unachievable by other means.

Genome (gene) editing is the most recent and significant genetic engineering technique targeting specific DNA sequences in the crops' genome (Scheben et al. 2017; Yin et al. 2017). Targeted mutagenesis of specific genes for knock-out, deletion, or allelic changes are now possible with a final product free of foreign DNA. Potato has already been shown to be amenable to genome editing and even to develop novel useful products (Butler et al. 2015, 2016; Clasen et al. 2016; Nicolia et al. 2015; Wang et al. 2015; Andersson et al. 2017; Ma et al. 2017). The two editing tools, TALEN and CRISPR/Cas9, must access the genome without integration of foreign DNA since it cannot be eliminated by crossing without losing most of the qualities of the original commercial variety. Transient expression by PEG-mediated protoplast transfection or Agrobacterium-mediated leaf infiltration generated the intended mutation, but the absence of a selectable agent and somatic variation of plants regenerated from protoplasts can make these strategies labor-intensive. Delivery of the editing reagent may be mediated by virus vectors, but their spread and elimination pose additional difficulties. Hence, genome editing technology offers great opportunities in potato but is still at its first stage of development.

True hybrid potato is a new potato breeding strategy which is increasingly been regarded as the game-changing solution to many of the pitfalls of conventional tetraploid breeding. The sparking step goes two decades back with the discovery of a self-incompatibility inhibitor gene in the wild species *S. chacoense* (Hosaka and Hanneman 1998a, b). The *S* locus inhibitor gene (*Sli*) was introgressed into diploid cultivated potato and shown to confer self-compatibility

(Phumichai et al. 2005). Soon after, few breeding programs started to introgress the *Sli* gene into their diploid potato lines and obtained S3 diploid lines with 80% homozygosity and good agronomic performance including yield (Lindhout et al. 2011). In parallel, an inbred line of *S. chacoense* (M6) was developed to produce recombinant inbred line populations (Jansky et al. 2014). As stated in the title of an opinion paper by a large community of US potato geneticists and breeders, it is proposed to "reinvent the potato as a diploid inbred-based line crop" (Jansky et al. 2016). New sources of self-compatibility system are needed to circumvent the use of the wild species *S. chacoense*. Within the diploid cultivated potato germplasm, self-compatible landraces exist, though rare, but can be used to develop inbred lines from distinct gene pools such as the Stenotomum group and Phureja group (Haynes and Guedes 2018). Recently, an even more promising new system has been developed by knocking out the self-incompatibility gene *S-RNase* using the CRISPR–Cas9 gene editing system (Ye et al. 2018).

In addition to obtaining quick genetic gain by fixing major genes for disease resistance or other important traits and exploiting heterosis by hybridization, the hybrid variety propagation is via true seeds. The use of botanical seeds has long been known to be an extremely interesting alternative to tuber seeds because of it low weight, lower content of pathogens, good storability, option for beneficial coating, and high multiplication rate. Previous work by CIP and other potato research organizations on the concept of True Potato Seed (TPS) aimed to complement traditional seed systems by the use of botanical seed as a mean to propagate potato. However, its actual adoption by farmers has been much less than originally expected. Breeding for good parental clones from tetraploid breeding lines led to the development of several varieties but adoption remained conditioned to reduced or scarcity of seed tuber supply at affordable prices (Almekinders et al. 2009). Recently, a TPS variety, Oliver F1, was developed by the Dutch breeding company Bejo Zaden B.V. (http://www.bejo.com/magazine/bejo-introduces-its-first-true-potato-seed-variety) and is now under deployment in some African countries where quality seed availability is rare. Many years of breeding to develop superior parental inbred lines with disease-resistance genes adapted to the various agroecologies and markets are needed but the potential benefits that could be derived from true hybrid potato seeds are immense. The next decades will tell us whether reinventing the potato as hybrid varieties from diploid inbred parental lines will be adopted by small-holder farmers in developing countries who are the likely first adopters of this new technology.

5.2 The Genome of the Potato

5.2.1 Cultivated, Wild Potato Genome Sequences Towards a Pan-Genome

The identification of the first cultivated potato genome sequence is and will remain a turning point in the history of potato science. Prior to its discovery, genetic markers were associated to genetic determinants of traits breeders and geneticists had been working on. A fraction of the genes was known while transcriptomes were describing their expression in tissues, at different times, and under various environmental situations. Candidate genes were tested for association with these quantitative trait loci but for the most, genes underlying QTL remained unknown. The potato genome sequence brought together all this genetic knowledge into a physical perspective for the first time. Eighty-six percent of the 844 Mb genome was assembled into 12 pseudomolecules where 39,031 protein-coding genes were predicted (The Potato Genome Sequencing Consortium 2011). The potato whose genome was sequenced is a homozygote diploid plant obtained after chromosome doubling of a monoploid derived by anther culture of a heterozygous diploid potato from the *Solanum tuberosum* Group Phureja (Paz and Veilleux 1999). This cultivar groups are diploid short-day adapted cultivars producing tubers lacking dormancy. They occur throughout the eastern slope of the Andes from western Venezuela to central Bolivia at elevation between 2000 and 3400 masl (Ochoa 1990). Other genome sequence from cultivated potato, in particular from the Group Andigenum and Tuberosum including modern cultivars are still missing and expected to reveal insight into the domestication/wild species contributions to the various groups of cultivars. The difficulties lie in the presence of four genome sequences derived by auto-ploidization and multiple introgression of chromosome segments from wild species (Rodríguez et al. 2010; Spooner et al. 2014). This makes assembly and phasing particularly difficult and only recently claims of successful assembly of all four genome sequences from modern potato cultivars were made (*NRGene* at http://www.nrgene.com). The comparison of 99 Mb of genome from a potato of the group Tuberosum with the DM potato genome sequence revealed collinearity and high sequence identity (The Potato Genome Sequencing Consortium 2011). A year after the release of the potato genome sequence, the tomato genome sequence was published together with its closest wild relative and compared to the potato genome sequence (The Tomato Genome Consortium 2012). As known from previous cytological and comparative genetic mapping and synteny studies, the tomato genome presents very similar chromosomal organization but nine large and several smaller inversions. The euchromatic, gene rich regions diverge by 8.7%, whereas the intergenic and repeat-rich heterochromatic regions diverge by 30%. The potato genome sequence was greatly improved by ordering and reordering 93% of the previously assembled genome into 12 pseudomolecules representing the 12 chromosomes of the potato (Sharma et al. 2013). This genome sequence continues to be the sole publicly available genome sequence of a cultivated potato. It is accessible through a friendly

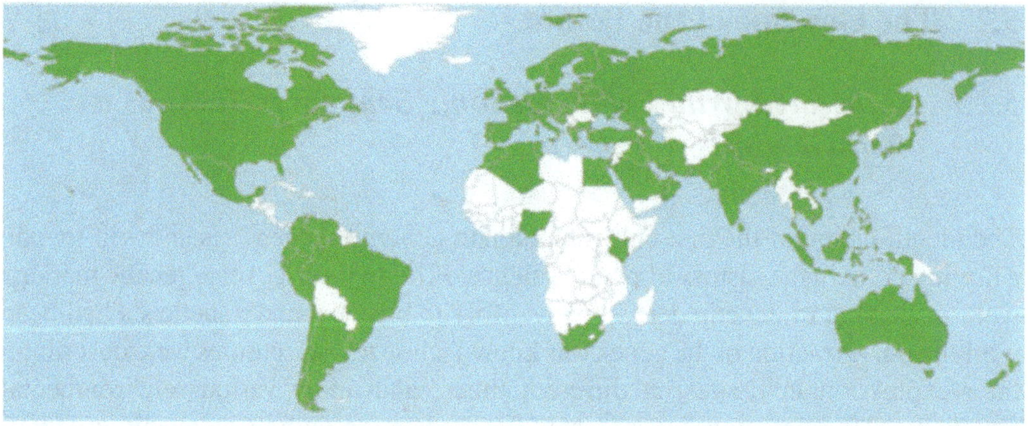

Fig. 5.1 Geographic distribution of SpudDB users. Filled in countries were the source of ten or more unique visits to SpudDB (http://solanaceae.plantbiology.msu.edu/) from October 2017 to October 2018 (Courtesy from John Hamilton, Robin Buell Michigan State University)

web genome browser hosted and maintained by the Buell Lab at Michigan State University in United States and includes annotation datasets, phenotypic and genotypic data from a diversity panel of 250 potato clones (Hirsch et al. 2013). This genomic resource is actively used by potato scientists worldwide (Fig. 5.1).

The initial efforts of the potato sequencing consortium were on resolving the two genome sequences of the dihaploid clone *S. tuberosum* Group Tuberosum RH89-039-16 (RH), but these could not be fully assembled in spite of the availability of the DM sequence. Higher level of heterogeneity was found among the two RH genomes than between RH and DM genomes (The Potato Genome Sequencing Consortium 2011). About 5% of the RH genome sequence (free of repetitive sequences) were aligned with the DM genome sequence and found to be mainly collinear with 97.5% sequence identity, whereas the two RH genome sequences presented 96.5% sequence identity. However, when larger RH genome sequences were obtained, loss of collinearity was frequently observed for the euchromatic region and the three highly diverged pericentric heterochromatin haplotypes of the chromosome 5 (de Boer et al. 2015). These findings stress the importance of sequencing other cultivated potato genome and perform de novo assembly.

Being a diploid and polyploid crop with frequent inbreeding depression and wild species introgression, genome sequence diversity is expected and has contributed to the difficulties of assembling more genome sequences from the cultivated potato.

The genome sequence of a wild species, *Solanum commersonii* was assembled using the potato genome sequence as reference (Aversano et al. 2015). This species has interesting sources of resistance to important diseases of the potato and is known for its freezing resistance and cold acclimation. The species has been used recently in breeding potato for resistance to bacterial wilt in potato and increased levels of resistance were observed (Carputo et al. 2009; Boschi et al. 2017). Flow cytometry estimated a total genome size of 830 Mb. The genome appears to be slightly smaller mainly due to differences in the intragenic regions, to have lower amount of repetitive DNA, and to have 126 cold-related genes not present in the *S. tuberosum* genome.

Recently, the genome sequence of another wild species was resolved using the M6 inbred clone of *Solanum chacoense* (Leisner et al. 2018). Flow cytometry estimated the genome to be 882 Mb. Using a de novo assembly procedure, 508 Mb of the genome assembly could be used to construct 12 pseudomolecules. These were compared to those of the first published genome sequence and shown with concordance for all of them. Interestingly, the genotype used was a 7-generation selfed *S. chacoense* plant but retained residual heterozygosity on all chromosome with three of them with significantly higher proportion. It is too early to assume that heterozygosity in some region is due either to deleterious alleles or to regions with reduced recombination. Genome annotation for gene-models revealed the presence of 37,740 genes. The *S. chacoense* genome sequence is a new resource for identifying important genes of key traits in population derived from M6.

The pan-genome of the cultivated potato covering traditional landraces (diploids to pentaploids), and modern potato cultivars of the Andigenum and Chilotanum gene pools represents today a huge endeavor due to its extraordinary diversity. When available, it would be a powerful resource for breeders to understand the genome structure of the cultivated potato between the core genome with genes present in all cultivars and the dispensable genome made of genes present only in some cultivar groups. The concept is not restricted to modern cultivars but can include wild relatives, or higher taxonomic level (Vernikos et al. 2015). Clearly, more genomes of wild species are also needed to be assembled to improve our understanding of the interspecific genome variation. Ten years after the beginning of sequencing the potato genome, it is worth noting that only one cultivated potato genome is publicly available unlike maize or rice. This highlights the complexity of resolving uneven heterozygosity of the two or four genomes present in wild and cultivated potato. New sequencing technologies and genome assembly software are about to deliver the genomes sequences from heterozygous potatoes. This is highly desirable due to the diversity of species that have contributed to the potato.

5.2.2 The Genome Plasticity of the Cultivated Potato

Comparative analysis of genome sequences in a small panel of closely related potatoes revealed extensive genome plasticity in potato (Hardigan et al. 2016). This study used a panel of doubled monoploid potatoes derived from *S. tuberosum* Group Phureja landraces with limited introgression from Group Stenotomum, Group Tuberosum, and *Solanum chacoense*. Large regions of the potato genome bearing stress-related gene families are duplicated or deleted revealing a possible evolutionary adaptation response to environmental stresses. Copy number variation (CNV) assessed with a minimum 100-bp size revealed that about 30% of the genes are duplicated or deleted in this panel of 12 closely related potatoes. The duplicated regions varied from 500 bp to 575 kb, with total CNV calls per individual varying from 2,978 to 10,532 located preferentially in intergenic sequences in pericentromeric region of the chromosomes. This genome plasticity concerns >7000 genes referred to as dispensable genes. A remarkably high level of genome heterogeneity is found in diploid potato, which is retained through clonal propagation.

Genome heterogeneity is responsible for differential gene expression observed among the genes of tetraploid cultivars (Pham et al. 2017). Genome-wide study of genomic variation and transcription in a panel of six North American tetraploid cultivars revealed the importance of preferential allele expression often associated with evolutionarily conserved genes. Additive allele expression genes in leaves and tubers were only slightly more abundant than preferred-allele expression genes. This can be due to the differential presence of regulatory sequences (promoters, enhancers) but also to structural differences (chromatin structure, epigenetic control). Copy number was frequent; about 40% of the genes from each cultivar were in variable copy number. Again here, copy number variation seemed to be more recent and concerning genes involved in response to biotic and abiotic stresses.

Resequencing of the genomes of a representative sample of cultivated potatoes revealed about 2622 genes under domestication selection, with only 14–16% shared by the North American modern potato cultivars and the Andigenum landraces (Hardigan et al. 2017). This relatively small original gene set suggests a relatively short original common domestication of cultivated potato which diverged into two geographically distinct and long-day adapted cultivar groups by the contribution of wild species. An equally plausible interpretation is two independent domestication events from distinct wild species. This hypothesis has been debated since the early days of potato taxonomy at the beginning of the twentieth century opposing the Russian and the English taxonomist schools advocating respectively multiple and single origin of the cultivated potato (reviewed in Spooner et al. 2014). The absence of extant wild species closely related to the ancestor species of the Southern domestication is the weakest support to this hypothesis (Spooner et al. 2012). The Hardigan study revealed the role of specific wild Solanum species in the evolution of the long-day adapted *S. tuberosum* cultivar group and adaptation to upland and lowland distinguishing the Andigenum and Chilotanum groups. However, both cultivated groups presented a significant contribution from the domestication progenitor *Solanum candolleanum* suggesting the differential contribution from wild species occurred after the domestication from the *S. candollearum* progenitor. Considering variants from the regions of introgression of wild species DNA, the nuclear phylogeny resolved the Chilotanum group and modern cultivars as deriving from the Andigenum group. This study brings closer to closure of a century-old controversy on the independent domestication event leading to the Chilotanum group of cultivars.

5.2.3 *New Genomic Tools for Potato Improvement*

Potato genomic resources are gradually expanding since the availability of the first potato genome sequence from an Andean potato landrace of the Phureja group (Hirsch et al. 2014). Partial genome sequences are available from dihaploid from modern cultivars and fully resolved haplotypes from tetraploid potato cultivars have been recently achieved. Transcriptomes corresponding to these genome sequences and similar ones have been produced under many important developmental and stress conditions.

The exploitation of potato genomic resources in modern cultivar development is mostly exemplified by the use of the Single Nucleotide Polymorphisms (SNP) arrays developed by the potato community (Douches et al. 2014). Several generations of SNP arrays were generated building on the original Infinium 8303 SNP array (Felcher et al. 2012). As listed by Hirsch et al. (2014), the SolCAP array was used to understand variation for glycoalkaloid biosynthesis in wild and cultivated potato, genotype several diversity panels for a retrospective view of North American potato breeding, for a taxonomic alignment, and for genetic structure of European potato cultivars. Since then, it has been used for genetic mapping in populations derived from a diploid inbred parent (Endelman and Jansky 2016; Peterson et al. 2016), genetic mapping of agronomic traits (Manrique-Carpintero et al. 2015), combined with other SNPs to extend its use to European potato breeding germplasm (Vos et al. 2015), assess linkage decay and testing GWAS models (Sharma et al. 2018), and test genetic identity of accessions in genebanks (Ellis et al. 2018).

5.3 From Genomes to the Genes of the Potato

5.3.1 *Gene Discovery Facilitated by the Genome Sequence*

Genomics-derived strategies for gene discovery have emerged with the availability of high density markers, decrease in sequencing costs, and the increasing power of bioinformatics.

GWAS has the potential to associate markers with regions, genes, underlying the phenotypic variation of trait of interest, and therefore to increase the effectiveness of potato breeding efforts. Unlike marker association studies based on biparental populations, GWAS is not constrained by the performance of one single genotype as the sole source of an allele of interest, and instead it exploits the power of large populations to identify marker-trait associations. A recent review of GWAS in potato highlighted the importance of understanding the structure (kinship) of the population under study (Sharma et al. 2018). Potato populations made of varieties and breeding lines have been studied to establish Linkage Disequilibrium (LD) between adjacent makers. This is an important parameter of the population under study because the shorter it is the higher is the significance of the association. LD decay in earlier studies were found to present large variation (1–10 cM until equilibrium) depending on the population, the locus, and the type of markers (reviewed in Spooner et al. 2014). The first whole-genome scan of LD decay on a large European potato cultivar population estimated LD decay to 5 cM (D'Hoop et al. 2010), concluding that association studies can be performed at moderate marker densities. Since the advent of SNP arrays, new GWAS have been conducted and revealed the power of this mapping approach over the biparental mapping (Stich et al. 2013). An extended SNP array of the 8303 SolCAP (SolSTW) was used to genotype 569 potato cultivars with 20k SNP markers (Vos et al. 2017). Although this study used a different estimator of LD decay than previous studies, it was found to be in the range of 1.5 Mb for old potato cultivars and 0.6 Mb for those of the second half of twentieth century, values which are compatible with the known limited number of meiosis (5–10) having taken place in the development of these European cultivar populations (Gebhardt et al. 2004; van Berloo et al. 2007). The most recent study using the SolCAP SNP array on a large European cultivar population of 351 tetraploid potatoes estimated LD decay in different regions (short and long arms, and pericentromeric heterochromatin) of each chromosome (Sharma et al. 2018). Again here, their estimates were in the range of 2.73 Mb for euchromatin and 3.27 Mb for whole chromosomal regions. Hence, most studies of LD decay report a modest decay of LD in European potato cultivar populations ranging from 0.6 to 20 Mb depending on the region and chromosome. Interestingly, smaller values of 0.3 Mb in chromosome 4 to 8 Mb in chromosome 8 were estimated for a population of 652 Andigenum cultivars (Berdugo-Cely et al. 2017). The lower distance for LD decay in these native cultivars is expected though a much lower distance could have been anticipated for a population from cultivar domesticated between 8,000 BC and 11,500 BC based on fossil evidence from the dry coast of Peru and south-central Chile (Spooner et al. 2014). It does appear that GWAS in potato can be successful at a modest marker density conferred by current SNP arrays in particular for traits with large QTL effects. However, GWAS alone will not be sufficient to associate markers directly to a specific gene contributing to the trait of interest in potato cultivar populations due to the limited number of meiotic recombination.

Annotation of the potato genome revealed the large family of plant resistance (R) genes discovered by motif sharing (nucleotide-binding site and leucine-rich repeat domain, NB-LRR) with an estimated number per haploid genome of 438 (Jupe et al. 2012). By rescreening the potato genome for NB-LRR target sequences, a total of 755 *R* gene homolog were identified (Jupe et al. 2013). This *R* gene enrichment and sequencing (RenSeq) method was applied to identify markers co-segregating with *R* genes for LB resistance and rapidly clone them (Jupe et al. 2013; Witek et al. 2016; Chen et al. 2018). A derived application of this genome-wide gene discovery is the diagnostic resistance gene enrichment sequencing (dRenSeq) which identifies full *R* genes and their homologs in breeding materials (Armstrong et al. 2018). Combined strategies to identify or clone, multiple resistance genes for diseases such as late blight, viruses, and nematodes will speed up the development of new cultivar with stacked resistance genes.

5.3.2 *Progress Toward Next Generation of Potato Varieties*

The exploitation of the pan-genome of the potato for varietal improvement will increase as more genomes are sequenced and traits phenotyped more accurately in broad germplasm. However, genetic gain will continue to be low in tetraploid breeding though faster and more predictable by the application of genomic selection. New breeding technologies and diploid hybrid breeding can generate unachievable genetic gains by tetraploid breeding.

Direct gene transfer in potato has been successful in generating disease resistance varieties since the early days of genetic engineering (Halterman et al. 2016). Existing widely grown varieties were genetically upgraded by addition of transgenes conferring resistance to pest and diseases, improved processing qualities, and consumer preferences (Table 5.1).

These transgenes produced new pest and pathogen toxins, silenced incoming viruses or endo-genes, or new enzymes for metabolite engineering. After a short life, the first generation of biotech potatoes were withdrawn as reported above. However, a renewed interest of the industry lead to the release of new biotech potatoes in the US (Waltz 2015). A long awaited released came about the same time in Argentina with a PVY virus-resistant potato variety (Bravo-Almonacid and Segretin 2016). With the exception of the latter, all biotech potato released so far were developed by the private sector. Efforts towards future release of late blight-resistant varieties have increased in the last years (Table 5.1). A 10-year research project in The Netherlands developed transgenic and cisgenic potatoes from four varieties with single and multiple *R* genes (Haverkort et al. 2016). *R* gene stacking was shown to confer high levels of resistance in the filed over several seasons (Zhu et al. 2012; Haesaert et al. 2015). One of these was even fully tested for regulatory approval (Storck et al. 2012). This biotech variety, Fortuna, was unfortunately withdrawn from regulatory approval because of the unfavorable European environment. In the US, a 5-year project aimed at the release in Indonesia and Bangladesh of late blight-resistant local varieties with three *R* genes (https://www.canr.msu.edu/biotechpp/index). These three-*R*-gene biotech potatoes are also the focus of a project aiming at release in sub-Saharan African countries potato varieties with an extremely high level of genetic tolerance to late blight, the most devastating disease in potato, caused by *Phytophtora infestans*, unrivalled by the genetic tolerance achieved to date through conventional breeding(Ghislain et al. 2018). Biotech potatoes have been field tested under natural infection for five seasons and have not shown any lesions caused by *P. infestans* (Fig. 5.2).

The latter two projects are benefiting from the release in the US of the Innate potato with late blight resistance for which the regulatory dossier is publicly available (Clark et al. 2014). One of the important costs in regulatory dossier development is the toxicity assessment of the new proteins for which the 3*R* gene technology can build a weight of evidence instead of costly purification, stability, and gavage testing (Habig et al. 2018). Therefore, when both projects estimated their regulatory costs, these were found to be reasonable unlike those reported by

Table 5.1 Traits of biotech potato from potatoes approved for food and cultivation (source ISAAA GM crop database)

Trait(s)	Trade name	Developer	First approved for food	First approved for cultivation
Colorado Potato Beetle resistance[a]	New Leaf™ Russet Burbank potato	Monsanto Co.	CAN USA (1995); AUS JPN NZL (2001); PHL (2003); KOR (2004)	USA (1994); CAN (1995)
	Atlantic NewLeaf™ potato	Monsanto Co.	CAN MEX USA (1996); AUS NZL (2001)	USA (1995); CAN (1997)
	Superior NewLeaf™ potato	Monsanto Co.	CAN (1995); USA (1996); MEX (1996); AUS JPN NZL (2001); PHL (2003); KOR (2004)	USA (1995)
Colorado Potato Beetle and PVY resistance[a]	New Leaf™ Y Russet Burbank potato	Monsanto Co.	USA (1998); CAN (1999); AUS JPN MEX NZL (2001); PHL KOR (2004)	USA (1997); CAN (1999)
	Shepody NewLeaf™ Y potato	Monsanto Co.	USA (1998); CAN (1999); AUS MEX NZL (2001); JPN PHL (2003); KOR (2004)	USA (1997); CAN (2001)
	Hi-Lite NewLeaf™ Y potato	Monsanto Co.	USA (1998)	
Modified starch (high amylose)[a]	Amflora™	BASF	EU (2010)	EU (2010)
	Starch Potato	BASF	USA (2014)	
Low asparagine (acrylamide), low black spot bruise	Innate® Cultivate	JR Simplot Co.	USA (2014); CAN (2016); AUS JPN MEX MYZ NZL (2017)	USA (2014); CAN (2016)
	Innate® Generate	JR Simplot Co.	USA (2014); CAN (2016); AUS MEX NZL (2017)	USA (2014); CAN (2016)
	Innate® Accelerate	JR Simplot Co.	USA (2014); CAN (2016); AUS MEX NZL (2017)	USA (2014); CAN (2016)
Low asparagine (acrylamide), low black spot bruise, late blight resistance	n/a (Russet Burbank)	JR Simplot Co.	USA (2015); AUS CAN NZL (2017)	USA (2015); CAN (2017)
	Innate® Acclimate	JR Simplot Co.	USA (2016); AUS CAN NZL (2017)	USA (2015); CAN (2017)
	Innate® Hibernate	JR Simplot Co.	USA (2016); AUS CAN NZL (2017)	USA (2015); CAN (2017)
PVY resistance	n/a (Spunta)	Technoplant Argentina	ARG (2018)	ARG (2018)

Countries are represented by three letter codes

[a]Refers to products phased out of the market

Fig. 5.2 Confined field trial conducted in Uganda with transgenic potato with three *R* genes (dark green plots), developed as described by Ghislain et al. (2018), and nontransgenic potato varieties (severely damaged plots)

larger players for commodities like maize (Kalaitzandonakes et al. 2006; Schiek et al. 2016). However, the adoption of biotech potatoes remains challenging due to negative perception by a large part of the public unfamiliar with the challenges and potential solutions to improve agriculture production. The long-standing opposition to industrialization of agriculture, the concerns about multinational corporate dominance, the lack of trust in risk assessment of regulatory agencies, the growing conflict of interest of the organic industry, and the fear of unknown manipulations of our food, have delayed the approval and adoption of biotech crops. The release of biotech potatoes addressing a major long-lasting threat on its production which calls back bad memories to Europeans and North Americans, may well result in a perception change provided strong public education is developed (Hallerman and Grabau 2016).

Gene editing in potato has already passed the stage of proof-of-concept as reviewed above. There are yet no potato products on the market, but gene-edited varieties will soon be released with traits governed by known existing genes whose regulation and allele structure determine the trait value (Table 5.2).

It is important to realize that gene editing is a complement to transgenesis, not replacement, because it is limited to the existing endogenous genes of the potato. Disruptive news came up recently when the European Court of Justice passed a judgment that genome edited crops should be regulated using the same regulatory framework as the transgenic crops (Callaway 2018). This decision is reminiscent of a previous one in 2012 when the European Food Safety Authority concluded that cisgenic crops should be regulated as transgenic crops (EFSA 2012). This European

Table 5.2 Traits targeted by genome editing in potato and opportunities for improving pest and disease resistance as well as nutritional qualities of the potato

Trait	Target gene	Expected impact	References
Heat tolerance (high yield under higher temperature)	Heat-shock cognate 70 (HSc70)	Enhanced yields of potato grown under lowland tropics	Trapero-Mozos et al. (2018)
Virus resistance (PVY potyvirus resistance)	Eukaryotic translation initiation factor 4E (eIF4E)	Reduce yield loss due to PVY and enhance tuber seed quality	Arcibal et al. (2016)
Reduced accumulation of reducing sugars	Vacuolar invertase (VInv)	Improved qualities of processed potato	Clasen et al. (2016)
Reduced acrylamide in processed products	Vacuolar invertase gene VInv and the asparagine synthetase genes StAS1 and StAS2	Reduction of acrylamide formation under extreme cooking temperature	Zhu et al. (2016)
Decreased accumulation of glycoalkaloids	Sterol side chain reductase 2 (St-SSR2)	Release of advanced breeding potato lines with elevated SGA	Sawai et al. (2014)
Inbreeding tolerance	S-RNase alleles (Sp3 and Sp4)	Generation of self-compatible diploid potato for developing hybrid potato varieties	Ye et al. (2018)
LB resistance	Ethylene response factor StERF3; 6 susceptibility genes; DND1 gene	Reduction of production losses and reduced costs of production	Tian et al. (2015), Sun et al. (2016a, b)
VitA biofortification	beta carotene hydroxylase *b-ch* gene	Enrichment in beta carotene in potato (precursor of VitA)	Van Eck et al. (2007)

decision will impact agricultural biotechnology innovation negatively not only in Europe but also in developing countries.

Hybrid breeding in potato has already been tested by farmers in developing countries and has received great excitement by the potato crop improvement actors in spite of the initial skepticism (Lindhout et al. 2017). The first yield assessment of hybrid varieties was conducted in two locations; the Netherlands and the Democratic Republic of Congo (de de Vries et al. 2016). In the latter, the best hybrid variety yielded three to four times the national average in sub-Saharan Africa (SSA) countries, whereas it yielded only half of the yield of conventional varieties in the Netherlands. A confounding factor is the type of seeds and health status that will need to be factored out for more precise yield comparison between hybrid and conventional potato. Nevertheless, the possibility of combining complementary traits from the parents, obtaining heterosis from hybridization of inbred parents, avoiding pathogen load of seed tubers, and facilitating transport of true seeds leaves no doubt that hybrid varieties will attract a lot of interest in the developing world.

5.4 Concluding Remarks

Despite early optimism, and unlike in other crops, the vast insight gained from its genes and genome has not been steadily translated into substantial genetic progress in potato, through either molecular breeding or transgenic approaches. Among the issues behind the above, the genetic complexity of tetraploid potatoes, issues related with public acceptance of transgenic crops, and a critical mass smaller than in other crops stand out as the most salient ones. Regardless, we remain confident that recent scientific developments, such as an increased focus on developing hybrid varieties at the 2X level, are one of the main factors that will change the above-described trend, since a main advantage of dealing with 2X instead of 4X genetics is a much more straightforward application of molecular approaches, as demonstrated already by the routinary use of such technologies in other Solanaceous crops such as potato, and to a lesser extent, pepper. In addition, early reports on the use of genomic selection in potato have demonstrated its ability to circumvent many of the pitfalls observed when QTL were used to attempt increasing the effectiveness of potato breeding efforts. The continuous reduction of DNA sequencing will enable collecting sequencing data on a larger scale than before, further facilitating both the identification of genomic regions associated with traits of economic importance, and a better understanding of quantitative traits in potato. Regarding the use of gene editing approaches, although they provide a much more targeted ability to modify the potato's genome, the full realization of its potential to facilitate the development of varieties carrying genetic alleles not hitherto found in the germplasm available will, by and large, depend on how the public acceptance of genetic modification evolves, both in developed and developing countries.

References

Almekinders CJM, Chujoy E, Thiele G (2009) The use of true potato seed as pro-poor technology: the efforts of an International Agricultural Research Institute to innovating potato production. Potato Res 52(4):275–293. https://doi.org/10.1007/s11540-009-9142-5

Andersson M, Turesson H, Nicolia A, Fält AS, Samuelsson M, Hofvander P (2017) Efficient targeted multiallelic mutagenesis in tetraploid potato (*Solanum tuberosum*) by transient CRISPR-Cas9 expression in protoplasts. Plant Cell Rep 36(1):117–128. https://doi.org/10.1007/s00299-016-2062-3

Arcibal E, Gold KM, Flaherty S, Jiang J, Jahn M, Rakotondrafara AM (2016) A mutant eIF4E confers resistance to potato virus Y strains and is inherited in a dominant manner in the potato varieties Atlantic and Russet Norkotah. Am J Potato Res 93(1):64–71. https://doi.org/10.1007/s12230-015-9489-x

Armstrong MR, Vossen J, Lim TY, Hutten RCB, Xu J, Strachan SM, Harrower B, Champouret N, Gilroy EM, Hein I (2018) Tracking disease resistance deployment in potato breeding by enrichment sequencing. Plant Biotechnol J 17(2):540–549. https://doi.org/10.1111/pbi.12997

Aversano R, Contaldi F, Ercolano MR, Grosso V, Iorizzo M, Tatino F, Delledonne M (2015) The *Solanum commersonii* genome sequence provides insights into adaptation to stress conditions and genome evolution of wild potato relatives. Plant Cell 27:954–968. https://doi.org/10.1105/tpc.114.135954

Berdugo-Cely J, Valbuena RI, Sanchez-Betancourt E, Barrero LS, Yockteng R (2017) Genetic diversity and association mapping in the Colombian Central Collection of *Solanum tuberosum* L. *Andigenum* group using SNPs markers. Plos One 12(3):e0173039. https://doi.org/10.1371/journal.pone.0173039

Boschi F, Schvartzman C, Murchio S, Ferreira V, Siri MI, Galván GA, Dalla-Rizza M (2017) Enhanced bacterial wilt resistance in potato through expression of *Arabidopsis* EFR and introgression of quantitative resistance from *Solanum commersonii*. Front Plant Sci 8:1642. https://doi.org/10.3389/fpls.2017.01642

Bradshaw JE (2009) Potato breeding at the Scottish Plant Breeding Station and the Scottish Crop Research Institute: 1920–2008. Potato Res 52:141–172. https://doi.org/10.1007/s11540-009-9126-5

Bradshaw JE (2017) Review and analysis of limitations in ways to improve conventional potato breeding. Potato Res 60:171–193. https://doi.org/10.1007/s11540-017-9346-z

Bradshaw JE, Ramsay G (2005) Utilisation of the commonwealth potato collection in potato breeding. Euphytica 146(1–2):9–19. https://doi.org/10.1007/s10681-005-3881-4

Bradshaw JE, Stewart HE, Wastie RL, Dale MFB, Phillips MS (1995) Use of seedling progeny tests for genetical studies as part of a potato (*Solanum tuberosum* subsp tuberosum) breeding programme. Theor Appl Genet 90:899–905

Bradshaw JE, Todd D, Wilson RN (2000) Use of tuber progeny tests for genetical studies as part of a potato (*Solanum tuberosum* subsp tuberosum) breeding programme. Theor Appl Genet 100:772–781

Bravo-Almonacid FF, Segretin MF (2016) Status of transgenic crops in Argentina. In: Collinge DB (ed) Biotechnology for plant disease control. Wiley Blackwell, New York, pp 275–283

Butler NM, Atkins PA, Voytas DF, Douches DS (2015) Generation and inheritance of targeted mutations in potato (*Solanum tuberosum* L.) using the CRISPR/Cas system. PloS One 10(12):e0144591. https://doi.org/10.1371/journal.pone.0144591

Butler NM, Baltes NJ, Voytas DF, Douches DS (2016) Geminivirus-mediated genome editing in potato (*Solanum tuberosum* L.) using sequence-specific nucleases. Front Plant Sci 7:1045. https://doi.org/10.3389/fpls.2016.01045

Callaway E (2018) CRISPR plants now subject to tough GM laws in European Union. Nature 560:16. https://doi.org/10.1038/d41586-018-05814-6

Carputo D, Aversano R, Barone A, Di Matteo A, Iorizzo M, Sigillo L, Zoina A, Frusciante L (2009) Resistance to *Ralstonia solanacearum* of sexual hybrids between *Solanum commersonii* and *S. tuberosum*. Am J Potato Res 86:196–202. https://doi.org/10.1007/s12230-009-9072-4

Chen X, Lewandowska D, Armstrong MR, Baker K, Lim T-Y, Bayer M, Harrower B, McLean K, Jupe F, Witek K, Lees AK, Jones JD, Bryan GJ, Hein I (2018) Identification and rapid mapping of a gene conferring broad-spectrum late blight resistance in the diploid potato species *Solanum verrucosum* through DNA capture technologies. Theor Appl Genet 131(6):1287–1297. https://doi.org/10.1007/s00122-018-3078-6)

Clark P, Habig J, Ye J, Collinge S (2014) Petition for determination of nonregulated status for Innate™ potatoes with late blight resistance, low acrylamide potential, reduced black spot, and lowered reducing sugars: Russet Burbank event W8. JR Simplot Company Petition JRS01 (USDA Petition 14-093-01p)

Clasen BM, Stoddard TJ, Luo S, Demorest ZL, Li J, Cedrone F, Coffman A (2016) Improving cold storage and processing traits in potato through targeted gene knockout. Plant Biotechnol J 14(1):169–176. https://doi.org/10.1111/pbi.12370

de Boer JM, Datema E, Tang XM, Borm TJA, Bakker EH, van Eck HJ, van Ham RCHJ, de Jong H, Visser RGF, Bachem CWB (2015) Homologues of potato chromosome 5 show variable collinearity in the euchromatin, but dramatic absence of sequence similarity in the pericentromeric heterochromatin. BMC Genomics 16(1):374. https://doi.org/10.1186/s12864-015-1578-1

D'Hoop BB, Paulo MJ, Kowitwanich K, Sengers M, Visser RGF, van Eck HJ, van Eeuwijk FA (2010) Population structure and linkage disequilibrium unravelled in tetraploid potato. Theor Appl Genet 121(6):1151–1170. https://doi.org/10.1007/s00122-010-1379-5

Douches D, Hirsch CN, Manrique-Carpintero NC, Massa AN, Coombs J, Hardigan M, Bisognin D, De Jong W, Buell CR (2014) The contribution of the Solanaceae Coordinated Agricultural Project to potato breeding. Potato Res 57:215–224. https://doi.org/10.1007/s11540-014-9267-z

EFSA Panel on Genetically Modified Organisms (GMO) (2012) Scientific opinion addressing the safety assessment of plants developed through cisgenesis and intragenesis. EFSA J 10(2):2561[33p]. https://doi.org/10.2903/jefsa20122561

Ellis D, Chavez O, Coombs JJ, Soto JV, Gomez R, Douches DS, Anglin NL (2018) Genetic identity in genebanks: application of the SolCAP 12K SNP array in fingerprinting and diversity analysis in the global in trust potato collection. Genome 61(7):523–537. https://doi.org/10.1139/gen-2017-0201

Endelman JB, Jansky SH (2016) Genetic mapping with an inbred line-derived F2 population in potato. Theor Appl Genet 129(5):935–943. https://doi.org/10.1007/s00122-016-2673-7

Felcher KJ, Coombs JJ, Massa AN, Hansey CN, Hamilton JP, Veilleux RE, Buell CR, Douches DS (2012) Integration of two diploid potato linkage maps with the potato genome sequence. PloS One 7(4):e36347. https://doi.org/10.1371/journal.pone.0036347

Finkers-Tomczak A, Bakker E, de Boer J, van der Vossen E, Achenbach U, Golas T, Suryaningrat S, Smant G, Bakker J, Goverse A (2011) Comparative sequence analysis of the potato cyst nematode resistance locus *H1* reveals a major lack of co-linearity between three haplotypes in potato (*Solanum tuberosum* ssp.). Theor Appl Genet 122(3):595–608. https://doi.org/10.1007/s00122-010-1472-9

Gatto M, Hareau G, Pradel W, Suárez V, Qin J (2018) Release and adoption of improved potato varieties in Southeast and South Asia. International Potato Center (CIP), Lima, Peru, 45p. ISBN 978-92-9060-501-0. Social Sciences Working Paper 2018-2, 42p

Gavrilenko T, Antonova O, Shuvalova A, Krylova E, Alpatyeva N, Spooner DM, Novikova L (2013) Genetic diversity and origin of cultivated potatoes based on plastid microsatellite polymorphism. Genet Resour Crop Evol 60(7):1997–2015. https://doi.org/10.1007/s10722-013-9968-1

Gebhardt C (2013) Bridging the gap between genome analysis and precision breeding in potato. Trends Genet 29:248–256. https://doi.org/10.1016/j.tig.2012.11.006

Gebhardt C, Ballvora A, Walkemeier B, Oberhagemannand P, Schuler K (2004) Assessing genetic potential in germplasm collections of crop plants by marker-trait association: a case study for potatoes with quantitative variation of resistance to late blight and maturity type. Mol Breeding 13(1):93–102

Ghislain M, Byarugaba A, Magembe E, Njoroge A, Rivera C, Roman ML, Tovar JC, Gamboa S, Forbes G, Kreuze J, Barekye A, Kiggundu A (2018) Stacking three late blight resistance genes from wild species directly into African highland potato varieties confers complete field resistance to local blight races. Plant Biotech J 17(6):1119–1129. https://doi.org/10.1111/pbi.13042

Habig JW, Rowland A, Pence MG, Zhong CX (2018) Food safety evaluation for R-proteins introduced by biotechnology: a case study of VNT1 in late blight protected potatoes. Regul Toxicol Pharmacol 95:66–74. https://doi.org/10.1016/j.yrtph.2018.03.008

Haesaert G, Vossen JH, Custers R, De Loose M, Haverkort A, Heremans B, Hutten R, Kessel G, Landschoot S, Van Droogenbroeck B, Visser RGF, Gheysen G (2015) Transformation of the potato variety Desiree with single or multiple resistance genes increases resistance to late blight under field conditions. Crop Prot 77:163–175. https://doi.org/10.1016/j.cropro.2015.07.018

Hallerman E, Grabau E (2016) Crop biotechnology: a pivotal moment for global acceptance. Food Energy Secur 5(1):3–17. https://doi.org/10.1002/fes3.76

Halterman D, Guenthner J, Collinge S, Butler N, Douches D (2016) Biotech potatoes in the 21st century: 20 years since the first biotech potato. Am J Potato Res 93(1):1–20. https://doi.org/10.1007/s12230-015-9485-1

Hamilton JP, Hansey CN, Whitty BR, Buell CR (2011) Single nucleotide polymorphism discovery in elite North American potato germplasm. BMC Genomics 12(1):302. https://doi.org/10.1186/1471-2164-12-302

Hardigan MA, Crisovan E, Hamilton JP, Kim J, Laimbeer P, Leisner CP, Manrique-Carpintero NC, Newton L, Pham GM, Vaillancourt B, Yang X, Zeng Z, Douches DS, Jiang J, Veilleux RE, Buell CR (2016) Genome reduction uncovers a large dispensable genome and adaptive role for copy number variation in asexually propagated *Solanum tuberosum*. Plant Cell 28:388–405. https://doi.org/10.1105/tpc.15.00538

Hardigan MA, Laimbeer FPE, Newton L, Crisovan E, Hamilton JP, Vaillancourt B, Wiegert-Riningera K, Wooda JC, Douches DS, Farréa EM, Veilleux RE, Buell CR (2017) Genome diversity of tuber-bearing Solanum uncovers complex evolutionary history and targets of domestication in the cultivated potato. Proc Nat Acad Sci 114(46):E9999–E10008. https://doi.org/10.1073/pnas.1714380114

Haverkort AJ, Boonekamp PM, Hutten R, Jacobsen E, Lotz LAP, Kessel GJT, Visser RGF (2016) Durable late blight resistance in potato through dynamic varieties obtained by cisgenesis: scientific and societal advances in the DuR*Ph* project. Potato Res 59(1):35–66. https://doi.org/10.1007/s11540-015-9312-6

Haverkort AJ, Struik PC, Visser RGF, Jacobsen E (2009) Applied biotechnology to combat late blight in potato caused by *Phytophthora infestans*. Potato Res 52:249–264. https://doi.org/10.1007/s11540-009-9136-3

Haynes KG, Guedes ML (2018) Self-compatibility in a diploid hybrid population of *Solanum phureja–S. stenotomum*. Am J Potato Res 95(6):729–734. https://doi.org/10.1007/s12230-018-9680-y

Hirsch CD, Hamilton JP, Childs KL, Cepela J, Crisovan E, Vaillancourt B, Hirsch CN, Habermann M, Neal B, and Buell CR (2014) Spud DB: A resource for mining sequences genotypes and phenotypes to accelerate potato breeding. Plant Genome 7(1):1–12. https://doi.org/10.3835/plantgenome2013.12.0042

Hirsch CN, Hirsch CD, Felcher K, Coombs J, Zarka D, Van Deynze A, De Jong W, Veilleux RE, Jansky S, Bethke P, Douches DS, Buell CR (2013) Retrospective view of North American potato (*Solanum tuberosum* L.) breeding in the 20th and 21st centuries. G3: Genes, Genomes, Genetics 3(6):1003–1013. https://doi.org/10.1534/g3.113.005595

Hosaka K, Hanneman RE (1998a) Genetics of self-compatibility in a self-incompatible wild diploid potato species *Solanum chacoense*: I detection of an *S* locus inhibitor (*Sli*) gene. Euphytica 99:191–197

Hosaka K, Hanneman RE (1998b) Genetics of self-compatibility in a self-incompatible wild diploid potato species *Solanum chacoense*: 2 localization of an *S* locus inhibitor (*Sli*) gene on the potato genome using DNA markers. Euphytica 103:265–271

Jansky SH, Charkowski AO, Douches DS, Gusmini G, Richael C, Paul C, Spooner DM, Novy RG, De Jong H, De Jong WS, Bamberg JB, Thompson L, Bizimungu B, Holm DG, Brown CR, Haynes KG, Vidyasagar R, Veilleux RE, Miller JC, Bradeen JM, Jiang JM (2016) Reinventing potato as a diploid inbred line-based crop. Crop Sci 11:1–11. https://doi.org/10.2135/cropsci2015.12.0740

Jansky SH, Chungand Y, Kittipadukal P (2014) M6: a diploid potato inbred line for use in breeding and genetics research. J Plant Regist 8:195–199. https://doi.org/10.3198/jpr2013.05.0024crg

Jansky SH, Spooner DM (2017) The evolution of potato breeding. Plant Breed Rev 41:169–211

Jupe F, Pritchard L, Etherington GJ, MacKenzie K, Cock PJA, Wright F, Sharma SK, Bryan GJ, Jones JDG, Hein I (2012) Identification and localisation of the NB-LRR gene family within the potato genome. BMC Genomics 13(1):75. https://doi.org/10.1186/1471-2164-13-75

Jupe F, Witek K, Verweij W, Śliwka J, Pritchard L, Etherington GJ, Maclean D, Cock PJ, Leggett RM, Bryan GJ, Cardle L, Hein I, Jones JDG (2013) Resistance gene enrichment sequencing (RenSeq) enables reannotation of the NB-LRR gene family from sequenced plant genomes and rapid mapping of resistance loci in segregating populations. Plant J 76(3):530–544. https://doi.org/10.1111/tpj.12307

Kalaitzandonakes N, Alston JM, Bradford KJ (2006) Compliance costs for regulatory approval of new biotech crops. In: Just RE, Alston JM, Zilberman D (eds) Regulating agricultural biotechnology: economics and policy. Natural resource management and policy, vol 30. Springer, Boston, pp 37–57. https://doi.org/10.1007/978-0-387-36953-2_3

Lawson C, Kaniewski W, Haley L, Rozman R, Newell C, Sanders P, Tumer NE (1990) Engineering resistance to mixed virus infection in a commercial potato cultivar: resistance to potato virus X and potato virus Y in transgenic Russet Burbank. Nat Biotechol 8(2):127

Leisner CP, Hamilton JP, Crisovan E, Manrique-Carpintero NC, Marand AP, Newton L, Pham GM, Jiang J, Douches DS, Jansky SH, Buell CR (2018) Genome sequence of M6, a diploid inbred clone of the high-glycoalkaloid-producing tuber-bearing potato species *Solanum chacoense* reveals residual heterozygosity. Plant J 94(3):562–570. https://doi.org/10.1111/tpj.13857

Lindhout P, Meijer D, Schotte T, Hutten RCB, Visser RGF, Eck HJ (2011) Towards F1 hybrid seed potato breeding. Potato Res 54:301–312. https://doi.org/10.1007/s11540-011-9196-z

Lindhout P, De Vries M, Ter Maat M, Ying S, Viquez-Zamora M, Van Heusden S (2017) Hybrid potato breeding for improved varieties. In: Achieving sustainable cultivation of potatoes, vol. 1. Breeding, nutritional and sensory quality. Burleigh Dodds Science Publishing, Cambridge, p 24. https://doi.org/10.19103/AS.2016.0016.04

Ma J, Xiang H, Donnelly DJ, Meng FR, Xu H, Durnford D, Li XQ (2017) Genome editing in potato plants by agrobacterium-mediated transient expression of transcription activator-like effector nucleases. Plant Biotechnol Rep 11(5):249–258. https://doi.org/10.1007/s11816-017-0448-5

Maluszynski M, Ahloowalia BS, Sigurbjörnsson B (1995) Application of *in vivo* and *in vitro* mutation techniques for crop improvement. Euphytica 85(1–3):303–315

Manrique-Carpintero NC, Coombs JJ, Cui Y, Veilleux RE, Buell CR, Douches DS (2015) Genetic map and QTL analysis of agronomic traits in a diploid potato population using single nucleotide polymorphism markers. Crop Sci 55(6):2566–2579. https://doi.org/10.2135/cropsci2014.10.0745

Mihovilovich E, Aponte M, Lindqvist-Kreuze H, Bonierbale M (2014) An RGA-derived SCAR marker linked to PLRV resistance from *Solanum tuberosum* ssp. andigena. Plant Mol Biol Rep 32(1):117–128

Nicolia A, Proux-Wéra E, Åhman I, Onkokesung N, Andersson M, Andreasson E, Zhu LH (2015) Targeted gene mutation in tetraploid potato through transient TALEN expression in protoplasts. J Biotechnol 204:17–24. https://doi.org/10.1016/j.jbiotec.2015.03.021

Ning Y, Liu W, Wang GL (2017) Balancing immunity and yield in crop plants. Trends Plant Sci 22(12):1069–1079. https://doi.org/10.1016/j.tplants.2017.09.010

Ochoa CM (1990) The potatoes of South America: Bolivia. Cambridge University Press, Cambridge

Paz MM, Veilleux RE (1999) Influence of culture medium and *in vitro* conditions on shoot regeneration in *Solanum phureja* monoploids and fertility of regenerated doubled monoploids. Plant Breed 118:53–57

Perlak FJ, Stone TB, Muskopf YM, Petersen LJ, Parker GB, McPherson SA, Fischhoff DA (1993) Genetically improved potatoes: protection from damage by Colorado potato beetles. Plant Mol Biol 22(2):313–321

Peterson BA, Holt SH, Laimbeer FPE, Doulis AG, Coombs J, Douches DS, Hardigan MA, Buell CR, Veilleux RE (2016) Self-fertility in a cultivated diploid potato population examined with the Infinium 8303 potato single-nucleotide polymorphism Array. Plant Genome 9(3):1–13. https://doi.org/10.3835/plantgenome2016.01.0003

Pham GM, Newton L, Wiegert-Rininger K, Vaillancourt B, Douches DS, Buell CR (2017) Extensive genome heterogeneity leads to preferential allele expression and copy number-dependent expression in cultivated potato. Plant J 92(4):624–637. https://doi.org/10.1111/tpj.13706

Phumichai C, Mori M, Kobayashi A, Kamijimaand O, Hosaka K (2005) Toward the development of highly homozygous diploid potato lines using the self-compatibility controlling *Sli* gene. Genome 48:977–984. https://doi.org/10.1139/G05-066

Plaisted RL, Hoopes RW (1989) The past record and future prospects for the use of exotic potato germplasm. Am Potato J 66:603–627

Ramakrishnan AP, Ritland CE, Sevillano RHB, Riseman A (2015) Review of potato molecular markers to enhance trait selection. Am J Potato Res 92(4):455–472. https://doi.org/10.1007/s12230-015-9455-7

Rodríguez F, Ghislain M, Clausen AM, Jansky SH, Spooner DM (2010) Hybrid origins of cultivated potatoes. Theor Appl Genet 121(6):1187–1198. https://doi.org/10.1007/s00122-010-1422-6

del Rosario Herrera M, Vidalon LJ, Montenegro JD, Riccio C, Guzman F, Bartolini I, Ghislain M (2018) Molecular and genetic characterization of the *Ryadg* locus on chromosome XI from Andigena potatoes conferring extreme resistance to potato virus Y. Theor Appl Genet 131(9):1925–1938. doi: https://doi.org/10.1007/s00122-018-3123-5

Sawai S, Ohyama K, Yasumoto S, Seki H, Sakuma T, Yamamoto T, Takebayashi Y, Kojima M, Sakakibara H, Aoki T, Muranaka T, Kazuki Saito K, Umemotog N (2014) Sterol side chain reductase 2 is a key enzyme in the biosynthesis of cholesterol, the common precursor of toxic steroidal glycoalkaloids in potato. Plant Cell 26:3763–3774. https://doi.org/10.1105/tpc.114.130096

Scheben A, Wolter F, Batley J, Puchta H, Edwards D (2017) Towards CRISPR/Cas crops–bringing together genomics and genome editing. New Phytologist 216(3):682–698. https://doi.org/10.1111/nph.14702

Schiek B, Hareau G, Baguma Y, Medakker A, Douches DS, Shotkoski F, Ghislain M (2016) Demystification of GM crop costs: releasing late blight resistant potato varieties as public goods in developing countries. Int J Biotechnol 14:112–131

Schönhals EM, Ortega F, Barandalla L, Aragones A, De Galarreta JR, Liao JC, Sanetomo R, Walkemeier B, Tacke E, Ritter E, Gebhardt C (2016) Identification and reproducibility of diagnostic DNA markers for tuber starch and yield optimization in a novel association mapping population of potato (*Solanum tuberosum* L.). Theor Appl Genet 129(4):767–785. https://doi.org/10.1007/s00122-016-2665-7

Sharma R, Bhardwaj V, Dalamu D, Kaushik SK, Singh BP, Sharma SK, Umamaheshwari R, Baswaraj R, Kumar V, Gebhardt C (2014) Identification of elite potato genotypes possessing multiple disease resistance genes through molecular approaches. Sci Hortic 179:204–211. https://doi.org/10.1016/j.scienta.2014.09.018

Sharma SK, Bolser D, de Boer J, Sønderkær M, Amoros W et al (2013) Construction of reference chromosome-scale pseudomolecules for potato: integrating the potato genome with genetic and physical maps. G3 (Bethesda) 3(11):2031–2047. https://doi.org/10.1534/g3.113.007153

Sharma SK, MacKenzie K, McLean K, Dale F, Daniels S, Bryan GJ (2018) Linkage disequilibrium and evaluation of genome-wide association mapping models in tetraploid potato. G3 (Bethesda) 8(10):3185–3202. https://doi.org/10.1534/g3.118.200377

Slater AT, Cogan NOI, Forster JW (2013) Cost analysis of the application of marker-assisted selection in potato breeding. Mol Breeding 32(2):299–310. https://doi.org/10.1007/s11032-013-9871-7

Slater AT, Cogan NOI, Hayes BJ, Schultz L, Dale MFB, Bryan GJ, Forster JW (2014a) Improving breeding efficiency in potato using molecular and quantitative genetics. Theor Appl Genet 127:2279–2292. https://doi.org/10.1007/s00122-014-2386-8

Slater AT, Wilson GM, Cogan NOI, Forster JW, Hayes BJ (2014b) Improving the analysis of low heritability complex traits for enhanced genetic gain in potato. Theor Appl Genet 127:809–820. https://doi.org/10.1007/s00122-013-2258-7

Slater AT, Cogan NOI, Forster JW, Hayes BJ, Daetwyler HD (2016) Improving genetic gain with genomic selection in autotetraploid potato. Plant Genome 9(3):1–15. https://doi.org/10.3835/plantgenome2016.02.0021

Sood S, Bhardwaj V, Pandey SK, Chakrabarti SK (2017) History of potato breeding: improvement diversification and diversity. In: Chakrabarti SK, Xie C, Tiwari JK (eds) The potato genome. Springer, Cham, pp 31–72

Spooner DM, Núñez J, Trujillo G, del Rosario Herrera M, Guzmán F, Ghislain M (2007) Extensive simple sequence repeat genotyping of potato landraces supports a major reevaluation of their gene pool structure and classification. Proc Nat Acad Sci 104(49):19398–19403. https://doi.org/10.1073/pnas.0709796104

Spooner DM, Jansky SH, Clausen A, del Rosario Herrera M, Ghislain M (2012) The enigma of *Solanum maglia* in the origin of the Chilean cultivated potato *Solanum tuberosum* Chilotanum Group. Econ Bot 66(1):12–21

Spooner DM, Ghislain M, Simon R, Jansky SH, Gavrilenko T (2014) Systematics, diversity, genetics, and evolution of wild and cultivated potatoes. Bot Rev 80:283–383. https://doi.org/10.1007/s12229-014-9146-y

Stich B, Urbany C, Hoffmann P, Gebhardt C (2013) Population structure and linkage disequilibrium in diploid and tetraploid potato revealed by genome-wide high-density genotyping using the SolCAP SNP array. Plant Breeding 132(6):718–724. https://doi.org/10.1111/pbr.12102

Storck T, Böhme T, Schultheiss H (2012) Fortuna et al. Status and perspectives of GM approaches to fight late blight. In: Thirteenth EuroBlight workshop St. Petersburg (Russia), 9–12 Oct 2011. PPO-Special Report 15:45–48.

Sun K, Wolters AMA, Loonen AE, Huibers RP, van der Vlugt R, Goverse A, Bai Y (2016a) Down-regulation of *Arabidopsis DND1* orthologs in potato and tomato leads to broad-spectrum resistance to late blight and powdery mildew. Transgenic Res 25(2):123–138. https://doi.org/10.1007/s11248-015-9921-5

Sun K, Wolters AMA, Vossen JH, Rouwet ME, Loonen AE, Jacobsen E, Bai Y (2016b) Silencing of six susceptibility genes results in potato late blight resistance. Transgenic Res 25(5):731–742. https://doi.org/10.1007/s11248-016-9964-2

The Potato Genome Sequencing Consortium (2011) Genome sequence and analysis of the tuber crop potato. Nature 475:189–195. https://doi.org/10.1038/nature10158

The Tomato Genome Consortium (2012) The tomato genome sequence provides insights into fleshy fruit evolution. Nature 485:635–641. https://doi.org/10.1038/nature11119

Thiele G, Hareau G, Suarez V, Chujoy E, Bonierbale M, Maldonado L (2008) Varietal change in potatoes in developing countries and the contribution of the International Potato Center: 1972–2007. Social Sciences Working Paper (2008-6)

Thornton, M., 2003. The rise and fall of Newleaf potatoes. Biotechnology: Science and society at a crossroads. Nat. Agric. Biotech. Council Rep. 15:235–243.

Tian Z, He Q, Wang H, Liu Y, Zhang Y, Shao F, Xie C (2015) The potato ERF transcription factor StERF3 negatively regulates resistance to *Phytophthora infestans* and salt tolerance in potato. Plant Cell Physiol 56(5):992–1005. https://doi.org/10.1093/pcp/pcv025

Tiwari JK, Devi S, Ali N, Luthra SK, Kumar V, Bhardwaj V, Chakrabarti SK (2018) Progress in somatic hybridization research in potato during the past 40 years. Plant Cell Tiss Org 132:225–238. https://doi.org/10.1007/s11240-017-1327-z

Trapero-Mozos A, Morris WL, Ducreux LJ, McLean K, Stephens J, Torrance L, Taylor MA (2018) Engineering heat tolerance in potato by temperature-dependent expression of a specific allele of *HEAT-SHOCK COGNATE 70*. Plant Biotechnol J 16(1):197–207. https://doi.org/10.1111/pbi.12760

Uitdewilligen JG, Wolters AML, D'hoop BB, Borm TJA, Visser RGF, van Eck HJ (2013) A next generation sequencing method for genotyping-by-sequencing of highly heterozygous autotetraploid potato. PLoS One 8(5):e62355. https://doi.org/10.1371/journalpone0062355

van Berloo R, Hutten RCB, van Eck HJ, Visser RGF (2007) An online potato pedigree database resource. Potato Res 50(1):45–57. https://doi.org/10.1007/s11540-007-9028-3

Van Eck J, Conlin B, Garvin DF, Mason H, Navarre DA, Brown CR (2007) Enhancing beta-carotene content in potato by RNAi-mediated silencing of the beta-carotene hydroxylase gene. Am J Potato Res 84(4):331

Vernikos G, Medini D, Riley DR, Tettelin H (2015) Ten years of pan-genome analyses. Curr Opin Microbiol 23:148–154. https://doi.org/10.1016/j.mib.2014.11.016

Visser RGF, Bachem CWB, Borm T, de Boer J, Van Eck HJ, Finkers R, van der Linden G, Maliepaard CA, Uitdewilligen JGAML, Voorrips R, Vos P, Wolters AMA (2014) Possibilities and challenges of the potato genome sequence. Potato Res 3–4:327–330. https://doi.org/10.1007/s11540-015-9282-8

Vos PG, Uitdewilligen JGAML, Voorrips RE, Visser RGF, van Eck HJ (2015) Development and analysis of a 20K SNP array for potato (*Solanum tuberosum*): an insight into the breeding history. Theor Appl Genet 128:2387–2401. https://doi.org/10.1007/s00122-015-2593-y

Vos PG, Paulo MJ, Voorrips RE, Visser RGF, van Eck HJ, van Eeuwijk FA (2017) Evaluation of LD decay and various LD-decay estimators in simulated and SNP-array data of tetraploid potato. Theor Appl Genet 130(1):123–135. https://doi.org/10.1007/s00122-016-2798-8

de Vries M, ter Maat M, Lindhout P (2016) The potential of hybrid potato for East-Africa. Open Agric 1(1):151–156. https://doi.org/10.1515/opag-2016-0020

Walker TS (1994) Patterns and implications of varietal change in potatoes. Working paper 1994-3 Social Sciences Department, International Potato Center (CIP) Lima, Peru

Waltz E (2015) USDA approves next-generation GM potato. Nat Biotechnol 33:12–13

Wang S, Zhang S, Wang W, Xiong X, Meng F, Cui X (2015) Efficient targeted mutagenesis in potato by the CRISPR/Cas9 system. Plant Cell Rep 34(9):1473–1476. https://doi.org/10.1007/s00299-015-1816-7

Watanabe K, Arbizu C, Schmiediche PE (1992) Potato germplasm enhancement with disomic tetraploid *Solanum acaule* I. Efficiency of introgression. Genome 35(1):53–57. https://doi.org/10.1139/g92-009

Witek K, Jupe F, Witek AI, Baker D, Clark MD, Jones JD (2016) Accelerated cloning of a potato late blight–resistance gene using RenSeq and SMRT sequencing. Nat Biotechnol 34:656–660. https://doi.org/10.1038/nbt.3540

Ye M, Peng Z, Tang D, Yang Z, Li D, Xu Y, Huang S (2018) Generation of self-compatible diploid potato by knockout of *S-RNase*. Nat Plants 4:651–654. https://doi.org/10.1038/s41477-018-0218-6

Yin K, Gao C, Qiu JL (2017) Progress and prospects in plant genome editing. Nat Plants 3:17107. https://doi.org/10.1038/nplants.2017.107

Zhu S, Li Y, Vossen JH, Visser RG, Jacobsen E (2012) Functional stacking of three resistance genes against *Phytophthora infestans* in potato. Transgenic Res 21:89–99. https://doi.org/10.1007/s11248-011-9510-1

Zhu X, Gong H, He Q, Zeng Z, Busse JS, Jin W, Bethke PC, Jiang J (2016) Silencing of vacuolar invertase and asparagine synthetase genes and its impact on acrylamide formation of fried potato products. Plant Biotechnol J 14(2):709–718. https://doi.org/10.1111/pbi.12421

Chapter 6
Potato Breeding

Meredith W. Bonierbale, Walter R. Amoros, Elisa Salas, and Walter de Jong

Abstract The breeding of crop plants is a highly effective means of increasing agricultural productivity in a sustainable and environmentally safe way. Prebreeding and population improvement not only capture essential genetic resources and move desired traits along variety development pipelines but also help assure the creation of broad and dynamic gene pools to meet future, unanticipated needs. To efficiently meet multiple breeding objectives requires both interdisciplinary collaboration and a grasp of a wide range of scientific knowledge and expertise. This chapter addresses a range of topics that define and govern potato breeding, drawing from the experiences of both international and regional potato breeding programs, to orient readers to the interlinked components of population improvement and variety development. Using a case study approach to discuss breeding objectives together with respective implications for breeding needs, methods, and awareness-raising approaches for impact, we detail some key research and achievements contributing to current state of the art. Major populations under improvement at the International Potato Center along with breeding objectives and trait levels selected are described in terms of the agroecologies or uses they address in developing country national programs; these are contrasted with a discussion of the Cornell University program that is oriented to the northeastern US. A sample stage gate process, accelerated multi-trait selection schemes, heritability and heterosis exploitation, genomic selection, data management, and end user consultations are introduced in the contexts of these two programs. The topic of this chapter is supported and augmented with further details on subjects closely related to potato breeding, provided in chapters contributed to

M. W. Bonierbale
International Potato Center, Lima, Peru

Duquesa Business Centre, Manilva, Malaga, Spain

W. R. Amoros · E. Salas (✉)
International Potato Center, Lima, Peru
e-mail: CIP-Breeding1@cgiar.org; e.salas@cgiar.org

W. de Jong
Cornell University, Ithaca, NY, USA
e-mail: wsd2@cornell.edu

H. Campos, O. Ortiz (eds.), *The Potato Crop*,
https://doi.org/10.1007/978-3-030-28683-5_6

this volume by Ortiz and Mihovilovich, Ghislain and Douches, Burgos et al., and Ellis et al. The authors hope that the content serves to orient researchers and managers in countries with different degrees of development to plan and succeed in impactful potato improvement programs.

6.1 Implications of Genetics, Genepools, and Biology for Potato Breeding

6.1.1 Key Features

Commercial potato of world importance is a heterozygous, autotetraploid, clonal crop (2n = 4x = 48). Modern varieties are the products of extensive breeding between different cultivar groups and wild species. Potato varieties grown outside of South America since the end of the sixteenth century, as well as landrace (indigenous) cultivars grown in lowland Chile and in the high Andes are referred to as *Solanum tuberosum*, within which several groups are recognized (Spooner et al. 2014).

The genetic resources available for potato improvement comprise a polyploid series (2n = 2x = 24 to 2n = 6x = 72) with genetic features that facilitate gene transfer across ploidy levels. Farmers' landrace varieties dominate potato production in the Andean center of origin and diversity; as cultivars they offer tremendous trait diversity in readily useable form. Landrace cultivars, improved varieties and wild potato species in *Solanum* section *Petota* comprise GPs (gene pools) 1 and 2, according to the gene pool concept of Harlan and de Wet (1971) and are relatively straightforward to use in breeding. The wild tuber-bearing potatoes (section *Petota*) include some 200 species (see Spooner et al. 2014 for a review of taxonomic treatments). These typically produce only small tubers, which often contain high levels of toxic glycoalkaloids, and many require short days for tuberization. Domestication of wild potatoes for use as food likely involved selection for increased tuber size and reduced glycoalkaloid content, probably simultaneously in multiple locations in Andean and coastal regions of South America (Ugent et al. 1982, 1987).

Unreduced gametes, self-incompatibility, and inter-specific reproductive barriers played key roles in polyploidization and the maintenance of species boundaries during the evolution and domestication of potato, and knowledge of their genetics is useful in germplasm enhancement. Diploid potatoes are out-crossing due to a system of gametophytic self-incompatibility (Pushkarnath 1942; De Nettancourt 1977), which prevents inbreeding and thereby promotes intraspecific genetic variation. Tetraploid potato, on the other hand, is self-compatible. The breakdown of the gametophytic self-incompatibility system that operates in diploids is a common phenomenon in angiosperm polyploids (Frankel and Galun 1977; Levin 1983), but the molecular mechanism is not known (Comai 2005). Selfing results in severe inbreeding depression in most potatoes. Self-incompatibility at the diploid level, and inbreeding depression in both diploids and tetraploids, make it difficult to eliminate unfavorable alleles or drive favorable alleles to fixation. Several genes governing reproductive isolation, crossability, and ploidy are further discussed in the chapter contributed to this volume by Ortiz and Mihovilovic.

Potato varieties are maintained by clonal (vegetative, asexual) reproduction. Potato tubers are modified stems and comprise the vegetative "seed" used to propagate a variety. In potatoes with an even number of chromosomes (2x, 4x, 6x) it is possible to produce sexual (botanical) seed which provides for the generation of new genotypes. Once variation has been created in the form of sexual seed, any seedling has the potential to become a new variety via clonal propagation. Although most potatoes produce a large number of botanical seeds per fruit (~200) as well as many fruits per plant, clonal propagation results in a low propagation coefficient (five to tenfold per generation). Clonal breeding facilitates intentional and unintentional exposure of candidate varieties to pathogens. This helps breeders eliminate undesirable genotypes, but also necessitates steps to reduce exposure to detrimental viruses that are transmitted through vegetative (tuber) seed. In common with other root and tuber crops the limited ability to phenotype and potato for desirable morphological or developmental features is challenged by the underground location of the harvested product.

6.1.2 Genome Constitution and Variation

Maximum heterozygosity has been considered essential for performance of tetraploid potato, with inbreeding leading to reduced vigor and yield, flower bud abortion, lack of flower bud formation, and sterility (De Jong and Rowe 1971; Mendiburu and Peloquin 1977). However, empirical research with hybrid families has suggested that poor performance may be due to the expression of recessive alleles (De Jong and Rowe 1971) or that the presence of certain alleles may be more important for high yield (Bonierbale et al. 1993). Genomic studies have revealed that tri-allelic and tetra-allelic single nucleotide polymorphisms (SNPs) are rare in potato cultivars (Hirsch et al. 2013; 2014), though these might be expected in outstanding clones if yield or vigor were associated with maximum heterozygosity. Nevertheless, when the products of individual genes are amplified by PCR (polymerase chain reaction) it is not uncommon to find three alleles at a locus.

Xu et al. (2011) proposed heterozygosity as the key feature enabling the frequent occurrence of gene presence/absence variants and other potentially deleterious mutations in the genome of the heterozygous diploid potato clone "RH." Next-generation (short read length) sequencing of 807 genes from 83 potato cultivars revealed a tremendous amount of genetic variation in potato. On average, there is one variant (SNP or indel) every 24 base pair (bp) in exons, and one variant every 15 bp in introns (Uitdewilligen et al. 2013). The average minor allele frequency of a variant is low, though, at 0.14, and 61% of variants have minor allele frequencies less than 0.05 (Uitdewilligen et al. 2013). Given the sequence variation, and keeping in mind that tetraploid potato can contain up to four alleles at each locus, it is easy to understand why so much phenotypic variation results every time two heterozygous potato clones are crossed. Considerable effects due to dominance and epistasis are possible, and experience with such out-crossing polyploids has shown that the genetic variation due to dominance and epistatic effects is large compared to that seen in diploid crops (Gruneberg et al. 2009). This suggests that heterosis largely determines the performance of out-crossing, clonally propagated crops, although its basis in potato has not been fully elucidated.

6.1.3 The Cost of Increasing Genetic Variation

Most plant breeding training programs place heavy emphasis on the need to increase genetic variation. There is value in this, of course, as wild species contain many useful traits, genes/alleles not present in modern cultivars. But there is also a considerable cost that is not appreciated by those unfamiliar with potato breeding. Wild potato species contain countless alleles that are undesirable for potato production, and when a cultivated potato is crossed with a wild accession, oftentimes the offspring is quite poorly adapted. Many further generations of crossing are needed to eliminate the undesirable alleles, while keeping the desired ones. When embarking on such a venture it is important to realize upfront that the process typically takes decades. It may be possible to use molecular markers to speed the process up, by selecting against the donor genome (and for the gene of interest) after each cycle of crossing, but the process will still be much slower than adding a gene to cultivated potato by *Agrobacterium*-mediated transformation, or by editing an existing cultivated allele to a desired wild species allele with CRISPR-Cas9 (https://www.yourgenome.org/facts/what-is-crispr-cas9), once the respective target genes are known. It is worth noting that the "adaptation gap" between wild species germplasm and cultivated germplasm is growing over time, as breeders continually work to improve cultivated germplasm by increasing the frequency of desirable alleles and decreasing the frequency of undesirable ones. Despite the challenges, support for germplasm enhancement programs that strategically bridge this gap and enable the continuous influx of valuable genes from crop wild relatives through improved populations and into varieties is critical to meeting the world's growing need for food in the face of climate change.

6.1.4 Genetic Enhancement

Potato breeding strategies frequently include research to efficiently access traits from beyond the variety-ready germplasm base. Such trait research or pre-breeding is expected to result in new materials, methods, tools, knowledge, and approaches to support the breeding process rather than in finished products or varieties. Genetic and biochemical research often provides insights and tools that enable gains in traits toward new breeding objectives or improved program results. Pre-breeding is conducted in parallel with mainstream breeding activities, such as by developing support populations that are upgraded for trait levels or improved for agronomic traits so that new types of diversity can be introduced from un-adapted to adapted germplasm without impeding advance toward established breeding goals.

Before embarking on breeding for new traits or using uncharacterized germplasm, it is important to assess and consider positive and negative correlations among traits. Trait correlation influences the success of cross combinations and can determine breeding progress when multiple traits are concerned. Negative trait associations can be critical in achieving breeding progress and influence the choice of parents as well as the selection approach that will be most successful. Embarking on medium- to long-term population development does not result in new varieties after a single recombination and selection cycle, but should result in better parents that will help meet the medium- to long-term objectives of breeding programs.

Support populations are useful for enhancing diversity from un-adapted germplasm to avoid introducing undesirable features into advanced breeding populations. In the case of potato, wild and landrace relatives often carry undesired agro-morphological traits like deep eyes, small tuber size and late maturity, a requirement for short days, or high glycoalkaloid content, in addition to untapped resistance to biotic and abiotic factors and nutritional traits. Those undesired traits may be eliminated from hybrid populations by backcrossing to improved types, or source populations may be enhanced for agronomic traits before desired traits are transferred to improved genepools. The large majority of potato's genetic resources are diploid, and breeding at this level results in faster genetic gains than breeding at the tetraploid level. Thus, particularly for multigenic traits, pre-breeding in diploid source germplasm before incorporating new traits into tetraploid breeding populations can be very effective.

Introgressing novel traits from distant wild species such as those in the tertiary gene pool (GP3) may require the use of bridge species to circumvent interspecific reproductive barriers, and unreduced gametes to transfer traits across ploidy levels. When interspecific crossing is possible, backcross schemes are modified from those used for inbred crops, such that a different genotype of the recipient germplasm (adapted type) is used in each cycle of crossing with the trait donor or selected hybrid to avoid inbreeding depression. This is illustrated by Gaiero et al. (2017) who introgressed partial resistance to bacterial wilt, caused by *Ralstonia solanacearum*, from the sexually incompatible GP3 species *S. commersonii* by bridge crosses with 2x *S. tuberosum* Phureja Group and successive backcrosses with different *S. tuberosum* Group Tuberosum genotypes. The resulting advanced backcross progenies are now being used by several breeding programs due to the acute need for resistance to bacterial wilt.

6.1.5 Case Study 1: Genetic Enhancement and Incorporation of Iron Content from Diploid into Tetraploid Cultivated Potatoes

With support from HarvestPlus (www.harvestplus.org) advocating the breeding of staple crops for micronutrient density, CIP has sought to increase the content of iron and zinc in potato through an inter-ploid breeding strategy. Recurrent selection in a base population of landrace potatoes took advantage of greater response to selection at the diploid level, and resulted in a population of diploid potato reaching 35 ppm iron and 30 ppm zinc from a baseline of 20 and 16 ppm, respectively. Elevated iron and zinc contents achieved in the source germplasm were incorporated into advanced, tetraploid populations via unreduced gametes. The resulting tetraploid potato population had iron and zinc concentrations twofold higher than baseline levels, which approaches the breeding targets for human populations with deficiencies of these minerals and high potato intake (Bonierbale et al. 2007; Section 6.3.4). Further work remains to be done to identify clones adapted to target agro-ecologies that maintain as favorable a package of traits as possible, including new levels of iron or zinc concentrations in resilient, consumer-accepted table potatoes.

Analyzing biofortified populations under recurrent selection in diploid potato germplasm, Paget et al. (2014) found moderate to high and positive correlation between iron (Fe) and zinc (Zn) contents from cycle I ($r^2 = 0.45$) and from cycle II ($r^2 = 0.72$), indicating that evaluation and selection for one of these traits will result in concomitant increase in the other. Negative genetic correlations were found between dry matter and Fe, Zn, Ca, and vitamin C contents (genetic correlation close to zero for vitamin C in Cycle 1) when analyzed on a dry-weight basis. In contrast, the same genetic correlation estimates were positive (but small) when analyzed on a fresh-weight basis (Table 6.1).

The genetic correlations in this example are strong enough that under multi-trait selection the breeding population is improved simultaneously for iron and zinc concentrations (Fig. 6.1a, b), whereas the average tuber weight of the population decreases, and the number of tubers per plant increases (Fig. 6.1c, d). A reduction

Table 6.1 Additive genetic correlations (are in bold) and confident intervals (are in brackets) for Fe, Zn, Ca, vitamin C, and tuber dry matter content from a multivariate analysis of Cycle1 data (Model 4) estimated on a dry and fresh-weight basis

Trait	Iron	Zinc	Calcium	Vitamin C
Cycle I (dry weight basis)				
Zinc	**0.45** [0.32 0.64]			
Calcium	**0.04** [−0.23 0.34]	**0.12** [−0.15 0.39]		
Vitamin C	**−0.01** [−0.18 0.29]	**0.10** [−0.15 0.30]	**0.05** [−0.27 0.33]	
Dry matter	**−0.23** [−0.42 −0.06]	**−0.24** [−0.41 −0.07]	**−0.19** [−0.36 0.07]	**−0.06** [−0.28 0.10]
Cycle II (dry weight basis)				
Zinc	**0.72** [0.42 0.88]			
Calcium	**0.35** [−0.04 0.61]	**0.57** [0.18 0.76]		
Dry matter	**−0.34** [−0.61 0.08]	**−0.38** [−0.66 0.10]	**−0.14** [−0.49 0.20]	
Cycle II (fresh weight basis)				
Zinc	**0.61** [0.33 0.84]			
Calcium	**0.07** [−0.32 0.52]	**0.45** [−0.02 0.77]		
Dry matter	**0.18** [−0.13 0.36]	**0.14** [−0.13 0.38]	**0.05** [−0.23 0.27]	

From: Paget et al. (2014)

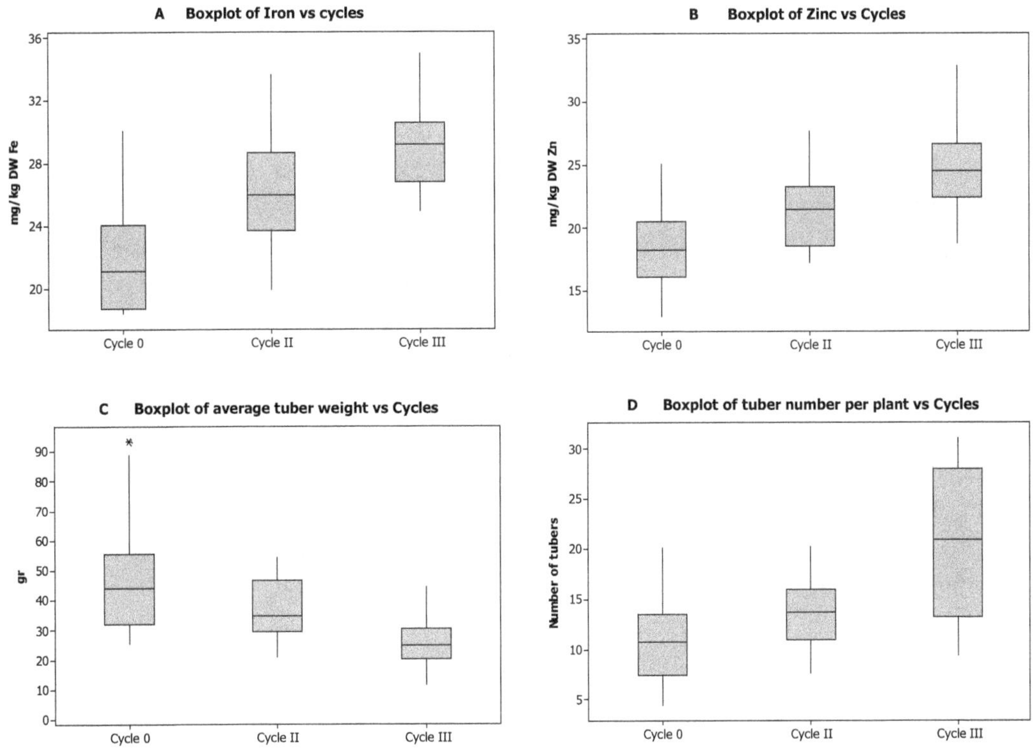

Fig. 6.1 Box plots of ranges of trait variation through three cycles of recurrent selection in diploid potato germplasm. (**a**) Iron concentration; (**b**) zinc concentration; (**c**) average tuber weight (reflecting size); (**d**) Tuber number per plant

in tuber size was encountered during population improvement at the diploid level even though it was considered in selection. This loss was mitigated, however, by returning to the tetraploid level via 4x–2x crosses by which the gains in the mineral contents realized at the 2x level were transferred to advanced populations.

6.2 Principles of Potato Breeding Methods and Approaches

The principal advantage of breeding clonally propagated crops is that each clonal variety is fixed and simple to maintain. Genetic purity is less of an issue in vegetatively—than in sexually—propagated crops. One substantial disadvantage of vegetative propagation though is that diseases are easily transmitted across clonal generations during propagation; another is that potato planting material is bulky and perishable by nature and the production of healthy material is expensive.

The single most challenging aspect of potato breeding is the identification of superior individuals that combine as many high priority traits, and as few weaknesses, as possible from a given cohort of F1 progeny in a reasonable time frame. Additional important challenges include the improvement of support populations, and the selection or construction of parents. The parents used for crossing are highly heterozygous—exceptions are inbreeding lines generated by self-fertilization or doubled-monoploid production. Each potato seed that results from a heterozygous × heterozygous cross differs, at many loci, from any other seed from the same cross. The resulting heterozygous genotypes are subject to selection after being fixed (stabilized in genetic terms) by clonal propagation.

Polyploidy, heterozygosity, and heterosis make the identification of good parents particularly challenging. At present, the performance of a parent can only be determined after the fact, that is, by looking at its progeny. Advances in genomic selection may make it possible to identify good parents in advance. Heterozygous tetraploid potato genotypes harbor great allelic diversity and interactions that are responsible for their performance as clones. As parental clones are not inbred, a genotype can never be reconstituted after sexual crossing or self-pollination. A practical consequence of this for potato breeding is that many, many, traits segregate in the progeny of any parent or parental combination. Directional breeding results in incremental changes in gene frequencies among progenies.

In the course of potato breeding an "F1 clone hybrid" is generally crossed with another "F1 clone hybrid," and the progeny is heterogeneous and has an extreme large segregation variance. A good parent generates a large genetic variation around a high family mean for a given trait. Once heterogeneous and heterozygous progenies are generated by crossing two potato clones, selection of clones within a given pool of genetic variation for variety development is conceptually, if not technically, straightforward. All the genetic advantages of clonally propagated crops can be used for variety development, and the genotype that will finally be released is among the progeny immediately after the initial crossing.

Inter-group crosses are important in population improvement of clonally propagated crops. Gruneberg et al. (2009) have suggested that this aspect of clone breeding is often neglected and may be the reason for low breeding progress in many clonally propagated crops when compared to the improvement of sexually reproducing ones. Complementary germplasm groups have not been identified in potato but strategies to maximize heterozygosity have been proposed to increase yield in tetraploid potatoes (e.g., Chase 1963). However, it is unlikely that the direct relationship of maximum heterozygosity and yield will extend to crosses involving unadapted germplasm (Bonierbale et al. 1993). Thresholds for heterosis were suggested upon finding of increased yield and vigor in two-way hybrids (cultivated × wild species) with respect to crosses within the cultivated genepool, but no additional increments in hybrids involving cultivated potato and two wild species (3-way hybrids) (Sanford and Hanneman 1982). Thus, it is unlikely that heterotic groups could be established in potato germplasm on the basis of genetic distance measures or maximum heterozygosity alone.

6.2.1 Population Improvement

Medium- to long-term genetic gain can be achieved over sequential cycles of crossing and selection. Recurrent selection is defined as reselection generation after generation, with inter mating of selected plants to produce the recombinant population for the next cycle of selection. The goal of recurrent selection is to improve the mean performance of a population of plants; a secondary goal, but nevertheless also important, is to maintain as much genetic variability as possible. Open recurrent selection is a method for improving the mean performance of a population while maintaining and increasing genetic variability by periodically introducing new sources of traits under selection. In genetic enhancement a few cycles of recombination facilitate the breakdown of linkage blocks so that desired traits from unadapted germplasm can be carried forward and undesirable ones left behind, minimizing the effects of linkage drag thus shortening the time required to introgress or incorporate new diversity into advanced populations. Depending on the stage of a population under improvement, selections may be considered for use as varieties, or as parents that will contribute to further gains toward complex breeding objectives. Figure 6.2

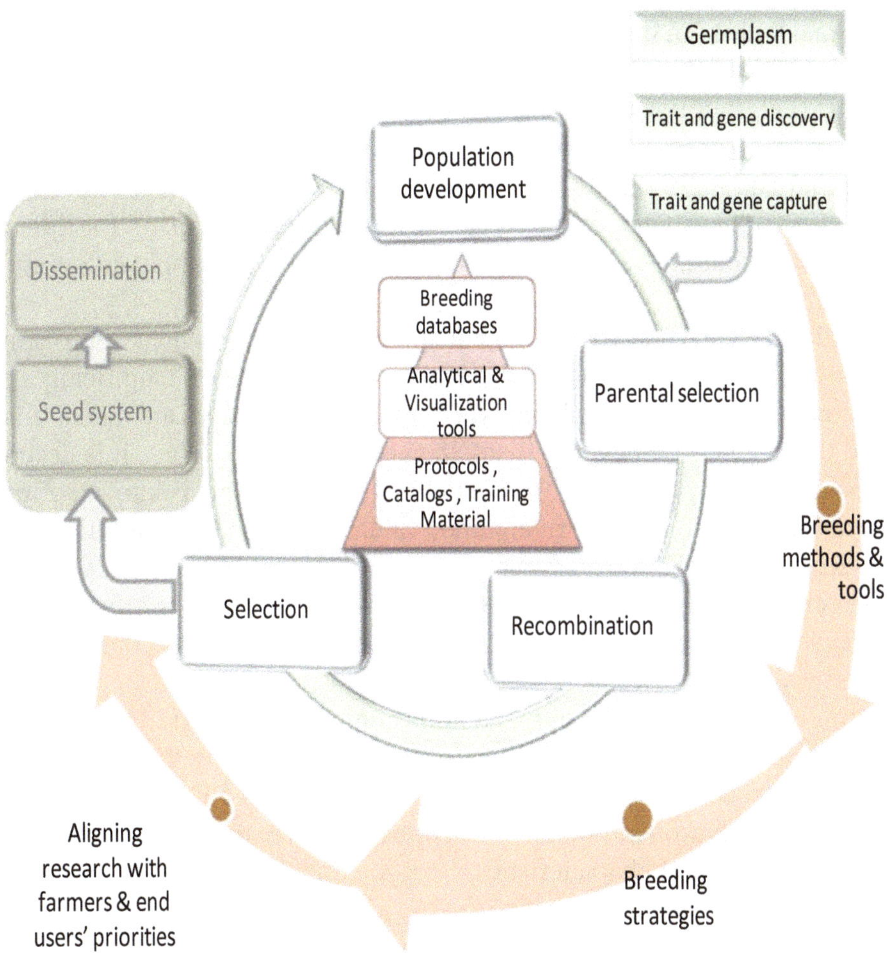

Fig. 6.2 Schematic representation of components of a potato breeding program. (Courtesy of Elisa Mihovilovich)

depicts simultaneous and integrated practices of recurrent selection, genetic enhancement, and variety selection in a potato breeding program.

In polyploid potato, more than one allele per locus can be transferred to the next generation in gametes, and thus, in contrast to diploids, the genetic variation due to dominance influences the response to selection in population improvement as long as the population is not in equilibrium. In tetraploid potato, a population is usually not in equilibrium after recombining parental material in controlled crossings, and 1/3 of the dominance variance is exploitable for selection progress when selection takes place on the female and male side [for further details see Wricke and Weber (1986) and Gallais (2004). The exploitation of the dominance variance in population improvement, in combination with faster genetic gains in diploids than in tetraploids, and the inheritance of 2n gametes as described by Ortiz (1998) provide great diversity and flexibility for potato improvement.

6.2.2 Crossing Parents

The choice of parents is an important step in any breeding program, especially for medium- and long-term breeding progress. The identification and number of crosses to perform at a given breeding stage is a factor of the knowledge available on potential parents (i.e., trait constitution and breeding value), the breeding objective, and facilities and resources available for crossing and selection. As mentioned above, the polyploid and highly heterozygous nature of potato mean that dominance and epistatic effects contribute considerably to clone performance, and for this reason, little is known about the value of a parent or specific cross combination until it has been tested.

Parental value can be assessed strategically through the conduct of progeny tests in appropriate breeding designs, the evaluation of pedigrees, or empirically through observation of selection ratios. Even when hundreds of crosses are made, it is often observed that the best clones trace back to very few crosses. Thus, it is desirable to predict which genotypes will be the best parents, since if this is known, efforts can be concentrated on the generation and the evaluation of the most promising combinations. Regardless of how much information is available to support the choice of parents, a practical approach is to sow around 200 seeds of every cross to be evaluated, and then sow more seed or emphasize the parents of those combinations that performed well in future years.

Parental value is a function of the genetic constitution of a trait donor. Complex traits are often comprised of several components, and defining those that contribute reliably to influence trait values is a prerequisite to successful identification of parents. Accurate evaluation of progeny is the most effective means to identify superior progenitors of inherited characteristics. Sprague and Tatum (1941) introduced the concepts of general combining ability (GCA) and specific combining ability (SCA) to distinguish between the average performance of parents in crosses (GCA) and the deviation of individual crosses from the average of crosses (SCA).

Studies of combining ability make it possible to identify parents (those with good GCA) that will perform well in most crosses, and also facilitate the development of superior hybrids through the use of parents with high SCA. Genetic analyses to measure combining ability further allow breeders to estimate genetic effects and parameters such as trait variance, covariance, correlations among traits, heritability, and the relative importance of additive and dominance variation. GCA represents mainly additive and additive × additive types of genetic variance. Thus, when a potato clone is selected as having good GCA, it means that the algebraic sum of the additive and additive × additive epistasis effects it passed on to its offspring produced a favorable result in excess of the average of all the offspring tested. GCA has a conceptual implication that each line being evaluated is tested against a large random sample of some specified population (Plaisted et al. 1962). In actual tests for GCA, however, the testers (usually males) often represent only a limited sample of a population.

6.2.3 Mating Designs

The term "mating design" refers to the mating of parents in a systematic plan of crosses to determine genetic parameters and/or parental value. These procedures are particularly useful for the identification of plants that will be the most effective parents in pre-breeding or recurrent selection programs. However, systematic mattings can be difficult to achieve because of sterility and incompatibility encountered during crossing. Mating designs have been classified according to the number of factors to be analyzed, the parents, and modalities of combination. Some mating designs are used more extensively than others, but each has its advantages and disadvantages depending on the reference population under consideration and the information desired. Following are descriptions of some of the most important designs for identifying superior progenitors to help assure genetic gains in a potato breeding program. Figure 6.3a–c, illustrates some of these mating designs.

Diallel Analysis is the best way to determine the combining ability of parents. It consists of the analysis of a set of crosses produced involving "*n*" lines in all possible combinations, a so-called diallel cross. The diallel mating design has been used extensively in potato germplasm enhancement and can be very useful if properly analyzed and interpreted. The analysis of diallels provides information on GCA and SCA of parents and their crosses and makes it possible to determine if reciprocal crosses give equivalent results.

The most commonly used *methods of diallel designs* are those proposed by:

- Griffing (1956), by which general and specific combining abilities are estimated;
- Gardner and Heberhart (1966), by which the variety and heterosis are evaluated; and
- Hayman (1954), in which information regarding additive and dominance effects for a characteristic and the genetic values of the parents is used.

Fig. 6.3 Schematic representation of different mating designs in which parent lines are indicated with numbers, and the cross combinations between them are represented with checkmarks in shaded cells. (**a**) Partial dialel, (**b**) Line × tester, and (**c**) Design II

A

Partial diallel									
		Males							
		1	2	3	4	5	6	7	8
Females	1				√	√	√		
	2					√	√	√	
	3						√	√	√
	4							√	√
	5								√
	6								
	7								
	8								

B

Line x Tester				
		Testers		
		1	2	3
Lines	1	√	√	√
	2	√	√	√
	3	√	√	√
	4	√	√	√
	5	√	√	√
	6	√	√	√
	7	√	√	√

C

Design II					
		Males			
		5	6	7	8
Females	1	√	√	√	√
	2	√	√	√	√
	3	√	√	√	√
	4	√	√	√	√

One of the most used is *Griffing's Method 2*, which estimates GCA and SCA) relating mainly to additive and non-additive gene effects (dominance and epistasis).

6.2.3.1 Partial Diallel

Analyses involving "*n*" parents in all possible combinations become unmanageable as the number of lines (*n*) increases. On the other hand, if only a small number of parents are tested, the estimates of combining ability tend to have a large sampling error. These difficulties led to the development of sampling crosses produced by large numbers of parents, without affecting the efficiency of the diallel technique. In a normal diallel, each line is involved in ($n - 1$) crosses. Kempthorne and Curnow (1961) presented the concept of the partial diallel design in which only a random sample of crosses, say of size'*s*, is analyzed where "*s*" is less than $n - 1$.

6.2.3.2 Line × Tester Design

The Line × Tester Design also provides information about the GCA and SCA of parents, and is also helpful in estimating various types of gene effects. The *crossing plan* of this design is as follows:

"*l*" lines are crossed to "*t*" testers so that $l \times t$ full-sib progenies are produced;

These progenies, with or without parents (i.e., testers and lines), are then tested in a replicated trial using a suitable experimental design, say randomized block design.

If there are three testers and seven lines, there are 7 × 3 = 21 crosses. For evaluation, the 21 crosses along with 10 parents, for a total of 31 entries, might be tested in a randomized complete block design with four replications. For this case, uniform planting material must be produced to enable inclusion of parental clones and progeny in the same trial.

6.2.3.3 Design II

In mating Design II (or Factorial Design), described by Comstock and Robinson (1948), the genetic information is similar to that obtained with Diallel Analysis.

Different sets of parents are used as males and females. If a set of eight parents is included in the design II, 16 crosses will be obtained. This design is advantageous when not all clones to be tested are male or female fertile.

6.2.4 Breeding Values

The estimated breeding value (EBV) of an individual can be calculated on the basis of pedigrees and performance in the course of a breeding program, and does not rely on the conduct of mating designs. Using appropriate statistical analysis, breeding value predicts how useful each individual would be as a parent; it expresses the ability of a parent to pass on superior trait levels to its offspring and is used for ranking breeding performance of an individual as a parent relative to the population average. The use of information on the individual and all relatives greatly increases the accuracy of selection in a breeding program (Lynch and Walsh 1998).

The calculation of breeding value goes beyond the typical estimation of genotypic or parental values with models based on fixed effects, by enabling the estimation of random effects of a mixed model. Mixed linear models are able to model different covariance structures and thus provide an improved representation of the underlying random and error components of variance (Oakey et al. 2007). The application of mixed models to estimate breeding values uses pedigree information to model and exploit genetic correlation among relatives and applies flexible variance–covariance structures for genotype-by-environment interaction to accurately predict performance (Piepho et al. 2008). The improved accuracy afforded by the use of all data in a breeding program with mixed models allows the analysis of repeated measures, unbalanced design experiments, spatial data, and multi-environment trials.

Breeding value has been successfully applied in several crops (sugarcane, eucalyptus, soybean, maize, and even potato). Slater et al. (2014) illustrate selection in potato based on breeding values. They conclude that using best linear unbiased prediction (BLUP) and pedigree to estimate breeding values can result in increased genetic gains for low heritability traits in auto tetraploid potato. Theory and applicability of breeding values in quantitative trait improvement are illustrated in Bernardo (2002) Relevant statistical packages that fit linear mixed models to large data sets using the Residual Maximum Likelihood (REML), approach, such as the Asreml-R reference manual (Butler et al. 2009) provide powerful software for the use of breeding values in plant breeding.

In animal breeding, EBVs are the basis for marketing breeding parents and they provide breeders with critical information for selection decisions. In applying EBVs, it is important to achieve a balance between the different groups of traits and to place emphasis on those traits that are important to the objective population, markets, and environment. It is not feasible, nor always desirable to seek high EBVs for all traits in a single progenitor. In fact, a comprehensive range of EBVs has the advantage that it is possible to avoid extremes in particular traits and select for animals/genotypes with balanced overall performance. The method is particularly useful in non-inbred populations and potato breeders should pay careful attention to this analytical approach.

6.2.5 *Early Versus Late Generation Selection*

In early generations it is possible to select for highly heritable traits for which accurate assessment of a genotype can be carried out on one or a few plants, but for more genetically complex traits, and for traits where interplant competition is an issue, it is necessary to evaluate clones in multi-plant plots planted as blocks under homogenous field conditions. The aim is an unbiased comparison of genotypes within blocks. The number of plants per plot and the number of replications or blocks depends on the breeding stage. The low propagation coefficient of potato (about ten depending on the propagation method used) limits the amount of planting material available at each stage of selection and is one reason that potato breeding is relatively slow (at least 8–10 years and usually more, from crossing until variety release). Large numbers of genotypes can be assessed for simple traits in early stages when small amounts of seeds are available; while more complex traits are assessed in later stages when larger quantities of seeds are needed for replicated trials, though less genetic diversity is represented.

Multistage selection can be managed in subsequent steps from early to later stages, or with indices that use a weight assigned to each trait. In practice, some characters are selected sequentially especially where there is clearly a lowest acceptable value (tuber size, shape, and color as well as pest and disease resistance), while others are selected simultaneously by aggregating characters into an index (often an intuitively formed index such as score values of overall performance.

6.2.6 *Case Study 2: The Use of a Selection Index in Potato Breeding*

An early-generation (seedling stage) selection method was applied at CIP for the identification of families and individuals that tuberize well under long days and high temperatures. Five groups of families generated by intercrossing CIP's best advanced parents from two populations and long day-adapted varieties underwent a greenhouse test during summer, applying two photoperiods: natural short days of 12 h and simulated long days using lamps to extend the photoperiod to 16 h. Four plant morphology prototypes were identified according to patterns of above- and below-ground growth including branching, stolon, and tuber formation (Fig. 6.4). Individual plants of each family were evaluated at harvest taking into account parameters including breeder's preference, a tuberization score based on the four patterns, tuber uniformity and physiological disorders including sprouting tubers, knobbiness, chain tubers, and cracking.

A selection index (SI) was built assigning a weight to each of these parameters, three for breeders' preference (BP), two for tuberization score (TS) and tuber uniformity (TU), and one for physiological disorders: sprouting (Sp), knobbiness (Kn), chained tubers (Ch), and cracks (Cr)

$$SI_n = BP_n(3) + TS_n(2) + TU_n(2) + Sp_n + Kn_n + Ch_n + Cr_n$$

Families with a high frequency of progenies exceeding an estimated selection index of 54 were selected. Seven of these selected families and three unselected were used in validation studies under field conditions.

Fig. 6.4 Plant morphology and growth patterns observed among families in early generation screening for adaptation to warm, long day environments. (**a**) Typical morphology under 12:12 h light:dark regime; (**b**) Typical morphology under 16:8 h light:dark regime; (**c**) Above-ground morphology of representative plants with growth patterns I, II, III, and IV; Below-ground morphology of representative plants with growth patterns I, II, III, and IV

6.2.7 Stability and Adaptation

Breeding programs typically breed for several locations rather than just one. Hence field evaluations that underpin selection must simulate a range of environments. For this reason, within limits of the propagation coefficient, clones are tested in plots within homogenous blocks at several locations over several years. The objective of multilocation trials is to assess promising genotypes with respect to narrow and broad adaptation. In practical potato breeding, "broad adaptation," the ability to perform well in a range of environments, appears essential for a clone to have any meaningful commercial success.

The wide range of quality preferences and the numerous pests and diseases addressed by potato breeding programs dictates the exposures required for selection. Decentralized selection is generally required by a program with multiple target environments and is best designed to address sets of traits that can logistically be assessed together in target environments or sites that represent them well. Along the course of selection, a combination of on-station and on-farm trials are usually performed.

National programs requesting potato breeding materials from CIP are provided with germplasm from different selection stages depending on their capacity to evaluate clones and the relative suitability of advanced germplasm to their production targets. Advanced clones and parents are available as in vitro plantlets. The genotypes provided meet multiple selection criteria for the given target environment, as assessed in similar testing sites. Upon receipt they are propagated and subject to multilocational trials before use as parents in national programs or testing as varieties.

6.2.8 Case Study 3: Breeding and Variety Development in Bangladesh

Breeding for stable yields and multiple disease resistance at CIP takes advantage of germplasm collections, broad-based advanced populations, and environmental diversity in Peru. One such population is the advanced lowland tropics virus resistant population, which is adapted to dry arid regions where virus pressure is high. With support from GIZ/BMZ and follow-up support from USAID, CIP provided 35 potato breeding lines of this population as in vitro plants to the Tuber Crop Research Centre of the Bangladesh Agricultural Research Institute in 2009. After local evaluation over a period of 7 years, a salt tolerant variety (BARI Alu-72) and a heat tolerant variety (BARI Alu-73) adapted to Bangladesh were released.

When distant environments can only be partially simulated in CIP's breeding program, samples of true seed families (TSF) from good parents can be distributed. TSF are available for national programs with the capacity to carry out early generation selection as well as variety identification trials. One advantage in working with TSF is the considerable reduction in time-to-release compared to that required when receiving advanced clones. Conducting selection with true seed in target environments can save 6 years or more, i.e., time that would otherwise be required for CIP breeders to conduct selection from the same TSF at their trial sites and recommend elite clones as in vitro plants.

When national programs test CIP germplasm as TSF, valuable information is returned to CIP regarding family performance (i.e., frequency of selected clones in each family), and this serves to refine the concept of adaptation and best bet materials for that location. National programs that invest in identifying parents based on local performance, and then utilize these parents for crossing and recurrent selection, benefit from knowing which breeding materials will serve those best and are likely to experience significantly faster genetic gains over the medium to long term.

6.2.9 Case Study 4: Collaborative Breeding in Vietnam

Advanced germplasm from CIP's lowland tropic virus resistant (LTVR) and late blight heat tolerant (LBHT) breeding populations developed in Peru were dispatched to Vietnam in the form of in vitro plants from 2005 to 2010. The materials were used in two ways: (i) for direct variety testing, and (ii) in cross-breeding to improve adaptation to local conditions.

Results of the collaboration demonstrated both the direct utility of clones from Peru in Vietnam—through the identification of two outstanding clones recommended for variety testing, and the advantage of local breeding, yielding, after three rapid cycles of recurrent selection, a new generation of elite clones with higher yields than the elite clones provided from Peru. The yield increase from local breeding and selection was realized in a considerably less time (9 years) than would have been required for the development, introduction, and testing of a new cycle of elite clones from CIP (at least 12 years) (Fig. 6.5a, b).

There are several advantages of testing early generation potato progenies at more than one location. Information from contrasting environments can be combined if the breeder tests a clone at two or more locations; those few clones that are broadly adapted can be identified early in the selection process. In addition, if one location experiences low disease pressure, daughter tubers from that site can be used to provide clean planting material for the next generation of selection. The healthy seed plot provides for an assessment of performance in the absence of major production constraints, while various pressures can be applied at additional sites. Alternatively, a protected environment (e.g., quarantine screen house) can be used to maintain a healthy copy of each selected genotype, but unlike the maintenance of a healthy

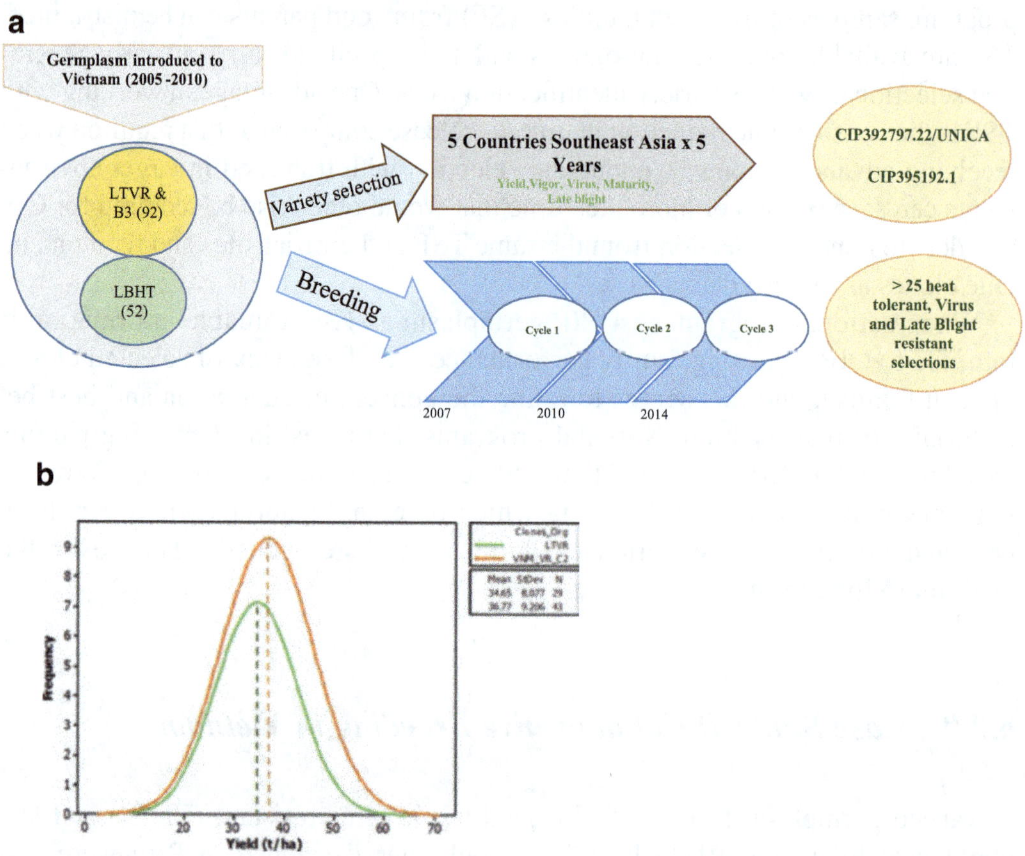

Fig. 6.5 Collaborative approach for potato breeding in Vietnam. (**a**) Two avenues for decentralized evaluation and improvement; (**b**) Frequency distributions for yield of selected potato clones in three cycles of recurrent selection conducted in Vietnam

seed plot, this typically does not allow for significant production of seed to support evaluation along the breeding cycle.

Rapid and real-time exchange of information on selection percentage between breeders and selectors helps inform breeders on the future crossing plans most likely to meet the requestor's needs. This simple data can confirm the breeders' concept of GCA and suitability of test environments with respect to the targeted ones. Breeders can use the information returned from selectors regarding percentage of selection in controlled cross families to exploit heterosis through cross combinations. Selection of parents and cross prediction can probably be improved by systematic collection of data on the percent of clones selected in families by the assignment of parents with good GCA or mutual SCA to complementary gene pools. This would benefit both CIP's and the national programs' breeding efforts, if it resulted in increased frequency of high-yielding clones with needed traits, or potentially, in yield jumps.

6.2.10 Importance and Relation of GCA and Heterosis in Potato

Heterosis was examined in a new tetraploid potato hybrid population obtained by crossing members of two advanced breeding populations developed at CIP for adaptation to the highlands (B3) and subtropical lowlands (LTVR). Significant positive heterotic values ranged from 18 to 60% for mid-parent heterosis and from 16 to 162% for best parent heterosis or heterobeltiosis. Positive heterotic values were found for tuber yield and tuber number in each environment. Significant GxE interaction on heterosis showed a differential effect of the environment on the magnitude of heterosis expression.

The expression of heterobeltiosis within B3 and LTVR confirms the effectiveness of the population breeding method on maintaining a broad genetic base (Mendoza and Sawyer 1985). Mid-parent heterosis can be the result of the combination of good levels of GCA of parents and some level of SCA of the cross. A reciprocal recurrent selection scheme (RRS) has been initiated to benefit from intercrossing between best parents of two complementary populations, B3 and LTVR (RRS), also known as recurrent reciprocal half sib selection, is a form of recurrent selection used to improve both GCA and SCA of a population for a character or characters.

6.3 Potato Breeding Procedures Overview

6.3.1 Hybridization

Heterozygous parents are recombined in controlled biparental combinations or by use of bulk pollen to create new variation in the form of sexual seed. The crossing block is established after selecting the parents to be used as females and as males, and the "planting on a brick method" or modifications of it can be practiced to encourage flowering and fruit set over a period of about 10 weeks. In the "brick method," potato tubers are placed on bricks partially buried in the soil. The tubers sprout and stems are allowed to grow. After some time the soil around the base of the full-size plant is washed away with a spray of water, which exposes emerging stolons and small tubers, but does not disrupt the root system that has penetrated in and around the brick. Thereafter, stolons and tuber that emerge from the base of the plant are removed. Potato has a complete flower with five anthers and a stigma that becomes receptive as pollen is shed from open flowers. Emasculation is practiced to prevent selfing when genetic studies will be performed, when parental value will be assessed in progeny tests and for best control of pedigrees (record-keeping).

Crossing at CIP is enhanced by flowering induction. Under long day conditions most potato clones flower to some extent. Typically, *S. tuberosum Phureja* Group flowers under both long and short-day photoperiods. Clones of Group *tuberosum*, however, usually will not flower under short days. In addition, there are many clones, especially those of early maturity, that rarely flower under any conditions or that flower sparsely and over a very short time period. For these reasons it is often desirable to induce flowering. This can be done by either of two methods: "grafting" or "planting on a brick." In the grafting method shoots of potato are grafted onto the stems of tomato plants. In both methods the idea is to prevent potato from developing beyond the vegetative flowering phase to the tuberization-senescense phase. In addition to these mechanical methods, other traditional flowering-enhancing practices may be followed such as long-day lighting, sprays of 40 ppm gibberellic acid at 4-day intervals, and heavy nitrogen fertilization. Temperature and humidity are also important. Ideal temperatures are 20 °C day and 16 °C night. Humidity should be 80% or higher. Procedures and crossing techniques for potato breeding as practiced at CIP are documented at https://research.cip.cgiar.org/potatoknowledge/proceduretechniques.php. A modification of "planting on a brick" using peat pots is shown. The open source software CIPCROSS (see Section 6.3.3) can be used to document all aspects of the crossing block, including pedigrees, storage, and inventory using bar coding for accuracy and efficiency.

6.3.2 Selection Schemes

Selection of potato varieties from botanical seed is conducted in several steps. Figure 6.6 shows how the crossing of two parents might be followed by five sequential steps in time (one selection step conducted with seedling plants and four subsequent selection steps conducted with cloned plants from tubers of the seedling genotypes). This scheme is an oversimplification since many crosses or families are developed simultaneously and many seeds of each family are subject to evaluation and selection at the same time, while here only three seedlings in each of 11 families are shown.

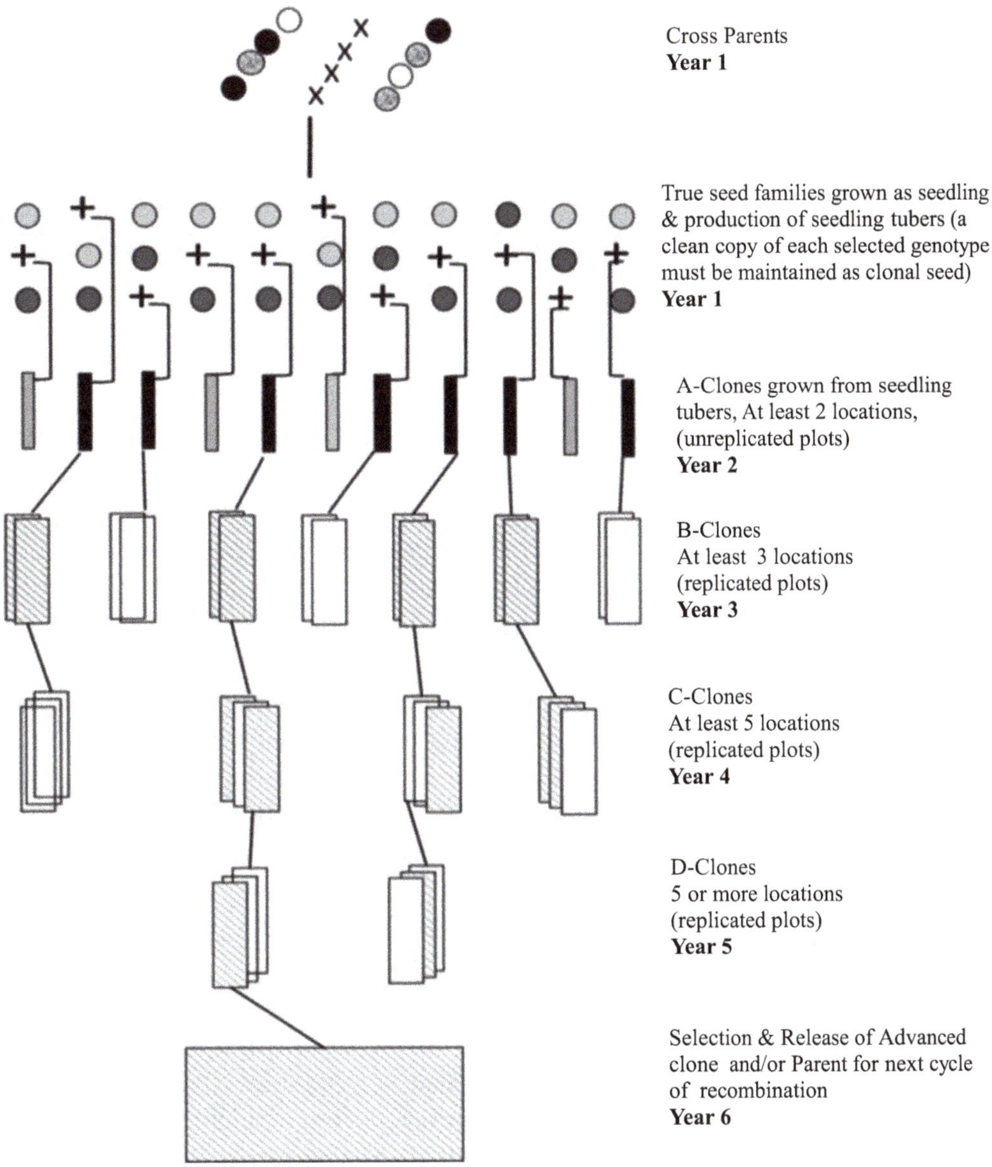

Fig. 6.6 Multilocation Clonal selection Scheme. (Modified from Gruneberg et al. 2009)

In principle, if the true seed plant could be cloned in large quantities and the population could be assessed with adequate accuracy, it is theoretically possible to select the "best" genotype in the first year, since maximum variation is present at the start of each selection cycle and no genetic changes are realized during it. In practice though, the individual genotypes selected in early stages of each clonal selection cycle are subject to propagation via tubers to permit adequate plot size and replication for later-stage assessments.

The overall selection scheme should consider the vulnerability of clonal planting material to disease. At least some tubers of every clone should be planted in a disease-free location, or time and resources should be allocated for the eventual elimination of pathogens from successful clones before they can be distributed among geographic regions.

Many breeding programs use only one location at the early selection stage, preferring to evaluate in favorable conditions such as on station, while also producing healthy seed for use in later selection stages. This location should be as free from virus pressure as possible to help keep seed clean over the years. For example, the New York breeding program makes all evaluations for the first 2 clonal years in a single location, going to a second location (for an unreplicated evaluation of yield) in the third clonal generation, and multiple locations (for replicated yield trials) starting in the fourth clonal year. Typical numbers of genotypes per cycle in this program are: 20,000 seedlings in year 1, 18,000 four hill plots in year 2, 1500 twenty-hill plots in year 3, and 250 hundred-hill plots in year 4.

Nevertheless, the first clonal generation can be evaluated at more than one location if sufficient clonal seed can be produced. For simultaneous evaluation at up to three locations in the first clonal generation, true seed families (TSF) are first converted to tuber families (TF) so that at least four copies of each genotype from each family are produced. Methods for evaluating TF include (1) planting three to five tubers of each genotype in a single row at a single location or (2) equivalent samples of each tuber family are evaluated in two or more locations, using one tuber per genotype and location. Multilocational testing of first clonal generation tuber families can be done with, or without the identification and labeling of individual genotypes. Detailed procedures for the production of tuber families followed at CIP are illustrated at: https://research.cip.cgiar.org/potatoknowledge/tuberfamilies.php

6.3.3 Data Management and Analytical Tools

Standardized information on the performance of progenies and selected clones across environments is necessary in order for breeders to efficiently make decisions about selection and variety release. The methods of data generation and processing that are utilized in plant breeding have radically changed in recent years. With the advancement of new high throughput technologies, data have grown in terms of quantity as well as complexity. However, the significance of the information that is hidden in newly generated experimental data can often be deciphered only by linking it to other data, collected previously and/or by others. Collaboration that makes it possible to connect disparate data sources and analyze them in meaningful ways with other researchers requires robust but practical data management solutions. CIP has developed an on line Global Trial Data Management System: https://research.cip.cgiar.org/gtdms/. Three key components are CIPCROSS, HIDAP, and Field Book Registry.

CIPCROSS is a botanical seed inventory tracking system for clonal crop breeding that enables tracking of breeding materials from crossing blocks and botanical seed inventories through to seed distribution. CIPCROSS is open source software, available online, and comprises two main tools: (a) The Crossing Management

System (CIPCROSS Tool v1.1 for Pocket PC) used the main development platforms Visual Basic 2008 for Windows Mobile OS and Microsoft SQL Server Compact 3.5 for Windows Mobile software. This tool facilitates the barcode labeling of parents and crosses, storing the information in a database and generating automated reports and (b) The Botanical Seed Inventory System (CIPSIS Tool v1.1) was designed as an easy-to-use web application. It uses PHP, Yii Booster, Bootstrap, CSS, Java Script, HTML and MySQL database programming languages. This tool facilitates the recording and accessing of information on the location of botanical seed in the storage facility, documentation of germination tests, updates on seed stocks, information about crosses and breeder's name. The system helps breeders develop, integrate, and organize their information in a database, avoid typing mistakes and saving work time (24 person–hours are saved per 14,250 labels). The Roots & Tubers Base centralized data repository facilitates access to the information with queries and filters for advanced searches (Fig. 6.7—Users can download these tools through the web page of "The Global Trial data Management System from CIP" https://research.cip.cgiar.org/gtdms/)

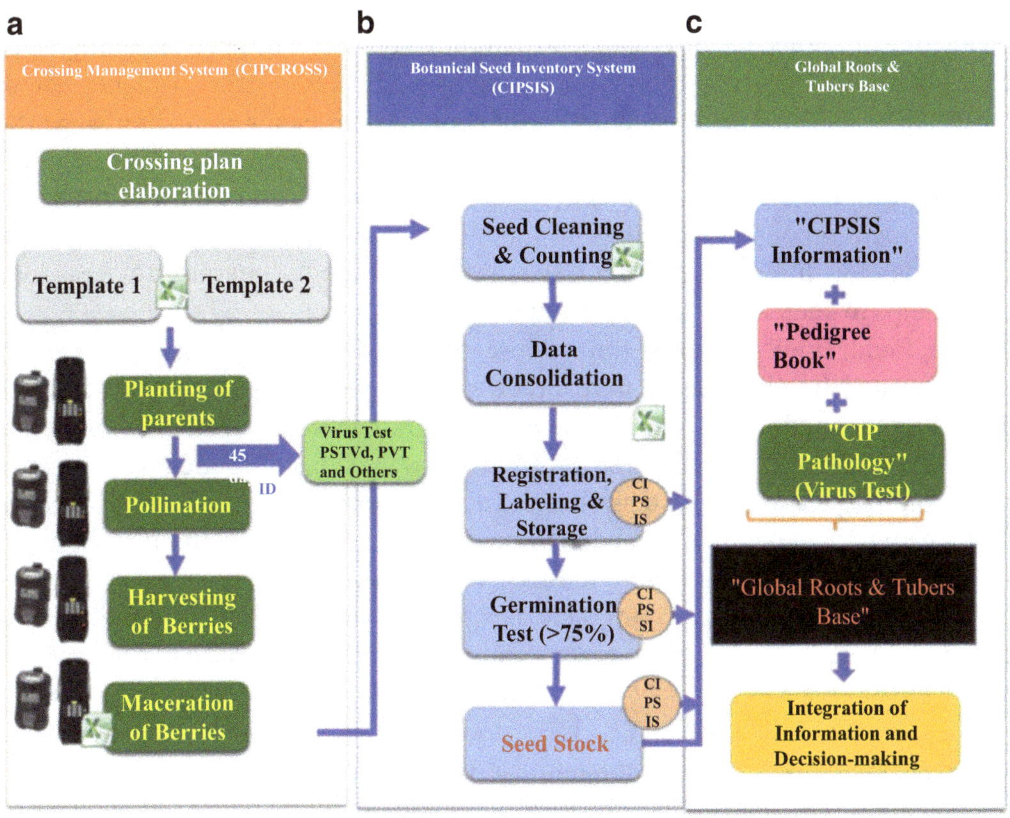

Fig. 6.7 CIPCROSS workflow: (**a**) The Crossing Management System (CIPCROSS Tool v1.1 for Pocket PC); (**b**) The Botanical Seed Inventory System (CIPSIS Tool v1.1) and c) Global Roots & Tubers Base as a centralize data repository; (**c**) Global Roots & Tubers Base as a centralize data repository

Data collection in potato breeding requires capacity to use and share standard protocols for the selection of clones from seedlings through to observational trials and on to preliminary and advanced yield trials, each accompanied by field books and structures for data collection, analysis, and reporting.

The highly integrated data analysis platform called HIDAP was developed by CIP's breeding program and Research Informatics Unit to facilitate and unify data collection, quality control, and data analysis for clonally propagated crops. HIDAP provides a single platform for use by potato and sweetpotato breeders. It supports compliance with Open Access, open standards such as the potato and sweetpotato crop ontologies and linkages with relevant corporate and community databases such as CIP's BioMart (https://research.cip.cgiar.org/gtdms/biomart) and SweetPotato Base (www.sweetpotatobase.org). HIDAP builds on the statistical platform R. This includes the R shiny tools, the knitr package, the Agricolae package also developed by CIP (https://cran.r-project.org/web/packages/agricolae/agricolae.pdf), and more than 100 other R packages. The R shiny package enables implementation of interactive web pages that are usable online and offline. The knitr package enables the creation of reproducible reports. Numerous statistical analyses can be performed using R and R functions developed at CIP. The software is available for download at https://research.cip.cgiar.org/gtdms/hidap/.

HIDAP is connected to the institutional pedigree and corporate database at CIP facilitating the tracking of clones and families generated through breeding. This connectivity enables verification and maintenance of the identity of clones across the different selection stages and tracing of pedigrees in selection and breeding. The HIDAP network enables researchers to share field books with colleagues, regional breeding programs, and/or partners. To use this network, users must register and create a login account. Once logged in, the user can share, download, and receive field books for different selection stages in a user-friendly interface. A download count helps to keep track of users and uses of this tool.

The Field Book Registry (https://research.cip.cgiar.org/cipfieldbookregistry/) facilitates updating field books in real time and viewing their status in the database. The data generated in CIP's potato breeding program is stored in the *"Global Roots & Tubers Base"* utilizing the free BioMart software https://research.cip.cgiar.org/gtdms/biomart/. This database has been structured for storage of phenotypic, genotypic, pedigree, geographical, and environmental data. Through the metadata and the search function using filters, the user can retrieve data from the experiments conducted by CIP or partner programs using CIP materials. The availability of the data is managed in conjunction with Dataverse following CGIAR open access guidelines (Fig. 6.8).

The use of the database and software tools enables analysis of phenotypic and genotypic data. A key goal is the identification of effective models that predict phenotypic traits and outcomes, elucidating important biomarkers and generating important insights into the genetic underpinnings of heritability.

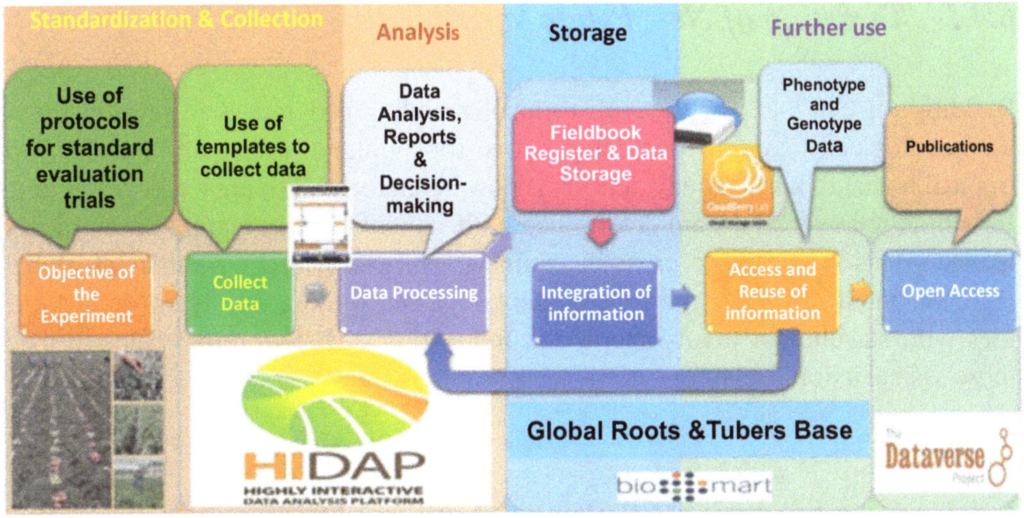

Fig. 6.8 Flowchart of CIP's breeding information system: Standardized evaluation protocols and ontologies implement templates to collect and analyze data with HIDAP software; Subsequent data storage through the field book register allows incorporation of breeding information into the Roots & Tubers Base in which phenotypic and genotypic data are integrated and made available for further use

6.4 Setting Objectives for Potato Breeding

Successful breeding relies fundamentally on having a clear set of prioritized breeding objectives. Key enabling information and tools for setting breeding objectives include: knowledge of the opinions and needs of stakeholder groups, knowledge of the production constraints and cropping systems of target populations and environments, and standardized means to measure and describe traits. Knowledge of the strengths and weaknesses of predominant varieties provides an important baseline for setting breeding objectives.

Having a good understanding of desired end-user traits requires ongoing, interdisciplinary communication and collaboration. This helps increase the chances that a new variety meets actual needs and is thus adopted, as well as minimizing the chance that new varieties (and the traits that define them, their management and marketing or use) might marginalize or disadvantage social or gender groups such as poor farmers or women, who often have access to different resources than men do. There is also a need to consider social and economic dimensions and client demography such as market access, youth, and urban/rural factors. A CGIAR system-wide initiative is currently underway to assess and encourage breeding that responds to social diversity including often-neglected or marginalized groups with emphasis on gender (http://www.rtb.cgiar.org/gender-breeding-initiative/?lang=en).

Potato breeding is characterized by having a large number of objectives, which include traits specific to the intended market and use of the crop (end-user traits), as well as traits related to productivity and protection against local diseases and insect pests (grower traits).

6.4.1 Targeting of New Potato Varieties

Building on CIP's experiences of a global breeding program that developed two agro-ecologically adapted populations (one for highland tropics and one for subtropical lowlands), increased attention is now given to prioritizing packages of traits for which relative values are determined with stakeholders in a more decentralized manner. This involves the cross-disciplinary estimation of trait values at global, regional, and local scales, the development of corresponding variety or product profiles with stakeholder involvement, and the setting of breeding priorities to meet those via selection decisions.

6.4.2 Case Study 5: RTB Priority Setting

An example of assessing trait values can be seen in the priority setting exercise conducted by the Root, Tubers and Bananas (RTB) Program of the CGIAR (http://www.rtb.cgiar.org/resources/impact-assessments/) Kleinwechter et al. (2014). Through a global survey carried out from 2012 to 2013, the RTB program sought to rank production constraints of each RTB crop (i.e., cassava, yam, banana, sweetpotato, and potato) and research options to alleviate them. This exercise was led by agricultural economists specialized in impact assessment. It provided an overview of problems affecting the potato sector and potential solutions including, but not limited to, breeding. The potato survey (Hareau et al. 2014) provided respondents with a list of 91 research options, organized around the areas of crop improvement, crop and resource management, seed management, genetic resources, value chains, postharvest utilization, and marketing, as well as socioeconomic research and extension.

Five of the top ten ranked research options for potato were directly related to breeding. Breeding for late blight resistance, drought tolerance, earliness, and high yield ranked second, third, fourth, and eighth of the 91 options provided, while the generic "germplasm enhancement" ranked seventh. An ex-ante assessment of the selected potato technologies revealed significant differences in terms of net present value (NPV) and internal rate of return (IRR) on investments across the different research options. Late blight- and virus-resistant varieties had the largest expected net benefits ($US 4.7 billion and $US 3.9 billion), and high rates of return of 87% and 104%, respectively, in the high-adoption scenario.

6.4.3 Case Study 6: Adjusting and Ranking Priority Traits in New York (NY) State

Staff of the breeding program in New York State U. S. meet many times each year with stakeholders, in both formal and informal settings, to continually discuss what the program's priorities should be. The frequent interaction helps stakeholders feel comfortable expressing their views, and allows the program to detect changes in industry priorities more quickly, as breeding goals always change with time.

Two examples can illustrate how stakeholder feedback has changed the NY program over the past 5 years. (1) NY growers recently made repeated requests to develop earlier maturing potatoes, to solve a problem the NY breeding program had unwittingly helped create. NY breeders have always put a high priority on yield, and selection for yield tends to select for later maturity. Unfortunately, a suite of late maturing varieties also shortened the timeframe in which NY growers could harvest their crop. The NY program now pays much more attention to maturity, and is prepared to select potatoes with lower yield if they mature early. (2) Potato chip factories recently began to ask for smaller chipping potatoes, as more and more of their product is now sold in small bags. In response, the program now selects smaller potatoes than it (or regional chipping factories) would have been willing to process in the not-so-distant past.

One approach that the NY breeding program has found useful when prioritizing traits is to compare what stakeholders ask for ("what stakeholders say") with the attributes of widely grown varieties ("what stakeholders do"). When there is an apparent disparity between words and action, there is an opportunity for deeper understanding. Fifteen years ago growers in NY kept asking for new varieties with resistance to common scab. What made the request unusual is that growers already had several resistant varieties to choose from, and that the variety they grew most each year was highly susceptible. What the NY program eventually realized is that the popular scab susceptible variety had two quality attributes—outstanding fry color out of cold storage and high specific gravity—that were far more important to the chip factories than the scab resistance growers kept asking for. As a result, NY re-ranked selection criteria, placing fry color and specific gravity above resistance to common scab.

6.4.4 Product Profiles

A product profile establishes a set of targeted attributes that a new plant variety or animal breed is expected to meet for successful release in a given market segment. Attributes must be understood as traits reaching a specified level; this level being defined either in absolute or relative terms (Ragot et al. 2018). Thus, a product profile may list yield (25 tons/ha or more; or 15% over variety V1 across a range of soil fertility conditions), tolerance to potato leaf roll virus (same as or better than variety V2), or dry matter content (no less than 18%, or no less than variety V3). The development of product profiles may best be done in collaboration with, for example, pathologists, agronomists, or nutritionists, as well as user communities who can contribute specific knowledge, tools, and approaches for setting trait levels, which in turn become the breeding targets within a profile.

6.4.5 Tools and Metrics

CIP's potato breeding program targets low input conditions by relying heavily on endemic disease pressure, poor soils and a series of intentional exposures to stress for screening and selection. The product profiles incorporate quantitative breeding objectives for productivity, protection, and utilization traits with emphasis on resilience, the setting of quantitative breeding objectives relies on knowledge of baseline variety characteristics and available genetic resources, the expected effects of changes in trait levels, and means to measure gains toward them.

Metrics for disease resistance: One special tool used by CIP is an interval scale for expressing potato resistance to late blight that indicates resistance levels required for satisfactory control of the disease in agro-ecologies with varying degrees of pathogen pressure. The use of this interval scale facilitates setting quantitative resistance breeding objectives in a robust manner within a breeding program's trait improvement framework. The late blight susceptibility scale of Yuen and Forbes (2009) enables assessment of resistance levels with a reduced coefficient of variation among trials as compared to other semi-quantitative metrics like AUDPC or rAUDPC. The scale uses reference cultivars in regression analysis and helps breeders to measure and describe resistance of genotypes independent of environment or inoculum level, which can vary from site to site and year to year. This approach gives breeders a simple numerical metric for quantitative traits that is useful for setting baselines, breeding goals, and calculating genetic gains.

Application of the susceptibility scale to setting breeding targets is based on the understanding that a variety with level 5 (more susceptible) would be sufficiently tolerant to provide the same level of protection in low-pressure agro-ecologies as a variety with level 2 (less susceptible) would provide in a agro-ecology with high disease pressure. The use of this scale is illustrated in Field assessment of resistance in potato to *Phytophthora infestans* at https://research.cip.cgiar.org/potatoknowledge/lateblight.ph.

Metrics for nutritional traits: To help reduce the health burden of iron and zinc deficiencies, quantitative targets for levels of these two elements in potato (a biofortification breeding goal) were set through collaboration between potato breeders at CIP and nutritionists of Harvest Plus. Consideration was given to (1) nutritional status of the target population, (2) dietary increments of iron and zinc known to be effective from other approaches (i.e., food fortification or nutrient supplementation), and (3) features affecting feasibility of nutrition impact such as consumption (potato intake), heritability, bioavailability, and retention of these minerals in potato prepared for use as food.

Nutritionists consider that dietary increments of 0.4 and 0.2 mg/day of iron and zinc, respectively, can have a positive biological effect on the health status of populations at risk of micronutrient malnutrition associated with deficiencies in these minerals. The provision of these increments through biofortified varieties is feasible when mineral concentration of a variety, consumption levels, bioavailability, and retention in the diet are sufficient. This food-based approach to reducing micronutrient malnutrition relies on variety change but not on modification of consumption patterns, i.e., high versus low potato intake. Logically, a population that consumes twice as much potato as another one would ingest twice as much iron from that source. And the percent of the Estimated Average Requirement (EAR) of the nutrients it provides would be double.

Quantification of these features for a given varietal and dietary context assumes conservative estimates of 10% and 25% bioavailability of iron and zinc, respectively, from potato, and that minerals are not lost in cooking. Figure 6.9 illustrates how different levels of iron and zinc concentrations in potato (referred to as biofortification levels: Base line, 1, 2, and 3, where base line represents current potato varieties) contribute to the EAR of iron and zinc for women of fertile age who consume 100, 200, or 400 g of potato per day. For reference, women in parts of Rwanda consume over 500 g of potato per day (Personal communication, Harvest Plus), while women in parts of Peru, may consume 800. The iron and zinc levels included in potato product profiles for populations of the Andes or the central African highlands are 45 mg/kg Fe and 35 mg/kg Zn (just over biofortification level 2 in Fig. 6.9) which can be expected to provide 50% of the EAR of both minerals for women consuming 400 g potato a day. This is a considerable increment over the baseline, but has been assessed as feasible by CIP's breeding program after evaluating genetic diversity, estimating heritability, and realizing significant gains (subsection 6.1.5) in cultivated potato germplasm. Achievement of such genetic gains would only provide 12–15% of the EAR for women who consume 100 g of potato per day or less, as in much of the subtropical lowlands of Asia. The inclusion of a biofortification breeding target to increase Fe or Zn intake from potato by 0.4 or 0.3 mg/day, or reach 50% EAR for this market segment, would require greater genetic gains than presently estimated to be feasible by the interploid breeding strategy undertaken by CIP. Nevertheless, high iron and zinc potatoes have been requested by the national programs of both India and Bangladesh. In such cases, a

A

Biofortification Level	Iron (mg/Kg DW)	Zinc (mg/Kg DW)
Baseline (non-biofortified)	20	16
1	27	24
2	42	32
3	50	40

B

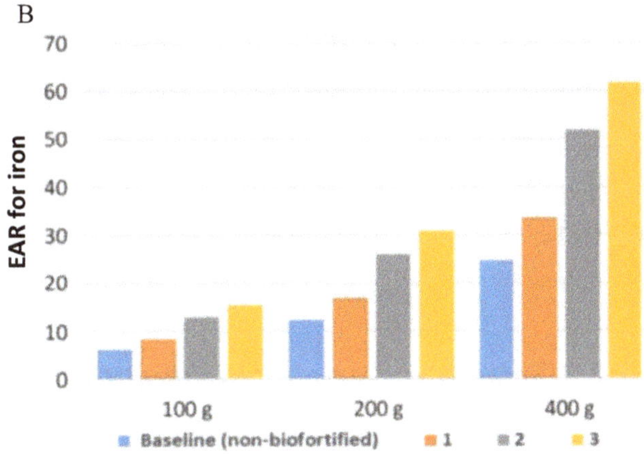

Fig. 6.9 Influence of Fe and Zn concentration and intake of potato on the estimated average requirement for women of fertile age. (**a**) Definition of baseline (non-biofortified) and incremental concentrations (biofortification levels 1, 2, and 3) of iron and zinc in potato tubers. (**b**) Percent of EAR for iron for women of fertile age met by consuming 100, 200, or 400 g/day of potatoes with concentration levels 0, 1, 2, and 3

food systems intervention seeking overall increments of micronutrients from co-staple crops should be considered.

Biofortification targets between level 2 and 3 in Fig. 6.9a have been set as part of the product profiles for table potatoes oriented to tropical highland agro-ecologies and populations with significant levels of anemia, who also consume potato as a main food or staple crop. The value of nutritional traits can be assessed with the disability adjusted life years (DALY metric). DALY is ordinarily used to assess the impact of public health burdens like human disease or illness. It extends the concept of potential years of life lost due to premature death to include equivalent years of "healthy" life lost by virtue of being in a state of poor health or disability. In so doing, mortality and morbidity are combined into a single, common metric (Meenakshi et al. 2007).

6.4.6 Setting of Breeding Priorities

Once product profiles are defined, breeding programs can proceed to refine trait lists and allocate resources to address those that are most feasible. While goals and priorities are important for breeding programs and the institutions that host and sponsor them, the list of traits a breeder can work with is limited, and in this light, he/she will select a subset of those in the profile for the day-to-day efforts of the program. Breeders may also be aware of a few "must have" traits that do not come up in stakeholder consultation, and since the inclusion of any trait in a breeding effort (whether to change or even maintain trait levels) requires investment, and thus has a cost, these should be made explicit and, ideally, added to the product profile.

Experience and knowledge of existing germplasm are applied to identify the set of attributes to be addressed in breeding. For each attribute, a quantified description of the desired result (maximize, reach specified level, maintain a certain minimum level, etc.) and a unique rank or priority is assigned. Ranking represents the relative effort required to develop the set of traits that will enable the new variety to respond to the opportunity or constraint identified in the targeting exercise and taken on by the breeding program. Relative effort required to address each objective are determined by the proximity of the program's genetic materials to the needed level of each trait, and the genetic control and relationships among various traits.

6.4.7 Selection Decisions in Potato Breeding

Because potato is highly heterozygous, a large number of traits segregate every time a cross is made. Every offspring receives a unique combination of desirable alleles at some loci and undesirable alleles at others. No offspring is ever "perfect." An issue of considerable practical importance, therefore, is deciding which phenotypes (clones; genotypes) are worth keeping, and which should be discarded at each stage of the selection scheme.

Selection decisions are the decisions a breeder makes to use, advance, or discard a selection unit (plant, progeny, progenitor, clone, experimental variety), guided by the breeding priorities, using specific decision-making tools (e.g., assessments of the targeted traits, estimates of breeding value, presence/absence of specific markers, selection indices, combining ability).

In early stages of selection, when the number of plants of each genotype is small, it is important to select stringently only for highly heritable traits. Meaningful assessments of yield, on the other hand, can only be performed in later stages. As selection progresses, and the individuals not meeting breeders' criteria have been eliminated, it becomes especially important to keep product profiles in mind, i.e., to select not just on the basis of individual traits, but on the suite of traits that collectively define a product, with attention to specific traits at specific stages of the selection scheme.

6.4.8 Case Study 7: Profile and Selection Decisions for Chipping Potato for Northeast USA

As nations develop there tends to be a shift in potato consumption, away from fresh and more towards processed potatoes, primarily as chips or French fries. Varieties intended for fries or chips ideally have higher levels of starch, so that they absorb less oil when fried. In addition, if potatoes need to be stored in the cold for a long time—typically to ensure a year-round supply of raw product—then it is useful if they are resistant to cold-induced sweetening, as the presence of high levels of glucose and fructose will cause potato slices to turn dark brown when fried.

To improve resistance to cold-sweetening, the New York program has adopted a low-tech approach that could be adopted anywhere. The essence of the approach is this: store potatoes at a temperature where only a small percent will fry to a light color. Intercross those that do, and after 5 years or so, once the population has reached a point where many progeny fry well, lower the cold storage temperature by 0.5–1.0 °C the following year. Intercross the few that fry well under the new, colder regime, and repeat the process. Recurrent selection has proven very effective at improving fry color in the New York program; many clones now fry well out of 3–9 months of storage at 6 °C. Ideally the breeding program will fry from slightly colder storages than those currently in use by regional industry, to help drive down the temperatures that industry can use over time. In general, the colder the storage, the longer that potatoes can be stored.

The two most important attributes in chipping potatoes are starch content (strongly correlated with specific gravity) and fry color, followed by tuber shape and size. About 70% of the New York breeding program effort is dedicated to developing new chipping varieties.

The New York program practices very little selection on seedling tubers, as the performance of plants grown in pots correlates poorly with performance of plants grown in the field. When the first field generation is harvested, the principal selection criteria are for tuber yield, tuber shape, and size (all visually assessed). The ideal chipping potato is round and the size of a baseball. Clones with low yield, or shapes and sizes too far from baseballs, are not selected.

The New York breeding program chips (out of cold storage) every clone that survives 1 or more years of selection in the field, discarding all that do not chip as well as, or better than, the industry standard chipping variety "Snowden." Absolute fry color can vary from season to season, although the ranking of fry color between clones does not vary greatly.

In the USA a specific gravity of 1.080 is deemed the minimum for processing. Nevertheless, because starch content can vary considerably from season to season (warm nights result in lower starch content than cool nights, all other things being equal), in New York the level of starch is always evaluated relative to that in the widely grown chipping variety, "Atlantic." The specific gravity of Atlantic is typically above 1.090 in New York, but in some seasons it is less. The New York breeding program will not continue to evaluate a chipping clone whose gravity averages 0.010 or more less than Atlantic, and prefers clones to average 0.005 or fewer points less. New York begins to evaluate specific gravity after a clone's second year in the field.

Replicated, multisite yield trials begin in the fourth field year, where clones that yield less than 90% of Atlantic are discarded. Clones that exhibit considerably more internal or external defects than "Atlantic" or "Snowden" in yield trials are also discarded, regardless of their yield, fry color, or specific gravity. Clones that survive several years of field trials are then evaluated on commercial farms, and processed in commercial chipping plants, where growers and chipping plants make the final decisions about which clones merit release as new varieties.

6.4.9 Selection Decisions: Marker-Assisted Selection

6.4.9.1 How the New York Program Uses Molecular Markers

Even though current NY breeder WDJ was trained as a molecular geneticist, the New York breeding program does not make extensive use of marker assisted selection, primarily because there are not yet many publicly available markers linked to traits of high priority.

One of the two markers currently used in New York is 57R (Finkers-Tomczak et al. 2011), tightly linked to the H1 gene, which confers resistance to race Ro1 of *Globodera rostochiensis*. Developing potatoes resistant to race Ro1 has been the highest priority of the New York breeding program for the past four decades; almost every cross made in NY has at least one Ro1-resistant parent. All offspring that survive 2 years of visual selection for appearance and yield in the field (about 200 clones, out of 20,000 initially planted) are screened with 57R over the winter. The vast majority of clones lacking the marker are discarded, the only exceptions being those that look especially promising for other high value traits (e.g., excellent chip color combined with high specific gravity).

RYSC3 (Kasai et al. 2000) is the other marker currently used in New York. This marker is tightly linked to Ry_{adg}, which confers immunity to potato virus Y (PVY), and like 57R, is used to screen clones that have survived 2 years of selection in the field. At present the presence/absence of the RYSC3 marker is just one of many data points NY uses to make a decision about each clone, with other data points including chip color, specific gravity, resistance to common scab, and visual assessments of yield and tuber appearance.

Although markers tightly linked to potato virus X (Gebhardt et al. 2006), wart (Gebhardt et al. 2006), verticillium wilt (Bae et al. 2008), Columbia root-knot nematode (Zhang et al. 2007), the pale cyst nematode (Sattarzadeh et al. 2006), and late blight resistance genes (e.g., Colton et al. 2006) have been reported, none of these traits are important enough in New York to influence selection decisions, and thus none of these markers are used routinely.

6.4.10 Breeding Objectives at CIP

Priority setting exercises (Fuglie 2007; Hareau et al. 2014), complemented by evolving requests from national programs for elite germplasm and candidate varieties, both contributed to CIP's profiles for new potato varieties tailored to the contrasting agroecologies, cropping, and food systems its program addresses. Two key products conceived in a recent exercise were (1) "Agile Potato" for intensification and diversification of cereal-based systems of Asia and, (2) mid-elevation tropical highlands late blight resistant potato.

CIP's major breeding populations are oriented to develop table and processing varieties for sustainable potato production in tropical highland and lowland areas faced with either of two basic sets of biotic and abiotic constraints: those encountered in the tropical and subtropical highlands, where late blight, cyst nematodes and frost are limiting, and farmers are increasingly faced with drought; and warmer, drier production areas, that are challenged by viruses, bacterial wilt, drought, root knot nematode and other pests, Increasing requests to support production in mid-elevation tropics of Sub Saharan Africa and Asia as well as temperate regions of Asia have led to the inclusion of heat tolerance and long day adaptation into our breeding objectives, requiring the introduction of new germplasm, screening methods, and trial sites. Present emphasis is on: earliness to enable potato's full potential to contribute to incomes and productive cropping systems; heat and drought tolerance to enable expansion to mid-elevation regions nearer to population centers, at the same time helping farmers adapt to climate change, and; nutritional value with specific attention to iron and zinc biofortification. CIP aims to reach new thresholds for these traits, while maintaining and broadening levels of disease resistance achieved in the past, and also responding to regional preferences for culinary characteristics and market opportunities.

6.4.10.1 Case Study 8: Agile Potato for Asia

Product Profile

The "Agile Potato" project seeks to develop potato varieties for fresh and processing use in short crop production windows of Asia. The product profile includes 70–90 day maturity; resistance to multiple viruses (PVY and PLRV, with PVX resistance desirable); stress tolerance (heat, drought, and salinity); resistance to late blight; and market-oriented quality (high dry matter and chip processing). The expectation is that varieties with short crop duration that resist disease, require less-frequent seed replacement, thrive in warm seasons, and store well under rustic conditions, can readily be inserted into diverse cropping patterns dominated by rice or wheat. Agile potato varieties would fit into cropping windows currently left fallow in the cereal-based systems of subtropical lowland and temperate regions of Asia. They should help increase system productivity through intensification as well as help meet growing demand for processed potatoes, thus contributing to income generation and employment. Breeding to meet this profile is backed up by trait research to enhance and accelerate breeding methods, and inter-disciplinary research to promote fast-track systems for variety identification and release. With this profile, CIP seeks to develop strategies for sustainable intensification of farming systems, raise awareness and expand consumer demand for potato while diversifying cereal-based diets with potato as a nutritious co-staple or vegetable food. A gamut of research on ecological and socioeconomic consequences of intensification is also planned in order to identify appropriate indicators of sustainability and monitor effects of benefits to intended sectors, including the poor whose livelihoods would be affected by the proposed intervention.

The target is to select 70–80 day heat-tolerant potatoes with competitive yields and more than 18% dry matter and resistance to potato viruses for subtropical lowlands of Southwest Asia. Commercial varieties and elite selections from previous cycles of the lowland tropics virus resistant population (LTVR) are comparators for performance in representative or target selection sites. In the following figures we show a group of selected early clones that performed significantly better than past generations and local varieties for yield and quality at 70 days after planting in warm environments, yielding in a range of 20–40 tons/ha, compared to the best local varieties such as Unica and Reiche (Fig. 6.10) and adequate dry matter content over 18% under the abiotic stress conditions of lowlands (Fig. 6.11).

Dry matter (DM) content is an important component to be taken into account in the breeding process especially when selection is focused on lowland tropical areas. In addition to the loss of fresh weight yield of tubers associated with high temperatures, the DM content of tubers are also reduced by 1% as the air temperature rises by 1 °C over a temperature variation average of 15–25 °C. This is due to the rapid

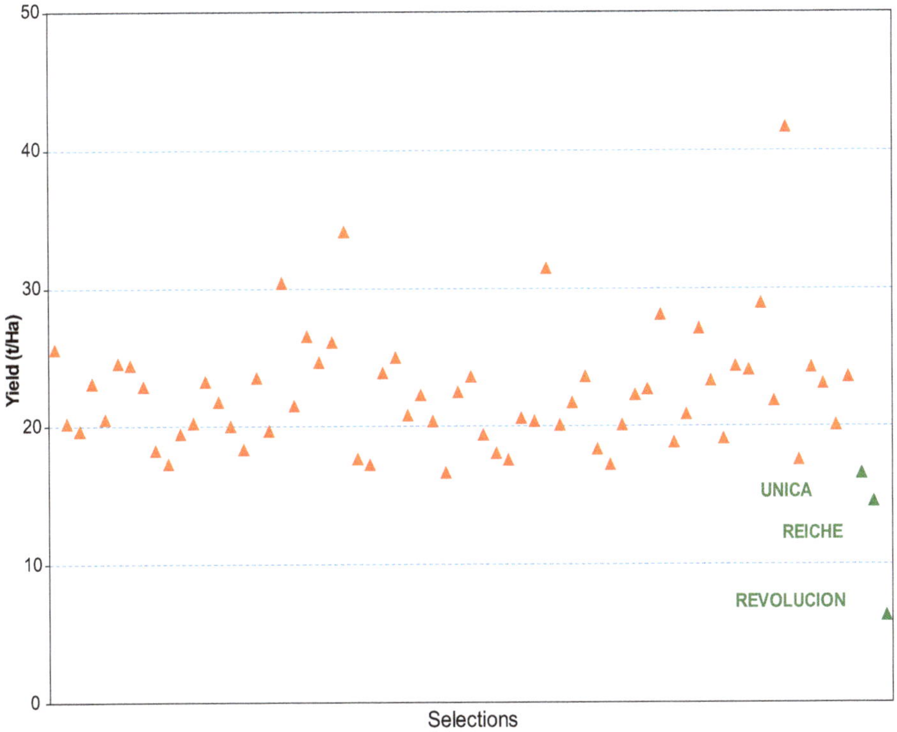

Fig. 6.10 Scatter plot showing yield of selected clones in a new generation of LTVR population that surpass levels of varieties and clones from previous generation grown under hot conditions of San Ramon, Junin, Peru, and harvested at 70 days after planting

decline of photosynthesis (Midmore and Rhoades 1988). This tendency of decrease in DM of clones in the LTVR population can be appreciated in Fig. 6.11, showing decreasing DM contents with decreasing altitude and thus increasing temperature of the evaluation sites. However, breeding by recurrent selection for yield as well as DM content in this population has maintained this important trait at between 11 and 19% even in the warmest season, and 15–21% in the high humid jungle site of San Ramon, Peru. This means that genetic diversity is sufficient to permit selection for heat tolerance and genotypes with good yield and adequate DM content can be identified in LTVR in spite of the critical negative correlation between temperature and DM in potato.

Crossing with long day adapted parents and evaluation under simulated long day conditions in Peru permitted the identification of LTVR families and advanced clones with adaptation to higher latitudes, for example of Vietnam where day lengths are intermediate, and Uzbekistan and Tajikistan where they are long combined with high summer temperatures. The same screening approach also helped provide pressure for improving the earliness (70–80 days) of LTVR at low latitudes. The compensatory effect of temperature on day length provided by warm mid-latitude conditions of southern Peru has also been key in breeding for early maturity, which can also be referred to as a relaxed requirement for short days. The identification of a moderate frequency of genotypes as well as superior parents with good performance under higher latitudes suggests that the LTVR breeding strategy

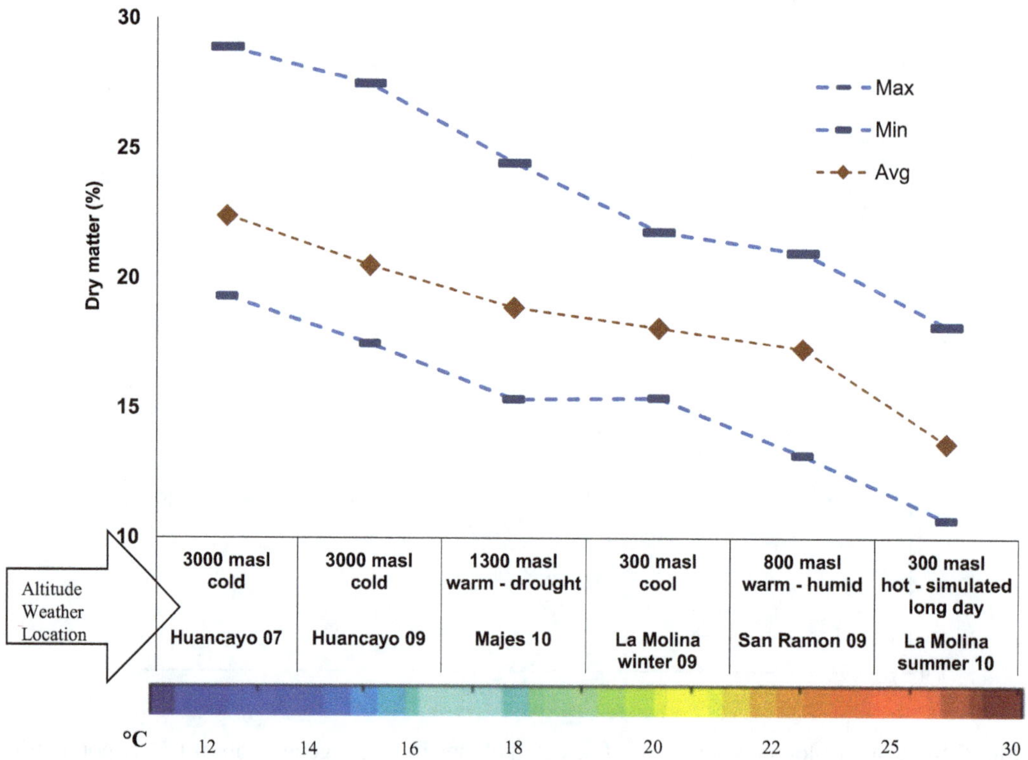

Fig. 6.11 Variation for dry matter content of potato selections in LTVR population across environments. Three hundred LTVR clones were grown in ten hill plots in six locations in Peru and evaluated for dry matter content within 10 days of harvest

is likely to be successful in developing the agile potato for both of its target environments—subtropical lowland and temperate low and mid-elevation regions of Asia.

Source Population LTVR

From the 1980s through the late 1990s, CIP endeavored to strengthen its lowland tropics (LT) population with multiple virus resistance developing the lowland tropics virus resistant population (LTVR). Development of the broad-based LTVR population used an open-recurrent selection strategy to incorporate foreign elite and advanced clones into its lowland tropics adapted breeding population. Lowland tropical environments were characterized as having day lengths of 10–14 h, minimum night-time temperatures of 18–20 °C, mean maximum temperatures greater than 25 °C and mean annual soil temperatures at 50 cm depth of 22 °C or more. LTVR population combines the heat tolerance and early bulking ability of *S. tuberosum* germplasm bred under the summer conditions of the northern Hemisphere (Cubillos and Plaisted 1976) on one hand, with virus resistance from native *Andigenum Group* germplasm and long day adapted *Andigenum Group* germplasm, namely Neotuberosum, on the other (Plaisted 1987; Mendoza 1990). The Neotuberosum and *Andigenum Group* germplasm contributed major genes for extreme resistance to potato virus Y (PVY) and potato virus X (PVX) (Muñoz et al. 1975; Gálvez et al. 1992). Early generation screening was routinely performed to discard PVX and PVY susceptible individuals in any trial in which LTVR families were evaluated.

Testing sites included CIP's experimental stations in Peru (see Section 6.4.6) which are relatively dry, humid, hot, and hot and humid lowland environments. Experiments comparing the yield stability of diverse clones across these locations revealed that hybrids from *S. tuberosum* × Neotuberosum crosses performed significantly better than those from *S. tuberosum* × *S. tuberosum or S. tuberosum* × *Andigenum Group*. It is possible that the higher level of heterozygosity together with prior selection of the divergent parental materials for adaptation accounted for the better performance of crosses between *S. tuberosum* and Neotuberosum clones (Mendoza and Estrada 1979). These authors also reported that *S. tuberosum* × neotuberosum hybrids are adapted to hot and humid conditions and able to produce tubers at 60–70 days after planting, albeit under short days. These hybrids made a strong contribution to the LTVR population, and thus may account in part for its adaptation to warm conditions. The subsequent cycles of genotypic recurrent selection led to the selection of a number of multiplex PVY and/or PVX extreme resistant progenitors (Mendoza et al. 1996) some of which have since been shown to also possess good GCA for yield and dry matter content. Multiplex refers to the genetic constitution of a polyploid individual having more than one copy of an allele, in this case RRrr or RRRr, assuring the transmission of resistance conferred by a major gene to nearly 100% of its cross progeny. In parallel to population improvement and parental line development, high yielding clones with excellent tuber quality were selected during each recurrent selection cycle.

A second stage of the breeding process aimed to incorporate foreign elite and advanced clones by means of an open-recurrent selection strategy. Assessment of genetic parameters in a North Carolina Design II estimated heritability for resistance to PLRV infection to be 0.54–0.69. Higher estimates were obtained when greenhouse-reared infective aphids were used to conduct controlled inoculations with sprouted tubers before tuber families were planted in the field, suggesting that controlled inoculations minimize errors or escapes because of a more even distribution of inoculum. Additive genetic effects were shown to contribute the most to resistance variability (74%), followed by dominance effects (10%) and additive × environment (location) effects (5%), suggesting that breeding advances could be expected in further cycles of recurrent selection (Salas 2002).

Multilocation Testing and Selection Decisions

A stage plan can be used to communicate selection decisions made in early and late stages of breeding. The stage-gate process portrays stepwise decisions with thresholds for each characteristic that govern decisions to keep or discard individuals in the course of a given breeding cycle. In this process, the "gates" describe the collection of thresholds or criteria that must be met if a clone is to pass from one stage of selection to another along the clonal selection scheme outlined in Fig. 6.12, presents a stage plan for selection in the LTVR population to meet the Agile Potato profile.

Yield and processing quality traits of LTVR clones generated during open-recurrent selection were evaluated in replicated experiments in a range of lowland environments of Peru. Dry matter content was consistently higher in the cool arid lowlands of the coast under irrigation (average = 21%) than in the warm humid area of the central jungle (average = 19%), highlighting the detrimental effect of high temperatures on this trait. As expected, significant G × E interaction was encountered for dry matter content, but also for glucose content and chip color. High yielding clones with good table or processing quality combining virus resistances and heat tolerance have been released by national programs and public institutions in target countries. Varieties of LTVR origin include: Tacna, Costanera, Unica, Reiche, Maria Bonita, and Maria Tambeña released in Peru; Kinga, Meva, Kinigi, Muziranzara, Muruta, Yayla Kizi, Baseko, Enfula, Chamak, Dheera, and IRA-92 in Africa; and Raniag, which was named in the Philippines. Evidence of the wide adaptation of bred materials from this population is provided by the success of the variety "Tacna" in the lowland subtropics of China where it is known as "Jizhangshu 8" and planted on an estimated 133,215 ha in 2015 from which 70,533 ha in Hebei, 26,000 ha in Shanxi, 22,667 in Inner Mongolia, and further 3000 has in the increasingly drought-prone northwest province of Gansu with more than 3000 ha (Gatto et al. 2018). Additional evidence of broad adaptation comes from the variety Unica that is now grown in at least six countries.

Breeding Priorities

Following CIP's multidisciplinary conceptualization of the agile potato, the potato breeding program refreshed its strategy by setting priorities for combining the needed traits in subgroups of the LTVR population. The inclusion of resistance to PVY in the LTRV population and selection for highland tropical as well as subtropical lowland conditions to enable seed production has had a cost in terms of earliness even after three to four cycles of recurrent selection. The infusion of Tuberosum germplasm and selection under heat and mid-long days, for plant type and stability from early generations is strategy to improve earliness in the LD (long day) and HT (heat tolerant) groups. In this interest, tuber quality becomes an issue and attention is needed to keeping dry matter high and glycoalkaloids, which provide a bitter flavor to potato and can be toxic, low.

Often unrecognized and rarely controlled, the accumulation of viruses in seed and ware tubers is perhaps the most serious constraint to economical potato production in the tropics. Breeding for resistance to the world's most important potato virus PVY is possible due to the availability of major genes (Ry_{adg}, Ry_{sto}) that confer extreme resistance, and for which selectable markers are available. On the other hand, developing effective levels of resistance to PLRV is one of the more challenging components of the Agile Potato profile. Most sources of resistance to PLRV are quantitatively inherited and provide partial resistance to infection and/or accumula-

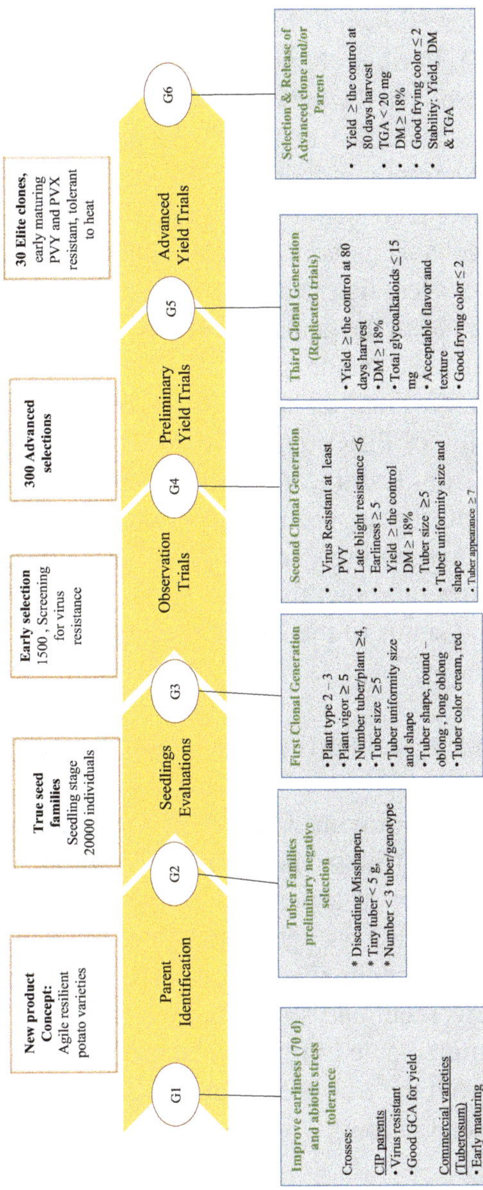

Fig. 6.12 Stage plan for selection in the LTVR population to meet the Agile Potato profile

tion of the virus. Despite moderate heritability, the frequency of resistant clones in advanced populations has been low. Thus, a relatively high priority is given to incorporating a new, highly heritable source of high levels of resistance to PLRV, designated Rl_{adg} recently identified in Andigenum Group germplasm into advanced populations that already carry extreme resistance to PVY at high frequency.

New sources of PLRV resistance, namely, accessions LOP-868, HUA-332 and OCH-7643, confirmed following mass selection in *S. tuberosum* Andigenum Group germplasm held at CIP, were shown to have exceptionally high GCA for this trait (Mihovilovich et al. 2007). Further characterization of these sources has been the basis of supporting trait research, the benefits of which are expected to boost levels and frequencies of PLRV resistance in the LTVR and other potato populations. Using a population of dihaploid individuals derived from LOP-868, Rl_{adg} was mapped to potato chromosome V, with high heritability demonstrated to be due to its multiple resistance mechanisms and a duplex allelic conformation of a major dominant gene in the donor accession (Velásquez et al. 2007). A PCR based marker (RGASC850) located 2 cM from the gene has been developed to facilitate the introgression and tracking of this major gene in advanced breeding populations (Mihovilovich et al. 2014). This marker has been incorporated into a multiplex PCR system for assisted selection of PLRV and PVY resistance using previously available information on Ry_{adg} from Kasai et al. (2000).

Table 6.2 presents the attributes, profile and breeding priorities contributing to the Agile Potato program and outlines some of the key research that supports the objective. Breeding priorities represent the relative effort dedicated to each attribute of the profile, given present status of the LTVR population.

Multi Trait Selection for Increased Earliness and Tolerance to Abiotic Stress

Most recently, emphasis in improving CIP's LTVR population toward the agile potato has targeted increased earliness and tolerance to abiotic stress. A new earlier maturing and abiotic stress tolerant generation that combined superior LTVR progenitors with tuberosum varieties was generated. Accelerated selection based on multilocation trials and simultaneous screening for abiotic stress (heat, drought and simulated long days) in divergent and stressful environments as well as screening for PVY resistance was implemented in Peru. The scheme requires only 4 years of selection and evaluation from the true seed progeny generation until the identification of advanced clones (see Fig. 6.6).

Evaluation and Selection Sites

CIPs proximity to significant agroecological diversity in Peru is an asset for the identification of patterns and trends of potato genotypes' response to environments. Potato is grown in coastal, high jungle and highland ecologies ranging from sea level to 3500 masl, providing wide variation in weather patterns including tempera-

Table 6.2 Agile potato profile, breeding objectives, and supporting research

Attribute	Product profile	Breeding priority
1. Yield	1. >25 tons/ha	1. Earliness
2. Earliness	2. 70 day maturity	2. High and stable yield across day lengths
3. Table quality	3. TBD (flavor appealing to consumers, more than 17% dry matter, <20 ppm glycoalkaloid content)	3. PLRV resistance
4. PVY resistance	4. Extreme resistance to PVY	4. Heat tolerance
5. PLRV resistance	5. Tuberization at >22 °C night temperature; bulking under warm day temperatures	5. Flavor and dry matter
6. Heat tolerance	6. Drought tolerance-TBD (ratio of fresh tuber yield to applied water expressed as kg/ha/mm)	6. PVY resistance
7. Good storability	7. Bacterial wilt resistance-TBD % plants wilted to degree *x* and % latent infection	7. Water productivity
8. Late blight resistance	8. Resistance to PLRV as high or higher than Granola	8. Storage quality
9. PVX resistance	9. TBD (*x* days to sprouting, low rottage, low water loss at given temperature)	9. Chip quality
10. Water productivity	10. Late blight resistance <6 on susceptibility scale	10. PVX resistance
11. Cold chipping ability	11. 20% selections with chip color score 1–2 after cold storage; dry matter >18%	11. Bacterial wilt resistance
12. Bacterial wilt resistance	12. Extreme resistance to PVX	

Supporting research:
Earliness: implement protocol and validate markers to develop MAS for bulking-based maturity. Complete GWAAS for day neutrality in subset of Trait Observation network panel across environments
Introgress and increase frequency of Rl_{adg} using MAS for PLRV resistance
Implement proximal sensing phenotyping for drought tolerance (water productivity). Develop inexpensive assay for glycoalkaloid contents
Validate pre-bred *S. commersonii* source of bacterial wilt resistance in Kenya and Ethiopia
Refine genome estimated breeding values (GEBV) for early bulking and heat tolerance extend pilot study of genomic selection for early bulking to LTVR in Asia
Optimize field trials and statistical analysis to make full use of trial data; genetic gains and heterotic groups at tetraploid and diploid levels

ture, humidity and rainfall. Most environments have short daylengths of around 11–12 h but the southern extreme at 17°S, provide daylength of about 14 h, which considering high summer temperatures approach threshold conditions for long day adaptation. Three experiment stations (San Ramon, Huancayo and La Molina) and additional sites accessed by local agreements (Table 6.3 provide evaluation conditions representative of the tropical lowland and highland potato environments targeted by agile potato. Indeed, potato seed for the subtropical lowlands is often produced at higher, vector-free, locations, requiring adaptation of the same variety to both highland and lowland conditions. Moving potato populations through the biotic and abiotic stresses present among these sites, and particularly the combination of temperature and day length, permit the identification of potato genotypes with wide adaptability and stability for yield.

Table 6.3 Location and conditions of major selection sites used for potato breeding by CIP in Peru

Location	Season	Climate	Latitude (°S)	Altitude (masl)	Geography	Main traits assessed	Agroecology
San Ramon (SR), Junín	Winter	Warm, humid	11	800	Central jungle	Heat tolerance; glycoalkaloid accumulation	Subtropical lowland humid
La Molina (LM), Lima	Spring-summer	Warm, semi-arid	12	133	Central coast	Yield, drought and heat tolerance, processing	Subtropical lowland arid
Tacna (Tac),	Summer	Hot, saline soils	18	500	Southern coast	Drought, soil salinity and adaptation to long days	Subtropical lowland arid
Huancayo, Junín	Winter–autumn	Cold	11	3200	Central highland	Yield potential, processing quality	Tropical highland cold
Oxapampa	Spring	Rainy, warm	10	1000	Central high jungle	Late blight	Tropical highland humid
Majes	Spring–summer	Warm, arid	16	1300	Southern coast	Drought tolerance, glycoalkaloid accumulation	Subtropical lowland arid

Trait Research in Support of the Agile Potato

Proximal sensing for efficient development of the agile potato: Remote sensing technologies are increasingly being used to assess crops by breeders and physiologists. These technologies have the potential to simplify evaluation of large breeding populations and improve selection efficiency over traditional field measurements. Different vegetative indices combined together may, for example, improve yield prediction (Rodriguez Junior et al. 2014). CIP's work in this area has aimed to explore genetic variation for the canopy temperature differential (CTD) and Normalized Difference Vegetation Index (NDVI) measurements and their relations with tuber yield components using 27 segregating families of the LTVR population under water stress, toward identifying families with tolerance to drought at early stages of selection. A high negative correlation (−0.66) of NDVI with Harvest index-fresh weight was observed, as were a high positive correlation (0.74) with plant wilting, a moderately high (0.50) correlation with plant vigor under drought conditions, and a significant negative correlation (−0.5) of CTD with tuber yield (tons/ha). Remote sensing techniques may provide a way to save resources and accelerate genetic gains through the early identification of families with tolerance to abiotic stress in a breeding program.

Trait research for understanding of complex, underground development: In breeding for a short vegetative cycle, CIP emphasizes bulking-based maturity, as opposed to foliar maturity or senescence. Tuber bulking results from two basic processes, tuber initiation and tuber growth (Radley et al. 1961). Like Andigenum Group potatoes, most varieties and clones from CIP's program yield significantly before physiological maturity, and development and harvest date cannot be assessed on foliar senescence alone. Indeed, if CIP-bred varieties are harvested when foliage has senesced, they will generally be well beyond the optimum time for harvest. Timing and duration of bulking vary with location, environmental factors, and genetics. Early tuber initiation and growth are necessary for acceptable production in areas or varieties in which potatoes are often harvested prior to physiological maturity.

To help provide early-maturing, resilient potatoes to farmers in medium and high latitudes of Asia, a panel of 162 CIP breeding lines was characterized for bulking behavior under four environmental conditions in field plots in La Molina, Peru. The environmental conditions for this study were: simulated long (16 h), and short (12 h) day lengths, in each of two seasons—cool winter and hot summer. The traits evaluated in the four field plots include yield (number and weight of tubers on two harvest dates) and dry matter content, in addition to which tuber induction was assessed by the cutting method of Ewing (1992). Tuber bulking ratio was calculated as the number of medium-sized and large tubers over the total number of tubers (tiny + small + medium + large) multiplied by the ratio of tuberized plants. Phenotypic assessment according to CIP's Protocol for tuber bulking maturity assessment of elite and advanced potato clones (https://research.cip.cgiar.org/potatoknowledge/bulking.php) enabled the classification of genotypes by maturity through cluster analysis according to Evanno and Regnaut (2005), and thereby, the identification of clones for which tuberization and bulking were more or less influenced by heat and photoperiod, including a small set of selections classified as day neutral.

Genotyping of the panel with version 1 of the SolCAP infinium array (8303 single nucleotide polymorphism (SNP) markers) enabled the identification of genomic regions and SNP markers influencing potato tuberization and bulking related traits affected by photoperiod and high temperatures. Prerequisites for this genome-wide association study (GWAS) included the calculation of the population membership matrix and SNP-based kinship relationship matrix and the estimation of linkage disequilibrium (LD decay to $r^2 = 0.2$ over 0.3 Mb) for the panel. GWAS-associated SNP jointly explained up to 60% of the phenotypic variation for the various traits, Therefore, MAS could be efficiently used for pre-screening of seedlings before further field evaluation, and contribute to shift the mean of the selection population for adaptation to warm, long day environments.

The same information was also used in a pilot study of genomic selection (GS). The underlying concept in GS is to model the entire complement of quantitative trait loci (QTL) effects across the genome to produce a genomic estimated breeding value (GEBV) from which progenitors or clones can be selected after genotyping. This pilot study tested several models proposed by Endelman (2011) to assess the feasibility of applying GS to improve selection of parental clones for early tuberization and bulking under long day-length. Prediction accuracies of $\geq$0.3 can be regarded as sufficiently high to merit implementation of GS (Heffner et al. 2010). Sufficiently high values were attained for tuberization induction, stolon length, bulking ratio at 75 and 90 days, and marketable tuber number at 75 days. As all markers are simultaneously fitted to capturing most of the trait heritability, GS can be expected to help reduce the length of selection cycles aimed at improving traits for which accurate phenotyping can be performed.

6.5 Prospects for True Hybrid Potato Breeding

Several countries and institutions have conducted research proposing the production of potato by sexual or botanical seed. Since the vegetative (tuber) planting material of clonally propagated potatoes is typically called seed potato, the term true potato seed (TPS) was introduced for tetraploid potato varieties produced from botanical seed. TPS aimed to overcome difficulties associated with production of vegetative seed including slow multiplication rate, accumulation of pathogens, and the bulky and perishable nature of tuber seed (Kidane-Mariam et al. 1985; Golmirzaie et al. 1994; Almekinders et al. 2009). As potato is a cross-fertilizing crop developed from heterozygous parents that do not "breed true," TPS is heterogeneous and thus variable for a large number of characteristics. While this variability is the source of variation for a clonal variety program, it is generally undesirable in varieties per se.

True hybrid potato breeding, on the other hand, aims at developing uniform true seed (TS) hybrid varieties from complementary superior inbred stocks developed at the diploid level. It would combine the advantages of TPS, (large numbers of disease-free propagules), diploid breeding (simplified genetics), and homogeneity (by way of combining inbred instead of heterozygous parents) in a new potato breeding system akin to that of maize.

Preliminary reports of productive diploid hybrids (de Vries et al. 2016), and the advantages of "re-inventing potato" as an inbred line-based crop (Jansky et al. 2016) make this effort one of the most intriguing developments in potato breeding over the past decade. Diploid inbreeding-tolerant potatoes would allow for considerably more systematic incorporation of new genes and traits, such as by backcrossing to fixed genotypes, as well as the possibility of substantial yield gains by crossing between well-defined heterotic groups. Working with diploid potatoes would also greatly facilitate the use of desirable recessive alleles. In principle, the low multiplication rate of clonally propagated potato can be circumvented by moving to diploid hybrids, where true seed from inbred × inbred crosses can produce vast numbers of genetically identical offspring in a short period of time.

Several steps need to be taken before diploid potato inbred line breeding might become routine. Most important is eliminating the self-incompatibility system, so that inbred lines can be readily developed. Recently, the dominant self-incompatibilty locus inhibitor allele (*Sli*) identified in certain self-compatible (SC) variants of the diploid wild potato *Solanum chacoense* (Hosaka and Hanneman 1998) has been used to overcome self-incompatibility (SI) and develop SC inbreds directly from the original source or by introgression to SI 2x cultivated germplasm (Phumichai and Hosaka 2006; Jansky et al. 2014). Additional research may be needed in order to manage SC as a trait in potato breeding, to improve the vigor of potato seedlings and to assure male fertility in diploid male parents.

Investment in inbreeding seems to be greater so far in the private than in the public sector. Nevertheless, some public breeding programs including CIP's have begun to seek some of the advantages that this novel strategy can offer to breeding and genetics for potato improvement. CIP has undertaken research to develop (1) a self-compatible) 2x breeding platform for broadening the genetic base of potatoes by hybridization across genepools, and (2) inbred genetic stocks for gene discovery and trait capture. Initial stocks and experience gained in the process of this research will contribute to longer term goals of developing complementary superior diploid inbred parental lines that will produce uniform, heterotic true seed 2x varieties when combined by crossing.

6.6 Variety Testing and Seed Links

6.6.1 How the Breeding Program Is Integrated with Seed Production in the State of New York in the United States

In addition to a public sector breeding program, the state of New York also maintains a public sector nuclear seed farm (the Uihlein Farm, in Lake Placid NY), which produces clean seed for commercial seed growers in the region. Each year the NY breeding program identifies up to five promising selections—typically clones that have been evaluated for at least 6 years in the field—and sends them to the Uihlein Farm. The Uihlein Farm first tests each clone for viruses, removing any that are present, and then, after a few years, depending on the multiplication scheme chosen, offers minitubers or limited-generation field grown tubers of these selections for sale.

Over time the breeding program has learned that it is important not to discuss promising clones with local industry until seed will soon be, or already is, available. If a clone elicits excitement, and growers ask for seed yet find none, they quickly lose interest, and it is very difficult to rekindle interest once seed does become available. When a clone is discussed before seed is available from the Uihlein Farm, the breeding program strives to provide up to 1000 kg of seed for commercial ware and chipstock growers to evaluate on their farms in the interim.

The NY seed farm bears some risk in this arrangement, as the breeding program typically discovers, over the course of additional years of evaluation, serious weaknesses in some of the clones being multiplied, rendering the clone unmarketable. On the other hand, the NY seed farm also benefits when it can be the first supplier of a new variety.

6.6.2 Regional Trials for Variety Assessment

Procedures, interfaces and responsibilities for trialing elite potato clones for variety release vary in degree of organization or formality across and sometimes within Countries.

As the aim of *plant breeding* is to produce new, improved *varieties, breeders should be aware of and anticipate meeting the several steps required for variety recognition, release and registration.*

Depending on program structure and target territory) and the complexity of the selection criteria, potato variety breeding typically involves a minimum of 6–7 years (and often more) of crossing and selection work, followed by a year or 2 of early testing to identify superior clones. The next step is the entry of candidate varieties into official variety registration recommendation trials. Programs in which breeding is done in the territory of release are more straight-forward in this respect than international programs in which additional time is required for quarantine exchange and bulking of seed in a new location. Depending on access to infrastructure and communication about objectives, this can take 2–5 years. The public and private sector differ in this respect since while public programs generally provide elite materials as true seed families or a small number of in vitro plants for research and breeding that may lead to local selection, breeding companies provide a reasonable quantity of seed of known varieties along with support and incentives for their testing and release.

6.6.3 Case Study 9: Regional Trials for Potato Variety Development in the United States

Northeast Regional Potato Trials—trialing system is a network for testing elite clones as candidate varieties. At present public-sector potato breeding in the US is divided into four regions—east, central, west and south. Each region has a regional trialing network, in which the agronomic performance of advanced clones is evaluated across a wide geographical range. The trial serves both to identify those relatively uncommon clones with broad adaptation, which are most likely to become commercially acceptable, and to identify varieties that may serve the needs of narrow geographic areas that have no breeding program of their own. Taking the eastern US as an example: the system consists of four public sector breeding programs—2 based in Maine, one in North Carolina, and 1 in New York—and 11 evaluation sites, 2 in NY and 1 each in Maine, North Carolina, Pennsylvania, Florida, Virginia, Ohio, Maryland and occasionally the Canadian provinces of New Brunswick and Prince Edward Island. Evaluators at each site rate clones using standardized scoring scales for yield, specific gravity, shape, internal and external defects, maturity, texture of tuber skin, and overall appearance, and upload the data to a common web portal (potatoes.ncsu.edu/NE.html). The originating breeding programs use the data to help make selection decisions, while regional evaluators use it to help inform local growers about which new varieties will perform best in their region. Breeders and evaluators also meet once a year to discuss, in person, the performance of clones in the trial the preceding year, and to make incremental adjustments in how the trial data is collected, analyzed and reported. Over the past decade separate national programs in the US, focused on chip and french fry variety development, have built upon the standardized framework of the eastern region trial network, with that data housed at potatoesusa.mediusag.com.

6.6.4 How Does the New York Program Decide When to Release a Variety?

In many countries a new potato variety cannot be legally grown until it has been formally evaluated in registration trials, found to be distinct, uniform and stable (DUS), and shown to have "value for cultivation and use." Decisions on all of these characters are made by an independent regulatory body, not the originating breeder.

In contrast, in the United States there is no legal requirement for comparable tests before a potato variety can be grown. (Although there is a requirement for DUS data, if a US breeder wants to procure plant breeders rights.) While some US universities have panels of breeders that evaluate candidate varieties, and can deny release of a variety by a colleague, others, like Cornell University in New York, have no formal potato variety release process at all.

Is either of these approaches preferable? Coauthor Walter de Jong is comfortable to operate in an unregulated environment. In his view the uniformity and stability tests are essentially meaningless for a clonally propagated crop. In addition, while he recognizes that evaluating distinctness is important to prevent anyone from re-releasing varieties developed by others, as long as two highly heterozygous autotetraploids have been crossed, all offspring will be distinct, as so many loci segregate. In other words, it is trivial to create a distinct potato. Variety release decisions in New York are made, instead, by gauging stakeholder interest. Prior to formal release, seed growers are permitted to grow candidate quantities under an evaluation license. When a candidate variety reaches a threshold, typically 10 ha or more of seed across all seed growers, and the seed growers ask for the variety to receive a formal (nonexperimental) name, it is released. One specific situation where the New York program will not release a variety, even if there is industry interest, is if tuber glycoalkaloids are too high (>20 mg/100 g tuber fresh weight).

6.6.5 *Case Study 10: All India Coordinated Crop Improvement Project and Release of Kufri-Lima from LTVR Population*

In India, state and national systems for verity testing and release coexist, and to some degree depend on each other, and these policies serve as a model for other countries in south Asia. Formal testing of a candidate variety by the All India Coordinated Crop Improvement Projects (AICCIP), and the subsequent recommendation of the variety for release by the AICCIP are required before variety can be released by the Central Sub-Committee that permits notification and thereby enables seed production and trade. In the case of potato, a minimum of 3 years of multilocational trials and assessment for Value for Cultivation and Use (VCU) take advantage of 16 locations across the country. These locations use standardized design and data templates and report on performance as well as meteorological conditions. Experiments are designed for Rabi (winter) season for irrigated areas such as Central, Northern, and Eastern regions (Indo-Gangetic Plains). Plateau regions (700–1200 masl) and hilly regions (1500–3000 masl) are represented in the multilocational trial series.

A new variety should be suitable for specified agro-climatic and soil conditions, have an ability to withstand typical stress conditions, and have tolerance/resistance to pests and diseases. It should also show distinct advantages over the existing equivalent released varieties, a process facilitated by standard formats for data collection. The breeders or Principal Investigators and Zonal Co-ordinators attend the meeting to provide wider information on the variety. The Director of Crop Development Programme is invited to provide information on the response of farmers to promotion and demonstration trials if they were conducted. The release proposal proforma requires the breeder to ensure availability of enough seed stock for seed multiplication on at least 10 ha.

An interesting feature of the All India trials project is that it reports on potato varieties as components of cropping systems, and not only on potato as a commodity. The pertinent proforma includes information on the productivity and profitability of potato and each of different crops in intercropping systems. Long-term experiments address issues of nutrient cycling and crop management that offer valuable methodology and possibly valuable data sets and recommendations for CIP's concern for the resilience of its Agile Potato proposition, proposing intensification of cereal-based systems with attention to sustainability.

Recent emphasis of collaboration between CIP and the Central Potato Research Institute (CPRI) of India have been on identifying early-maturing, heat tolerant, virus-resistant varieties suitable for intensifying and diversifying rice-based production systems. CIP introduces elite clones as in vitro plants as well as true seed from crosses between parents with good GCA for features of performance in subtropical lowland environments for used in breeding or direct identification of varieties for the vast Indo-Gangetic Plains and hilly regions of southwest Asia. The release of Kufri-Lima in October 2017, bred as CIP397065.28, is an exciting result of this collaboration. Potato virus are a serious problem for farmers in India, particularly in Gujarat, Rajasthan, Madhya Pradesh, Chhattisgarh, Odisha, West Bengal, and East Uttar Pradesh, where temperatures are higher in the winter season. Crop losses due to viruses range between 20 and 50%, and farmers from these areas traditionally need to import seed from Northern India (Punjab). "Farmers usually have to wait until temperatures drop to plant potatoes, but because of its tolerance to heat, Kufri Lima can be planted a full 20–30 days earlier than other local varieties." Earlier planting means earlier harvests, giving Kufri Lima farmers the ability to sell their potatoes at a premium price before other varieties hit the market. Farmers can expect to be paid 40–50% higher prices than those who harvest potatoes during the normal season. Heat tolerance and early maturity mean that Kufri Lima farmers can invest their earnings into a second winter harvest helping them to improve the overall productivity of the cropping system, while virus resistance may enable seed production closer to home. https://www.potatopro.com/companies/central-potato-research-institute-cpri.

Kufri-Lima was introduced to CPRI as CIP397065.28 in 2006 after initial selection in Peru. It was evaluated for adaptation, quality, and resistance to degeneration along with several other selections from the LTVR population with the support of ICAR (International Council of Agricultural Research), Central Potato Research Institute, GIZ/BMZ and USAID.

6.6.6 Variety Release and Registration

Keen awareness of timelines, procedures and actors involved in decisions on the release of varieties from elite clones and of policies and practices that permit or enable seed production is essential for the uptake of new varieties and subsequent benefit of their component and composite traits. Country specific systems for variety testing, release and dissemination of seed to farmers are governed by sets of national laws, scientific guidelines, norms, and standard practices which together can be termed "Regulatory Frameworks."

Registration is often required for seed to be multiplied and marketed in a country or territory or for exportation outside the territory. Release is intended to make a newly developed variety available to the public for general cultivation in the regions for which it is adapted. Even though they are not directly responsible, breeders should be aware of the designated service responsible for recommending the registration of new varieties in their target locations, as well as the Basic information required for application. Additional efficiencies can be realized when breeding programs are aware of and contribute to the availability of Basic material required for application and variety release. Knowledge of the agency with which a reference sample has to be deposited, up to date reference to the format for information required and the quantity of seed to be available at time of release vary be country or state. Depending on the territory in question, the designated service responsible for approving the registration of new varieties is usually a ministry or department of agriculture, which may have crop specific rules and regulations for variety development. In India and Bangladesh, for example, potato is among the crops for which a special category called "notified" is required for varieties to enter seed production schemes.

Additional policies, processes and agencies are those responsible for demonstration or promotion trials for pre-released and released varieties where required; maintenance of certification standards; and the production and storage of foundation seed and provision of seed of new varieties. Policy regulating seed trade is of less direct concern, though also important to breeding programs. An excellent reference for procedures and policy for variety testing and release can be found at http://www.coraf.org/wasp2013/wp-content/uploads/2013/07/ECOWAS_VAR_REGIST_MANUAL_SEP_081.pdf.

6.6.7 Regional Frameworks for Variety Release

Beyond national seed regulatory systems, several regional frameworks are being developed or implemented across countries. Advantages of regional frameworks include improved access by farmers to seed of improved varieties at affordable prices, avoidance of repetitive national testing, making seed trade easier, faster, and less expensive for introduction of new varieties, timely availability of quantity and high quality of seed and choice of variety needed. Regional frameworks encourage individual countries to adopt national seed legislation and establish national variety catalogues that contribute to growers' awareness of new varieties. Regional registration can result in a larger market share for varieties and reduced time and cost for national variety release when second and third country releases are simplified by admissibility of available performance trial data.

Fast track options for variety registration at regional level rely on policies and communication practices that recognize prior registration in one member country and one season testing as sufficient for variety release in a second one. Similarly relevant are federal and state or provincial procedures that complement each other, enabling the benefits of new varieties to be realized across states or countries. Mechanisms are also needed whereby release proposal on the basis of data from state and farmers' field trials could be used for the zonal release of a variety, across an agro-ecological zone that covers more than one state or one country (Tripp et al. 1997). These same authors conclude that adjustment to seed regulatory frameworks is necessary because of significant changes in national seed systems, including: reductions in budget for public agricultural research; the failure of many seed parastatals; increasing concern about plant genetic diversity; pressure for the establishment of plant variety protection; the increasing contributions of commercial seed enterprises; and the emergence of innovative local level variety development and seed production initiatives.

A report of test cases in Africa supported by Syngenta documented the inclusion of three new potato varieties in the COMESA (Common Market for Eastern & Southern Africa) regional catalogue in 2016 (https://www.syngentafoundation.org/

sites/g/files/zhg576/f/seeds_policy_regional_variety_release_test_cases_317.pdf). Notably, these are the first non-maize varieties to be included in the region's list.

6.7 Concluding Remarks

Potato breeding is at a crossroad. Gene editing and large scale genotyping are at hand, and true hybrid breeding appears feasible, even for this complex polyploid and traditionally clonally propagated crop. Together with pre- and post-Mendelian methods, these advances promise to help accelerate genetic gain to nourish a growing population, help preserve the environment, and confront economic and climate constraints. Sustained breeding and supporting trait research will assure potato's role in a more comprehensive and integrated agriculture in which quality and resilience, in addition to the historical emphasis on yield, are major goals. Community efforts including the development of cross-species tools and research aimed at understanding the principles underlying crop performance will be needed to meet growing demands on agriculture. The programs illustrated here use crop ontologies to support understanding of traits and attempt to exploit heterosis, inter-ploidy breeding, and accelerated selection schemes to reach their goals. Beyond this, research on root systems, G × E interaction and photosynthetic efficiency merit attention. The lengthy 12–20 years required to develop and release a new potato variety, coupled with climate variability, call for intensified research on selection strategies and choice of breeding method. The use of marker-assisted and genomic selection, the possibilities offered by inbreeding coupled with gametophyte selection and hybrid development, "big data" and efficient real-time phenotypic data collection and analysis are most compelling at present. Finally, collaboration across sites and disciplines and the utilization of statistical models for assessing genotypic adaptation and breeding value are critical to the efficient development and deployment of better potato varieties.

Acknowledgements We gratefully acknowledge Elisa Mihovilovich and Thiago Mendes for valuable discussions toward this chapter.

References

Almekinders CJM, Chujoy E, Thiele G (2009) The use of true potato seed as pro-poor technology: the efforts of an international agricultural research institute to innovating potato production. Potato Res 52:275–293

Bae J, Halterman D, Jansky S (2008) Development of a molecular marker associated with Verticillium wilt resistance in diploid interspecific potato hybrids. Mol Breed 22:61–69

Bernardo R (2002) Breeding for quantitative traits in plants. Stemma Press, Woodbury

Bonierbale M, Plaisted RL, Tanksley S (1993) A test of the maximum heterozygosity hypothesis using molecular markers in tetraploid potatoes. Theor Appl Genet 86:481–491

Bonierbale M, Amoros W, Burgos G, Salas E, Suarez H (2007) Prospects for enhancing the nutritional value of potato by plant breeding. African Potato Association conference proceedings, Alexandria, Egypt, vol 7, pp 26–46

Butler DG, Cullis BR, Gilmour AR, Gogel BJ (2009) ASReml-R reference manual. v. 3. Queensland Department of Primary Industries & Fisheries, Brisbane

Chase SS (1963) Analytic breeding in *Solanum tuberosum* L.—a scheme utilizing parthenotes and other diploid stocks. Can J Genet Cytol 5:359–363

Colton L, Groza H, Wielgus S, Jiang J (2006) Marker-assisted selection for the broad-spectrum potato late blight resistance conferred by gene RB derived from a wild potato species. Crop Sci 46:589–594

Comai L (2005) The advantages and disadvantages of being polyploid. Nat Rev Genet 6:836–846

Comstock RE, Robinson HF (1948) The components of genetic variance in populations of biparental progenies and their use in estimating the average degree of dominance. Biometrics 4:254–266

Cubillos A, Plaisted R (1976) Heterosis for yield in hibrids between *S. tuberosum* spp. *Tuberosum* and *S. tuberosum* spp. *Andigena*. Am Potato J 53:143–150

De Jong H, Rowe PR (1971) Inbreeding in cultivated diploid potatoes. Potato Res 14:74–83

De Nettancourt D (1977) Incompatibility in angiosperms. Springer, Berlin

Endelman JB (2011) Ridge regression and other kernels for genomic selection with R package rrBLUP. Plant Genome 4:250–255

Evanno G, Regnaut S (2005) Detecting the number of clusters of individuals using the software STRUCTURE: a simulation study. Mol Ecol 14:2611–2620. https://doi.org/10.1111/j.1365-294X.2005.02553.x

Ewing EE (1992) Critical photoperiods for tuberization: a screening technique with potato cuttings. Am Potato J 55:43–53

Finkers-Tomczak A, Bakker E, de Boer J, van der Vossen E, Achenbach U, Golas T, Suryaningrat S, Smant G, Bakker J, Goverse A (2011) Comparative sequence analysis of the potato cyst nematode resistance to locus H1 reveals a major lack of co-linearity between three haplotypes in potato (*Solanum tuberosum* ssp.). Theor Appl Genet 122:595–608

Frankel R, Galun E (1977) Pollination mechanisms, reproduction, and plant breeding. Springer, New York

Fuglie K (2007) Priorities for potato research in developing countries: results of a survey. Am J Potato Res 84:353–365

Gaiero P, Mazzella C, Vilaró F, Speranza P, de Jong H (2017) Pairing analysis and in situ hibridisation reveal autopolyploid-like behaviour in *Solanum commersonii* × *S. tuberosum* (potato) interspecific hybrids. Euphytica 213:137. https://doi.org/10.1007/s10681-017-1922-4

Gallais A (2004) Quantitative genetics and selection theory in autopolyploid plants. INRA, Paris

Gálvez R, Mendoza HA, Fernández-Northcote EN (1992) Herencia de la inmunidad al virus Y de la papa (PVY) en clones derivados de *Solanum tuberosum* spp. Andigena. Fitopatología 27:8–15

Gardner CO, Heberhart SA (1966) Analysis and interpretation of the variety cross diallel and related populations. Biometrics 22:439–452

Gatto M, Hareau G, Pradel W, Suarez V, Qin J (2018) Release and adoption of improved potato varieties in Southeast and South Asia. International Potato Center (CIP) Lima, Peru, 42p. Social Sciences Working Paper No. 2018-2. ISBN 978-92-9060-501-0

Gebhardt C, Bellin D, Henselewski H, Lehmann W, Schwarzfischer J, Valkonen J (2006) Marker-assisted combination of major genes for pathogen resistance in potato. Theor Appl Genet 112:1458–1464

Golmirzaie A, Malagamba P, Pallais N (1994) Breeding potatoes based on true seed propagation. In: Bradshaw J, MacKay G (eds) Potato genetics. CAB International, Wallingford, pp 499–512

Griffing B (1956) Concepts of general and specific combining ability to relation to diallel crossing system. Anst J Biol Sci 9:463–493

Grüneberg WJ, Mwanga R, Andrade M, Espinoza J (2009) Chapter 5: Selection methods Part 5: Breeding clonally propagated crops. In: Ceccarelli S, Guimarães EP, Weltzien E (eds) Plant breeding and farmer participation. FAO, Rome, pp 275–322

Hareau G, Kleinwechter U, Pradel W, Suarez V, Okello J, Vikraman S (2014) Strategic assessment of research priorities for potato. CGIAR Research Program on Roots, Tubers and Bananas (RTB), Lima (Peru). RTB Working Paper 2014-8. Available online at: www.rtb.cgiar.org

Harlan IR, de Wet IMI (1971) A rational classification of cultivated plants. Taxon 20(4):509–517

Hayman BI (1954) The theory and analysis of diallel crosses. Genetics 39:789–809

Heffner EL, Lorenz AJ, Jannink J l, Sorrells ME (2010) Plant breeding with genomic selection: Gain per unit time and cost. Crop Sci 50(5):1681–1690

Hirsch C, Hisrch CD, Felcher K, Coombs J, Zarka D, VanDeynze A, DeJong W, Veilleux RE, Jansky S, Bethke P, Douches DS, Buell CR (2013) Retrospective view of North American potato (*Solanum tuberosum* L.) breeding in the 20th and 21st centuries. G3 Genes Genomes. Genetics 3:1003–1013

Hirsch CD, Hamilton J, Childs K, Cepela J, Crisovan E, Vaillancourt B, Hirsch CN, Habermann M, Neal B, Buell CR (2014) Spud DB: a resource for mining sequences, genotypes, and phenotypes to accelerate potato breeding. Plant Genome 7(1):1–12

Hosaka K, Hanneman RE (1998) Genetics of self-compatibility in a self-incompatible wild diploid potato species *Solanum chacoense*. 1. Detection of an S locus inhibitor (Sli) gene. Euphytica 99:191–197. https://doi.org/10.1023/A:1018353613431

Jansky SH, Chung YS, Kittipadukal P (2014) M6: a diploid potato inbred line for use in breeding and genetics research. J Plant Regist 8:195–199. https://doi.org/10.3198/jpr2013.05.0024crg

Jansky SH, Charkowski AO, Douches DS, Gusmini G, Richael C, Bethke PC, Spooner DM, Novy RG, De Jong H, De Jong WS, Bamberg JB, Thompson AL, Bizimungu B, Holm DG, Brown CR, Haynes KG, Sathuvalli VR, Veilleux RE, Miller JC, Bradeen JM, Jiang JM (2016) Reinventing potato as a diploid inbred line–based crop. Crop Sci 56:1–11

Kasai K, Morikawa Y, Sorri VA, Valkonen JP, Gebhardt C, Watanabe KN (2000) Development of SCAR markers to the PVY resistance gene Ryadg based on a common feature of plant disease resistance genes. Genome 43:1–8

Kempthorne O, Curnow RN (1961) The partial diallel cross. Biometrics 17:229–250

Kidane-Mariam HM, Mendoza H, Wissar R (1985) Performances of true potato seed families derived from intermating tetraploid parental lines. Am Potato J 62:643–652. https://doi.org/10.1007/BF02853473

Kleinwechter U, Hareau G, Suarez V (2014) Prioritization of options for potato research for development—results from a global expert survey. CGIAR Research Program on Roots, Tubers and Bananas (RTB), Lima (Peru). RTB Working Paper 2014-7. Available online at: www.rtb.cgiar.org\

Levin DA (1983) Polyploidy and novelty in flowering plants. Am Nat 122:1–25

Lynch M, Walsh B (1998) Genetics and analysis of quantitative traits. Sinauer Associates, Sunderland

Meenakshi JV, Johnson N, Manyong VM, De Groote H, Javelosa J, Yanggen D, Naher F, Gonzalez C, Garcia J, Meng E (2007). How cost-effective is biofortification in combating micronutrient malnutrition? An ex-ante assessment HarvestPlus Working Paper No.2/. HarvestPlus, Washington, DC. www.harvestplus.org

Mendiburu AO, Peloquin SJ (1977) Bilateral sexual polyploidization in potatoes. Euphytica 26:573–583

Mendoza HA (1990) Mejoramiento poblacional. Una estrategia para la utilización del germoplasma de papa. En: Mejoramiento de papa en el Cono Sur, pp 47–62

Mendoza HA, Estrada NR (1979) Breeding potatoes for tolerance to stress: heat and frost. In: Mussel H, Staples RR (eds) Stress physiology in crop plants. Wiley, New York, pp 227–262

Mendoza HA, Sawyer RL (1985) The breeding program at the International Potato Center (CIP). In: Russell GE (ed). Progress in plant breeding, 1. London, Butterworths, 325 p

Mendoza HA, Mihovilovich EJ, Saguma F (1996) Identification of triplex (YYY) potato virus Y (PVY) immune progenitor derived from *Solanum tuberosum* spp. Andigena. Am Potato J 73(1):13–19. ISSN 0003-0589

Midmore DJ, Rhoades RE (1988) Applications of agrometeorology to the production of potato in the warm tropics. In: Rijks D, Stigter CJ (eds) Agrometeorology of the potato crop. International Society for Horticultural Science (ISHS); International Potato Center (CIP); World Meteorological Organization (WMO); European Association for Potato Research; Royal Netherlands Meteorological Institute, Wageningen, pp 103–136. Acta Horticulturae. no. 214

Mihovilovich E, Alarcón L, Pérez AL, Alvarado J, Arellano C, Bonierbale M (2007) High levels of heritable resistance to potato leafroll virus (PLRV) in *Solanum tuberosum* subsp. Andigena. Crop Sci 47(3):1091–1103

Mihovilovich E, Aponte M, Linqvist-Kreuze H, Bonierbale M (2014) An RGA-derived SCAR marker linked to PLRV resistance from *Solanum tuberosum* ssp. Andigena. Plant Mol Biol Report 32:117–128. https://doi.org/10.1007/s11105-013-0629-5

Muñoz FJ, Plaisted RL, Thurston HD (1975) Resistance to potato virus Y in *S. tuberosum* spp. Andigena. Am Portao J 52:107–115

Oakey H, Verbyla AP, Cullis BR, Wei XM, Pitchford WS (2007) Joint modeling of additive and non-additive (genetic line) effects in multi-environment trials. Theor Appl Genet 114:1319–1332. https://doi.org/10.1007/s00122-007-0515-3

Ortiz R (1998) Potato breeding via ploidy manipulations. Plant Breed Rev 16:15–85

Paget M, Amoros W, Salas E, Eyzaguirre R, Alspach P, Apiolaza L, Noble A, Bonierbale M (2014) Genetic evaluation of micronutrient traits in diploid potato from a base population of Andean Landrace Cultivars. Crop Sci 54:1949–1959

Phumichai C, Hosaka K (2006) Cryptic improvement for fertility by continuous selfing of diploid potatoes using Sli gene. Euphytica 149:251–258. https://doi.org/10.1007/s10681-005-9072-5

Piepho H-P, Möhring J, Melchinger AE, Büchse A (2008) Blup for phenotypic selection in plant breeding. Euphytica 161:209–222

Plaisted RL (1987) Advances and limitations in the utilization of Neotuberosum in potato breeding. In: Jellis GJ, Richardson DE (eds) The production of new potato varieties. Cambridge University Press, Cambridge, pp 186–196

Plaisted RL, Sanford L, Federer WT, Kehr AE, Peterson LC (1962) Specific and general combining ability for yield in potatoes. Am Potato J 39:185–197

Pushkarnath P (1942) Studies on sterility in potatoes. 1. The genetics of self- and cross-incompatibilities. Indian J Genet and Pl Br 2:11–36

Radley RW, Taha MA, Bremner PM (1961) Tuber bulking in the potato crop. Nature 191:782–783

Ragot M, Bonierbale M, Weltzien E (2018) From market demand to breeding decisions—a framework. RTB Working Paper.

Rodriguez Junior FA, Ortiz-Monasterio I, Zarco-Tejada PJ, Ammar K, Gérard B (2014) Using precision agriculture and remote sensing techniques to improve genotype selection in a breeding program. Conference Paper.International Conference on Precision Agriculture. http://quantalab.ias.csic.es/pdf/Rodrigues%20Jr.%20et%20al.%202014.pdf

Salas E (2002) Estudio de la variabilidad genética de la resistencia al Virus del Enrollamiento de la papa (PLRV) en una población tetraploide. Tesis para optar el Grado de Magister Scientiae. Universidad Nacional Agraria—La Molina. Lima, Peru

Sanford JC, Hanneman RE (1982) Intermating of potato haploids and spontaneous sexual polyploidization—effects on heterozygosity. Am J Potato Res 59:407–414

Sattarzadeh A, Achenback U, Lubeck J, Strahwald J, Tacke E, Hofferbvert H, Rothsteyn T, Gebhardt C (2006) Single nucleotide polymorphism (SNP) genotyping as basis for developing a PCR-based marker highly diagnostic for potato varieties with high resistance to Globodera pallida pathotype Pa2/3. Mol Breed 18:301–312

Slater AT, Wilson GM, Cogan NO, Forster JW, Hayes BJ (2014) Improving the analysis of low heritability complex traits for enhanced genetic gain in potato. Theor Appl Genet 127(4):809–820

Spooner DM, Ghislain M, Simon R, Jansky SH, Gavrilenko T (2014) Systematics, diversity, genetics, and evolution of wild and cultivated potatoes. Bot Rev 80:283–383. https://doi.org/10.1007/s12229-014-9146. —Toward a Rational Classification of Cultivated Plants—Jstor https://www.jstor.org/stable/1218252

Sprague GF, Tatum A (1941) General vs. specific combining ability in single crosses of corn. J Am Soc Agron 34:923–932

Tripp R, Louwaars NP, Van Der Burg WJ, Virk DS, Witcombe JR (1997) Alternatives for Seed regulatory reform: an analysis of variety testing, variety regulation and seed quality control, Agricultural Research and Extension Network Paper, no. 69. http://hdl.handle.net/10535/4581

Ugent DL, Pozorski S, Pozorski T (1982) Archaeological potato tuber remains from the Casma Valley of Peru. Econ Bot 36:182–192

Ugent DL, Dillehay T, Ramirez C (1987) Potato remains from a Late Pleistocene settlement in southcentral Chile. Econ Bot 41:17–27

Uitdewilligen JGAML, Wolters A-MA, D'hoop BB, Borm TJA, Visser RGF, van Eck HJ (2013) A next-generation sequencing method for genotyping-by-sequencing of highly heterozygous autotetraploid potato. PLoS One 8(5):e62355. https://doi.org/10.1371/journal.pone.0062355

Velásquez AC, Mihovilovich E, Bonierbale M (2007) Genetic characterization and mapping of major gene resistance to potato leafroll virus in *Solanum tuberosum* ssp. *Andigena*. Theor Appl Genet 114(6):1051–1058

De Vries MI, ter Maat M, Pim Lindhout P (2016) The potential of hybrid potato for East-Africa. Open Agric. 1: 151–156

Wricke G, Weber WE (1986) Quantitative genetics and selection in plant breeding. de Gruyter, Berlin

Xu X et al (2011) Genome sequence and analysis of the tuber crop potato. Nature 475(7355):189–195

Yuen J, Forbes G (2009) Estimating the level of susceptibility to *Phytophthora infestans* in potato genotypes. The American Phytopathological Society. Phytopathology 99(6):2

Zhang LH, Mojtahedi H, Kuang H, Baker B, Brown CR (2007) Marker-assisted selection of Columbia root-knot nematode resistance introgressed from *Solanum bulbocastanum*. Crop Sci 47:2021–2026

Chapter 7
Genetics and Cytogenetics of the Potato

Rodomiro Ortiz and Elisa Mihovilovich

Abstract Tetraploid potato (*Solanum tuberosum* L.) is a genetically complex, polysomic tetraploid ($2n = 4x = 48$), highly heterozygous crop, which makes genetic research and utilization of potato wild relatives in breeding difficult. Notwithstanding, the potato reference genome, transcriptome, resequencing, and single nucleotide polymorphism (SNP) genotyping analysis provide new means for increasing the understanding of potato genetics and cytogenetics. An alternative approach based on the use of haploids ($2n = 2x = 24$) produced from tetraploid *S. tuberosum* along with available genomic tools have also provided means to get insights into natural mechanisms that take place within the genetic load and chromosomal architecture of tetraploid potatoes. This chapter gives an overview of potato genetic and cytogenetic research relevant to germplasm enhancement and breeding. The reader will encounter findings that open new doors to explore inbred line breeding in potato and strategic roads to access the diversity across the polyploid series of this crop's genetic resources. The text includes classical concepts and explains the foundations of potato genetics and mechanisms underlying natural cytogenetics phenomena as well as their breeding applications. Hopefully, this chapter will encourage further research that will lead to successfully develop broad-based potato breeding populations and derive highly heterozygous cultivars that meet the demands of having a resilient crop addressing the threats brought by climate change.

7.1 Introduction

The most grown potatoes are tetraploid ($2n = 4x = 48$), but farmers in the Andes grow diploid ($2n = 2x = 24$), triploid ($2n = 3x = 36$), and pentaploid ($2n = 5x = 60$) cultivars (Watanabe 2015). The basic chromosome number of these tuber-bearing

R. Ortiz (✉)
Swedish University of Agricultural Sciences (SLU), Alnarp, Sweden
e-mail: rodomiro.ortiz@slu.se

E. Mihovilovich
Independent Consultant, Lima, Peru

H. Campos, O. Ortiz (eds.), *The Potato Crop*,
https://doi.org/10.1007/978-3-030-28683-5_7

Solanum species is 12. Diploid cultivars along with tetraploid cultivars are used in potato breeding through ploidy manipulations with haploids and $2n$ gametes (or gametes with the sporophytic chromosome number), and chromosome engineering using aneuploidy (Ortiz 1998). Improvement of cultivated potato is challenged by its high heterozygosity (Bradshaw et al. 2006; Hirsch et al. 2013) and complex polysomic tetraploid inheritance (Howard 1970; Ortiz and Peloquin 1994; Ortiz and Watanabe 2004). The tetraploid *Solanum tuberosum* has four homologues, which include 12 unique chromosomes each, thus showing tetrasomic inheritance (Bradshaw 2007). Epistasis and heterozygosity are key for succeeding in $4x$ potato breeding because multi-allelic quantitative trait loci showing high-order genic interactions, while additivity also contributes to quantitative traits with high heritability.

The genetics of tetrasomic potato depends on four sets of homologous chromosomes (instead of two as in diploid potato). Three genotypes (*AA*, *Aa*, *aa*) are expected after selfing a heterozygous diploid (*Aa*), while the selfing of a comparable tetraploid (*AAaa*) gives five different genotypic classes in its offspring: *AAAA* (quadruplex), *AAAa* (triplex), *AAaa* (duplex), *Aaaa* (simplex), and *aaaa* (nulliplex). Double reduction—related to tetrasomic inheritance—occurs when two chromosomes in a gamete derive from two sister chromatids; i.e., the sister chromatids end in same gamete. Quadrivalent formation, a single crossing over between the centromere and the locus to allow sister chromatids to attach to two different centromeres, that these centromeres with sister chromatids move to the same pole in anaphase I, and sister chromatids go the same pole in anaphase II are necessary for double reduction. The probability of double reduction to occur is noted as α, which is equalto $\frac{qea}{2}$, where q is the quadrivalent frequency, e is the frequency of equational separation that depends on the gene–centromere map distance, and a is the frequency of non-disjunction (often $\frac{1}{3}$). Chromosome segregation arises when $\alpha \neq 0$,thus indicating that a locus of interest lies close to the centromere; while if $\alpha = \frac{1}{7}$ or $\frac{1}{6}$ then chromatid segregation or maximal equational division (MED), respectively, occurs. MED is rarely found because of the requirements for its occurrence (Burham 1984). DNA-aided marker analysis confirmed the occurrence of double reduction and that it increases with distance from the centromeres (Bourke et al. 2015). As noted by Gálvez et al. (2017), the potato reference genome and transcriptome, research on both gene expression and regulatory motif, plus resequencing and SNP genotyping analyses, provide new means for increasing the understanding of potato genetics. For example, DNA resequencing allows assembling a genome reference for each cultivar or landrace, which may provide useful knowledge regarding structural differences between the various potato groups. In this regard, the resequencing of diversity panel including wild species, landraces, and cultivars demonstrated that a limited gene set accounts for early improvement of the potato cultigen, while distinct loci seems to be involved on the adaptation *S. tuberosum* group Andigenum (upland potato) and *S. tuberosum* groups Chilotanum and Tuberosum (lowland potato) populations (Hardigan et al. 2017). Signatures of selection in genes regulating pollen development/gametogenesis reduced fertility. Introgression of truncated alleles of wild species, particularly *S. microdontum*, was noted in long day cultivars, thus showing how wild tuber-bearing *Solanum* species are key sources of variation for breeding.

7.2 Haploids and Disomic Inheritance

Tetraploid potato shows significant inbreeding depression (De Jong and Rowe 1971) because polyploidy and heterozygosity mask deleterious recessive mutations and buffer genomic imbalance (Comai 2005; Henry et al. 2010; Tsai et al. 2013). These characteristics led to an alternative breeding approach based on the use of haploids ($2n = 2x = 24$) produced from tetraploid *S. tuberosum*. The homozygosity/heterozygosity of methylated DNA may be, however, involved in inbreeding depression/heterosis in self-compatible diploid potatoes because DNA methylation may suppress gene expression (Nakamura and Hosaka 2010).

The induction of haploid plants is generally referred to as "haploidization." There are two main pathways by which haploid formation can be induced in potato: androgenesis and gynogenesis. Androgenesis is through in vitro culturing of whole anthers or free microspores on a nutrient rich medium to induce plantlet regeneration from single gametic cells or haploid calli (Veilleux 1996), while gynogenesis is haploidization via the "maternal" or seed parent's genome. Potato haploids are routinely obtained by gynogenesis, a process in which specific *S. tuberosum* Group Phureja ($2n = 2x = 24$) selections, known as "haploid inducers," contribute the paternal gametes for pollination of the desired haploid progenitor. The formation of a haploid embryo begins when the egg is either induced into parthenogenesis (Hermsen and Verdenius 1973) or when the zygote experiences spontaneous abortion of the pollen donor's set of chromosomes (Clulow et al. 1991). Evidence for the latter has been the identification of Phureja-specific molecular markers in aneuploids ($2n = 2x + 1 = 25$) among the offspring of some haploid induction crosses (Clulow et al. 1991; Clulow and Rousselle-Bourgeois 1997; Samitsu and Hosaka 2002; Straadt and Rasmussen 2003; Ercolano et al. 2004). Ortiz et al. (1993a) suggested that the genetics of the ability to induce haploids is relatively simple.

Putative haploids are identified firstly by the lack of a dominant morphological marker for anthocyanin pigmentation on developing embryos (embryo spots) or on seedling shoots (nodal bands). This marker that allows early haploid selection is present in homozygosis in certain pure *S. tuberosum* Group Phureja clones or has been bred to homozygosity in its derived hybrids. The ploidy of the resultant seedlings is confirmed by counting chromosomes in mitotically dividing root cells (Sopory 1977), counting chloroplasts in stomatal guard cell pairs (Singsit and Veilleux 1991) or through flow cytometric analysis (Owen et al. 1988). The haploid-inducing clones often used owing to their relatively superior haploid-induction frequency and homozygosity for the seed marker "embryo-spot" are the following

Table 7.1 Haploid induction ability of "IVP 35," "IVP 101," and "PL-4"

Character[a]	"IVP 35"	"IVP 101"	"PL-4"
Number of haploids per 100 berries	69	62	96
Number of haploids per 1000 seeds	52	85	103

[a]Pooled data per haploid inducer from 13 seed parents

Group Phureja clones: "IVP 35," "IVP 48", and "IVP 101" (Ross 1986). "IVP 101" has been derived from the cross [(G609 × "IVP 48") × ("IVP 10" × "IVP 1")]. G609 is a haploid from Group Tuberosum cultivar "Gineke" that combines its own haploid induction ability with a high degree of male fertility, profuse flowering, and vigor.

The efficiency of haploid production is determined by both, production ability of the tetraploid seed parent and induction ability of the diploid pollinator (Hougas et al. 1964; Frandsen 1967; Hermsen and Verdenius 1973). However, interaction between seed parents and pollinators were also noted (Frandsen 1967). Despite this interaction, haploid induction ability of "IVP 101" has proved to be higher than "IVP 35" and "IVP 48" (Hutten et al. 1993). By the end of the 1990s a promising haploid inducer named "PL-4" (CIP596131.4) was selected at the International Potato Center (CIP, Lima, Perú) as a transgressive genotype from the cross between "IVP 35" × "IVP 101" due to its highest haploid inducer ability, degree of flowering, shedding and pollen viability relative to its parents (M. Upadhya and R. Cabello, CIP, unpublished data). Historical data accumulated from 2001 to 2009 from haploid induction crosses between 37, 4x breeding lines with both "IVP 101" and "PL-4" showed that the latter produced twice the amount of seeds without embryo spot (putative haploids) of "IVP 101." A more comprehensive study to determine the haploid inducer ability of "PL-4" relative to their parents was performed during 2015 and 2016 at CIP. This involved haploid induction of 13, 4x breeding clones with the three haploid inducers. Haploid confirmation was made by counting chloroplasts in stomatal guard cell pairs and flow cytometric analysis in seedlings grown from seeds without embryo spot. "PL-4" produced a significantly higher number of haploids than its parents. Meanwhile, "IVP 101" outperformed "IVP 35" in number of haploid per 1000 seeds (Table 7.1). There were also differences in haploid production ability between seed parents (Table 7.2). Two breeding clones, CIP 300056.33 and CIP 392820.1, showed the highest number of haploids.

7.2.1 Further Research and New Evidence on Haploid Origin

Previous research in potato haploids originated by gynogenesis detected aneuploids ($2n = 2x + 1 = 25$ and $2n = 2x + 2 = 26$) instead of the expected 24-chromosome karyotypes with concurrent appearance of Group Phureja-specific molecular markers (Clulow et al. 1991, 1993; Waugh et al. 1992; Wilkinson et al. 1995; Clulow and Rousselle-Bourgeois 1997; Ercolano et al. 2004). Moreover, there was one case in which translocation of a Group Phureja chromosomal segment to the Group Tuberosum genome was detected by genomic in situ hybridization (GISH; Wilkinson

Table 7.2 Haploid production ability of 13, 4*x* breeding clones from CIP potato breeding program

CIP-number[a]	Breeding code	Number of haploids per 100 berries	Number of haploids per 1000 seeds
CIP 300056.33	LR00.014	141	169
CIP 300072.1	LR00.022	68	101
CIP 300093.15	LR00.027	82	58
CIP 301023.15	C01.020	40	63
CIP 388615.22	C91.640	75	141
CIP 388676.1	Y84.027	40	32
CIP 390478.9	C90.170	2	2
CIP 390637.1	93	76	107
CIP 391931.1	458	11	17
CIP 392780.1	C92.172	9	21
CIP 392820.1	C93.154	258	175
CIP 397073.16	WA.104	39	30
CIP 397077.16	WA.077	100	68
Average		72	76

[a]Pooled data per breeding clone from three haploid inducers

et al. 1995). The cause, frequency, and nature of these introgressions remain unknown, though they may affect performance of the haploids (Allainguillaume et al. 1997). A similar phenomenon has been observed in CenH3-based haploid induction in *Arabidopsis thaliana* (Ravi and Chan 2010; Ravi et al. 2014; Tan et al. 2015). In this system, mis-segregation of the haploid inducer chromosomes leads to genome elimination. In a fraction of the haploid progeny, one or few of the haploid inducer chromosomes were retained, resulting in aneuploid progeny. This DNA introgression was identified readily by low-pass sequencing and single nucleotide polymorphism (SNP) analysis (Tan et al. 2015). These findings added evidence that DNA introgression from a haploid inducer is expected to involve large contiguous segments, and often whole chromosomes. Lately, K.R. Amundson et al. (unpublished) surveyed a haploid segregating population for aneuploids by low-pass sequencing. The population was previously developed at CIP for tetraploid genetic mapping of a major gene controlling *Potato leaf roll virus* (PLRV) resistance in the Group Andigena cultivar “Alca Tarma” (Velásquez et al. 2007). They identified 19 haploids (11.4%) that displayed elevated relative sequence read coverage of a single chromosome consistent with 25-chromosome karyotypes in root tip metaphase spreads in these putatively aneuploid clones. By sequencing parental genotypes to higher depth (40–66×) and identifying homozygous SNP between “Alca Tarma” and either of the two haploid inducers, “IvP-101” or “PL-4,” plus assuming to have sired each haploid (Velásquez et al. 2007), they found nearly 0% haploid inducer

SNP for all chromosomes of these aneuploids. Thus, the additional chromosomes observed likely did not originate from the haploid inducer genome, but were maternally inherited. This lack of Group Phureja SNP in haploids was previously reported in another study that employed DNA markers (Samitsu and Hosaka 2002). The production of aneuploid gametes is a common property of polysomic polyploids (Comai 2005), and "Alca Tarma" was not an exception. Admusson et al. (unpublished) concluded that for haploid inducers "IvP-101" and "PL-4," either the mechanism of haploid induction does not involve egg fertilization or genome elimination in "Alca Tarma" was very efficient.

7.3 Relevance of Haploids in Plant Breeding and Genetics

Haploids showed disomic inheritance, which means that each chromosome paired with its homolog, thus providing means for simplifying genetic research in potato. They can also be efficiently used for research on chromosome pairing and natural mutation accumulated at the tetraploid level. Initially, potato haploids were envisioned as a tool to simplify the breeding of *S. tuberosum* cultivar production by reducing tetraploid germplasm to a diploid breeding level (Chase 1963). A second early reason for the production of haploids was to acquire a "genetic bridge" between the various genomes of *Solanum* species. Ploidy barriers between the cultivated and wild *Solanum* species could be circumvented by crossing haploids to the wild diploid *Solanum spp.* and novel hybrid germplasm incorporated back into tetraploid breeding programs through $4x \times 2x$ crosses using $2n$ gamete formation, or by colchicine-doubling of the novel diploid hybrid (Ross 1986). In addition to their use in breeding, diploid potato hybrids represent a powerful tool for genetic analysis due to its much simpler segregation ratios compared to tetraploid cultivated potatoes (Ortiz and Peloquin 1994). Thus, diploid potato has been used to determine the inheritance of economically important traits such as tuber shape (De Jong and Burns 1993; Van Eck et al. 1994b), tuber flesh and skin pigmentation (De Jong 1987; Van Eck et al. 1994a), and tuber skin texture (De Jong 1981). The genetic basis of some physiological mutants has also been analyzed with the use of diploids (De Jong et al. 1998, 2001). Haploids have been convenient for trait mapping (Kotch et al. 1992; Pineda et al. 1993; Freyre et al. 1994; Simko et al. 1999; Naess et al. 2000; Velásquez et al. 2007) and in the development of an online catalogue of amplified fragment length polymorphisms (AFLP) covering the potato genome (Rouppe van der Voort et al. 1998). On the other hand, many breeders have extracted haploids from superior parents and also maintain a diploid gene pool composed of hybrids between haploids and diploid wild species carrying specific quality and host plant resistance genes not found in cultivars (Carputo and Barone 2005; Ortiz et al. 2009).

Currently, there are some ongoing efforts towards genetically restructuring potato as a diploid inbred line-based crop (Jansky et al., 2016; Lindhout et al. 2011). Here, the vision is a diploid potato crop composed of a series of inbred lines that capture the favorable genetic diversity available in the potato cultigen. This diploid genepool with a broad suite of traits represent a valuable stock for fixing desirable gene combinations and realized breeding gains. Last but not least, haploids have been regarded as a tractable ploidy state for copy number variation (CNV) analysis in tetraploid potatoes as these variants are meiotically transmissible and diploid gametes can likely shield deficiencies such as recessive lethal and dosage sensitive loci (Comai 2005; Lovene et al. 2013; Henry et al. 2015). CNV is defined as stretches of DNA from 1 kilobase (kb) to several megabases (Mb) that display different copy numbers in populations (Feuk et al. 2006). Analyses of multiple genotypes in *Arabidopsis* and maize suggest that CNV may play a significant role in phenotypic diversity and hybrid heterosis in plant species (Swanson-Wagner et al. 2010; Cao et al. 2011). Moreover, they can affect phenotype impacting important agronomic and host plant resistance traits (Maron et al. 2013; Díaz et al. 2012; Zhu et al. 2014; Cook et al. 2012). Direct CNV detection in tetraploid potato is challenging. The dosage increase associated with a duplication is subtler in tetraploids (25% increase) than in diploids (50% increase), and high heterozygosity impedes haplotype assembly (Potato Genome Sequencing Consortium 2011). As a consequence, CNV analysis is often limited to few loci of interest and is costly to be practical in breeding programs. Hence, sampling the gametophyte genome of tetraploids by ploidy reduction through haploidy is an alternative approach.

7.4 2*n* Gametes

Gametes with the sporophytic chromosome number should be named as 2*n* gametes and not as "unreduced" gametes as wrongly dubbed. They result from pre-meiotic, meiotic, or post-meiotic abnormalities during gametogenesis. The modes of formation are pre-meiotic doubling, first division restitution (FDR), chromosome replication during meiotic interphase, second division restitution (SDR), post-meiotic doubling, and apospory (diploid sac formed from nucellus or integument cell). FDR and SDR mechanisms are the most common for 2*n* pollen and 2*n* egg formation in potato (Ortiz 1998). Heterozygous 2*x* parents transmit 80% and 40% of their heterozygosity to their 4*x* hybrid offspring after sexual polyploidization with FDR or SDR 2*n* gametes, respectively.

The parallel orientation of the spindles in the second meiotic division accounts frequently for FDR 2*n* pollen, while omission of the second division after a normal first division seems to be often involved in SDR 2*n* eggs. The abnormal meiosis leading to these 2*n* gametes are under the genetic control of recessive mutants: *ps* for 2*n* pollen and *os* for 2*n* egg, both of which appear to be ubiquitous in *Solanum* species. The finding of genes whose mutations led to a high frequency of 2*n* gametes in the model plant species *Arabidopsis thaliana* provided further means for understanding their formation in plants (Brownfield and Kölher 2011). It appears to be very likely that a mechanism related to a loss of protein function leads to the formation of 2*n* gametes.

The frequency of 2*n* gametes may be affected by incomplete penetrance and variable expressivity, which is under minor modifier genes and influenced by plant age and the environment. Phenotypic recurrent selection could be effective for increasing the frequency of FDR 2*n* pollen, while recurrent selection with progeny testing may raise SDR 2*n* egg expressivity.

There are synaptic mutants affecting gametogenesis in haploids, *Solanum* species and haploid-species hybrids. They may cause poor pairing, reduced chiasma, or both, thus reducing recombination. For example, the synaptic mutant sy_3—found in Group Phureja–haploid hybrids—along with *ps* produces FDR 2*n* pollen without crossing over (FDR-NCO), while the desynaptic mutant *ds-1* generates sterile *n* eggs and fertile FDR 2*n* eggs owing to a direct equational division of univalent chromosomes at anaphase I, i.e., pseudohomotypic division. Desynaptic gametes may transfer about 95% of the 2*x* genotype to their 4*x* hybrid offspring.

7.5 Cytoplasm Diversity and Male Sterility

There are six distinct cytoplasmic genome types in potato, namely, M, P, A, W, T, and D. Many clones bred worldwide show a genetic bottleneck in cytoplasmic diversity due to the continuous use of cytoplasmic-based male sterility "lineages" derived from *S. demissum* or *S. stoloniferum* that are often used as sources of host plant resistance. For example, T (45%), D (38%), and W (11%) are the most frequent types in CIP bred germplasm (Mihovilovich et al. 2015); while the most popular among EU cultivars and breeding clones are T (59%), D (27%), and W (12%) (Sanetomo and Gebhardt 2015), and cultivars and breeding lines from Japan plus a sample of landraces and foreign cultivars show 73.9% T, 17.4% D, and 2.4% W (Hosaka and Sanetomo 2012).

Cytoplasmic factors and nuclear alleles are involved in indehiscence, shriveled microspores, sporad formation, anther-style fusion, ventral-styled anthers, and thin anthers (Grun et al. 1977). Group Andigena and its ancestors, Group Stenotomum and Phureja, share most plasmon factors, of which many differ from those found in Group Tuberosum. Hence, cytoplasmic-genetic male sterility is often noted in hybrid offspring among some tuber-bearing *Solanum* species because interactions between sensitive factors in the cytoplasm of one species and nuclear genes from the other species. Hybrids derived from crossing Group Tuberosum haploids as female and Group Phureja or Stenotomun as males are very often male sterile, but the reciprocal cross show male fertile offspring. Male sterility ensues from the interaction of a dominant gene (*Ms*) from the Group Phureja or Stenotomun with Group Tuberosum sensitive cytoplasm. Diploid (2EBN) wild species such as *S. chacoense, S. berthaultii,* and *S. tarijense* do not carry genes that interact with the Group Tuberosum cytoplasm as shown by the high male fertile hybrid offspring between them.

The frequency of male fertile offspring in hybrids between Group Tuberosum and groups Stenotomum or Phureja may vary because some tetraploid cultivars bear a dominant male fertility restorer (*Rt*) gene (Iwanaga et al. 1991b). The *Ms* and *Rt* genes, which are independently inherited, are very distal from the centromere, thus showing both loci chromatid segregation (Ortiz et al. 1993b). The *Rt* gene allows to partially bypass male sterility using crossbreeding, e.g. by crossing Group Tuberosum haploids bearing *Rt* with 2× species carrying *Ms* because ½ (*rt*/*rt* × *Ms*/*ms*) ¾ (*Rt*/*rt* × *Ms*/*ms*) or 100% (*Rt*/*Rt* × *Ms*/*Ms* or *Rt*/*Rt* × *Ms*/*ms*) of the resulting hybrid offspring will be male fertile.

7.6 Self-Incompatibility and *s* Locus Inhibitor Mechanism

Diploid potatoes and their related wild tuber-bearing *Solanum* species are self-incompatible due to a gametophytic self-incompatibility (GSI) system controlled by the interaction of a pollen *S* gene with pistil *S* gene(s). GSI inhibits fertilization by self-pollen or pollen from closely related (sibling) plants (Hanneman Jr 1999). In this system, compatibility is controlled by the *S* locus, which consists of two closely linked genes: *S-RNase* and *S-Locus F-box* (*SLF/SFB*) that control the female and male specificity, respectively. S-RNase-based self-incompatibility systems are widespread mechanisms for controlling selfing (Hancock et al. 2003). *S* locus variants, currently known as S-haplotypes, determine self-incompatible pollen rejection when there is a match between the single S-haplotype in the haploid pollen and either of the two haplotypes in the diploid system. S-RNases are the determinants of *S*-specificity in the pistil and act in pollen recognition as well as in direct pollen growth inhibition by degrading pollen RNA in incompatible pollinations (McClure et al. 1990). A distinct S-RNase protein is expressed from each functional S-haplotype and upon recognition the protein enters intact pollen tubes retaining its potentially cytotoxic enzyme activity (Gray et al. 1991). Conversely, pollen RNA is stable in compatible pollinations as interaction between S-RNase and *SLF* confers resistance to the cytotoxic effects of S-RNase (Fig. 7.1; Golz et al. 2001). The *SLF/SFB* is a family of F-box protein genes whose most well-known role is ubiquitin-mediated protein degradation. Pollen modifier genes encoding proteins that form complexes with SLF provides for ubiquitylation and degradation of non-self S-RNase as a necessary step to overcome the cytotoxicity of S-RNase. On the other hand, self S-RNases fail to bind productively and thus escape degradation (Hua et al. 2008; Zhang et al. 2009).

Modifier genes encoding putative pistil self-incompatibility factors were found in potato wild relatives. *HT-A* and *HT-B* are two similar genes expressed in the genus *Solanum*, but only *HT-B* has shown to be strongly suppressed in self-incompatibility breakdown in *S. chacoense*. HT-B proteins appear to be degraded in pollen tubes after compatible pollination, while this is not the case in incompatible pollen tubes where substantial amounts of HT-B reactive protein were found (Fig. 7.1; Goldraij et al. 2006).

Self-compatibility can be obtained by converting a self-incompatibility diploid (e.g., S_1S_2) to a tetraploid (de Nettancourt 1977). Here the defect occurs only in the pollen due to the so-called heteroallelic pollen (HAP) effect. Thus, $S_1S_1S_2S_2$ pistils

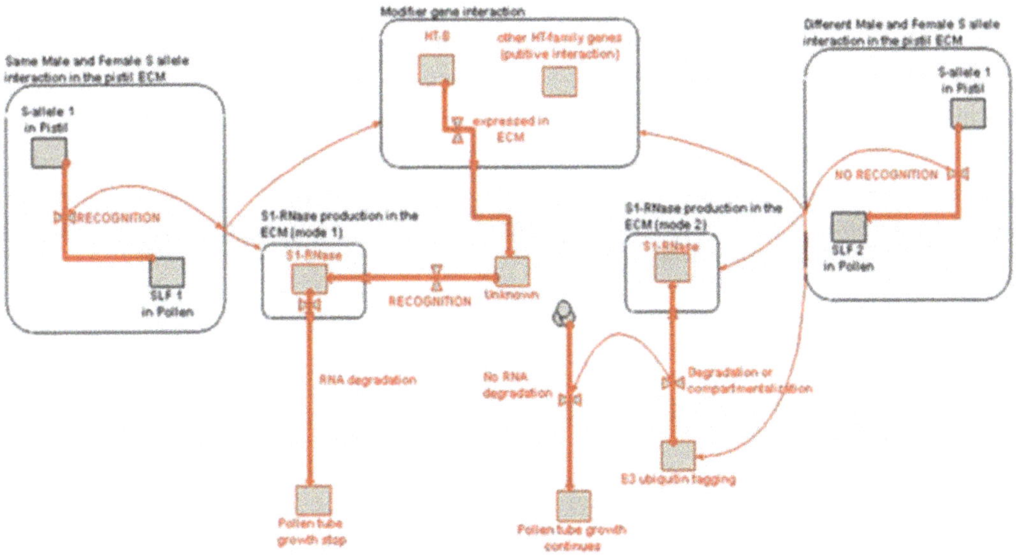

Fig. 7.1 A model for S-RNase-based self-incompatibility. When the pollen tube enters the extracellular matrix (ECM) the proteins HT-B, S-RNase, and 120K (not represented) are expressed. On a positive recognition of self-pollen (or a self *SLF* gene) the pollen tube RNA is broken down by the production of S-RNase and fertilization is unlikely. In the opposite reaction, when an unrecognizable match is made, S-RNase is degraded after being tagged by an E3 ubiquitin ligase complex. (Courtesy: Dr. Philippe Kear, International Potato Center)

reject S_1- and S_2-pollen normally, but diploid pollen is not rejected. Self-incompatibility breakdown only occurs in the HAP case, S_1S_2. Pollen *S* functions to provide resistance to S-RNase. Self-compatible variants have often been described among genotypes of self-incompatible potato species, such as, *inter alia*, *S. chacoense*, *S. kurtzianum*, *S. neohawkesii*, *S. pinnatisectum*, *S. raphanifolium*, *S. sanctaerosae*, *S. tuberosum* Groups Phureja and Stenotomum (Cipar et al. 1964), and *S. verrucosum* (Eijlander 1998). SC variants were also noted in haploids ($2n = 2x = 24$ chromosomes) from Group Tuberosum (De Jong and Rowe 1971; Olsder and Hermsen 1976) and in hybrids between Group Phureja and haploids ($2n = 2x = 24$ chromosomes) of Group Andigenum (Cipar 1964).

Self-compatible variants in *S. chacoense* show a single dominant gene "*Sli*" that is expressed in a sporophytic fashion (Hanneman 1985; Hosaka and Hanneman 1998a). The existence of self-incompatible progeny segregating from *S. chacoense* selfed plants having *Sli* gene in a heterozygous condition shows that fully functional S-haplotypes are transmitted through pollen even when the *Sli* factor is not. *Sli* was mapped to the end of chromosome 12 (Hosaka and Hanneman 1998b). Since the *S* locus has been localized on chromosome 1 (Gebhardt et al. 1991; Jacobs et al. 1995; Rivard et al. 1996), it is evident that the *Sli* gene is independent of the *S* locus. *Sli* can be regarded as a dominant gain-of-function (GOF) pollen-part mutant (PPM) that interacts in some way with the GIS system in pollen and results in self-compatible plants. This pollen side gene may inhibit S-RNase uptake, break down the pollen-stigma recognition system, overcome the cytotoxic activity of S-RNase independent of its pollen *S*-genotype or through the action on a non-S-specific factor (Hosaka and Hanneman 1998b; McClure et al. 2011). Likewise, self-compatible variants in *S. verrucosum* (*ver*) has been assumed to be a pistil side nonfunctional S-RNase haplotype (*Sv* allele) that allows *ver* plants be pollinated by its own pollen as well by the pollen of other self-incompatible potato species (Eijlander 1998). Absence of pistillate S-RNases seems to be a characteristic feature of this species (Makhan'ko 2011).

7.6.1 Self-Compatibility in Breeding

Most cultivated tetrasomic polyploid or self-incompatible diploid potatoes have not realized breeding gains due to low recombination, long generation cycles, polyploidy, inbreeding depression, and poor adaptation of wild potato germplasm (Visser et al. 2009; Lindhout et al. 2011). New breeding methods that involve the development of diploid inbred lines in potato were proposed as a strategy to address many perceived limitations faced by potato breeders (Birhman and Hosaka 2000; Phumichai et al. 2005). In the last few years a trend emerged in a group of potato breeders to reconsider the crop as a diploid species composed of a series of inbred lines that capture the favorable genetic diversity available in cultivated and wild potatoes. Inbreeding due to selfing may be efficient for organizing the whole gene pool into various favorably interacting and stable epistatic systems (Allard 1999).

The self-incompatibility inhibitor (*Sli*) gene opens new doors to explore inbred line breeding in potato (Lindhout et al. 2011). Highly inbred *S. chacoense* lines such as M6, which has been self-pollinated for seven generations, are vigorous and fertile (Jansky et al. 2014). CIP has incorporated *Sli* into diploid cultivars of *S. tuberosum* groups Stenotomum and Phureja and selected a panel of 20 *Sli* bearing self-compatible hybrids to provide a more desirable self-compatibility source than wild *S. chacoense* for the development of inbred lines. These self-compatible hybrids denoted BSLi, are being used to incorporate novel diversity from wild species and take full advantage of modern genetics and genomics tools to generate inbred genetic resources such as recombinant inbred lines (RILs), for fundamental gene discovery and gene mapping.

Self-compatibility has been identified in five diploid cultivated potatoes of *S. tuberosum* Phureja (phu) and Stenotomum (stn) Groups held in CIP's genebank. Selfing three of these self-compatible diploid cultivars; i.e., CIP705468 (goniocalyx), CIP703320 (stn), and CIP701165 (stn) yielded progenies that segregated for self-compatible and self-incompatible individuals. The segregation ratio 2 self-compatible: 1 self-incompatible was significantly skewed from the expected ratio of 3 self-compatible: 1 self-incompatible for a mutant factor in heterozygosis because of the small population size ($N < 60$) analyzed in each selfed cultivar. Self-compatible plants due to pistil side mutations that compromised S-RNase or HT factors produce only self-compatible plants. The same is true for GOF mutations of the *SLF* gene

(McClure et al. 2011). On the other hand, self-compatible plants with a *Sli* mutation in heterozygocity produce 3 self-compatible: 1 self-incompatible plants regardless of its haplotype condition in the *SLF* locus; i.e., *SxSx*, *SxSy*, *SySy*. Hence, the presence of self-compatible plants in the offspring of self-compatible 2*x* cultivars suggests a dominant pollen-side mutation similar to the *Sli* gene since this is the only scenario that yields self-incompatible offspring. Further research will be required to elucidate whether this *Sli*-like phenotype is a novel pollen-side GOF factor or *Sli* gene variant. Whatever its nature, self-compatible 2*x* cultivars would provide a more desirable self-compatible source than *S. chacoense* as they will avoid the undesirable linkage drag associated with the use of a wild species in the development of 2*x* inbred lines.

7.6.2 Interspecific Crosses and Incompatibility

Interspecific reproductive barriers (IRBs) complicate using wild germplasm species for crop improvement (Zamir 2001; Jansky et al. 2013). Therefore, the rich trait diversity available in CIP's extensive collection of tuber-bearing *Solanum* accessions requires overcoming IRBs to be introgressed into cultivated *Solanum* taxa. These barriers include incompatibility between pollen and pistil, male sterility resulting from interactions between nuclear and cytoplasmic genes, and endosperm failure (Camadro et al. 2004). Post-zygotic IRBs are due to EBN incompatibilities that lead to endosperm failure, whereas pre-zygotic IRBs are typically associated with pollen tube growth inhibition (Camadro and Peloquin 1981; Fritz and Hanneman 1989; Novy and Hanneman 1991; Camadro et al. 1998; Erazzú et al. 1999; Hayes et al. 2005).

Genetic and molecular research shows that some self-incompatibility factors also function in prezygotic IRBs. However, self-incompatibility and IRBs differ in terms of specificity and the precise factor requirements. IRBs show broad specificity, and a single S-RNase can cause rejection of pollen from species or groups of species (Murfett et al. 1996; Tovar-Méndez et al. 2014). *S-RNase* and *HT* genes dual roles in self-incompatibility and interspecific pollen rejection points out pleiotropic effects and hence linkage between these two mechanisms.

IRB mechanisms' complexity is such that multiple redundant mechanisms can contribute to interspecific incompatibility, even between a single pair of species (Murfett et al. 1996; McClure et al. 2000). This may complicate experiments because defects in one rejection mechanism do not necessarily result in compatibility. For example, HT-proteins previously found implicated only in S-RNase-dependent self-incompatibility and IRBs have also been involved in S-RNase-independent pollen rejection in tomato (McClure et al. 2011).

Selective pressures like reproductive assurance make self-incompatible to self-compatible mating system transitions (MSTs) common in nature (Barrett 2002; Goldberg et al. 2010; Goldberg and Igic 2012). Self-compatibility has not been extensively investigated in the potato clade and few self-compatible species have been recognized. Loss of S-RNase function is a common route to self-compatibility. This is the case of self-compatible variants of *S. verrucosum* in which pollen tubes of other potato species, including those having 1EBN can grow without inhibition in their pistils reaching ovules in great quantity (Hermsen and Ramanna 1976; Makhan'ko 2011). Furthermore, dominant gain-of-function (GOF) pollen-part mutants such as *Sli*, which results in self-compatible variants in *S. chacoense*, increase seed set in interspecific crosses to *S. pinnatisectum* in addition to suppressing self-incompatibility (Sanetomo et al. 2014).

Phumichai et al. (2006) were able to introduce after crossing the *S*-locus inhibitor gene (*Sli*), which can inhibit gametophytic self-incompatibility, in diploid potatoes and alter self-incompatible to self-compatible plants, into 32 diploid genotypes. *Sli* has been also successfully introduced to diploid cultivars from *S. tuberosum* Phureja and Stenotomum Groups using SC *S. chacoense* variants as male parents at CIP. Assuming that this pollen side mutant acts either breaking down pollen-stigma recognition system or overcoming S-RNAse cytotoxic activity (Hosaka and Hanneman 1998b), then a change in interspecific compatibility may occur after crossing *Sli*-bearing self-compatible hybrids as pollen parents and self-incompatible sources of late blight resistance from wild *S. piurae* and *S. chiquidenum* species. Previous attempts at CIP to cross these wild species with 2*x* *S. tuberosum* cultivars produced very few seeds with *S. chiquidenum* and no seed set at all with *S. piurae*. Embryo rescue was often required to save the few hybrids produced from *S. chiquidenum*. Linkage between self-incompatibility and IRBs results in MSTs with significant implications in germplasm enhancement programs particularly when decisions between different crop wild relatives have to be made in interspecific crosses.

7.7 Unilateral Compatibility

This is a very common IRB pattern that refers to crosses that are compatible in only one direction (Lewis and Crowe 1958). Most IRBs conform to the self-incompatible × self-compatible rule, in which pollen from the self-compatible species is rejected on pistils of related self-incompatible species but the reciprocal pollination is compatible (Nathan Hancock et al. 2003; Bedinger et al. 2011). In potato only some IRBs conform to this rule (Hermsen and Ramanna 1976; Eijlander et al. 2000). For example, crosses between the self-compatible 2*x* (1EBN) species *S. pinnatisectum* and the 2*x* (1EBN) self-incompatible species *S. cardiophyllum* were successful only when self-compatible *S. pinnatisectum* was used as the female parent (Chen et al. 2004). A similar pattern was observed in crosses between the self-compatible 2*x* (1EBN) species *S. commersonii* and the self-incompatible 2*x* (2EBN) species *S. chacoense* (Summers and Grun 1981). Although it is not absolute, the consistency of the "self-incompatible × self-compatible" rule suggests a link between inter- and intraspecific pollen rejection. The suggested linkage is that the *S*-locus controls unilateral incompatibility as well as self-incompatibility. Conversely, there are many exceptions to the self-incompatible × self-compatible rule in potato such as the occurrence of both unilateral and bilateral self-incompatible × self-incompatible conflicts (Camadro et al. 1998, 2004; Kuhl et al. 2002; Raimondi et al. 2003). Genetic systems entirely independent of the *S*-locus have been proposed to explain cross incompatibility, which is a term used to emphasize potato clade cross distinctive complexity. Cross incompatibility includes both post- and pre-zygotic mechanisms.

An important phenomenon worth mentioning is compatibility encountered in crosses between self-compatible *S. verrucosum* plants (2*x*, 2EBN) and a wide range of accessions of various 1EBN potato species. Success on producing these novel sexual hybrids was achieved with several 1EBN diploid species, such as *inter alia*, *S. bulbocastanum*, *S. pinnatisectum*, *S. polyadenium*, *S. commersonii*, and *S. circaeifolium* (Yermishin et al. 2014). Absence of S-RNases in pistils in self-compatible *S. verrucosum* allowed growth of pollen tubes from these 1EBN wild species and fertilization of egg cells. In addition, "rescue pollination" was used to improve hybridization effectiveness. "Rescue pollination" also known as "double pollination" is a technique that reduces premature fruit drop and involves the application of pollen from the incompatible species, followed a day or 2 later by that of a compatible species, denoted as "mentor pollen" (Singsit and Hanneman Jr 1990). The "mentor pollen" fertilizes several ovules, stimulating the development of fruit and pollen tubes from the incompatible parent to reach the ovules and effect fertilization (Singsit and Hanneman Jr 1990; Yermishin et al. 2014). Pollen of Group Phureja pollinators are used as "mentor pollen" because of their typical dominant seed spot marker, so its offspring can be visually identified and eliminated (Brown and Adiwilaga 1991; Iwanaga et al. 1991a).

A feature of inter-EBN interspecific hybridization using self-compatible *S. verrucosum* as "bridge species" is male sterility in most resulting hybrids (Abdalla and Hermsen 1972–Abdalla and Hermsen 1973; Yermishin et al. 2014). Cytoplasmic male sterility factors (CMS) from *S. verrucosum* have been assumed to interact with dominant nuclear genes from the 1EBN male parents, resulting in male sterility of hybrids. CMS has also been suggested to account for male sterility of hybrids produced when 2*x* (2EBN) Group Tuberosum haploids were used as male parents in crosses with cultivated 2*x* (2EBN) Phureja and Stenotomum Groups (Grun et al. 1962; Ross et al. 1964; Carroll 1975) as well as with a wide range of wild species (Hermundstad and Peloquin 1985; Tucci et al. 1996; Santini et al. 2000). Male fertile hybrids obtained when the haploids were the male parent corroborated this assumption (Tucci et al. 1996; Novy and Hanneman 1991). Consequently, this type of barrier can be overcome by carrying out reciprocal crosses.

Male sterility of hybrids from *S. verrucosum* did not represent a drawback since crosses with Tuberosum haploids were successful when the (*S. verrucosum* × 1EBN) hybrids were used as females (Yermishin et al. 2014). However, male fertile Tuberosum haploids are not widely available (Makhan'ko 2008). Breeders may overcome this drawback by extracting haploids from 4*x S. tuberosum* cultivars or breeding clone bearing the male fertility dominant restorer gene *Rt* that gives fertility to plants that contain the dominant male sterility gene *Ms* in the presence of sensitive cytoplasm (Iwanaga et al. 1991b). Selection of Group Tuberosum haploids carrying a restorer of fertility (*Rt*) gene can be used to pollinate (*S. verrucosum* × 1EBN) hybrids and produce male fertile hybrids for further backcrossing with cultivated potato.

An alternative appealing approach for creating opportunities to solve the problem of the prezygotic interspecific incompatibility with a number of 1EBN wild species was proposed by Polyukhovich et al. (2010). These investigators transferred the nonfunctional S-RNase *Sv* haplotype from self-compatible *S verrucosum* directly to Group Tuberosum haploids using some rare pollen receptive haploids. Further development of homozygote *SvSv* hybrids were achieved after selfing or sib-mating of F_1 self-compatible hybrids with high functional pollen fertility. These *SvSv* hybrids were identified based on their good penetration of pollen tubes in the styles of 1EBN *S. bulbocastanum* and *S. pinnatisectum* species. These *SvSv* Tuberosum hybrids may produce fertile hybrids in direct crosses with 1EBN wild diploid potato species.

According to available knowledge, 2*x* (2EBN) self-compatible *S. verrucosum*, which is used as a "bridge species", provides potato breeders an ideal route in germplasm enhancement aimed at introgressing valuable genes from 1EBN wild diploid *Solanum* species into breeding populations. Efficiency of hybrid production using self-compatible *S. verrucosum* is greater than other methods based on ploidy manipulations such as somatic hybridization or embryo rescue, which require considerable experience and expenditure of time and resources (Jansky 2006).

Group Tuberosum haploids are the most promising recipients for introgression of novel genes for traits of interest from diploid wild species germplasm (Peloquin et al. 1989; Jansky et al. 1990). Smaller population size in comparison with tetraploids is needed for selecting recombinants that meet breeding requirements due to their disomic inheritance. Accumulation of desirable genes and elimination of undesirable ones flows faster at this ploidy level. In addition, the presence of naturally occurring meiotic mutations in the potato gene pool, which leads to the production of 2*n* gametes, allows chromosome doubling for the transfer of valuable gene combinations to the tetraploid level (Carputo et al. 2000; Yermishin et al. 2014).

7.8 Endosperm Balance Number (EBN) and Interspecific Reproductive Barriers

Various genetic mechanisms account for the success or failure of seed development in flowering plants, which undergo double fertilization during sexual reproduction (Kinoshita 2007). Failure for producing triploids after tetraploid × diploid crosses—also known as "triploid block"—led to studying as a reproductive barrier the endosperm, which provides nourishment to the seed embryo. A first concept was to consider a 2:3:2 ploidy balance between maternal tissue, endosperm and embryo but further research demonstrated that normal endosperm development depends on having a 2:1 maternal to paternal genome dosage in the endosperm (Ehlenfeldt and

Ortiz 1995 and references therein). This endosperm dosage system for both intra- and inter-specific crossing seems to be multigenic in *Solanum* species, in which is known as the endosperm balance number (EBN) that also explains some aspects of species evolution therein. For example, Ortiz and Ehlenfeldt (1992) indicated the role of EBN in the origin of both diploid and polyploid potato species or how it becomes a hybridization barrier for speciation among sympatric *Solanum* species with same ploidy.

The EBN is a unifying concept that may predict endosperm function in intraspecific, interploidy, and interspecific crosses in potato and wild crop relatives (Johnston et al. 1980). Each species carries an EBN value that is constant across interspecific crossing, thus determining the effective ploidy in the endosperm, which must be in a 2 maternal:1 paternal ratio that is a necessary for successful endosperm development. The EBN, which is in itself an arbitrary value, is given to a species based on its crossing behavior with known EBN standards. The EBN does not reflect directly the ploidy of a species (Hanneman Jr 1999). For example, there are 2*x* (1EBN), 2*x* (2EBN), 3*x* (2EBN), 4*x* (2EBN), 4*x* (4EBN), 5*x* (4EBN), and 6*x* (4EBN) *Solanum* species.

The EBN concept was useful to elucidate the nature of the pollinator effect in haploid extraction (Peloquin et al. 1996). Haploid embryos appear to be associated with hexaploid endosperms as a result of having the union of 2-chromosome sets from the pollinator with the polar nuclei and lack of fertilization of the egg, thus having a 2 maternal: 1 paternal EBN in the endosperm that normal seed development requires.

The "triploid block" is a reproductive barrier resulting from endosperm malfunction due to the epigenetic phenomenon of genomic imprinting (Ehlenfeldt and Ortiz 1995; Köhler et al. 2009). Evidence shows that the endosperm dosage systems are imprinted within the gametes, thus the same gene being functionally different in maternal and paternal chromosomes. The maternally and paternally imprinted genes often carry a DNA methylation or histone modification in their vicinity (Kinoshita 2007). The position of these epigenetic modifications determines how the gene will be expressed; i.e., either from the maternal or paternal inherited allele (Köhler et al. 2012). Maternally imprinted genes appear to repress endosperm proliferation, which seems to be promoted by paternally imprinted genes. Abnormal endosperm development results from the imbalance of these imprinted genes. This "parental conflict" is consistent with the differential maternal and paternal genome effects in interploidy mating (Köhler et al. 2009). Products of imprinted PcG genes such as *Medea* (*MEA*) and *Fertilization Independent Seed2* (*FIS2*) could be limiting factors affecting endosperm and seed development. For example, FIS PcG proteins repress fertilization-independent seed formation and restrict endosperm proliferation (Köhler and Makaverich 2006).

The EBN provides useful prior knowledge for germplasm transfer from 2*x* (1EBN), 2*x* (2EBN), 4*x* (2EBN) and 6*x* (4EBN) species into 4*x* (4EBN) potato (Ortiz and Ehlenfeldt 1992). For example, chromosome engineering using 4*x* (2EBN) and 2*x* (2EBN) addition lines will allow to introduce specific chromosomes bearing the target allele into the 4*x* cultigen pool for further use in breeding.

Likewise, 2*x* chromosome addition lines from may result using 2*x* (1EBN) and 2*x* (2EBN) *Solanum* species. The first step will be to get a 3*x* (2EBN) interspecific hybrids after crossing both 2*x* species and thereafter using these hybrids by crossing them with 2*x* (2EBN) species to obtain 2*x* (2EBN) chromosome addition lines, which can be further used for breeding at 2*x* level. Haploids from 5*x* (4EBN) clones may be another method for producing 2*x* (2EBN) offspring with traits from the 2*x* (1EBN) parent because haploid extraction will allow screening for the right balance in the endosperm.

7.9 Trait Genetic Research: A Summary Prior to DNA Markers

Swaminathan and Howard (1953) provided the first summary of the inheritance of most important traits in potato, which was further updated by Howard (1960, Howard 1970), Bradshaw and Mackay (1994 and chapters therein), and Tiemens-Hulscher et al. (2013). Table 7.3 give some details about trait genetics and gene symbols as noted in some of these publications.

Further research using the candidate gene approach led to identifying diagnostic DNA-based markers for genes involved, *inter alia*, in quantitative resistance to late blight (*R1* gene family in chromosome V) or cyst nematode (major QTL in same resistance "hot spot" on potato chromosome V), and both chip color (e.g., co-localized with a cold-sweetening QTL in chromosome IX and other sugar QTL in chromosome X) and tuber starch content in tetraploid potato cultivars (Gebhardt et al. 2007). Other annotated loci in the potato genetic map are *Y* (yellow flesh color) in chromosome III, *H3* of *Gpa4* in chromosome IV, Rx_2 (extreme resistance to *Potato Virus X*) and Nx_{tbr} (hypersensitivity to *Potato Virus X*) in chromosome V, Nx_{phu} in chromosome IX, *Ro* (tuber shape) and anthocyanin (including flower and skin color) in chromosome X, Ry_{sto}, and *Ry-adg* in chromosome XI, and both *Gpa 2* (resistance to potato cyst nematode) plus Rx_1 in chromosome XII.

7.10 Cytogenetics for Crossing, Scaling Up and Down Ploidy, and Chromosome Engineering

Germplasm enhancement (mistakenly replaced by some as pre-breeding, Ortiz 2002) is the early component of sustainable plant breeding that includes identifying a useful character, "capturing" its genetic diversity, and putting those genes into a "usable" form (Peloquin et al. 1989). Potato provides an interesting example of using crop wild relatives in germplasm enhancement. The wild species *S. demissum* (6*x*, 4EBN), *S. stoloniferum* (4*x*, 2EBN), and *S. vernei* contributed host plant resistance genes to late blight, *Potato Virus Y*, and cyst nematode, respectively; while

Table 7.3 Gene symbols and their genetics of key traits for potato breeding

Trait	Gene symbol	Genetics (disomic inheritance unless indicated otherwise)
Plant growth type		Upright dominant to prostrate, and intermediate procumbent recessive to upright but unknown to prostrate
Dwarfism		Recessive phenotype producing compact, dark green, rosette plant
Flower color	*D* red	*F* involved in anthocyanin expression in flowers
	P blue	Purple = *D_P_F_*, Red to rose *D_ppF_*, Blue = *ddP_F_*, White = *D_P_ff* or *ddppF_*
Skin color	*D* red	*I* engaged in anthocyanin expression in tuber skin
	P blue	Purple = *D_P_I_*, Red *D_ppI_*, Blue = *ddP_I_*, White to yellow to brown = *D_P_ff* or *ddppF_*
Flesh color	*Y* yellow, *y* white	Yellow caused by carotenoids dominant to white
	Or allele at *Y* locus	Orange depends on two genes: one determining production and other (recessive) accounting for accumulation of zeazhantin. Orange dominant to yellow
	D red	Purple or red due to anthocyanins regulated by *B*.
	P blue	Purple = *D_ P_ B_*, Red = *D_ pp B_*, Blue = *dd P_ B_* (2*x*)
Tuber color pattern		Splashed and spotted only in eyes while speckled (dominant) reverses; i.e., eyes are either white or yellow
Tuber shape	*Ro* round	Round (*Ro__*) dominant to long (*roro*)
	ro long	
Russet skin		Three independent loci having an additive effect to each other; i.e., *AABBCC* show more russet skin that *AaBbCc*
Stem pubescence		Single gene, being pubescent dominant to glabrous stem
Early maturity		Earliness due to dominant allele with additive effects; i.e., duplex (*AAaa*) earlier than simplex (*Aaaa*) in 4*x*
Tuber dormancy		Dominant early sprouting with short dormancy
Tuberization under long days		High heritability but influenced by day length, light intensity, and temperature

(continued)

Table 7.3 (continued)

Trait	Gene symbol	Genetics (disomic inheritance unless indicated otherwise)
Chip color		3-Gene hypothesis for both reversion resistance (ability to produced light-colored chips after harvest or short storage) and reconditioning (controlled tuber warming after storing fall crop for 4–6 months) to eliminate reducing sugars: A dominant allele in each of three loci for good chipping, being one or two loci common to both traits
Late blight resistance	12 known *R* genes derived from *S. demissum, Rpi-blb3* from *S. bulbocastanum*, and *Rpi-abpt* derived from quadruple hybrid involving *S. acaule*, *S. bulbocastanum*, plus groups Phureja and Tuberosum	Dominant *R* for race-specific host resistance corresponding to virulence genes in oomycete *Phytophthora infestans* (*R8* and *R9* seem to be "durable") Many genes are involved in partial host plant resistance that slow down the development of all *Phytophthora infestans* races, thus being likely more durable than race-specific resistance
Early blight resistance		High heritability for partial resistance, thus additivity being the most important gene action
Potato leaf roll virus resistance	N_L for hypersensitivity	Dominance of resistance but major gene does not provide enough durable resistance to a cultivar
Potato virus X resistance	Rx_i for extreme resistance, while Nx_i controls hypersensitivity	Dominant inheritance
Potato virus Y resistance	Ny_{chc}, Ny_{dms}, $Ry_{sto}{}^{n1}$ and $Ry_{sto}{}^{n2}$ gives hipersensitivity, whereas *Ry*-adg, *Ry-chc*, *Ry-hou* and *Ry-sto* are for extreme resistance	Dominant extreme resistance from *S. stoloniferum*, *S. hougasii,* and Group Andigena, while multigenic for Group Phureja
Potato virus A	*Na* protects for infection through hypersensitivity	
Wart resistance		Dominant alleles in two loci necessary for resistance
Verticillium wilt		2 genes
Bacterial wilt resistance		3–4 dominant genes required
Black leg and bacterial soft rot resistance		Minor genes with additive effects increase resistance
Potato tuber moth resistance		Simple inheritance due to additivity

(continued)

Table 7.3 (continued)

Trait	Gene symbol	Genetics (disomic inheritance unless indicated otherwise)
Root-knot nematode resistance		High narrow-sense heritability for resistance in Group Phureja (susceptible)—*S. sparsipilum* (likely three major complementary genes in heterozygous state) segregating population, but reciprocal effects suggest maternal effects of *S. sparsipilum* or cytoplasmic-nuclear gene interaction
Potato cyst nematode resistance	*H1* (from Group Andigena), *H2* (from CPC2602), *Fa* and *Fb* (from *S. spegazzinii*), plus *B* and *C* from *S. verneii* provide host plant resistance against *Globodera rostochiensis*, while *Gpa2*, *Gpa5*, *GPa1*, *H2*, and *H3* of *Gpa4* give resistance against *G. pallida*	*H1* gives host plant resistance against Ro-1 and Ro-4 pathotypes

processing quality was derived from S. *chacoense* (2*x*, 2EBN) in various cultivars (Jansky et al. 2013).

Introgression and incorporation are the two approaches for using wild species in plant breeding (Simmonds 1993). Introgression refers to transferring one or a few alleles from wild germplasm to breeding populations that lack them, while a large-scale program for developing breeding populations using wild germplasm to broaden its genetic base is known as incorporation. Figure 7.2 illustrates both in potato breeding. "Bridge" species, double pollination, embryo rescue, 2*n* gametes, and the EBN knowledge allows using chromosome engineering, while haploids, wild 2*x* (2EBN) species, and 2*n* gametes are used in ploidy manipulations.

CIP released in the mid-1990s diploid bred-germplasm generated from using haploids from tetraploid cultivars and breeding clones, plus diploid landraces and wild species (Watanabe et al. 1994). These genetic resources are of value for potato breeding due to their genetic diversity, crossability with the 4*x* cultigen pool facilitated by 2*n* gametes (mostly FDR 2*n* pollen), and high host plant resistance to pathogens and pests derived mostly from wild relatives. Ploidy manipulations at CIP led to the transmission of host plant resistance to cyst and root-knot nematodes, bacterial wilt, early blight, and potato tuber moth, as well as producing high-yielding 4*x* (4EBN) breeding clones also having yield stability over environments (Ortiz et al. 1994). Burundi released in the 1990s the tetraploid potato cultivar "Nemared," which derived from this diploid breeding population (Fig. 7.3), because of both its host plant resistance to root knot nematode and desired agronomic traits.

Potato, as shown above, is the model crop species for germplasm enhancement of polysomic polyploids. Crop wild relatives and landraces are the diversity sources, while haploid derived from the tetraploid cultigen pool "capture" this diversity after crossing them with the former. The haploid-species hybrids producing 2*n* gametes

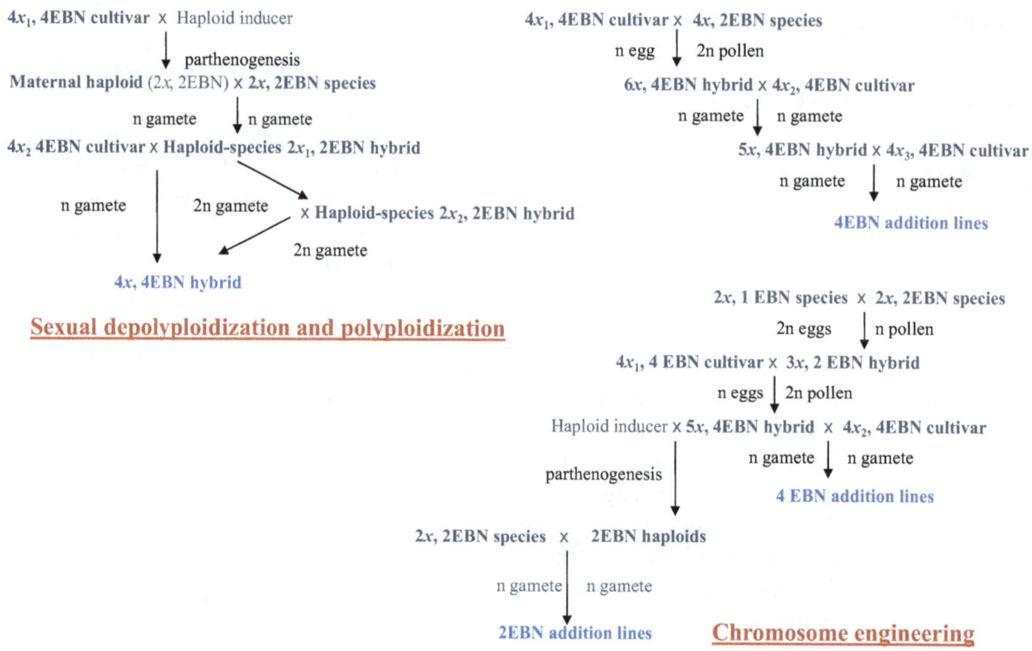

Fig. 7.2 Germplasm enhancement approaches in potato breeding: introgression through chromosome engineering and incorporation using sexual depolyploidization and polyploidization

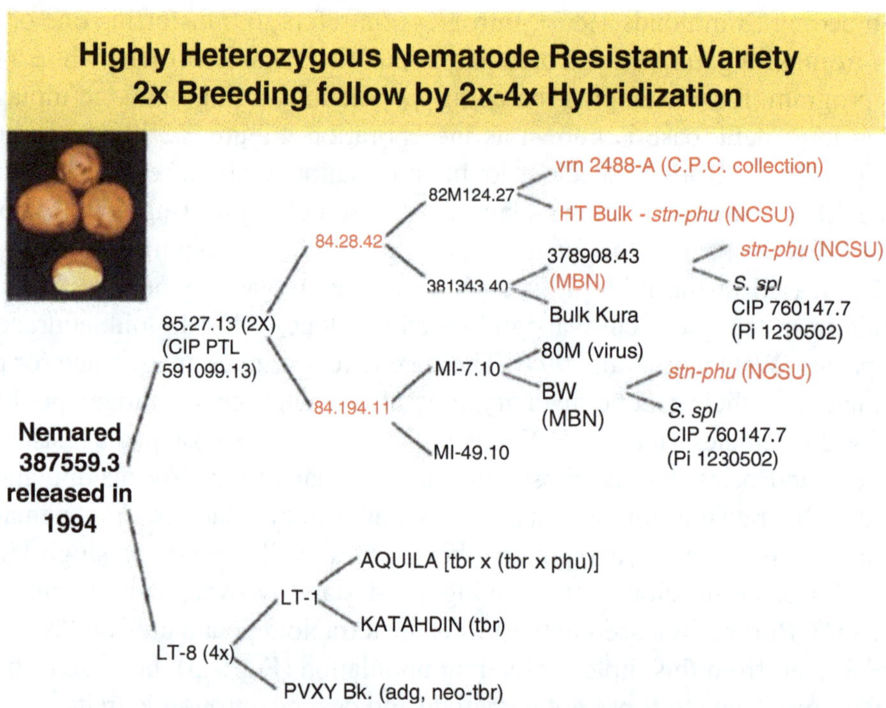

Fig. 7.3 Pedigree of tetraploid cultivar potato "Nemared" resulting from ploidy manipulations

transfer this diversity to the tetraploid breeding pool through sexual polyploidization in which the EBN ensures the resulting ploidy of the hybrid offspring.

7.11 Concluding Remarks

Despite its the biological characteristics which made genetic improvement more complex in potato than in other crops, potato breeders have at their disposal such a powerful and effective approach as the unique ability to conduct across ploidy and species-wide crosses in order to introgress relevant genetic variation from its genetic resources into potato breeding programs. Progress in the field of cytogenetics of potato enables a more effective transfer of relevant genetic variation from wild *Solanum* accessions kept in genebanks or through in situ approaches. The increasing demands to develop resilient potato varieties able to withstand the threats brought by climate change, as well as the substantial progress recently achieved with the development of potato hybrid cultivars at the 2*x* level, underpin the increasing contribution of cytogenetics to the genetic improvement of the potato crop. The increasing availability of vast amounts of genomic information, as well as the continuous reduction of expenses associated with the sequencing of whole genomes are expected to further increase the contribution of potato cytogenetics, in terms of accelerating the development of varieties which provide not only superior adaption to changing production environments but also increased food and nutrient security to the millions of people to whom potato represents a staple crop.

References

Abdalla MMF, Hermsen JGT (1972) Plasmons and male sterility types in *Solanum verrucosum* and its interspecific hybrid derivatives. Euphytica 21:209–220

Abdalla MMF, Hermsen JGT (1973) An evaluation of *Solanum verrucosum* Schlechtd. For its possible use in potato breeding. Euphytica 22:19–27

Allainguillaume J, Wilkinson MJ, Clulow SA et al (1997) Evidence that genes from the male parent may influence the morphology of potato dihaploids. Theor Appl Genet 94:241–248

Allard RW (1999) History of plant population genetics. Ann Rev Genet 33:1–27

Barrett SCH (2002) The evolution of plant sexual diversity. Nat Rev Genet 3:274–284

Bedinger PA, Chetelat RT, McClure B et al (2011) Interspecific reproductive barriers in the tomato clade: opportunities to decipher mechanisms of reproductive isolation. Sex Plant Reprod 24:171–187

Birhman RK, Hosaka K (2000) Production of inbred progenies of diploid potatoes using an S-locus inhibitor (Sli) gene and their characterization. Genome 43:495–502

Bourke PM, Voorrips RE, Visser RGF, Maliepaard C (2015) The double-reduction landscape in tetraploid potato as revealed by a high-density linkage map. Genetics 201:853–863

Bradshaw JE (2007) The canon of potato science: 4 tetrasomic inheritance. Potato Res 50:219–222

Bradshaw JE, Mackay GR (eds) (1994) Potato genetics. CAB International, Wallinford

Bradshaw JE, Bryan GJ, Ramsay G (2006) Genetic resources (including wild and cultivated *Solanum* species) and progress in their utilization in potato breeding. Potato Res 49:49–65

Brown CR, Adiwilaga KD (1991) Use of rescue pollination to make a complex interspecific cross in potato. Am J Potato Res 68:813–820

Brownfield L, Kölher C (2011) Unreduced gamete formation in plants: mechanisms and prospects. J Exp Bot 62:1659–1668

Burham C (1984) Discussion in cytogenetics. Burgess Publishing, Minneapolis

Camadro EL, Peloquin SJ (1981) Cross-incompatibility between two sympatric polyploid *Solanum* species. Theor Appl Genet 60:65–70

Camadro EL, Verde LA, Marcellán ON (1998) Pollen-pistil incompatibility in a diploid hybrid potato population with cultivated and wild germplasm. Am J Potato Res 75:81–85

Camadro EL, Carputo D, Peloquin SJ (2004) Substitutes for genome differentiation in tuber-bearing Solanum: interspecific pollen-pistil incompatibility, nuclear-cytoplasmic male sterility, and endosperm. Theor Appl Genet 109:1369–1376

Cao J, Schneeberger K, Ossowski S et al (2011) Whole-genome sequencing of multiple *Arabidopsis thaliana* populations. Nat Genet 43:956–963

Carputo D, Barone A (2005) Ploidy level manipulations in potato through sexual hybridization. Ann Appl Biol 146:71–79

Carputo D, Barone A, Frusciante L (2000) 2n gametes in the potato: essential ingredients for breeding and germplasm transfer. Theor Appl Genet 101:805–813

Carroll CP (1975) The inheritance and expression of sterility in hybrids of dihaploid and cultivated diploid potatoes. Genetica 45:149–162

Chase SS (1963) Analytic breeding in *Solanum tuberosum* L.—a scheme utilizing parthenotes and other diploid stocks. Can J Genet Cytol 5:359–363

Chen Q, Lynch D, Platt HW et al (2004) Interspecific crossability and cytogenetic analysis of sexual progenies of Mexican wild diploid 1EBN species *Solanum pinnatisectum* and *S. cardiophyllum*. Am J Potato Res 81:159–169

Cipar MS (1964) Self compatibility in hybrids between Phureja and haploid Andigena clones of *Solanum tuberosum*. Eur Potato J 7:152–160

Cipar MS, Peloquin SJ, Hougas R (1964) Variability in the expression of self-incompatibility in tuber-bearing diploid *Solanum* species. Amer Potato J 41:155–162

Clulow SA, Rousselle-Bourgeois F (1997) Widespread introgression of *Solanum phureja* DNA in potato (*S. tuberosum*) dihaploids. Plant Breed 116:347–351

Clulow SA, Wilkinson MJ, Waugh R et al (1991) Cytological and molecular observations on Solanum phureja-induced dihaploid potatoes. Theor Appl Genet 82:545–551

Clulow SA, Wilkinson MJ, Burch LR (1993) *Solanum phureja* genes are expressed in the leaves and tubers of ancusomatic potato dihaploids. Euphytica 69:1–6

Comai L (2005) The advantages and disadvantages of being polyploid. Nat Rev Genet 6:836–846

Cook DE, Lee TG, Guo X et al (2012) Copy number variation of multiple genes at *Rhg1* mediates nematode resistance in soybean. Science 338:1206–1209

De Jong H (1981) Inheritance of russeting in cultivated diploid potatoes. Potato Res 24:309–313

De Jong H (1987) Inheritance of pigmented tuber flesh in cultivated diploid potatoes. Amer Potato J 64:337–343

De Jong H, Burns VJ (1993) Inheritance of tuber shape in cultivated diploid potatoes. Am Potato J 70:267–283

De Jong H, Rowe PR (1971) Inbreeding in cultivated diploid potatoes. Potato Res 14:74–83

De Jong H, Kawchuk LM, Burns VJ (1998) Inheritance and mapping of a light green mutant in cultivated diploid potatoes. Euphytica 103:83–88

De Jong H, Kawchuk LM, Coleman WK et al (2001) Development and characterization of an adapted form of droopy, a diploid potato mutant deficient in abscisic acid. Am J Potato Res 78:279–290

de Nettancourt D (1977) The genetic basis of self-incompatibility. In: Incompatibility in angiosperms. Monographs on theoretical and applied genetics, vol 3. Springer, Berlin

Díaz A, Zikhali M, Turner AS et al (2012) Copy number variation affecting the Photoperiod-B1 and Vernalization-A1 genes is associated with altered flowering time in wheat (*Triticum aestivum*). PLoS One 7:e33234

Ehlenfeldt MK, Ortiz R (1995) On the origins of endosperm dosage requirements in *Solanum* and other angiosperma genera. Sex Plant Reprod 8:189–196

Eijlander R (1998) Mechanisms of self–incompatibility and unilateral incompatibility in diploid potato (*Solanum tuberosum* L.). Dissertation, Wageningen University

Eijlander R, Ter Laak W, Hermsen JGT et al (2000) Occurrence of self-compatibility, self-incompatibility and unilateral incompatibility after crossing diploid *S. tuberosum* (SI) with *S. verrucosum* (SC): I. Expression and inheritance of self-compatibility. Euphytica 115:127–139

Erazzú LE, Camadro EL, Clausen AM (1999) Pollen-style compatibility relations in natural populations of the wild diploid potato species *Solanum spegazzinii* Bitt. Euphytica 105:219–227

Ercolano MR, Carputo D, Li J, Monti L et al (2004) Assessment of genetic variability of haploids extracted from tetraploid ($2n = 4x = 48$) *Solanum tuberosum*. Genome 47:633–638

Feuk L, Carson AR, Scherer SW (2006) Structural variation in the human genome. Nat Rev Genet 7:85–97

Frandsen NO (1967) Haploidproduktion aus einem Kartoffelzuchtmaterial mit intensiver Wildarteinkreuzung. Der Zuchter 37:120–134

Freyre R, Warnke S, Sosinski B et al (1994) Quantitative trait locus analysis of tuber dormancy in diploid potato (*Solanum* spp.). Theor Appl Genet 89:474–480

Fritz NK, Hanneman RE (1989) Interspecific incompatibility due to stylar barriers in tuber-bearing and closely related non-tuber-bearing Solanums. Sex Plant Reprod 2:184–192

Gálvez JH, Tai HH, Barkley NA et al (2017) Understanding potato with the help of genomics. AIMS Agric Food 2:16–39

Gebhardt C, Ritter E, Barone A et al (1991) RFLP maps of potato and their alignment with the homeologous tomato genome. Theor Appl Genet 83:49–57

Gebhardt C, Li L, Pajerowska-Mukthar K et al (2007) Candidate gene approach to identify genes underlying quantitative traits and develop diagnostic markers in potato. Crop Sci 47:S106–S111

Goldberg EE, Igic B (2012) Tempo and mode in plant breeding system evolution. Evolution 66:3701–3709

Goldberg EE, Kohn JR, Lande R et al (2010) Species selection maintains self-incompatibility. Science 330:493–495

Goldraij A, Kondo K, Lee CB et al (2006) Compartmentalization of S-RNase and HT-B degradation in self-incompatible Nicotiana. Nature 439(7078):805–810

Golz JF, Oh HY, Su V et al (2001) Genetic analysis of Nicotiana pollen-part mutants is consistent with the presence of an S-ribonuclease inhibitor at the S-locus. Proc Natl Acad Sci U S A 98:15372–15376

Gray JE, McClure BA, Bonig I et al (1991) Action of the style product of the self-incompatibility gene of Nicotiana alata (S-RNase) on in vitro-grown pollen tubes. Plant Cell 3:271–283

Grun P, Aubertin M, Radlow A (1962) Multiple differentiation of plasmons of diploid species of *Solanum*. Genetics 47:1321–1333

Grun P, Ochoa C, Capage D (1977) Evolution of cytoplasmic factors in tetraploid cultivated potato (Solanaceae). Am J Bot 64:412–420

Hancock CN, Kondo K, Beecher B, McClure B (2003) The S-locus and unilateral incompatibility. Phil Trans R Soc Lond B 358:1133–1140

Hanneman RE Jr (1999) The reproductive biology of the potato and its implication for breeding. Potato Res 42:283–312

Hanneman RE (1985) Self-fertility in *Solanum chacoense*. Am Potato J 62:428–429. (abstract)

Hardigan MA, Parker F, Laimbeer E et al (2017) Genome diversity of tuber-bearing Solanum uncovers complex evolutionary history and targets of domestication in the cultivated potato. Proc Natl Acad Sci USA. https://doi.org/10.1073/pnas.1714380114

Hayes RJ, Dinu II, Thill CA (2005) Unilateral and bilateral hybridization barriers in interseries crosses of 4x 2EBN *Solanum stoloniferum*, *S. pinnatisectum*, *S. cardiophyllum*, and 2x 2EBN S. tuberosum haploids and haploid-species hybrids. Sex Plant Reprod 17:303–311

Henry IM, Dilkes BP, Miller ES, Burkart-Waco D et al (2010) Phenotypic consequences of aneuploidy in *Arabidopsis thaliana*. Genetics 186:1231–1245

Henry IM, Zinkgraf MS, Groover AT et al (2015) A system for dosage-based functional genomics in poplar. Plant Cell 27:2370–2383

Hermsen JGT, Ramanna MS (1976) Barriers to hybridization of *Solanum bulbocastanum* Dun. and *S. verrucosum* Schlechtd. and structural hybridity in their F1 plants. Euphytica 25:1–10

Hermsen JT, Verdenius J (1973) Selection from *Solanum tuberosum* group Phureja of genotypes combining high-frequency haploid induction with homozygosity for embryo-spot. Euphytica 22:244–259

Hermundstad SA, Peloquin SJ (1985) Male fertility and 2n pollen production in haploid-wild species hybrids. Am Potato J 62:479–487

Hirsch CN, Hirsch CD, Felcher K et al (2013) Retrospective view of North American potato (*Solanum tuberosum* L.) breeding in the 20th and 21st centuries. G3 3:1003–1013

Hosaka K, Hanneman RE (1998a) Genetics of selfcompatibility in a self-incompatible wild diploid potato species *Solanum chacoense*. I. Detection of an S locus inhibitor (*Sli*) gene. Euphytica 99:191–197

Hosaka K, Hanneman RE (1998b) Genetics of self-compatibility in a selfcompatible wild diploid potato species *Solanum chacoense*. 2. Localization of an S-locus inhibitor (Sli) gene on the potato genome using DNA markers. Euphytica 103:265–271

Hosaka K, Sanetomo R (2012) Development of a rapid identification method for potato cytoplasm and its use for evaluating Japanese collections. Theor Appl Genet 125:1237–1251

Hougas RW, Peloquin SJ, Gabert AC (1964) Effect of seed-parent and pollinator on frequency of haploids in *Solanum tuberosum*. Crop Sci 4:593–595

Howard HW (1960) Potato cytology and genetics, 1952-59. Biblphia Genet 19:87–216

Howard HW (1970) Genetics of the potato *Solanum tuberosum*. Logos Press, London

Hua ZH, Fields A, Kao TH (2008) Biochemical models for S-RNase-based self-incompatibility. Mol Plant 1:575–585

Hutten RCB, Scholberg EJMM, Huigen DJ et al (1993) Analysis of dihaploid induction and production ability and seed parent x pollinator interaction in potato. Euphytica 72:61–64

Iwanaga M, Freyre R, Watanabe K (1991a) Breaking the crossability barriers between disomic tetraploid *Solanum acaule* and tetrasomic tetraploid *S. tuberosum*. Euphytica 52:183–191

Iwanaga M, Ortiz R, Cipar M, Peloquin SJ (1991b) A restorer gene for genetic-cytoplasmic male sterility in cultivated potatoes. Am Potato J 68:19–28

Jacobs JME, Van Eck HJ, Arens P et al (1995) A genetic map of potato (*Solanum tuberosum*) integrating molecular markers, including transposons, and classical markers. Theor Appl Genet 91:289–300

Jansky S (2006) Overcoming hybridization barriers in potato. Plant Breeding 125:1–12

Jansky S, Peloquin SJ, Yerk GL (1990) Use of potato haploids to put 2x wild species germplasm in usable form. Plant Breed 104:290–294

Jansky S, Dempewolf H, Camadro EL, Simon R, Zimnoch-Guzowska E, Bisognin DA, Bonierbale M (2013) A case for crop wild relative preservation and use in potato. Crop Sci 53:746–754

Jansky S, Chung YS, Kittipadukal P (2014) M6: a diploid potato inbred line for use in breeding and genetics research. J Plant Reg 8:195–199

Jansky SH, Charkowski AO, Douches DS, Gusmini G, Richael C, Bethke PC, Spooner DM, Novy RG, De Jong H, De Jong WS, Bamberg JB, Thompson AL, Bizimungu B, Holm DG, Brown CR, Haynes KG, Sathuvalli VR, Veilleux RE, Miller JC, Bradeen JM, Jiang J (2016) Reinventing potato as a diploid inbred line–based crop. Crop Sci 56:1412–1422

Johnston SA, den Nijs TP, Peloquin SJ, Hanneman RE Jr (1980) The significance of genic balance to endosperm development in interspecific crosses. Theor Appl Genet. 57:5–9

Kinoshita T (2007) Reproductive barrier and genomic imprinting in the endosperm of flowering plants. Genes Genet Syst 82:177–186

Köhler C, Makaverich G (2006) Epigenetic mechanisms governing seed development in plants. EMBO Rep 7:1223–1227

Köhler C, Scheid OM, Erilova A (2009) The impact of the triploid block on the origin and evolution of polyplpid plants. Trends Genet 26:142–148

Köhler C, Wolff P, Spillane C (2012) Epigenetic mechanisms underlying genomic imprinting in plants. Annu Rev Plant Biol 63:331–352

Kotch GP, Ortiz R, Peloquin SJ (1992) Genetic analysis by use of potato haploid populations. Genome 35:103–108

Kuhl JC, Havey MJ, Hanneman RE (2002) A genetic study of unilateral incompatibility between diploid (1EBN) Mexican species *Solanum pinnatisectum* and *S. cardiophyllum* subsp. *cardiophyllum*. Sex Plant Reprod 14:305–313

Lewis D, Crowe LK (1958) Unilateral interspecific incompatibility in flowering plants. Heredity 12:233–256

Lindhout P, Meijer D, Schotte T et al (2011) Towards F_1 hybrid seed potato breeding. Pot Res 54:301–312

Lovene M, Zhang T, Lou Q, Buell CR et al (2013) Copy number variation in potato—an asexually propagated autotetraploid species. Plant J 75:80–89

Makhan'ko OV (2008) Interspecific incompatibility in diploid potato breeding. Zemlyarobstva i Akhova Raslin 1:11–14

Makhan'ko OV (2011) Incompatibility in interspecific and intraspecific hybridization of diploid potato and ways of its overcoming. Dissertation, Institute of Genetics and Cytology Belarusian National Academy of Sciences

Maron LG, Guimarães CT, Kirst M et al (2013) Aluminum tolerance in maize is associated with higher *MATE1* gene copy number. Proc Natl Acad Sci USA 110:5241–5246

McClure B, Cruz-Garcia F, Romero C (2011) Compatibility and incompatibility in S-RNase based systems. Ann Bot 108:647–658

McClure BA, Gray JE, Anderson MA et al (1990) Self-incompatibility in Nicotiana alata involves degradation of pollen rRNA. Nature 347:757–760

McClure BA, Cruz-Garcia F, Beecher BS, Sulaman W (2000) Factors affecting inter- and intraspecific pollen rejection in Nicotiana. Ann Bot 85:113–123

Mihovilovich E, Sanetomo R, Hosaka K et al (2015) Cytoplasmic diversity in potato breeding: case study from the International Potato Center. Mol Breed 35:137. https://doi.org/10.1007/s11032-015-0326-1

Murfett J, Strabala T, Zurek D et al (1996) S RNase and interspecific pollen rejection in the genus Nicotiana: multiple pollen-rejection pathways contribute to unilateral incompatibility between self-incompatible and self-compatible species. Plant Cell 8:943–958

Naess SK, Bradeen JM, Wielgus SM et al (2000) Resistance to late blight in *Solanum bulbocastanum* is mapped to chromosome 8. Theor Appl Genet 101:697–704

Nakamura S, Hosaka K (2010) DNA methylation in diploid inbred lines of potatoes and its possible role in the regulation of heterosis. Theor Appl Genet 120:205–214

Nathan Hancock C, Kondo K, Beecher B et al (2003) The S locus and unilateral incompatibility. Phil Trans R Soc Lond B 358:1133–1140

Novy RG, Hanneman RE (1991) Hybridization between Gp. Tuberosum haploids and 1EBN wild potato species. Am Potato J 68:151–169

Olsder J, Hermsen JT (1976) Genetics of selfcompatibility in dihaploids of Solanum tuberosum L. 1. Breeding behaviour of two selfcompatible dihaploids. Euphytica 25:597–607

Ortiz R (1998) Potato breeding via ploidy manipulations. Plant Breed Rev 16:15–86

Ortiz R (2002) Germplasm enhancement. In: Engels JMM, Ramanatha Rao V, Brown AHD, Jackson MT (eds) Managing plant genetic diversity. CAB International, Wallingford—International Plant Genetic Resources Institute, Rome, pp 275–290

Ortiz R, Ehlenfeldt M (1992) The importance of endosperm balance number in potato breeding and the evolution of tuber bearing Solanums. Euphytica 60:105–113

Ortiz R, Peloquin SJ (1994) Use of 24-chromosome potatoes (diploids and dihaploids) for genetic analysis. In: Bradshaw JE, Mackay GR (eds) Potato genetics. CAB International, Wallinford, pp 133–153

Ortiz R, Watanabe KN (2004) Genetic contributions to breeding polyploid crops. Recent Res Devel Genet Breed 1:269–286

Ortiz R, Iwanaga M, Camadro EL (1993a) Utilización potencial de progenie autofecundada de IvP-35 como inductor de haploides en papa por cruzamientos 4*x*-2*x*. Revista Latinoamericana de Papa 5/6:46–53

Ortiz R, Iwanaga M, Peloquin SJ (1993b) Male sterility and 2*n* pollen in 4*x* progenies derived from 4*x* × 2*x* and 4*x* × 4*x* crosses in potatoes. Potato Res 36:227–236

Ortiz R, Iwanaga M, Peloquin SJ (1994) Breeding potatoes for the developing countries using wild tuber bearing *Solanum* species and ploidy manipulations. J Genet Breed 48:89–98

Ortiz R, Simon P, Jansky S et al (2009) Ploidy manipulation of the gametophyte, endosperm and sporophyte: a tribute to Professor Stanley J. Peloquin (1921–2008). Ann Bot 104:795–807

Owen HR, Veilleux RE, Haynes FL et al (1988) Photoperiod effects on 2*n* pollen production, response to anther culture, and net photosynthesis of a diplandrous clone of *Solanum phureja*. Am Potato J 65:131–139

Peloquin SJ, Yerk GL, Werner JE, Darmo E (1989) Potato breeding with haploids and 2n gametes. Genome 31:1000–1004

Peloquin SJ, Gabert AC, Ortiz R (1996) Nature of "pollinator" effect in potato haploid production. Ann Bot 77:539–542

Phumichai C, Mori M, Kobayashi A et al (2005) Toward the development of highly homozygous diploid potato lines using the self-compatibility controlling Sli gene. Genome 48:977–984

Phumichai C, YIkeguchi-Samitsu Y, Fujimatsu M et al (2006) Expression of S-locus inhibitor gene (Sli) in various diploid potatoes. Euphytica 148:227–234

Pineda O, Bonierbale MW, Plaisted RL et al (1993) Identification of RFLP markers linked to the H_1 gene conferring resistance to the potato cyst nematode *Globodera rostochiensis*. Genome 36:152–156

Polyukhovich YV, Makhan'ko OV, Savchuk AV et al (2010) Production of bridge lines for overcoming interspecific incompatibility in potato. Vestsi Natsiyanalnai Akademii Navuk Belarusi. Seryia Biyalagichnykh Navuk 2:51–58

Potato Genome Sequencing Consortium (2011) Genome sequence and analysis of the tuber crop potato. Nature 475:189–195

Raimondi JP, Sala RG, Camadro EL (2003) Crossability relationships among the wild diploid potato species *Solanum kurtzianum*, *S. chacoense* and *S. ruiz-lealii* from Argentina. Euphytica 132:287–295

Ravi M, Chan SW (2010) Haploid plants produced by centromere-mediated genome elimination. Nature 464:615–618

Ravi M, Marimuthu MPA, Tan EH et al (2014) A haploid genetics toolbox for *Arabidopsis thaliana*. Nat Commun 5:5334

Rivard SR, Cappadocia M, Landry BS (1996) A comparison of RFLP maps based on anther culture derived, selfed, and hybrid progenies of *Solanum chacoense*. Genome 39:611–621

Ross H (1986) Potato breeding-problems and perspectives. Verlag Paul Parey, Berlin

Ross RW, Peloquin SJ, Hougas RW (1964) Fertility of hybrids from *Solanum phureja* and haploid *S. tuberosum* matings. Eur Potato J 7:81–89

Rouppe van der Voort JR, Van Eck HJ, Draaistra J et al (1998) An online catalogue of AFLP markers covering the potato genome. Mol Breed 4:73–77

Samitsu Y, Hosaka K (2002) Molecular marker analysis of 24-and 25-chromosome plants obtained from *Solanum tuberosum* L. subsp. *andigena* (2n = 4x = 48) pollinated with a *Solanum phureja* haploid inducer. Genome 45:577–583

Sanetomo R, Gebhardt C (2015) Cytoplasmic genome types of European potatoes and their effects on complex agronomic traits. BMC Plant Biol 15:162. https://doi.org/10.1186/s12870-015-0545-y

Sanetomo R, Akino S, Suzuki N et al (2014) Breakdown of a hybridization barrier between *Solanum pinnatisectum* Dunal and potato using the S locus inhibitor gene (Sli). Euphytica 197:119–132

Santini M, Camadro EL, Marcellán ON et al (2000) Agronomic characterization of diploid hybrid families derived from crosses between haploids of the common potato and three wild Argentinean tuber-bearing species. Am J Potato Res 77:211–218

Simko I, Vreugdenhil D, Jung CS et al (1999) Similarity of QTLs detected for in vitro and greenhouse development of potato plants. Mol Breed 5:417–428

Simmonds NW (1993) Introgression and incorporation: strategies for the use of crop genetic resources. Biol Rev 68:539–562

Singsit C, Hanneman RE Jr (1990) Isolation of viable inter-EBN hybrids among potato species using double pollinations and embryo culture. Am Potato J 67:578

Singsit C, Veilleux RE (1991) Chloroplast density in guard cells of leaves of anther-derived potato plants grown in vitro and in vivo. HortScience 26:592–594

Sopory SK (1977) Differentiation in callus from cultured anthers of dihaploid clones of *Solanum tuberosum*. Z Pflanzenphysiol 82:88–91

Straadt IK, Rasmussen OS (2003) AFLP analysis of *Solanum phureja* DNA introgressed into potato dihaploids. Plant Breed 122:352–356

Summers D, Grun P (1981) Reproductive isolation barriers to gene exchange between *Solanum chacoense* and *S. commersonii* (Solanaceae). Am J Bot 68:1240–1248

Swaminathan MS, Howard HW (1953) The cytology and genetics of the potato (*Solanum tuberosum*) and related species. Biblphia Genet 16:1–192

Swanson-Wagner RA, Eichten SR, Kumari S et al (2010) Pervasive gene content variation and copy number variation in maize and its undomesticated progenitor. Genome Res 20:1689–1699

Tan EH, Henry IM, Ravi M et al (2015) Catastrophic chromosomal restructuring during genome elimination in plants. Elife 4:e06516

Tiemens-Hulscher M, Delleman J, Eising J, Lammerts van Beuren ET (2013) Potato breeding. Aardappelwereld, The Hague

Tovar-Méndez A, Kumar A, Kondo K et al (2014) Restoring pistil-side self-incompatibility factors recapitulates an interspecific reproductive barrier between tomato species. Plant J 77:727–736

Tsai H, Missirian V, Ngo KJ et al (2013) Production of a high-efficiency TILLING population through polyploidization. Plant Physiol 161:1604–1614

Tucci M, Carputo D, Bile G et al (1996) Male fertility and freezing tolerance of hybrids *involving Solanum tuberosum* haploids and diploid Solanum species. Potato Res 39:345–353

Van Eck HJ, Jacobs JM, Van den Berg PM et al (1994a) The inheritance of anthocyanin pigmentation in potato (*Solanum tuberosum* L.) and mapping of tuber skin colour loci using RFLPs. Heredity 73:410–421

Van Eck HJ, Jacobs JM, Stam P et al (1994b) Multiple alleles for tuber shape in diploid potato detected by qualitative and quantitative genetic analysis using RFLPs. Genetics 137:303–309

Veilleux RE (1996) Haploidy in important crop plants—potato. In: Mohan Jain S, Sopory SK, Veilleux RE (eds) In vitro haploid production in higher plants, vol 3. Kluwer Academic, Dordrecht, pp 37–39

Velásquez AC, Mihovilovich E, Bonierbale M (2007) Genetic characterization and mapping of major gene resistance to potato leafroll virus in *Solanum tuberosum* ssp. *andigena*. Theor Appl Genet 114:1051–1058

Visser RGF, Bachem CWB, Boer JM et al (2009) Sequencing the potato genome: outline and first results to come from the elucidation of the sequence of the world's third most important food crop. Am J Potato Res 86:417–429

Watanabe K (2015) Potato genetics, genomics, and applications. Breed Sci 65:53–68

Watanabe K, Orrillo M, Iwanaga M et al (1994) Diploid potato germplasm derived from wild and landrace germplasm resources. Am Potato J 71:599–604

Waugh R, Baird E, Powell W (1992) The use of RAPD markers for the detection of gene introgression in potato. Plant Cell Rep 11:466–469

Wilkinson MJ, Bennett ST, Clulow SA et al (1995) Evidence for somatic translocation during potato dihaploid induction. Heredity 74:146–151

Yermishin AP, Polyukhovich YV, Voronkova EV et al (2014) Production of hybrids between the 2EBN bridge species *Solanum verrucosum* and 1EBN diploid potato species. Am J Potato Res 91:610–617

Zamir D (2001) Improving plant breeding with exotic genetic libraries. Nat Rev Genet 2:983–989

Zhang Y, Zhao Z, Xue Y (2009) Roles of proteolysis in plant self-incompatibility. Annu Rev Plant Biol 60:21–42

Zhu J, Pearce S, Burke A et al (2014) Copy number and haplotype variation at the VRN-A1 and central FR-A2 loci are associated with frost tolerance in hexaploid wheat. Theor Appl Genet 127:1183–1197

Part III
Pest and Diseases

Chapter 8
Insect Pests Affecting Potatoes in Tropical, Subtropical, and Temperate Regions

Jürgen Kroschel, Norma Mujica, Joshua Okonya, and Andrei Alyokhin

Abstract Ensuring the sustainable production of potato is an important challenge facing agriculture globally. Insect pests are major biotic constraints affecting potato yields and tuber quality. The high pesticide uses to control them is of high human and environmental health concern, and it is expected that this will be further exacerbated through impacts of climate change. The chapter provides an overview of the geographical distribution of potato insect pests and their importance in tropical, subtropical, and temperate potato production regions. Climate change will potentially contribute to expand their geographical range of distribution, and increasing populations will lead to greater crop and post-harvest losses. Good progress has been made in applying insect pest modeling in pest risk analysis of potato pests to inform and create better awareness of future pest risks under climate change. Potato pests include some of the species which have evolved resistance to a wide variety of chemicals; and potato growers have already experienced the situation that available chemicals failed to control their targets. This chapter emphasizes the development, use, and adaptation of Integrated Pest Management (IPM) across all potato-growing regions of the world. Ultimately, this will lead to sustainable and more resilient potato production systems not overly dependent on pesticides. IPM requires a good knowledge and understanding of individual potato production systems; identifying pest species, knowing their biology and symptoms of infestation is essential for making educated decisions on their integrated management. To address this need,

J. Kroschel (✉)
Independent Consultant for International Agricultural Research & Development, Filderstadt, Germany
e-mail: jurgen.kroschelconsult@outlook.com

N. Mujica
Independent Consultant, Lima, Peru

J. Okonya
International Potato Center, Kampala, Uganda
e-mail: J.Okonya@cgiar.org

A. Alyokhin
University of Maine, Orono, ME, USA
e-mail: alyokhin@maine.edu

H. Campos, O. Ortiz (eds.), *The Potato Crop*,
https://doi.org/10.1007/978-3-030-28683-5_8

the chapter provides detailed information for a total of 49 insect pests of potato and the status quo of their management around the world.

8.1 Introduction

The potato (*Solanum tuberosum* L.) is native to the Andean highlands of South America. Today, potato is produced in more than 149 countries in temperate, subtropical, and tropical agroecologies, demonstrating the versatility and adaptability of this crop to a wide range of environmental conditions. Potato provides humans with an abundant and relatively inexpensive source of high-quality nutrients. Historically, their incorporation into everyday diets commonly coincided with periods of rapid population growth in a variety of nations (Zuckerman 1999). Presently, potatoes continue playing a very important role in feeding the human population. The last 30 years were characterized by an explosive growth in their popularity in Asia and Africa, which were previously reliant on other staple crops. Therefore, ensuring sustainability of potato production is currently an important challenge facing agricultural professionals worldwide (Vincent et al. 2013). Insect pests are major biotic factors affecting potato yield and tuber quality. Globally, losses are estimated on average at 16% (Oerke et al. 1994). Locally, if not routinely controlled, reductions in tuber yield and quality can be between 30 and 70% for various pests (Mujica and Kroschel 2013; Kroschel and Schaub 2013). The high pesticide use in potato is of high human and environmental health concern, which needs to be addressed by developing and more widely implementing Integrated Pest Management (IPM) approaches.

8.2 Potato Insect Pests' Geographical Distribution and Invasiveness

Due its global geographical distribution, potato is affected by a wide range of insect pests. In this book chapter we listed and described a total of 49 species: nine major species occurring in tropical and subtropical regions; two major species affecting potato in temperate regions; six major and 32 minor species of temperate, subtropical, and tropical regions. Farmers in tropical and subtropical countries must contend with a higher number of pest species, and with some exceptions, a minimum of 2–4 pests often reach pest status requiring the application of control methods (Kroschel et al. 2012). Many pests have evolved in the center of potato origin, and farmers in the Andean region are confronted by a higher number of pests than farmers in Africa or Asia. Some species such as the potato tuber moth, *Phthorimaea operculella* (Zeller), and the leafminer fly, *Liriomyza huidobrensis* (Blanchard) have become invasive and occur today as serious pests in many tropical and subtropical regions.

In contrast, the strong adaptation of Andean potato weevils, *Premnotrypes* spp., to the climate of the Andean region and its monophagous feeding habitat on potato and its wild relatives have restricted its distribution. There are, however, still several other pests which could also gain global proportions. Just very recently in 2006, the tomato leaf miner, *Tuta absoluta* Meyrick, although a more minor pest in potato, was unintentionally introduced to Spain, from where it continued its devastating journey across Africa and into Asia where it reached India within less than 10 years. As farmers had not been prepared and no control measures had been in place, the pest caused large production losses in tomato (*Lycopersicon esculentum* Mill.); under certain conditions also potato was more heavily infested as known from South America. The bud midge, *Prodiplosis longifilia* Gagne, currently with a restricted distribution in Florida and Virginia, and South America (Colombia, Peru, and Ecuador) could become an invasive species supported by its very polyphagous feeding habit. The Colorado potato beetle, *Leptinotarsa decemlineata* (Say), native to Mexico, has spread across most of the United States, and was introduced into France in the 1920s from where it spread further reaching also parts of China (CABI 2017a).

8.3 Impacts of Climate Change on Potato Insect Pests

Climate, especially temperature, has a strong and direct influence on the development and growth of insect pest populations. Herbivorous insects—as all other arthropods—are exothermic organisms that cannot internally regulate their own temperature. Their development depends on the temperature to which they are exposed in the environment. A rise in temperature due to climate change may both increase or decrease pest development rates and related crop losses. Hence, an increase in temperature can potentially affect range expansion and outbreaks of many insect pests including pests of potato.

Innovative modeling approaches such as process-based climatic response phenology models (Orlandini et al. 2017; Mujica et al. 2017; Sporleder et al. 2004, 2017) have been used to assess the effect of temperature increase under projected changes in global temperature for the year 2050 (Kroschel et al. 2016a) for a wide range of potato pests: *P. operculella* (Kroschel et al. 2013, 2016b); *L. huidobrensis* (Mujica et al. 2016); Guatemalan potato tuber moth, *Tecia solanivora* (Povolny) (Schaub et al. 2016); Andean potato tuber moth, *Symmetrischema tangolias* (Gyen) (Sporleder et al. 2016); the White flies, *Bemisia tabaci* (Gennadius) and *Trialeurodes vaporariorum* (Westwood) (Gamarra et al. 2016a, b).

These predictions have clearly demonstrated that insect pests of potato will respond to climate change by expanding their geographical range of distribution and increasing population densities will lead to greater crop and post-harvest losses. There are, however, distinct differences among the different pest species. The damage potential of *P. operculella* for example will potentially progressively increase in all regions where the pest already prevails today, with a range expansion into temperate and tropical mountainous regions (Kroschel et al. 2013, 2016b). In comparison, *L. huidobrensis* is much less adapted to warmer climates and climate change will differently affect this species. The global predictions clearly indicate a slight to moderate decrease in the establishment potential of the pest in most of the tropical and subtropical potato production areas, but still with high pest risks. Further, a range expansion into temperate regions of Asia, North and South America, and Europe, as well as into subtropical and tropical mountainous regions is expected, with a moderate increase of its establishment and damage potential (Mujica et al. 2016). A further range expansion into tropical mountainous regions such as the Andes has been also predicted for *T. solanivora* and *S. tangolias* (Quiroz et al. 2018).

Potato production systems of tropical countries are highly susceptible to pest infestations due to often year-round favorable climatic conditions for pest population growth and host plant availability. Even smaller changes in temperature predicted for tropical regions compared to temperate regions will have stronger consequences on pest development due to already higher existing metabolism rates of organisms such as insects (Dillon et al. 2010). This does affect not only the general life cycle of an insect pest but also all other biological processes, including feeding rates, plant growth, and activity of biotic antagonist. Lessons have been learnt from the El Niño phenomena to better understand possible climate change effects on pest abundance and severity in tropical production areas. During the 1997 El Niño phenomena in Peru, the mean temperature on the Peruvian coast increased by about 5 °C above the annual average. While infestation of potato by *L. huidobrensis* decreased, infestations by all other pests increased in agricultural and horticultural crops. The farmers' only adaptive strategy to cope was to apply high doses of pesticides every 2–3 days (Cisneros and Mujica 1999a).

8.4 Insect Pest Control with Insecticides in Potato

Potato foliage contains a considerable amount of glycoalkaloids, which provide at least some protection from herbivory. Nevertheless, they are attacked by a robust complex of phytophagous insects, some of which can destroy the potato crop in the absence of adequate control measures. Just as with most other cultivated plants, management of insect pests of potato is achieved predominantly through application of pesticides. By some estimates, potatoes are the most chemically dependent crop in the world (Vincent et al. 2013). Insect pests are not the only factor responsible for this notoriety, as large amounts of fungicides are used to combat diseases caused by fungi and oomycetes. Still, insecticides remain to be a foundation of insect pest management in most potato fields around the world, and their use can be rather heavy at times.

Although insecticides have been largely successful in keeping potato production going, there are serious and well-known concerns about long-term sustainability of this approach. Nontarget effects of insecticides on a variety of organisms, including humans and beneficial insects, gained considerable notoriety since 1960s. The use of highly hazardous pesticides in potato in countries such as Ecuador and Peru has caused serious health risks to farmers (Orozco et al. 2009). Worldwide decline in beneficial pollinators documented in the early 2000s provided additional fuel to the fire of public apprehension of using toxic chemicals in agriculture. Furthermore, as discussed in the following section, more and more insecticides have lost their efficiency due to resistance development in insect populations (Alyokhin et al. 2013, 2015).

As more and more insecticides are becoming phased out due to environmental concerns or become ineffective due to resistance development in targeted insect populations, the number of options available to potato growers dwindles. Developing replacement insecticides is an increasingly difficult and expensive task, and it is highly questionable that a plethora of new active ingredients will regularly appear on the market in perpetuity (Alyokhin et al. 2015). Therefore, good stewardship of existing chemicals and, whenever possible, their replacement with nonchemical control alternatives become an increasingly important business strategy for the pesticide industry and potato farmers. Therefore, this chapter puts a considerable emphasis on describing various nonchemical management options for insect pests.

In the modern age of industrialized agriculture, a farmer field is often considered to be a type of a production facility. Although there is a certain element of truth to such an approach, it is important to remember that it is still comprised of living organisms that are interacting with each other and with their environment. In other words, it is an ecosystem, with the same basic characteristics as all other ecosystems on Earth. This includes evolutionary processes leading to adaptation to a particular set of environmental conditions through survival and reproduction of specific genotypes. In many cases, such an adaptation means developing an ability to survive exposure to insecticides, a phenomenon known as insecticide resistance (Alyokhin et al. 2015).

Potato pests include some of the species that are most prone to evolving resistance to a wide variety of chemicals. The Arthropod Pesticide Resistance Database (2018) lists 469 cases of green peach aphid (*Myzus persicae* (Sulzer)) resistance to a total of 80 active ingredients; 300 cases of Colorado potato beetle (*Leptinotarsa decemlineata* (Say)) resistance to a total of 56 active ingredients; 111 cases of greenhouse whitefly (*T. vaporariorum*) resistance to 27 active ingredients; and 501 cases of two-spotted spider mite (*Tetranychus urticae* C. L. Koch) resistance to rather impressive 95 active ingredients. The extent of resistance is likely to be underestimated because not every case of its development is entered into the database. It is possible that the ability to deal with toxic glycoalkoloids contained in potato foliage serves as a preadaptation to resisting chemical toxins made by humans (Alyokhin and Chen 2017). Not every population of a given pest species is resistant to all compounds that have been recorded to fail against that species. However, these statistics vividly illustrate the seriousness of the problem. On several occasions, potato growers already experienced the situation when virtually all commercially available chemicals failed to control their targets (Alyokhin et al. 2013). It is very likely that such a situation will arise again in the foreseeable future. Therefore, we need to be proactive to prevent this from happening.

Insecticide resistance is usually associated with several traits that may be taken advantage of when devising resistance management plans. Due to the pleiotropic effects of resistant alleles, in the absence of insecticides resistant insects often have lower reproductive output and/or suffer higher mortality. As a result, they are being outcompeted by susceptible insects. Also, in many cases resistance is inherited as an incompletely dominant or an incompletely recessive trait. Therefore, insects that are heterozygous at the resistant locus can be successfully controlled with sufficiently high dose of insecticide. So, resistance management techniques are generally directed towards preventing the situation when only highly resistant homozygotes survive in a population (Alyokhin et al. 2013, 2015). In practical terms, this can be achieved by doing the following:

- Monitoring insecticide efficacy. Larger than usual number of surviving pests may indicate that their population is becoming resistant.
- Avoiding applications of the same or related products repeatedly throughout a growing season. Instead, application schedule should consist of a sequence of insecticides with different modes of action. Information on a product's mode of action is available from its manufacturer.
- Applying insecticides at rates that are not lower than a recommended minimum. Otherwise, heterozygotes will survive and breed with each other. Following Mendelian laws of inheritance, some of their offspring will become homozygously resistant and capable of surviving even the recommended insecticide rates.
- Applying insecticides only when pest populations are sufficiently high to cause economically important damage. Therefore, the use of control thresholds developed for different potato pests is highly recommended. Trying to kill every single insect will, by definition, result in the survival of only highly resistant genotypes.
- Whenever possible, leaving parts of the field untreated to allow susceptible pests to survive and interbreed with resistant pests. Because resistance is almost never completely dominant, resulting heterozygotes will be killed by insecticides.
- The use of insecticide applications should be not the first but the ultimate control option in an IPM approach after all other management options could not prevent to keep a specific pest population under the economic threshold.

8.5 Integrated Pest Management in Potato

Integrated Pest Management (IPM) is defined as an "ecosystem approach to crop production and protection that combines different management strategies and practices to grow healthy crops and minimize the use of pesticides." It means "a careful consideration of all available pest control techniques and subsequent integration of appropriate measures that discourage the development of pest populations and keep pesticides and other interventions to levels that are economically justified and reduce or minimize risks to human health and the environment. IPM emphasizes the growth of a healthy crop with the least possible disruption to agro-ecosystems and encourages natural pest control mechanisms" (FAO 2018). It has been also defined as "a decision support system for the selection and use of pest control tactics, singly or harmoniously coordinated into a management strategy, based on cost/benefit analyses that consider the interests of and impacts on producers, society, and the environment" (Kogan 1998).

Although IPM has been widely promoted by researchers, policymakers, and agricultural practitioners, it is still very far from being universally adopted, but there have been large shifts towards bringing it into the mainstream of agricultural production. Since 2014, for example, a European Union (EU) Directive has obliged all professional plant growers within the Union to apply the general principles of IPM (European Parliament and the Council of the European Union 2009). To make IPM, however, successful, integrating several methods to combat pests requires interdisciplinary research and effective nonchemical control methods for all pests in a specific crop agroecology to fully avoid or minimize the pesticide use. But many research programs working on IPM still focus on single plant protection methods and not on a systematic study of the compatibility and optimization of simultaneously implemented pest management elements (Stenberg 2017). The International Potato Center (CIP) has developed a holistic working framework for potato pest management research and development and its application by farmers. This approach led, for example, to the development of IPM for different potato agroecologies in Peru under smallholder production considering ecological, economic, and environmental benefits (Mujica and Kroschel 2018; Kroschel et al. 2012). Further, Horn and Page (2008) reported from a successful potato IPM program for large-scale potato growers in Australia.

Individual potato IPM programs vary greatly depending on the specific agroecology and the socioeconomic conditions for which it is developed or adapted. However, they all share several important components, which distinguish them from the so-called conventional approaches that largely rely on calendar-based sprayings of pesticides.

- Application of best cultural practices, which includes the use of healthy seed, suitable crop rotations, and intercropping systems among others which also support natural biological control.
- Correct and timely pest identification and regular monitoring of population development. This applies the use of monitoring tools such as pheromone- or physical-based trapping methods (e.g., yellow sticky traps). Educated pest management takes advantage of specific aspects of pest biology that can be used to diminish its populations in one way or another. It should also be preventative, initiated before the occurrence of economic losses.
- Application of economic thresholds. Attempts on complete pest elimination are costly, often futile, and generally counterproductive. A control action should be

taken only when the cost of damage due to injury by a pest exceeds the cost of the control action. Control thresholds have been developed for several potato pests, but they need to be verified and adjusted for the conditions of a specific location and potato variety.

- Using diverse control tactics in a complementary or, whenever possible, synergistic ways. Pesticides are the ultimate option integrated with cultural, biological, regulatory, and other controls to result in a unified multi-pronged attack against pest populations.
- Evaluating results. This component is often overlooked, but very important. Efficiency of IPM control programs should be continuously monitored to make necessary adjustments when needed considering economic, ecological, and environmental assessments.

Among other benefits, IPM is likely to dramatically slow down evolution of insecticide resistance. Simultaneous adaptations to diverse and unrelated management techniques will require statistically unlikely genetic changes in pest populations (Alyokhin et al. 2015). As a result, integrating different techniques is likely to reduce our reliance on the "pesticide treadmill" of constantly replacing failed chemicals. With all its advantages, IPM should be made available to farmers across all potato-growing areas of the world. Practicing this approach requires, however, a good understanding of individual production systems down to a single-field level and knowing their specific components. Identifying pest species and knowing their biology is essential for making educated decisions on their management. To address this need, the following subchapters provide an overview of major and minor insect pests of potato and their management around the world.

8.6 Major Pests in Tropical and Subtropical Regions

8.6.1 Potato Tuber Moths

***Phthorimaea operculella* (Zeller, 1873),**
***Symmetrischema tangolias* (Gyen, 1913),**
***Tecia solanivora* (Povolny, 1973) (Lepidoptera: Gelechiidae)**

Distribution The potato tuber moth (PTM), *Phthorimaea operculella,* originated in the tropical mountainous regions of South America. Today it has a worldwide distribution and is considered the most damaging potato pest in the developing world (Fig. 8.1). It is present in almost all tropical and subtropical regions of the world, in North, Central, and South America, Africa, Asia, Australia, and Europe. The Andean potato tuber moth (APTM), *Symmetrischema tangolias,* is native to South America (Peru and Bolivia) but has in the last decades spread to other regions of the world (Fig. 8.2). Records include North America, Australia, New Zealand, and more recently Indonesia. Although this species is known as a pest of potato and tomato in South America and the Australian region, it does not attack any of these

Fig. 8.1 Adult female of *Phthorimaea operculella* (**a**) and symptoms of larvae infestation on leaves, stems, and tubers (**b**, **c**, **d**). (Photo credits: CIP)

Fig. 8.2 Adult male and female of *Symmetrischema tangolias* (**a**), and symptoms of larvae infestation on stems and tubers (**b**, **c**). (Photo credits: CIP)

two crops in North America but instead feeds on black nightshade (*Solanum americanum* Mill.). The Guatemalan potato tuber moth (GPTM), *Tecia solanivora,* probably originated from Guatemala and is endemic throughout Central America (Fig. 8.3). In 1983, the pest was unintentionally introduced into Venezuela and then invaded Colombia and Ecuador. In 2000, *T. solanivora* was introduced in the Canary Islands (Tenerife). Since then the pest has been considered as a major threat to potato throughout southern Europe and was listed as a quarantine pest by the European and Mediterranean Plant Protection Organization (EPPO 2005a; Kroschel et al. 2016a, b; Kroschel and Schaub 2013). In 2014, it was finally recorded in mainland Spain where efforts are going on to eradicate the pest (Jeger et al. 2018).

Host range *P. operculella* is an oligophagous pest (i.e., an insect feeding on a restricted range of food plants) of vegetable crops that belong mainly to the family Solanaceae: potato (*Solanum tuberosum* L.), tomato (*Lycopersicon esculentum* Mill.)*,* and tobacco (*Nicotana tabacum* L.). Also, wild species of the Solanaceae family, including important weeds, e.g., black night shade (*Solanum nigrum* L.) are hosts. In total, the host range comprises 60 species. Further, crops of the family Chenopodiaceae are attacked including eggplant (*Solanum melongena* L.), bell

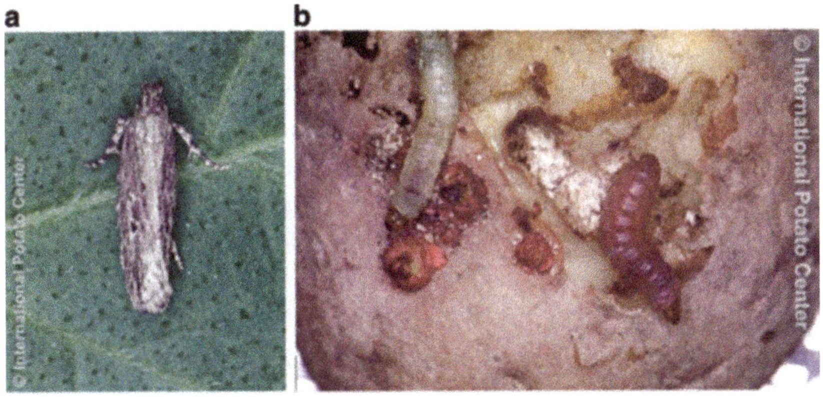

Fig. 8.3 Adult male of *Tecia solanivora* (**a**), and symptoms of larvae infestation on tubers (**b**). (Photo credits: CIP)

pepper (*Capsium annuum* L.), Cape gooseberry (*Physalis peruviana* L.), aubergine (S. *melongena* L.), and sugar beet (*Beta vulgaris* L.). *S. tangolias* has a more restricted host range comprising of potato, tomato, sweet cucumber (*Solanum muricatum* Aiton), poroporo (*Solanum aviculare* G. Forst), kangaroo apple (*Solanum laciniatum* Aiton), black nightshade, and bell pepper. *T. solanivora* is monopahagous attacking only potato.

Symptoms of infestation *P. operculella* attacks potato by mining the leaves and stems and by feeding on the tuber. Mines are the typical symptoms of leaf damage caused by the larvae eating the mesophyll without damaging the upper and lower epidermis. When the foliage dies, the larvae enter the soil through cracks where they may eventually find and feed upon tubers. Larvae enter potato tubers via the eyes and continue to bore or tunnel through the tuber just below the skin. Larval excreta are pushed out through the holes, which can be observed immediately after larvae start their mining activity. Larvae of APTM enter the potato stem making a small hole in the plant axils (between stem and lateral petioles). From this hole, galleries made by the larvae run downward within the stem. Excreta are pushed out through the initial hole made by the larva. When stems are severely damaged, the upper part of the stem wilts or the whole plant collapses. Young plants can suffer tip death from boring larvae. Eggs may be found in slits on the stem of a food plant. In tubers, larvae enter through potato eyes. Initially, the small hole can hardly be seen by the naked eye. As in stems, larval excreta are then pushed out through the hole, which becomes apparent after several days of mining activity. Inside the tuber, the larva tunnels just under the surface at first, but later penetrates more deeply. GPTM larvae feed exclusively on tubers during potato cultivation and during storage. Damage is caused by larvae that bore galleries into the tubers. After the larvae have left tubers, the exit hole is clearly visible. In potato fields, *T. solanivora* attack occurs from tuberization until harvest (Kroschel and Schaub 2013; Niño 2004). Tubers infested by either of the three species develop a bitter taste and are unsuitable for human or livestock consumption (Keller 2003; Kroschel and Schaub 2013).

Impacts on production losses

P. operculella. Under heavy field infestation, potato foliage can be destroyed, which can result in substantial yield loss of up to 70%. High infestations early in the season can directly affect tuber yield. Strong correlation exists between leaf and consequent tuber infestation, which suggests that reducing *P. operculella* population density during the growing period is key to reducing potato tuber infestation at harvest. Hence, the most devastating yield losses are largely a result of earlier tuber infestation in the field, generally where moths have laid eggs through soil cracks on the developing tubers, or when harvest is delayed. *P. operculella* also damages harvested potato tubers in storage. The damage to potatoes in rustic stores can be total within a few months if the tubers are left untreated. Infested tubers are unsuitable not only for human consumption but also for use as seed. Infested tubers produce fewer yields and initiate a fast development of a new field *P. operculella* population (Kroschel and Schaub 2013; Kroschel et al. 2012; Keller 2003; Kroschel 1994, 1995).

S. tangolias. This species has become an economically important pest in potato fields and in storage in mid-elevation regions of the Andes (Peru, Bolivia, and Ecuador); its status in Colombia is not well confirmed. In the Andes, losses in the field may reach up to 30%, but most economically significant damage occurs when infested tubers are transferred to potato stores where reinfestation takes place. Without adequate management, farmers can completely lose their house-stored potatoes within 3–4 months of storage. In Australia and New Zealand, where the pest was accidentally introduced from South America, it is more recognized as a local pest of tomato, poroporo, sweet cucumber and other Solanaceae crops, and is commonly referred to as "tomato stem borer." However, at national level it is considered a minor pest and its economic impact on these crops is not well reported in the literature (Kroschel and Schaub 2013; Keller 2003).

T. solanivora. Complete losses of harvested tubers have been observed occasionally after the invasion of the pest into new areas when farmers were not yet familiar with pest control. Generally, tuber damage rates at harvest vary between 2 and 15% in the Andean region. When infested potato tubers are stored without application of control methods *T. solanivora* can destroy, depending on the storage period and temperature, of whole potato stock (Kroschel and Schaub 2013; Niño 2004).

Methods of prevention and control

Control of the potato tuber moths must take place both in the field and in storage. Implementation of integrated pest management is recommended to reduce the pest problem in field and stores. *T. solanivora* is the most difficult potato moth to control as the larvae feed only inside potato tubers where they are hard to reach (Kroschel and Schaub 2013; Kroschel et al. 2012; Keller 2003; Pollet et al. 2003; Kroschel 1995).

Monitoring with pheromone traps. For all three potato tuber moth species sexual pheromones have been identified and synthesized. They are used for monitoring the flight activity of adult male populations to detect early the presence of the different moths in the field and store to take adequate control measures.

Cultural practices. Some common practices for potato tuber moths are the use of pest-free seed tubers, deep planting, regular irrigation to avoid soil cracking, high hilling to protect tubers, timely harvest, not leaving the tubers after harvest exposed in the field for a long time (especially throughout the night), i.e., harvest and store immediately, and removal of leftover tubers to reduce the overwintering field population. Also, early maturing varieties can contribute to reduced risk of infestation.

Biological control. Classical biological control can be an effective strategy in all those regions in which the pests has been unintentionally introduced to keep the pest population below economic threshold; for this approach, the species *Copidosoma koehleri* (Blanchard), *Apanteles subandinus* (Blanchard), and *Orgilus lepidus* (Muesebeck) have been widely and successfully used (Canedo et al. 2016a, b, c; Kroschel and Schaub 2013).

Biopesticides. Microbial biopesticides for *P. operculella* field control have been tested based on *Bacillus thuringiensis* subsp. *kurstaki* (*Btk*) and *P. operculella*-specific granulovirus (*Phop*GV, Baculoviridae). *Btk* was effective but required repeated applications because it is quickly degraded by UV light. Likewise, *Phop*GV has shown mixed results. To protect *Phop*GV against UV inactivation a variety of adjuvants (e.g., dyes, optical brighteners) have been tested but simple preparations of *Phop*GV-infected larvae macerated in water were superior. Applications of *Phop*GV doses sufficient to cause >95% mortality are considered not being economical, and low dose treatments are proposed for a relatively inexpensive partial suppression of the field population (Lacey and Kroschel 2009; Sporleder and Kroschel 2008; Kroschel and Sporleder 2006; Sporleder 2003; Kroschel et al. 1996).

Attract-and-kill. This approach has been developed to control of *P. opercullela* and *S. tangolias* under field and storage conditions. It consists of a co-formulation of the insect pest-specific sexual pheromone, which "attracts" males, and a contact insecticide at very low concentration which "kills" males getting in contact with the product. The oil formulation is applied at a droplet size of 100 μL using a special handheld applicator; it is applied at 2500 droplets/ha. It effectively reduces the male population and the number of offspring, hence controlling larvae damage in the crop. It provides pest-specific control, and is harmless to natural enemies, humans, and the environment (Kroschel and Zegarra 2010, 2013). In Peru, the two products AdiosMacho-*Po*® and AdiosMacho-*St*® have been registered to be commercialized in Peru and the Andean region.

Chemical control. Broad-spectrum insecticides have been commonly used to suppress potato tuber moth population and economic damage, but which has been associated with many negative effects causing resistance of the pests to various active ingredients and affecting farmers and the environment.

Integrated Pest Management. Effective IPM practices for potato tuber moths have been developed, which can be applied successfully if potato tuber moths are the only economically important pests in an agroecosystem (e.g., Republic of Yemen, Kroschel 1995). However, potato is often affected by several pest species which requires a system approach to manage all economically important potato pests (Kroschel et al. 2012). This means effective IPM practices are required for all pests to fully eliminate or minimize the use of insecticides.

Storage management

Potato tuber moth infestation occurs frequently in rustic farmer-managed potato stores in developing countries, especially if temperature is suitable for rapid population build up and the storage lasts for several months. Storage facilities should be cleaned thoroughly before potato tubers are stored. Fine netting at windows should protect adult moths from entering storage facilities. Only healthy tubers should be selected for storage. Infested potato tubers need to be destroyed. However, initial infestations cannot be easily observed and are the main reasons why potato tuber moths enter storage facilities and infest potato. Sex-pheromone-baited water traps or funnel traps (Delta) can be used for monitoring the moth presence but in known region of potato tuber moths' occurrences, potatoes should be treated before storage.

Biopesticides. Biopesticides based on *Btk* and *Phop*GV are used in potato tuber moth storage control. The microbials are formulated in inert materials (e.g., talcum) and dusted over potatoes before storage. Since *Phop*GV is only effective in *P. operculella* and *T. solanivora*, in regions where all three species occur simultaneously, the use of *Btk* has the advantage to control all three species. Further, *Btk* is mostly available as a commercial biopesticide while *Phop*GV has to be multiplied in potato tuber moth larvae. The product Matapol, e.g., is a co-formulation between *Btk* and *Phop*GV, commercialized in Bolivia. It has also been shown that inert materials (e.g., calcium carbonate, kaolin, talcum, silicium rich sand) can be used effectively without the addition of active biologicals (*Btk,* or *Phop*GV) as they control first instar larvae through desiccation (Kroschel and Koch 1996; Mamani et al. 2011; Sporleder and Lacey 2013; Schaub and Kroschel 2017).

Attract-and-kill. Attract-and-kill formulations (see above) can be applied at a density of one drop (100 μL)/qm of storage area to reduce the male population and hence tuber infestation in potato stores (Kroschel and Zegarra 2013).

Chemical control. Malathion dust (WHO Class III) is often observed to be sold in developing countries to treat stored tubers. This is especially critical if precautions are not taken properly by farmers and potatoes are stored in living areas. Pyrethroids (e.g., fenvalerate) have shown to be highly effective equally to *Btk* treatments described above (Kroschel and Koch 1996).

8.6.2 Pea Leafminer Fly

Liriomyza huidobrensis Blanchard (Diptera: Agromyzidae)

Distribution The pea leafminer *Liriomyza huidobrensis* is an agricultural pest endemic to South America (Fig. 8.4). Since the early 1980s, the pest has been also recorded in many other countries around the world, presumably associated with the global trade of ornamental plants (Mujica et al. 2016; CABI 2012).

Fig. 8.4 Adult of *Liriomyza huidobrensis* (**a**) and symptoms of the larvae infestation on potato leaves (**b**). (Photo credits: CIP)

Host range *L. huidobrensis* is highly polyphagous and has been recorded from plants of 14 families (Spencer 1973, 1990). The long list includes potato, bean, pea, alfalfa (*Medicago sativa* L.), and vegetables such as tomato, celery (*Apium graveolens* var. dulce (Mill.) Pers.), lettuce (*Lactuca sativa* L.), pepper (*Capsicum annuum* var. *longum* (DC.) Sendtn.), and spinach (*Spinacia oleracea* L.). In addition, leafminer flies infest many weed species and ornamental plants.

Symptoms of infestation Adults and larvae of *L. huidobrensis* damage the plant foliage. Adults cause damage by puncturing the leaf surface to feed on the leaf tissue, and to lay eggs. Newly hatched larvae mine into the leaf and feed on the chloroplast-rich mesophyll, making a serpentine mine whose diameter increases as the larva grows. A large proportion of grown larvae remain close to the midrib. Leaf tissue affected by larval mining becomes necrotic and brownish. Highly infested crop fields appear burned (Cisneros and Mujica 1999b).

Impacts on production losses *L. huidobrensis* is a serious pest of arable crops, vegetables, and ornamental plants under field and glasshouse conditions in many parts of the world. Plant injuries caused by adult and larval activities reduce photosynthesis activity and cause leaf wilting. For potato, yield losses of up to 100% were reported in Argentina, Chile, and Indonesia (Cisneros and Mujica 1999b). Pest intensity-crop loss relationships for the leafminer fly in different potato varieties indicated that the accumulated foliar injury up to the growth stages of flowering and berry formation produced the highest yield losses in the different potato varieties (Mujica and Kroschel 2013). Economic injury levels in Peru varied according to control costs and commodity values, and potato varieties with longer vegetation period can tolerate higher levels of foliar injury by the leafminer fly before control measures are needed (Desiree: 21–28%, Revolucion: 34–47%, Canchan: 31–40%, Maria Tambeña: 40–53%, Tomasa: 55–74%, and Yungay: 40–54% of foliar injury).

Methods of prevention and control

Ecological and economical sound control of the leafminer fly is best realized when based on IPM, which promotes natural enemies in combination with cultural practices and low-toxic insecticides (Mujica 2016; Mujica et al. 2016; Kroschel et al. 2012; Weintraub et al. 2017).

Monitoring pest populations. Counting the number of flies captured in yellow sticky traps monitors adult leafminer fly activity. Counting the number of larvae or fresh tunnels per leaflet by sampling the bottom, middle, and top parts of the plant is used to monitor larval infestation. Both methods can be adapted for decision making and applying an action threshold (AT) to avoid unnecessary applications of insecticides. The AT can be defined as the level of pest population at which control measures should start to prevent the pest population from reaching an economic injury level (EIL, point where economic losses will begin). The AT is typically set below the EIL accounting for the lag time to implement effective control measures. Preemptive insecticidal control is economically not justified until foliar injury exceeds these values (Mujica and Kroschel 2013; Mujica and Kroschel 2018).

Crop management. Healthy, vigorous growing potato plants can better tolerate leafminer damage, particularly during the vegetative phase. Balanced N-fertilization is important as high N-content in leaves promotes leafminer fly development. Continuous food availability by replanting host crops will favor the abundance of the leafminer fly. Rotation with nonhosts is therefore recommended.

Conserving beneficial insects. Leafminer flies are controlled by many beneficial insects, which are either predators or parasitoids. Strategies to conserve beneficial insects can be manifold and include diversified cropping systems, high structural floristic diversity in agricultural landscapes, special weed management practices, and reduced use of broad spectrum insecticides (Mujica et al. 2016).

Classical biological control. It can be an effective strategy in all those regions in which the pea leafminer fly has been unintentionally introduced and where natural enemies of leafminer flies are absent to keep the pest population below economic threshold. The endoparasitoids *Halticoptera arduine* Walker (Pteromalidae), *Chrysocharis flacilla* Walker (Eulophidae) and *Phaedrotoma scabriventris* (Nixon) (Braconidae) were successfully introduced and established in three agro-ecological regions (low, middle, and high altitude) in Kenya (Muchemi et al. 2014; Mujica et al. 2016). Yet, biocontrol of leafminer flies needs to be accompanied by additional, biocontrol-compatible control measures which need to be ultimately integrated into one holistic IPM concept for vegetables that addresses all major pests of the system.

Use of entomopathogenic nematodes. The entomopathogenic nematode *Heterorhabditis indica* (Rhabditida: Heterorhabditidae) caused 58.7% of leafminer larval mortality in potato leaves under semi-field conditions. It could be considered as biocontrol-compatible control measure as part of an IPM program for potato and vegetables (Mujica et al. 2013).

Physical control. Yellow attracts leafminer fly adults. The use of mobile (1 × 4 m bands of stick traps which are moved across the fields fixed at tractors or hand-carried) and stationary (50 × 50 cm, 60–80 traps/ha) yellow sticky traps can effectively reduce the leafminer fly adult population. In the Cañete valley of Peru, a cumulative capture of up to seven million adults/ha by using fixed and mobile yellow sticky traps (US$66.7/ha) resulted in a reduction of the control costs by 55.5% compared with chemical control ($200.0/ha), and an average use of six adulticide applications per season (Mujica et al. 2000).

Chemical control. Decisions to use insecticides should be made according to the monitoring results and when the leafminer population is expected to cause economic damage (Mujica and Kroschel 2013). Systemic insecticides with translaminar properties are most effective in controlling leafminer fly larvae. Such insecticides include abamectin and spinosad or cyromazine (Weintraub 2001).

Integrated Pest Management. An IPM strategy based on the use of seed treatment, action threshold, trapping devices and selective application of insecticides showed a higher efficacy to control potato pests including *L. huidobrensis* than the conventional application of insecticides by farmers in the Cañete valley of Peru. IPM reduced the total quantity of pesticides used per season by 56% compared to the conventional management, representing a decrease of 69.2% in the environmental impact. Further, IPM achieved 35% of higher marketable potato yield than conventional management (Mujica and Kroschel 2018). Leafminer fly is a polyphagous pest and biocontrol-compatible control measures need to be integrated into an IPM concept which considers also other economic pests in potato.

8.6.3 Andean Potato Weevils

***Premnotrypes suturicallus* Kuschel, *P. vorax* (Hustache), *P. latithorax* (Pierce) (Coleoptera: Curculionidae)**

The Andean potato weevil complex consists of at least 14 species with 12 in the genus *Premnotrypes* and two in the genus *Rhigopsidius* and *Phyrdenus*. The most important species attacking potato are *Premnotrypes vorax*, *P. latithorax* and *P. suturicallus* (Fig. 8.5). All *Premnotrypes* species show sexual dimorphism; the females (6.8–8.0 mm) are larger than males (5.6–7.5 mm) (Alcázar and Cisneros 1999).

Distribution The Andean potato weevils (*Premnotrypes* spp.) are the most serious pests of the potato in the Andean region above 2800 m above sea level. Its distribution extends from Argentina to Venezuela, covering a mountainous territory of about 5000 km in length. These weevils are native to the Andes where wild potato and cultivated species are their hosts (Alcázar and Kroschel 2008).

Host range The host range of Andean potato weevils includes only potato and its wild relatives.

Symptoms of infestation The Andean potato weevil, in the larval and adult stages, causes damage to the potato. Adults feed on leaves starting from their edges, marking a very characteristic form of semicircle. When the beetle population is very

Fig. 8.5 Adult of *Premnotrypes suturicallus* (**a**: male, left; female, right), and symptoms of weevil damage on leaves (**a**, **b**) and of larvae infestation in potato tubers (**c**). (Photo credits: CIP)

high, leaves are eaten up to the central leaf vein. Occasionally, adults can also damage on stolons, tubers during formation, and the stem bases. Serious damage to tubers begins when newly emerged larvae penetrate tubers. As the larvae develop, they build characteristic tunnels that are usually filled with excreta. Once the larval stage is complete, the larvae leave the tuber making characteristic circular exit holes (Alcázar and Cisneros 1999; Alcázar and Kroschel 2008).

Impacts on production (losses) Andean potato weevils cause substantial yield losses that seriously threaten Andean farmers' food security. Losses largely vary (16–45%) even when insecticides are applied. If weevils are not routinely controlled, losses can even reach 80–100% (Ortiz et al. 1996; Kroschel et al. 2012).

Methods of prevention and control
(Alcázar and Cisneros 1999; Alcázar and Kroschel 2008; Kroschel et al. 2012).

Cultural control. The main sources of weevil infestations are potato fields of the previous season. It is recommended to maintain community rotation system over distances of about 1 km between fields. Let the field rest for 3–5 years before planting potato and avoid planting potatoes for two consecutive seasons. Best candidate crops to rotate with include faba bean and barley. Elimination of volunteer potato plants in other crops should be practiced. Early harvest and elimination of crop residues using animals (pigs, sheep) reduces weevil populations in the field.

Mechanical control. Night collection of adults through plant shaking into buckets.

Physical barriers. It is a method that prevents the adult weevils from entering the field by using a plastic barrier placed around the potato field. It is recommended to install the barriers before or at the time of planting, to prevent the first adults from entering the new fields. Results from the use of plastic barriers studies revealed a higher efficacy for the control of the Andean potato weevil than applications of insecticides (Kroschel et al. 2009).

Plant barriers. Interrupting adult migration into potato fields through plant barriers of tarwi (*Lupinus mutabilis* Sweet) or mashua (*Tropaeolum tuberosum* Ruíz and Pavón) has also shown effects of reducing the weevil population.

Chemical control. Generally, the use of insecticides has not demonstrated a consistent efficacy and farmers still experience losses while spraying insecticides 2–3 times.

Natural enemies. Although the weevils are native to the Andes, no specific parasitoids have been identified but predators like carabids are wide-spread and affect the weevil population. Most common species are *Blennnidus* sp., *Notiobia schnusei* (Van Endem) and *Harpalus turmalinus* Er. In addition, entomopathogenic fungi (*Beauveria bassiana* (Balsamo) Vuillemin) and nematodes (*Heterorhabditis* sp., and *Steinernema* sp.) have been identified and used to develop biocontrol strategies (Kroschel et al. 2012; Alcázar and Cisneros 1999; Kaya et al. 2009).

Integrated Pest Management. An IPM strategy consists of the use of plastic barriers in combination with cultural practices; one insecticide application might be required if plastic barriers are used in potato–potato rotations (Kroschel et al. 2011, 2012).

8.6.4 Potato Psyllid

Bactericera (ex-Paratrioza) cockerelli (Sulc) (Hemiptera: Triozidae)

Distribution The potato psyllid *Bactericera cockerelli* is native to North America and occurs mainly in the Great Plains region of the United States, from Colorado, New Mexico, Arizona, and Nevada, north to Utah. It is also found in Mexico, Guatemala, Honduras, and Nicaragua, and is suspected to be present in other Central American countries, including El Salvador (Fig. 8.6). Following an accidental introduction, *B. cockerelli* has become widespread in New Zealand (CABI 2016; OIRSA 2016; Rehman et al. 2010).

Host range *B. cockerelli* is found primarily on plants of the family Solanaceae. The psyllid infests and develops on a variety of cultivated and weedy plant species including potato, tomato, pepper, eggplant, and tobacco as well as noncrop species such as black nightshade, groundcherry (*Physalis* spp.), and matrimony vine (*Lycium* spp.) (CABI 2016).

Fig. 8.6 Nymphs of *Bactericera cockerelli* infesting potato leaves (**a**) and potato plant affected with potato purple top (**b**). (Credits: Courtesy of CIP)

Symptoms of infestation Both adults and nymphs feed by sucking the sap of plants. Direct damage is caused by injecting substances that destroy plant cells (toxins). Chlorophyll production is interfered causing leaf yellowing and stuntedness, which together result in a condition known as psyllid yellows. Indirect damage is related to the transmission by nymphs and adults of the bacterium *Candidatus liberibacter solanacearum* (psyllaurous) that causes infectious diseases known as permanent tomato (PT), potato purple top (PM), and zebra chip. Typical symptoms of potato zebra chip include yellowing and curling of foliage, stunted growth, formation of aerial tubers, shortened and thickened internodes, leaf scorching, reduced tuber size and yield, and early plant death. Belowground, zebra chip is characterized by the presence of collapsed and necrotic stolons, and browning of internal vascular tissues, which, upon frying, exhibits dark brown streaks, hence the term "zebra chip" (Martin 2016; Rehman et al. 2010; Garzón 2002).

Impacts on production losses Potato zebra chip disease is a serious disorder of potatoes that has resulted in millions of dollars in losses to the potato industry. The characteristic symptom is dark striping in the tuber when fried as French fries or potato chips, rendering the infected potatoes unmarketable. It has been determined that *C. liberibacter* is transmitted to potato very rapidly by the potato psyllid, and that a single psyllid per plant can successfully transmit this bacterium to potato in as little as 6 h. This low psyllid density, coupled with a short inoculation access period, represents a substantial challenge for growers in controlling the potato psyllid and preventing zebra chip transmission. Just a few infective psyllids feeding on potato for a short period could result in substantial spread of the disease within a potato field or region (Garzón 2002; Rehman et al. 2010; Munyaneza et al. 2007, 2009).

Methods of prevention and control

Monitoring pest population. Early season management of this insect is crucial to minimize damage and psyllid reproduction in the field. Egg and nymphal stages require visual examination of the foliage. Adults can be sampled with yellow traps, which are effective in detecting immigrant populations and must be installed from the beginning of the cropping season. Low populations before or at the start of tuber formation reduce production significantly, but once the tubers are formed plants tolerate direct damage (Henne et al. 2010; Butler and Trumble 2012; Workneh et al. 2012).

Plant resistance. The most valuable and effective strategies to manage zebra chip would likely be those that discourage vector feeding, such as use of plants that are resistant to psyllid feeding or less preferred by the psyllid. Unfortunately, no potato variety has so far been shown to exhibit sufficient resistance or tolerance to zebra chip or potato psyllid.

Biological control. Several predators and parasites of *B. cockerelli* are known, though there is little documentation of their effectiveness. In New Zealand, fungal isolates of *Lecanicillium muscarium* (Petch) Zare & W., *Isaria fumosorosea* (Wize), and commercial formulations of *Beauveria bassiana* Balsamo-Crivelli (Vuillemin), *I. fumosorosea*, and *Metarhizium anisopliae* (Metchnikoff) Sorokin were successfully tested against nymphs of *B. cockerelli* in laboratory assays and adults and nymphs in greenhouse assays, outperforming a conventional spiromesifen insecticide (Mauchline and Stannard 2013). Commercial formulations of *Metarhizium anisopliae* (F52®, Novozymes Biologicals) and *Isaria fumosorosea* (Pfr 97®, Certis USA) significantly reduced plant damage and potato zebra chip symptoms in Weslaco, Texas (Lacey et al. 2011).

Chemical control. Good insecticide coverage or translaminar activity is important because psyllids are commonly found on the underside of the leaves. Because several generations often overlap, caution should be taken when selecting and applying insecticides targeted against the potato psyllid in relation to which life stages are present in the crop and timing of insecticide applications.

Integrated Pest Management. *B. cockerelli* is often associated with the pathogen *Candidatus liberibacter solanacearum* and this represents the main challenge in the search for IPM components that can be integrated into the potato crop. Several management strategies have been proposed such as cultural, biological, and chemical control. In New Zealand, the proposed IPM strategy includes careful timing of insecticide applications, incorporating alternative chemicals, integrating treatments with vector monitoring or sampling, establishing action thresholds (although this is difficult for vectors), determining the likelihood of infections throughout the year, and combining chemical applications with resistant or tolerant varieties (Vereijssen et al. 2018).

8.6.5 Bud Midge

Prodiplosis longifila **Gagne (Diptera: Cecidomyiidae)**

Distribution *Prodiplosis longifila* is a polyphagous species only in the Americas (Fig. 8.7). Its origin is unknown. It occurs in North America (first reports from Florida in 1934 and from Virginia in 1990) and South America (Colombia, Peru, and Ecuador). Some publications mention its presence in the "West Indies" but detailed records for individual countries in the Caribbean are not available and a previous record for Jamaica is now considered to be a misidentification (EPPO 2015; Gagné 1986).

Host range *P. longifila* is polyphagous and has been recorded from many plant species. However, economic damage is mainly reported on potato, tomato, pepper, asparagus and to a lesser extent on Tahiti lime (*Citrus* × *aurantifolia* (Christm.) Swingle, *Citrus* x *latifolia* (Yu. Tanaka) Tanaka). Lists of host plants include important crops such as onion (*Allium cepa* L.), watermelon (*Citrullus lanatus* (Thunb.) Matsum. & Nakai), melon (*Cucumis melo* L.), cucumber (*Cucumis sativum* L.), artichoke (*Cynara scolymus* var. *scolymus* L.), soybean (*Glycine max* L.), alfalfa (*Medicago sativa* L.), beans (*Phaseolus vulgaris* L.), castor bean (*Ricinus communis* L.), and grape (*Vitis vinifera* L.). Because the identity of the different populations of *P. longifila* has not yet been resolved, some records of hosts previously attributed to *P. longifila* may belong to another species (EPPO 2015).

Symptoms of infestation *P. longifila* is minute in size, but extremely injurious under warm environmental conditions. The larva passes through three stages, with the first and second larval stages being the most harmful and causing damage to flowers and buds of different crops. The larvae of *P. longifila* scrape the epidermal tissues of plant structures using piercing-sucking mouthparts. Feeding on buds distorts growth points and results in winding stems and black appearance. It causes

Fig. 8.7 *Prodiplosis longifila* adult (**a**), larvae (**b**) and damage on potato sprouts (**b**). (Photo credits: CIP)

flower buds to fall and results in distorted fruit of low quality (EPPO 2015; Hernandez et al. 2015; Sarmiento 1997).

Impacts on production losses *P. longifila* is one of the most important pests of Solanaceae crops (potatoes and other vegetables) and asparagus in the Neotropics causing yield losses >50%, if not controlled. It affects buds, flowers, and fruits, and causes severe damage to both open and protected crops. During the first 50 days after plant emergence, the bud midge is a critical pest that affects plant growth by destroying both the terminal and lateral buds. However, infestations after flower initiation do not affect potato yields. High temperatures favor the increase in midge population and hence crop damage, which could explain the reduced potato yields recorded during the El Niño phenomenon in Peru (EPPO 2015; Hernandez et al. 2015; Sarmiento 1997).

Methods of prevention and control
Crop management. Eggs and first instar larvae are in protected places (e.g. under the calyx) and are not visible with the naked eye (Gonzales-Bustamante 1996). Therefore, factors that favor vigorous potato plant growth (appropriate fertilization, healthy seed with vigorous sprouts, irrigation, and biostimulants) prevent pest damage by the rapid growth of buds and the exposure of larvae to dehydration and natural enemies. Management practices to reduce infestation include optimal planting density, orientation of the furrows from east to west to favor insolation, elimination of broadleaf weeds, avoidance of susceptible crops, management of the planting season, and rotation with nonsusceptible crops (Mujica and Kroschel 2018).

Seed treatment. Seed treatment with a systemic insecticide protects buds against initial bud midge infestation from emergence to formation of the first leaves and stems. This is a highly recommended practice since it ensures protection for a period of 35 days during a critical period of the crop, without affecting natural enemies (Mujica and Kroschel 2018).

Chemical control. For a chemical control of this pest, the following should be considered: (a) dusting of sulfur powder on young plants to reduce initial infestation, (b) the use of translaminar or systemic insecticides, (c) dusting of dry powder insecticides at the soil or lower third of the plants, and (d) a combined control of larvae and adults with various active ingredients such as imidacloprid, dimethoate, mixtures of chlorpyrifos and alphacypermethrin, and vegetable and mineral oils (Castillo 2010). Sulfur dust has repellent effects on the adults, but low toxicity to larvae. Treatments with sulfur are more effective in the late afternoon when adults become active and more exposed to the treatments.

Biological control. The most commonly occurring biological control agents are egg and larva parasitoids of the genus *Synopeas* (Hym.: Platygasteridae), reaching parasitism of 16–20% (Hernández 2014). Important predators of third instar larvae and pupa are *Chrysoperla asoralis* Banks, *Nabis capsiformis* Germar, and *Methacantus tenellus* Stal.

Biopesticides. In Peru, the entomopathogenic nematode *Heterorhabditis* sp. (native) showed 80% pathogenicity in larvae and pupae of *P. longifila* (Pacheco 2015). Mortality of *P. longifila* caused by *Metarrhizium anisopliae*, *Baeuveria bassiana*, *Paecilomices fumosoroseus,* and *Verticillium lecanii* under laboratory and greenhouse conditions was 75 and 69%, 71 and 73%, 62 and 50%, 59 and 44%, respectively (Reategui et al. 2009). The application of entomopathogens in commercial fields of asparagus showed a reduction of *P. longilifa* pupae by 79.2% (*B. bassiana*), 70.8% (*M. anisopliae*), 30.4% (*V. lecanii*) (Castillo 2010).

8.7 Major Pests in Temperate Regions

8.7.1 Colorado Potato Beetle

***Leptinotarsa decemlineata* (Say) (Coleoptera: Chrysomelidae)**

Distribution The Colorado potato beetle *Leptinotarsa decemlineta* is native to Mexico (Fig. 8.8), but has spread to Central America, most of the United States (except Alaska, California, Hawaii, and Nevada), and to southern Canada. This pest has also been introduced into Europe and parts of Asia (CABI 2017a; Jacques and Fasulo 2015).

Host range Colorado potato beetle is an oligophagous species feeding on about ten species of plants in the family Solanaceae (Alyokhin et al. 2013). These include potato, tomato, and eggplant.

Symptoms of infestation Both adults and larvae are voracious leaf feeders and can completely defoliate potato plants. In rare cases, tubers exposed at the soil surface can also be eaten. Characteristic black and sticky excreta is left on the stem and leaves by the larvae and adults.

Impacts on production losses *L. decemlineata* is one of the most economically important insect pests of potato. Adults and larvae reduce potato tuber yields by

Fig. 8.8 Colorado potato beetle, Leptinotarsa decemlineata, adults (**a**) and larvae feeding on potato leaves (**b**). (Photo credits: Andrei Alyokhin)

devouring foliage. Larvae may defoliate potato plants resulting in yield losses of up to 100% if the damage occurs prior to tuber formation. Losses in potato are most severe if the incidence of last instar larvae, the most voracious stage, peaks during the time of tuber formation. This species is a huge pest problem and farmers spend millions of dollars each year on insecticides to control it (CABI 2017a; Jacques and Fasulo 2015).

Methods of prevention and control
IPM programs against this potato pest emphasize reducing insect pest populations by using several control measures, including crop rotation, altered planting dates (to avoid peak pest populations during vulnerable stages of crop development), and use of *B. thuringiensis* toxins or other pesticides when economic injury levels are reached. IPM approaches seek to maximise the impact of naturally occurring biological control agents in suppressing beetle populations (CABI 2017a; Jacques and Fasulo 2015; Laznik et al. 2010; Capinera 2001; Ferro 2000).

Sampling. Entire plants should be examined for above-ground life stages of the insect. Adult beetles are attracted to the yellow color and can be captured with traps. However, traps are seldom used because the various life stages are clearly distinguishable from each other.

Cultural control. The Colorado potato beetle may be managed culturally by crop rotation. Distances of at least 0.5 km between potato fields are required to provide protection. Beetles initially disperse by walking, so crop rotation and/or trenching can significantly reduce field infestations. Destruction of crop debris after harvest may contribute to reducing beetle populations. Beetle densities are also often lower on mulched potato plots (Alyokhin 2009).

Biological control. While many natural enemies have been identified, they are usually not able to control Colorado potato beetle populations below the economic injury levels. The tachinid fly *Myiopharus doryphorae* (Riley) may produce high parasitism rates early in the season and prevent the beetles from becoming a serious pest. Also, the entomopathogenic nematodes (*Steinernema feltia* Filipjev) and the fungus *Beauveria bassiana* have been reported to offer some control against this pest (Alyokhin 2009).

Biopesticides. Abamectins and spinosins generally provide excellent control of this pest (but see cautionary note on insecticide resistance below). Bacterial insecticides based on delta endotoxin of bacterium *Bacillus thuringiensis* subsp. *tenebrionis* are also effective, but they must be applied against the first two instars.

Chemical control. Insecticides are commonly used to control populations of Colorado potato beetle. Resistance to insecticides develops very rapidly in this species. For example, for endrin the first failure was reported after 1 year of its use, while for oxamyl it was observed already within the first year of use (Forgash 1985). Overall, failures were reported to 51 compounds belonging to ten chemical groups and possessing eight modes of action (Alyokhin et al. 2013).

8.7.2 European Corn Borer

***Ostrinia nubilalis* (Hübner) (Lepidoptera: Crambidae)**

Distribution *Ostrinia nubilalis* is native to Europe and was introduced into North America near Boston (Massachusetts) in 1917 (Baker et al. 1949). It gradually spread from there to other parts of the United States and Canada. It has also been reported in Asia (China, India, Georgia and Indonesia, middle East (Syria, Israel, Lebanon, Iran, Turkey) and North Africa (Algeria, Egypt, Libya, Morocco and Tunisia) (EPPO 2014; Li et al. 2003). Two races, E and Z that are morphologically indistinguishable exist in both USA and Europe.

Host range The European corn borer is primarily a pest of maize (*Zea mays* L.), but very polyphagus feeding on host plants from over 17 families but not necessarily being a pest. Economic host crops include potato, pepper, celery (*Apium graveolens* L.), tomato, beans, hop (*Humulus lupulus* L.), oat (*Avena sativa* L.), millet (*Panicum* spp.), sorghum (*Sorghum bicolor* L. Moench), cotton (*Gossypium arboreum* L.), and fruits such as apple (*Malus pumilla* Miller) and peach (*Prunus persica* (L.) Batsch L.). Feeding also occurs in ornamental plants and weeds including hollyhock (*Alcea rosea* L.), mugwort (*Artemisia vulgaris* L.), pigweed (*Amaranthus* spp.), and others.

Symptoms of infestation In potato, the most obvious sign of infestation is stem wilt. The entrance hole is in the stem and is easy to locate because of the presence of the excreta that larvae expel while feeding inside the stem. Larval presence can be confirmed by cutting the stem (CABI 2017b). Young larvae tunnel into leaf petioles while older larvae tunnel into the main stem. Larval damage can, however, be confused with that of other stem borers.

Impacts on production losses Potato plants present high tolerance for this insect; fairly high levels of infestation often do not seem to significantly affect tuber yields (Kennedy 1983). Varieties with weak stems are more likely to suffer reduced yields, as affected stems will break more easily during heavy wind- or rain- storms. However, severe infestations may result in considerable crop losses. Also, larval entry holes may facilitate infection by fungal and bacterial pathogens such as *Erwinia carotovora*.

Methods of prevention and control

Sampling. Monitoring of adult moth flights is suggested to determine the timing for oviposition and hence if a field is at risk. Sticky traps baited with strain-specific sex pheromones can be used for this purpose. Early infestations can also be more damaging than later ones and scouting for egg masses and observation of damage by the first instar larvae on leaf petioles. It is prudent to be vigilant and to monitor populations in areas where this insect has been known to previously cause damage.

Cultural control. This can involve early planting to escape damage, intercropping with nonhost plants and proper field sanitation involving of ploughing the field at the end of the cropping season.

Chemical control. European corn borer larvae are not particularly difficult to kill, but pesticide applications, to be effective, must be made before the larvae enter the stem. Pesticide application is recommended only when more than 10% of the stems show evidence of tunneling (Nault and Kennedy 1996).

Biological control. *Bacillus thuringiensis* subsp. *kurstaki* (*Btk*) and *B. thuringiensis aizawai* can be successfully applied to target first instar larvae. Alternative control methods include release of commercially available parasitic wasps (*Trichogramma nubilale* Ertle and Davis and *T. brassicae* Bezdenko) (CABI 2017b). Other natural enemies that have been used to control the European corn borer include *Lydella thompsoni* (Herting), *Diadegma terebrans* (Gravenhorst) and *Macrocentrus grandi* Goidanich (Baker et al. 1949).

8.8 Major Pests Globally Present

8.8.1 Aphids

***Myzus persicae* (Sulzer, 1776),**
***Macrosiphum euphorbiae* (Thomas, 1878) (Hemiptera: Aphididae)**

Distribution The green peach aphid, *Myzus persicae*, is thought to have its origins in China, just as its overwintering host plant, the peach (*Prunus persica* (L.) Batsch) (Fig. 8.9). However, the green peach aphid is highly adaptable and is currently cosmopolitan in distribution. The potato aphid (*M. euphorbiae*) originated in North America but has spread to the temperate parts of Europe and Asia and is found in all potato growing areas globally (CABI 2017c, d).

Host range Both species are extremely polyphagous, being capable of feeding on several hundred-plant species (Blackman and Eastop 2000).

Symptoms of infestation Aphids can damage potato plants directly by feeding on sap, and indirectly by transmitting various viral diseases. *Direct damage*. Continuous

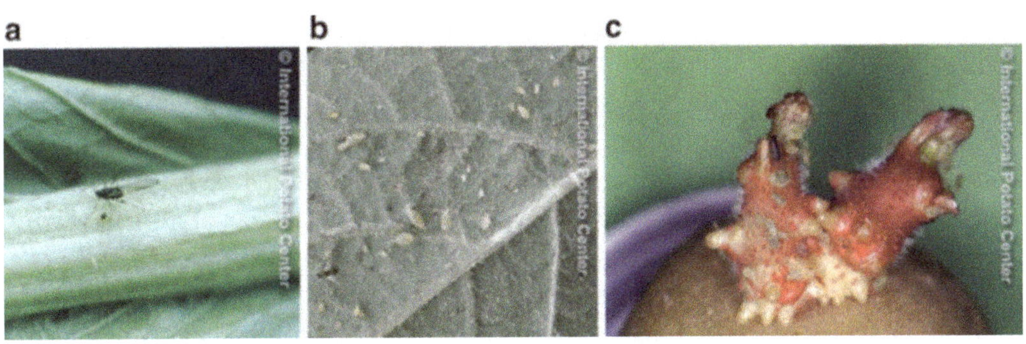

Fig. 8.9 Winged green peach aphid start infestations on potato (**a**), wingless green peach aphids start colonies under the leaf (**b**), aphid colonies on potato tuber sprouts in stores (**c**), where they transmit viruses to seed potatoes. (Photo credits: CIP)

sucking of sap by large numbers of aphids considerably weakens and slows plant development. Removing sap and injecting toxic saliva causes leaf deformation. The weakened plant produces low quality tubers. During feeding, aphids excrete honey dew which promotes growth of a sooty black mold on plant leaves or stems, hence reducing the photosynthetic area for the plant. *Indirect damage*. The most important damage caused by aphids in potato is virus transmission. Symptoms of virus infestation vary depending on the virus transmitted. For instance, potato leaf roll virus (PLRV) causes leaf-rolling and tuber stem necrosis in potato. Other PLRV symptoms include thickening, curling, chlorotic spotting and yellowing of the leaves (CABI 2017c, d; Larrain et al. 2003; Salazar 1995).

Impacts on production losses *Direct damage*: Large build-up in aphid population may result in death of heavily infested plants. Since aphid distribution within potato fields is usually clumped, this results in patches of dead plants, sometimes referred to as "aphid holes." However, this phenomenon is relatively uncommon. In most cases, direct damage by aphids does not affect potato yields. *Indirect damage*. Most aphid-related damage to potato crops is caused by virus transmission. Potato leafroll virus (PLRV) and potato virus Y (PVY) are the two most important potato-infecting viruses. Both severely diminish potato yield and quality in most potato-growing areas. PLRV is a persistent virus transmitted exclusively by potato colonizing aphid species. PVY is nonpersistently transmitted by at least 50 different aphid species, most of which probe potato plants with their proboscis, but do not settle nor reproduce on them. *M. persicae* is the most effective vector of both persistent and nonpersistent viruses. It is also a highly polyphagous species that often moves into potato fields from surrounding crops and noncrop vegetation (CABI 2017c, d; CIP 1996; Larrain et al. 2003; Salazar 1995).

Methods of prevention and control

Biological control. Various natural enemies of aphids act as efficient biological control agents; among those are different predatory insects (e.g., lady beetles and larvae of green lacewings) and parasitic wasps (*Aphidius* spp.). Other generalist predators include *Orius* spp., *Geocoris* spp., and *Nabis* spp. Guard rows of flowering vegetation planted within potato fields can provide a habitant for aphid natural enemies (Powell and Pell 2017). Entomopathogenic fungi commonly infect aphids under field conditions and may cause epizootics that significantly reduce aphid densities (Pell et al. 2001). Numerous biological control products that use seven species of entompathogenic fungi (mostly *Beauveria bassiana* and *Lecanicillium* spp.) are commercially available for aphid control. Proper timing is very important when using these products because fungal spores are strongly influenced by environmental conditions, such as temperature and relative humidity (Kim et al. 2013).

Biological insecticides. Extracts from garlic (*Allium sativa* L.), neem (*Azadirachta indica* A. Juss.), red chili (*Capsicum annum* L.), pyrethrum flowers (*Chrysanthemum* sp.), either singly or in mixtures provide some aphid control, especially at an early stage of infestation.

Cultural practices. Weeding and removal of alternative and overwintering hosts such as wild mustards (*Brassica* spp.), use of wheat straw or white plastic as mulch, and intercropping with onion, garlic or coriander (*Coriandrum sativum* L.) have shown to reduce aphid populations. Planting only high-quality disease-free potato tubers and rogueing out virus-infected plants is recommended for preventing virus damage even in the presence of substantial aphid populations (CIP 1996; Larrain et al. 2003; Sanchez and Vergara 2002).

Chemical control. Insecticides specific to the homopterans such as spirotetramat, flonicamid and pymetrozine have low effect on natural enemies and are therefore good candidates for IPM. To the contrary, use of broad-spectrum insecticides may flair up populations of green peach aphids that recover quicker than populations of their natural enemies. On a small scale, application of potassium-based liquid soap sprays may reduce numbers of potato-colonizing aphids. When practicing chemical control, it is important to remember that green peach aphid has a high propensity for developing insecticide resistance that rivals that of the Colorado potato beetle. Failures have been reported for at least 69 different active ingredients, including all commonly used chemical classes (Alyokhin et al. 2013).

8.8.2 Whiteflies

***Bemisia tabaci* (Gennadius 1989),**
***Trialeurodes vaporariorum* (Westwood 1856) (Homoptera: Aleyrodidae)**

Distribution *Bemisia tabaci* is a common pest in tropical and subtropical regions but is less prominent in temperate habitats (Fig. 8.10). *Trialeurodes vaporariorum* originated in tropical or subtropical America (probably Brazil or Mexico). Currently, it has become a cosmopolitan pest, with records in every zoogeographic region of the world (CABI 2017e, f).

Host range *B. tabaci* is a highly polypaghous pest. In addition to Solanaceae crops such as potato, tomato, and eggplant, it infests a wide range of other field crops and vegetables; among those are sweetpotato (*Ipomoea batatas* (L.) Lam.), cotton, beans, cucumber, melon, soybean, cassava (*Manihot esculenta* Crantz), and many others.

Symptoms of infestation Adult whiteflies can be easily seen on the underside of the leaves. Sucking sap from plant tissue is associated with several physiological plant disorders, such as chlorosis of new foliage. Heavy infestations with hundreds of adult greenhouse whiteflies on the lower surfaces of potato plant leaflets produce whitish spots that subsequently turn dark. As the population increases, sooty mold accumulates on the leaf surface, necrosis develops on the leaf margins, and some leaves curl upward. Necrotic leaf areas expand and coalesce, causing the whole leaf to dry-out. Producing sugar-rich honeydew promotes growth of sooty mold fungi, thus interfering with normal photosynthetic processes. Adult whiteflies transmit a number of viruses, each of which has its own suite of symptoms. For example, potato apical leaf curl virus (PALCV) causes upward or downward curling of leaves. Infection with potato yellow vein virus (PYVV) initially results in bright yellow

Fig. 8.10 *Bemisia tabaci* adults on potato leaves (**a**, **b**) and potato plant with symptoms of Potato yellow vein virus (PYVV) transmitted by *Trialeurodes vaporariorum* (**c**). (Photo credits: CIP)

veins. As disease progresses, the leaves become yellow, while the veins regain their green color. Infected plants may produce deformed tubers with abnormally large protruding eyes (CABI 2017e, f; Larrain et al. 2003; Sanchez and Vergara 2002).

Impacts on production losses Both direct and indirect damage may result in yield reduction of 40% or more (Anderson et al. 2005; CABI 2017e, f; Franco-Lara et al. 2013).

Methods of prevention and control

Monitoring pest populations. Direct observation and use of yellow sticky traps are useful methods for monitoring whiteflies and for early detection and the documentation of relative whitefly abundance over time.

Biological control. Despite abundant research on natural enemies and nonchemical control of *B. tabaci* few natural enemies have been identified and used in biological control. Nymphs are affected by entomopathogenic fungi such as *Verticillium lecanii* Zimmerman, *Paecilomyces farinosus* Holmsk, and *P. fomosoroseus* Wize. *V. lecanii* has been developed as a biopesticide (CABI 2017e, f).

Crop management. Maize can be planted as barrier and/or cucurbits as trap crop for controlling adults and to enhance the development of biological control agents (CABI 2017e, f).

Physical control. Yellow attracts whitefly adults. Stationary yellow stick traps are recommended to be installed at planting around fields to capture whitefly adults migrating from other crops. The use of mobile and stationary yellow sticky traps can effectively reduce white fly adult populations in potato (Cisneros and Mujica 1999c; Mujica 1998).

Chemical Control. Avoid use of broad spectrum insecticides in order not to affect natural enemies. The chitin inhibitor, buprofezin, has shown high efficacy (>90%) in controlling immature stages. Soaps (such as potassium soaps) can also be applied on a small scale. Chemical control should be based on using an action threshold of three nymphs per leaf (CABI 2017e, f; Cisneros and Mujica 1999c).

8.8.3 Ladybird Beetles

***Henosepilachna vigintioctomaculata* (Motschulsky), *Henosepilachna vigintioctopunctata* (F.) (Coleoptera: Chrysomelidae)**

Distribution *Henosepilachna vigintioctomaculata* and *H. vigintioctopunctata* are two related and morphologically similar species of phytophagous lady bird beetles that share a common name of a potato ladybird, hadda beetle, or 28-spotted lady bird beetle (Fig. 8.11). *H. vigintioctomaculata* is common in temperate areas of Asia, including China, Japan, Korea, and Russia. The ranges of both species overlap, particularly in China. However, *H. vigintioctopunctata* prefers warmer climates and is widely distributed in Southeast and South Asia, including Pakistan, and has been introduced to Australia, New Zealand, several Pacific islands, and South America (Katakura 1980; Naz et al. 2012; Xu et al. 2013).

Host range Potato ladybirds are polyphagous species feeding on a variety of plant species. However, most severe damage is usually reported for Solanaceae's crops, including potato (Xu et al. 2013).

Symptoms of infestation Both adults and larvae feed on potato leaves, causing their complete defoliation and eventual death in cases of severe infestation. Feeding by larvae usually starts on the lower sides of infested leaves. Lower epidermis and mesophyll get consumed, while higher epidermis and large veins remain relatively intact (Jackson 2016; Xu et al. 2013). Affected leaves eventually dry up and die. Furthermore, feeding damage may facilitate infections by gray mold, *Botrytis cinerea* Pers. ex Fr. (Yao et al. 1992).

Impacts on production losses Potato ladybirds can be serious pests of potato. Yield losses have been estimated to reach 10–15% in normal years, and 20–30% in the years of heavy infestation (Song et al. 2008). In especially severe cases, complete crop destruction is possible (Jackson 2016).

Fig. 8.11 28-spotted ladybird beetle, Henosepilachna vigintioctomaculata, adult (**a**) and larva (**b**). (Photo credits: Yulin Gao, Institute of Plant Protection, Chinese Academy of Agricultural Sciences)

Methods of prevention and control

Monitoring pest populations. All life stages of potato ladybirds are rather conspicuous, in large part due to their aposematic coloration. Therefore, they can be relatively easily detected by visual observations. Eggs are usually laid in clusters of 20–30 on lower leaf surfaces, thus requiring that scouted leaves are turned upside-down.

Biological control. The parasitic wasp *Pediobius foveolatus* (Crawford) can parasitize 50–60% of field potato ladybird populations. In some cases, this reduces damage to economically acceptable levels, and no further control is necessary (Puttarudriah and Krishnamurti 1954; Venkatesha 2006). However, such high parasitism rates should not be taken for granted, and a careful monitoring of ladybird populations is required.

Crop management. Removing and destroying crop residue after harvest deprive resident potato ladybirds of food, thus dramatically enhancing their mortality (Jackson 2016; Xu et al. 2013).

Physical control. Manual destruction of potato ladybird beetles is recommended for small potato plots (Xu et al. 2013). Success of this approach is facilitated by the tendency of ladybird adults to aggregate at overwintering sites outside of potato fields. Adult ladybird beetles are strongly phototactic (Zhou et al. 2015). Therefore, light traps can be used to reduce their numbers. However, feasibility of this approach remains to be investigated. Also, light traps attract a wide variety of insects, including beneficial species (Zhou et al. 2015).

Biological insecticides. Some botanical products, such as azadirachtin, have a good efficacy for ladybird management (Ghosh and Chakraborty 2012; Jeyasankar et al. 2014). Also, the entomopathogenic fungi *Beauveria bassiana* and *Metarrhizium anisopliae* (Vishwakarma et al. 2011) and *Bacillus thuringiensis* (Song et al. 2008) are used.

Chemical Control. Potato ladybird beetles can be easily killed by common broad-spectrum insecticides, which remains unfortunately the most common approach to controlling these pests (Xu et al. 2013).

8.9 Minor Pests Globally Present

8.9.1 Cutworms

***Agrotis ipsilon* (Hufnagel, 1776), *Peridroma saucia* (Hübner, 1808), *Agrotis segetum* (Denis & Schiffermüller, 1775) (Lepidoptera: Noctuidae)**

Distribution Cutworms are larvae of several noctuid moth species which are cosmopolitan (Fig. 8.12). Most widely distributed are the black cutworm (*Agrotis ipsilon*), the variegated cutworm (*Peridroma saucia*) and the turnip moth (*Agrotis segetum*). Their origin is uncertain but specimens which were used to describe *A. ipsilon* in 1766 were collected from Austria.

Fig. 8.12 Cutworm larval damage on potato tuber (**a**), and freshly cut potato plant (**b**). Cutworm larva (**c**), pupa (**d**) and adult (**e**) of *Agrotis ipsylon*. (Photo credits: CIP)

Host range Cutworms are polyphagous and have been reported as a pest on nearly all vegetable crops and some cereals. It's a pest in potato, maize, alfalfa, clover (*Trifolium* L.), cotton, rice, sorghum, strawberry, sugar beet, and tobacco. The larvae sometimes prefer to feed on weeds such as bluegrass (*Poa pratensis* L.), curled dock (*Rumex crispus* L.), lambsquarters (*Chenopodium album* L.), yellow rocket (*Barbarea vulgaris* W.T. Aiton); and redroot pigweed (*Amaranthus retroflexus* L.). Considerable damage has also been observed among shrubs and trees such as linden (*Tilia* sp.), wild plum (*Prunus* sp.), crabapple (*Malus* sp.), and lilac (*Syringa vulgaris* L.). Cereals such as oats (*Avena sativa* L.), barley (*Hordeum vulgare* L.), sorghum, maize, and wheat (*Triticum* spp.) are also affected.

Symptoms of infestation The cutworm larvae remain in the soil at the base of the plant during the day. At night, some species cut down the stems of young potato plants, while other species climb the plants and feed on their leaves. Old instar larvae can occasionally tunnel into potato stems disrupting plant growth. Tubers closer to the ground surface may suffer occasional damage. In a single night, as single larva can cut down several potato plants.

Impacts on production losses Cutworms if not detected early and during outbreaks can cause up to 100% crop loss through cutting down of potato plants.

Methods of prevention and control

Monitoring. Use of sex pheromones in white and yellow traps during spring are quite effective in predicting attacks. Baited traps can also be used to monitor larva populations in the field (Sanchez and Vergara 2002).

Cultural control. Deep tillage exposes larvae and pupae to the action of natural enemies and may crash them mechanically. Weed control also aids in reducing cutworm damage by reducing the sites for egg-laying because these insects infest a wide range of host plants. Properly covering tubers during hilling impedes larval access to the tubers. Late harvest results in a greater damage of tubers. Hand picking of the larvae very early in the morning is the most commonly used method for cutworm control among smallholder farmers in Africa. Fallowing and crop rotation with none host plants such as bluegrass, onion, garlic, peppermint (*Mentha piperita* L.) and coriander can also reduce cutworm populations as well as habitat management and farming practices that conserve populations of existing natural enemies. Trap cropping using susceptible host plants such a sunflower or cosmos (*Cosmos* sp.) has also been used in combination with daily killing the trapped cutworm larvae. Flooding of the infested field between crops can also kill the cutworm larvae. Sticky substances such as mollasses and saw dust can also be spread at the base of the plant to trap the cutworm larvae.

Transgenic plants. Several transgenic crops (maize, tobacco, wheat, cotton and maize) have been reported to be resistant to cutworm damage. The genes responsible for resistance against *A. ipsilon* include *Bacillus thuringiensis* cry1Ac gene (Bt), barley trysin inhibitor (Bti-cme) and cowpea trypsin inhibitor (CpTi). Preliminary results of genetically modified potatoes show that feeding larvae on leaves which expresses Cry3Aa affect larvae by being toxic, and also that Bt potatoes curb the growth and reproduction of the adults of *Agrotis ipsilon*.

Biological control. Cutworms are hosts for numerous parasitoid wasps and flies, including species of Braconidae *Cotesia ruficrus* (Haliday), *Snellenius manilae* (Ashmead), Ichneumonidae *Tenichneumon panzer* (Wesmael), Tachinidae *Bonnetia comta* (Fallen), *Euplectrus plathypenae* and Eulophidae, with rates of parasitism as high as 75–80%. Up to 60% of cutworm larvae have been reported to be killed by an entomopathogenic nematode, *Hexamermis arvalis* in central USA. Entomopathogenic nematodes that have been used in the control of *A. ipslon* larvae include *Steinernema glaseri, S. riobrave and Steinernema carpocapsae* (Sanchez and Vergara 2002).

Biopesticides. Extracts from the following plants have demonstrated to be toxic to the larvae of *A. ipslon*: *Nerium oleander* L. leaves, neem leaves and seeds, *Melia azedarach* L. fruits, *Bassia* muricate (L.) Asch., *Lantana* sp., *Parthenium* sp., *Hyptis* sp., *Ipomoea carnea* Jacq., *Tephrosia nubica* (Boiss.) Baker, *Bidens pilosa L.* and *Rumex nepalensis* Spreng. roots (CABI 2018a).

Chemical control. Spotted or localized field infestations by cutworms are typical, calling for focused subsurface soil insecticide applications. The use of toxic baits is recommended because of its specificity against these insects and because it does not markedly affect the biological control agents. Toxic baits prepared as a mix of bran, molasses, water, and insecticide should be applied at the base of plants at dusk (CIP 1996; Larrain et al. 2003).

8.9.2 Armyworms

***Spodoptera eridania* (Stoll), *Spodoptera frugiperda* (J E Smith), *Mamestra configurata* (Walker), *Spodoptera ornithogalli* Guenée, *Copitarsia decolora* (Guenée), *Feltia* spp., (Lepidoptera: Noctuidae)**

Distribution *Spodoptera frugiperda* is distributed in North, Central, and South America, Africa, and India (CABI 2018a) (Fig. 8.13). *S. frugiperda* is on the EPPO A1 list of quarantine pests and is intercepted occasionally in Europe on imported plant material (CABI 2018b). *S. eridania* occurs throughout southern USA, Central and South America, and the Caribbean. It is not established in Europe (CABI 2018c). *Mamestra configurata* is restricted to North America (Canada, USA, Mexico) (CABI 2018d).

Host range Army worms feed on a wide range of crops and are important pests especially in cereals. However. they have also been reported to feed on potato in absence of the primary hosts (Strand 2006).

Symptoms of infestation Armyworms are mainly forage feeders and feed during the day, do not make burrows, and may migrate in mass into potato fields from adjacent crops. They also feed on tubers that are exposed on the surface or accessible through cracks in the soil (CIP 1996; Larrain et al. 2003; Sanchez and Vergara 2002).

Impacts on production losses Armyworms can cause up to 100% crop defoliation of potato plants. Some defoliation from armyworms can be tolerated. Keeping defoliation between 10 – 15% will generally prevent yield losses.

Methods of prevention and control It is recommended to make suitable soil preparations that expose the pupa of the insect to environmental conditions and predators. There is biological control based on the use of natural enemies of eggs and larvae, such as Tachinid insects, Inchneumonids, Trichogramatids, the fungus

Fig. 8.13 *Spodoptera frugiperda* larvae infesting potato leaves (**a**). (Photo credits: CIP)

Zoophthora radicans affecting larvae, among others (Larrain et al. 2003). Spotted or localized field infestations are typical, calling for focused insecticide treatments. Toxic baits prepared as a mix of bran, molasses, water, and insecticide should be applied at the base of plants at dusk (CIP 1996).

8.9.3 Wireworms

***Agriotes lineatus* (L.), *A. vobscurus* (L.), *A. sputator* (L.)., *Athous haemorrhoidalis* (F.), *Conoderus rudis* (Brown), *C. vespertinus* (Fabricius), *Ctenicera pruinina* (Horn), *C. cuprea* (Fabricius), *Melanotus communis* (Gyllenhal) (Coleoptera: Elateridae)**

Distribution Wire worms are the larvae of click beetles (Fig. 8.14). Over 39 species in 21 genera of wireworms have been reported to attack potato. Wireworms are found throughout the world, and species vary greatly among regions. While in the UK, the three most important pest species in potato are *Agriotes lineatus*, *A. vobscurus,* and *A. sputator*, in the US, these are *Limonius canus* LeConte, *Limonius californicus* (Mannerheim), *A. lineatus*, *A. obscurus* L. and *A. sputator*.

Host range Grasses are the main host but they also attack potato, asparagus, carrot (*Daucus carota* subsp. *sativus* (Hoffm.) Schübl. & G. Martens), sugar beet (*Beta vulgaris* L.) and leek (*Allium ampeloprasum* L.)

Symptoms of infestation Wireworms are frequent pests in temperate climates, but less so in warm tropical areas. Thin, lustrous larvae with small thoracic legs live underground and may be up to 25 mm long. The larvae can borrow deeply inside tubers, and spend considerable time over there producing irregularly shaped tunnels in tubers (CIP 1996; Larrain et al. 2003; Sanchez and Vergara 2002).

Impacts on production losses Wireworms may injure potatoes by feeding on the seed piece resulting in weak stands, but most of their damage is caused by tunneling

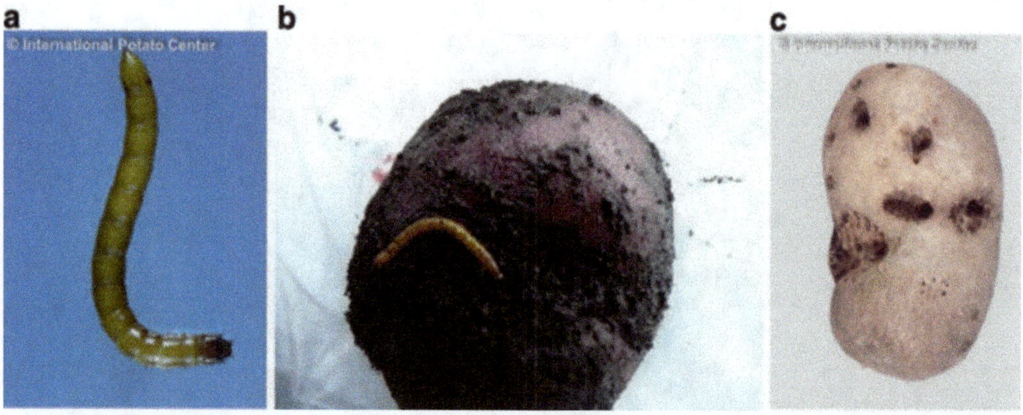

Fig. 8.14 Wireworm larvae (**a**, **b**) and larval damage on potato tuber (**c**). (Photo credits: CIP)

into tubers, which reduces yield quality. Wireworm tunneling also creates an entry point for certain plant pathogens, eventually leading to tuber rot. In some years and regions of the U.S., up to 45% of the total potato tuber harvest has been downgraded or rejected outright because of wireworm injury, resulting in substantial economic loss (Steele 2011).

Methods of prevention and control

Monitoring. Avoid growing potatoes in wireworm infested fields. Plan and utilize a range of risk assessment methods such as pheromone and bait trapping, as well as soil sampling to confirm the status of each field. In arable rotations, plough-based cultivation may help to reduce wireworm populations (Larrain et al. 2003).

Cultural control. Wireworms feed on the roots of various crops, particularly grasses. Consequently, before planting potatoes in pasture areas, the soil borne wireworm larvae population must be reduced by proper plowing and rotation with other crops that require frequent tilling. Keeping potato fields free from weeds has also been reported to reduce wireworm populations.

Biological control. Currently there are no commercial biological control agents available for controlling wireworms, although some strains of the insect-pathogenic fungus *Metarhizium anisopliae* have shown promising results under experimental conditions (Kabaluk et al. 2005).

Chemical control. Insecticides applied to soil may be required in certain circumstances. To be effective, insecticides for the control of wireworms need to be incorporated into the soil at planting and remain active during the cropping season (Parker and Howard 2001).

8.9.4 Flea Beetles

Epitrix spp. (Coleoptera: Chrysomelidae)

Distribution *Epitrix* spp. is a genus of many flea beetles that are known to feed upon members of the Solanaceae family. *Epitrix tuberis* Gentner (1944) and *E. cucumeris* (Harris 1851) are common pests of potato in North America (Malumphy et al. 2016) (Fig. 8.15). A nonnative pest of potato recently established and causing significant economic damage in Portugal and Spain was identified as *Epitrix papa*

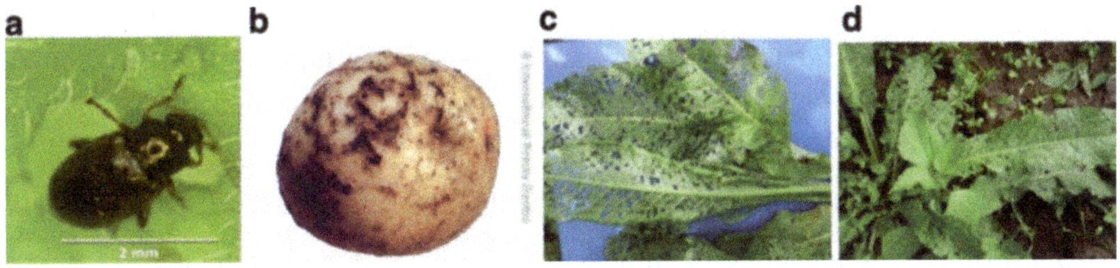

Fig. 8.15 Adult damage on potato leaves (**a**) and larval damage on potato tuber (**b**). Heavy adult damage and infestation on a weed in potato field in Kabale, Uganda (**c**, **d**). (Photo credits: CIP)

sp. n. and included in the list of quarantine pests in Europe (Eyre and Giltrap 2013; Malumphy et al. 2016; Sanchez and Vergara 2002). *Epitrix yanazara* Bechyné (1959) was the major economic important species in the central highlands of Peru (Kroschel et al. 2012) which is a region with a high diversity of flea beetles (Furth et al. 2015).

Host range The host range comprises potato, tomato, eggplant, tobacco as well as weeds of the family Solanaceae.

Symptoms of infestation Flea beetles are small beetles and jump easily in the foliage of plants. Their feeding results in characteristic circular holes less than 3 mm in diameter. Larvae that feed on roots, stolons, and tubers also cause damage. They bore the tubers superficially or scratch the skin, thus facilitating penetration of pathogenic fungi (CIP 1996; Larrain et al. 2003; Sanchez and Vergara 2002).

Impacts on production losses The flea beetles are considered one of the most serious pests threatening the entire EPPO region. In North America, they are ubiquitous. In cases of severe infestation, defoliation by adults may cause leaves to dry completely, thus affecting photosynthesis and plant yield. Larval damage makes tubers become unviable for sale, and the destruction of the roots can result in plant death (Eyre and Giltrap 2013; Malumphy et al. 2016; Sanchez and Vergara 2002). In Peru, populations increased severely above economic thresholds if no pesticides were applied (Kroschel et al. 2012).

Methods of prevention and control

Cultural control. Crop rotation. Removal of host weeds and appropriate soil management to ensure vigorous potato plants contribute to reducing flea beetle populations.

Chemical control. Many insecticides that are used to control other insect pests of potato are also effective against flea beetles. Potato plants withstand some foliage damage beyond which insecticides are required (CIP 1996; Larrain et al. 2003; Sanchez and Vergara 2002).

8.9.5 White Grubs

***Phyllophaga* spp., *Anomala* spp., *Popillia japonica* Newman, *Anomala orientalis* Waterhouse, *Holotrichia javana* Brenske, *Holotrichia oblita* Faldermann, *Temnorhynchus coronatus* (Fabricius), *Holotrichia longipennis* Blanchard, *Brahmina coriacea* Hope, *Melolontha melolontha Linnaeus* (Coleoptera: Scarabaeidae)**

White grubs are found in many genera in the family Scarabaeidae.

Distribution Scarab beetles and thus white grubs are found worldwide.

Host range White grubs are polyphagus and feed on over 1000 plant species including field crops such as potato, sweetpotato, maize, asparagus, and soybean;

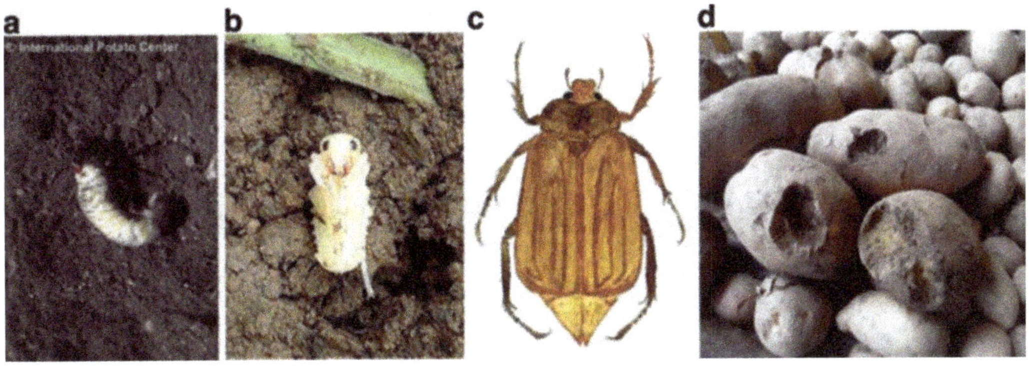

Fig. 8.16 Typical white grub larva (**a**) and pupa (**b**) of the genus *Phyllophaga*. Adult of *Melolantha melolantha* (**c**) and damage of larvae in potato tubers (**d**). (Photo credits: CIP)

fruits such as apple, peach, grape, as well as trees and ornamentals (Pathania and Chandel 2016; Visser and Stals 2012; Misra and Chandel 2003) (Fig. 8.16).

Symptoms of infestation White grubs are the larvae of relatively large beetles. The damage is done mainly by the grubs which remain in the soil. They damage the plant by feeding on roots, underground stem parts, stolons, and tubers. Earlier instars feed on the roots and may cause the plants to dry up. After tubers are developed, the grubs switch to feeding on them, leaving large holes (BioNET-EAFRINET 2011; CIP 1996; Sanchez and Vergara 2002; Visser 2012).

Impacts on production losses Economic damage results mostly from larval damage to potato tubers. Severe damage usually occurs when planting potato in former pasture or grazing fields, which are preferred habitats for white grubs, and on no-till or reduced-till land (CIP 1996; Larrain et al. 2003; Sanchez and Vergara 2002). Yield losses of 40–80% have been reported in India (Pathania and Chandel 2016; Misra 1995).

Methods of prevention and control

Cultural practices. Deep tilling exposes white grubs to adverse environmental conditions such as sunlight and frost, and to predatory vertebrates. Ensuring proper drainage in the field will reduce white grub populations since they prefer moist soil, especially with high content of decaying organic matter. Potato should not be planted directly following pasture, grass, or several successive years of cereals and/or corn. Crop rotation is an effective control method when potato is rotated with resistant or less susceptible crops, such as deep-rooted legumes (e.g., alfalfa, cowpea, and pigeon peas) and Allium crops (onions, garlic). Strips of African marigold (*Tagetes* sp.), sunflower (*Helianthus annuus* L.), and castor can be used as trap crops. Allium crops also repel adult beetles from laying eggs (Larrain et al. 2003; BioNET-EAFRINET 2011).

Biological control. Natural enemies that control white grubs include parasitic wasps in the genera *Tiphia, Myzinum* (Hymenoptera: Tiphiidae), and *Pelecinus polyturator* Drury (Hymenoptera: Pelecinidae), and the fly, *Pyrgota undata* Wiedemann (Diptera: Pyrgotidae). Spores of the pathogens *Bacillus popilliae*,

B. lentimorbus, and *Metarhizium anisopliae* can be used to inoculate the soil. Nematodes species such as *Steinernema* sp. and *Heterrorhabditis* sp. can also be effectively used against white grubs (Sanchez and Vergara 2002).

Chemical control. This pest is not easily controlled with insecticides. Chemical treatment on grubs must be done when they are young, as older larvae are more robust and move to a greater depth as they develop (Gupta and Gavkare 2014).

8.9.6 Mites

***Tetranychus urticae* C. L. Koch, *T. evansi* Baker & Prichard, *Polyphagotarsonemus latus* (Banks) (Arachnidae: Trombidiformis, Tarsonemidae)**

Distribution The origin of the tomato or tobacco red-spider mite *Tetranychus evansi* is Brazil. The pest was accidentally introduced into Southern Africa during the 1980s. The two spotted- or red-spider mite *Tetranychus urticae* and the broad mite *Polyphagotarsonemus latus* are widely distributed in both tropical and subtropical areas of the world (CABI 2017g, h) (Fig. 8.17).

Host range Mites are polyphagous. Major hosts are potato, tomato, tobacco, pepper, eggplant, pumpkins and squashes (*Cucurbita pepo* L.), cucumber, watermelon, celery, maize, beans, strawberry (*Fragaria* × *ananassa*), cotton, citrus, papaya (*Carica papaya* L.) (Hazzard 2013; Landis and Davis 1947; Goftishu et al. 2016).

Symptoms of infestation *T. urticae* and related mites are generally known as red spiders, although red is not always their characteristic color. Mites are extremely small, almost microscopic, and feed on the cellular matter of leaves. Chlorotic spots caused by mites give leaves a tan coloring, whereas high infestation will cause leaf and plant wilting. *P. latus* attacks sprouts and tender leaves, deforming them. Damage to growing plants is particularly severe. The white mite colonies are located on the underside of the young leaves and on the buds. The infested plants take a dark

Fig. 8.17 *Tetranychus urticae* feeding on a potato leaf (**a**), and symptoms of white mite feeding causing deformation of buds and tender leaves (**b**) and toasted appearance on the underside of a potato leaf (**c**). (Photo credits: CIP)

to moderate green coloration and develop a somewhat leathery appearance. Heavy infestation may result in plant death (CIP 1996; Larrain et al. 2003; Sanchez and Vergara 2002; Landis and Davis 1947).

Impacts on production losses Plants with severe damage of *P. latus* do not form tubers and remain very small. When infestations occur in an intense and violent way they can cause the destruction of entire crop field. In the same way, attack by hundreds or thousands of red spider mites can cause thousands of lesions, thus can significantly reduce the photosynthetic capability of plants (CIP 1996; Larrain et al. 2003; Sanchez and Vergara 2002).

Methods of prevention and control
Monitoring. Mite infestations normally begin at boundaries of the potato fields especially those neighboring fields of preferred host plants such as maize, alfalfa, and mint. Stressed potato plants tend to be more susceptible. Mites can build high populations in a very short time during hot (above 30 °C) dry seasons. Constant visual monitoring and early detection is key. Check on the underside of leaves that show symptoms of nutrient deficiency (CABI 2017g, h).

Cultural practices. Sprinkler irrigation helps to limit mite damage by increasing the humidity (above 60%) on plant leaves, thus making conditions less favorable for mites. Proper field sanitation is done through removal of infested plants and volunteer crops (CIP 1996; CABI 2017g, h).

Natural control. Often mites are kept under control by their natural enemies, including predatory thrips, lacewings, gall midges (*Feltiella acarisuga* Vallot) and ladybird beetles (*Stethorus punctillum* Weise) and predatory mites (*Phytoseiulus persimilis* Athias-Henriot, *Neoseiulus californicus* (McGregor), *Amblyseius andersonii* Berlese, *Galendromus occidentalis* Nesbitt, *Mesoseiulus longipes*, and *Hypoaspis miles* Berlese (Pundt 2007). Mites become primary pest when natural enemies are destroyed by insecticide applications to control other insect pests. So, avoid use of nonselective pesticides which reduce predatory mite populations.

Biopesticides. Plant extracts such as azadirachtin (neem oil), or from cotton seed, soybean, clover, garlic, rosemary (*Rosemarinus officinalis* L.), and pyrethrum have shown some efficacy against red spider mites (CABI 2017g, h).

Chemical control. When control measures are required, usually a single application of an acaricide is sufficient especially when populations are still low (<five mites per plant). In some cases, focal treatments are recommended using either soap, sulphur, Bifenazate, abamectin, spirotetramat, and spiromesifen (Hazzard 2008).

8.9.7 *Thrips*

Frankliniella occidentalis **(Pergande),** ***Thrips tabaci*** **Lindeman (Thysanoptera, Thripidae)**

Distribution The western flower thrips *Frankliniella occidentalis* originated in western North America and has since become a major pest on many crops across the

Fig. 8.18 Pale or brown nymphs and darker adults of *Frankliniella occidentalis* on the underside of potato leaves (**a**), and symptoms of thrips feeding causes silvering on underside of a potato leaf (**b**). (Photo credits: CIP)

US and around the world (Fig. 8.18). The onion thrip *Thrips tabaci* is thought to have originated in the Mediterranean region but is now found on all continents except Antarctica.

Host range *F. occidentalis* feeds on more than 200 crops in over 50 genera and this includes vegetables, fruit trees, and ornamentals.

Symptoms of infestation Thrips can cause major damage mainly during dry climatic conditions. Because of heavy feeding, potato leaves develop a silver or chlorotic dotting of the tissues and become deformed. The silver color is due to the emptying of the cells (CIP 1996; CABI 2017i, j; Learmonth 2017).

Impacts on production losses Thrips are minor pest of potato. Some species are vectors for tomato spotted wilt virus (TSWV), with western flower thrips (*F. occidentalis*) as the most important virus vector. The virus reduces crop yield and tuber quality (CABI 2017i, j).

Methods of prevention and control

Cultural practices. Thrips feed on a wide range of plants, including weeds from which they can invade potato crops. Therefore, weed control is a good management strategy for this pest. Susceptible crops (onion, tomato, corn, artichoke, and pumpkin) should not be planted next to a potato crop. Any plants showing tomato spotted wilt virus symptoms should be removed (CIP 1996; CABI i, j; Learmonth 2017).

Behavioral control. Yellow or blue (especially for *F. occidentalis*) sticky traps can be used to monitor the presence of thrips.

Biological control. Various species of the worldwide anthocorid genus *Orius* are used in biological control against thrips, and these bugs are important as predators in many natural populations. *Amblyseius swirskii* Athias-Henriot and *Neoseiulus cucumeris* Oudemans are two of the most widely used predatory mites in the biological control of *F. occidentalis*. Fungal pathogens and nematodes, such as *Beauveria bassiana* and *Steinernema feltiae*, are also being used commercially. Parasitoids include *Ceranisus menes* Walker for the western flower thrip (CABI 2017i, j).

Chemical control. Low to medium toxicity pesticides used against thrips include Spinosad, Spinetoram, and Abamectin (Bentley and Rice 2015).

8.9.8 Tomato Tuber Moth

Tuta absoluta **(Meyrick) (Lepidoptera: Gelechiidae)**

Distribution The tomato leaf miner, *Tuta absoluta*, was first identified in Peru by Meyrick in the year 1917 from samples collected in Huancayo (3200 m above sea level) (Fig. 8.19). Since then, it has been recorded in all South American countries. The pest is predominantly found in areas below 1000 m asl. It was first introduced into Europe in Spain in 2006, from where it is believed to have spread to other European countries. Its range has expanded considerably to Asian and African countries in recent years (CABI 2017k).

Host range Hosts of *T. absoluta* are the cultivated Solanaceae tomato, potato, eggplant, sweet cucumber (*Solanum muricatum* L.), and tobacco. Other cultivated host plants are slender amaranth (*Amaranthus viridis* L.), groundcherry, and the Johnson grass (*Sorghum halepense* (L.) Pers.). Noncultivated Solanaceae hosts include a wide range of weeds such as *Lycopersicon hirsutum* Dunal, *Lycopersicum puberulum* Ph, *Solanum americanum* Mill., *S. bonariense* L., *S. elaeagnifolium* L., *S. hirtum*, *S. lyratum*, *S. nigrum* L., *S. puberulum*, *S. sisymbriifolium* Lam., *S. saponaceum*, *Datura stramonium* L. (jimsonweed), *D. ferox* L., *Nicotiana glauca* Graham, *Lycium chilense* (Bertero), *Lycopersicon pennellii* var. *puberulum* (Correll), *Solanum dubium* Fresen, and *Solanum puberulum* Phil.

Symptoms of infestation On potato, *T. absoluta* attacks aerial parts and occasionally tubers. Foliar damage is a result of mine-formation within the mesophyll by the feeding larvae. The newly emerged larvae are quite mobile and begin their feeding by building galleries on the leaflets. Depending on the place of oviposition, a larva may penetrate leaf mesophyll, leaf petiole, and young shoots. The mines expand as

Fig. 8.19 Adult of *Tuta absoluta* (**a**), larvae infestation on potato leaf (**b**), and tuber (**c**). (Photo credits: CIP)

the larva develops, causing deformation or rot of the affected leaves. The larvae can migrate to another part of the plant, especially when they reach their maximum development or when the affected foliage withers and dries (CIP 1996; Notz 1992; EPPO 2005b).

Impacts on production losses *T. absoluta* is a major pest of tomato (*Lycopersicum esculentum*), but it also attacks potato. Larval feeding causes desiccation of damaged leaves and shoots, resulting in low tuber production. Larval damage is generally accentuated in young plants in the tuberization phase. Possibilities of *T. absoluta* infestation of potato tubers are very low and restricted to very high field infestation rates during the last stage of the crop (CIP 1996; CABI 2017k; OIRSA 2015b; SENASICA 2016).

Methods of prevention and control

Monitoring with pheromone traps. Pheromone traps can be used both for early detection and for monitoring the flight activity of the pest. They can also help to reduce the population when used for mass trapping, in which case a higher number of traps need to be deployed. For monitoring purposes, one trap should be placed per hectare (Niederwieser 2017; OIRSA 2015b).

Cultural practices. Good agricultural practices for the control of *T. absoluta* include crop rotation with non-solanaceous crops (preferably cruciferous crops), ploughing, adequate irrigation and fertilization, removal of infested plants, and complete removal of post-harvest plant debris.

Biological control. Nesidiocoris tenuis Reuter is an efficient predator of eggs and small larvae of the tomato tuber moth. Egg parasitoids of the family Trichogrammatidae are the natural enemies most used worldwide in biological control programs against Lepidoptera, through inoculative and inundative releases. Releases of *Trichogramma pretiosum* Riley reduced the population of *T. absoluta* to a maximum of 49%. Therefore, combinations with the application of a bioinsecticide based on *Bacillus thuringiensis* subsp. *kurstaki* have been recommended (Niederwieser 2017).

Botanicals/Biopesticides. Neem oil (Azadirachtin) acts as a contact and systemic insecticide and has been shown to be effective against low infestations of *T. absoluta* and for controlling first and second instar larva. The use of *Bacillus thuringensis* is recommended at low-medium infestation levels in combination with Azadirachtin (SENASICA 2016).

Chemical control. Several insecticides have been effectively used against *T. absoluta* in potato; however, this pest is known to develop resistance against effective insecticides.

8.9.9 Potato Leafhoppers

***Empoasca fabae* (Harris 1841) (Hemiptera: Cicadellidae)**

Distribution The leafhopper *Empoasca fabae* is widely distributed (Fig. 8.20).

Fig. 8.20 *Empoasca fabae* adult (**a**); and potato leave damage (**b**). (Courtesy by CIP)

Host range *E. fabae* is a polyphagous pest. Its main hosts are potato, alfalfa, eggplant, bean, celery, cucumber, cucurbits, groundnut (*Arachis hypogaea* L.), melon, rhubarb (*Rheum rhabarbarum* L.), strawberry, sweet potato, and tomato.

Symptoms of infestation The potato leafhopper *E. fabae* is a sucking insect that extracts sap directly from the vascular system of the leaflets, petioles and sometimes the stems. The attack on potato crops is sporadic and is favored by humid conditions. Insects live in the abaxial (lower) part of the leaflets. They inject toxic saliva while feeding, causing leaf necrosis and interfering with plant growth. Under severe attack, the attacked plants die prematurely. Leafhopper can transmit some viruses, although their occurrence is rare (CIP 1996; Cook et al. 2004; Larrain et al. 2003).

Impacts on production losses The complex of symptoms caused by the leafhopper leads to reduced growth and reproduction of plants. Depending on the stage of plant development, in heavily infested fields up to 75% the yield can be lost (Backus et al. 2005; Cook et al. 2004; Medeiros et al. 2004). *E. fabae* has become a major pest with the introduction of modern synthetic insecticides. In Minnesota, USA, annual losses to potato were estimated at $US 7 million (Noetzel et al. 1985). The relationship between yield loss and leafhopper numbers have been determined to be directly linear (Radcliffe and Johnson 1994). In south western Ontario, Canada, average losses of up to 85% for potatoes have been recorded (Tolman et al. 1986). *E. fabae* is also a serious pest of potato in some parts of India with severe hopper burn especially of early planted crops (Verma et al. 1994).

Methods of prevention and control

Most of the potato leafhopper management in potato and other crops is based on monitoring, cultural control, and the use of insecticides.

Monitoring. The presence of the leafhopper can be determined rapidly using entomological nets or yellow sticky traps located at the edges of the crop. Nymphs are best monitored by examining the leaves, especially the underside. Monitoring middle part of a plant gives a more precise estimate of the population of leafhopper nymphs than monitoring basal and apical parts. The economic threshold in potato is 10 or more nymphs per 100 leaves, or 10% of the leaves infested with nymphs (CIP 1996; Cook et al. 2004).

Cultural practices. Infestation of potato fields may be prevented by avoiding proximity to crops such as beans that host high leafhopper populations. Also, elimination of crop residues and appropriate irrigation help reduce populations of this pest (Larrain et al. 2003).

Natural control. Although the leafhopper has several natural enemies, such as predators and parasites, they play a very minor role in potato leafhopper control (CIP 1996; Cook et al. 2004; Larrain et al. 2003).

Plant resistance. Resistant or tolerant varieties should be considered in affected regions, but if leafhopper populations increase, systemic insecticides may become necessary (Backus et al. 2005).

Chemical control. Applications must be made only if populations reach economic thresholds.

8.9.10 Leaf Beetles

Diabrotica viridula **Fabricius,** ***D. speciosa*** **Erichson (Coleoptera: Chrysomelidae)**

Distribution Central America (Costa Rica and Panama) and South America (Argentina, Bolivia, Brasil, Columbia, Ecuador, French Guyana, Paraguay, Peru, Uruguay, and Venezuela (Fig. 8.21).

Host range South American Diabrotica species are presented in at least 116 species in 24 families. *Diabrotica speciosa* larvae developed well on maize, peanut (*Arachis hypogaea* L.), and soybean roots, and not so well on pumpkin (*Cucurbita maxima* Duchesne and *Cucurbita andreana* Naudin), beans (*Phaseolus* spp.), and potato roots. *Diabrotica viridula*, preferred maize as adult and larval food, and for oviposition (Cabrera 2003).

Symptoms of infestation Feeding on leaflets by adults interferes with photosynthesis and, therefore, reduces production of tubers. Larvae damage stolons and form galleries in tubers (Cabrera 2003; Lara et al. 2004).

Fig. 8.21 Adult of *Diabrotica speciosa*. (Photo credits: CIP)

Impacts on production losses In some areas, the most important economic damage to potato is caused by the subterranean larvae gnawing the surface of tubers, which lose quality and become susceptible to soil pathogens (CIP 1996). Eggs and larvae do not develop under dry conditions, so damage is most severe during wet seasons.

Methods of prevention and control
Cultural practices. Deep tillage leads to exposing of larvae and pupae to the action of predators and adverse environmental factors. Weeds that serve as alternating hosts should be eliminated (CIP 1996; Sanchez and Vergara 2002).

Chemical control. In cases of heavy infestation, especially when potato plants are small, insecticide applications are recommended against adult beetles (CIP 1996; Sanchez and Vergara 2002).

8.10 Concluding Remarks

Since potato is a major crop for humankind, it has a global distribution and it is attacked by a myriad pests which can substantially reduce its productivity and its quality. The increasing awareness about the nutritional, agronomic, and cash creating advantages potato provides is likely to further increase its status as a global crop, particularly in developing subtropical and tropical countries. The development, adaptation and use of integrated pest management will be an important area of future research crucial for a sustainable and more resilient and economic profitable potato production in all potato growing regions worldwide. Emphasis should be given to develop and use biological approaches in pest management. This will reduce the dependence on insecticides as well as will reduce the risk that insect populations develop resistance against insecticides.

Previous work has clearly shown that many insect pests of potato will respond to climate change by expanding their geographical range of distribution and increasing population densities will lead to greater crop and post-harvest losses, particularly in subtropical and tropical regions. Modeling of the response of potato pest populations to global warming will help to predict potential changes in pest distribution and severity in order to support potato growers in the adaptation of their pest management strategies.

References

Alcázar J, Cisneros F (1999) Taxonomy and bionomics of the Andean potato weevil complex: *Premnotrypes* spp. and related genera. In: Arthur C, Ferguson P, Smith B (eds) Impact on a changing world: program report 1997-98. International Potato Center, Lima, pp 141–151. http://www.cipotato.org/market/PgmRprts/pr97-98/17weevil. Accessed 11 Jun 2018

Alcázar J, Kroschel J (2008) Andean potato weevil. In: Wale SJ, Platt HW, Cattlin ND (eds) Diseases, pests and disorders of potatoes: a color handbook. Academic Press, Boston, pp 129–131

Alyokhin A (2009) Colorado potato beetle management on potatoes: current challenges and future prospects. In: Tennant P, Benkeblia N (eds) Potato II. Fruit, vegetable and cereal science and biotechnology, vol 3 (Special Issue 1), pp 10–19

Alyokhin A, Chen YH (2017) Adaptation to toxic hosts as a factor in the evolution of insecticide resistance. Curr Opinion Insect Sci 21(1):33–38

Alyokhin A, Chen YH, Udalov M, Benkovskaya G, Lindström L (2013) Evolutionary considerations in potato pest management. In: Giordanengo P, Vincent C, Alyokhin A (eds) Insect pests of potato: global perspectives on biology and management. Academic Press, Oxford, pp 543–571

Alyokhin A, Mota-Sanchez D, Baker M, Snyder WE, Menasha S, Whalon M, Dively G, Moarsi WF (2015) Red Queen on a potato field: IPM vs. chemical dependency in Colorado potato beetle control. Pest Manage Sci 71(3):343–356

Anderson PK et al (2005) Whitefly and whitefly-borne viruses in the tropics: building a knowledge base for global action. International Center for Tropical Agriculture (CIAT), Cali, Colombia, p 351

Arthropod Pesticide Resistance Database (2018). https://www.pesticideresistance.org. Accessed 3 Jul 2018

Backus EA, Serrano MS, Ranger CM (2005) Mechanisms of hopperburn: an overview of insect taxonomy, behavior, and physiology. Annu Rev Entomol 50:125–151

Baker WA, Bradley WG, Clark CA (1949) Biological control of the European corn borer in the United States. Tech Bull U S Dep Agri 983:1–185

Bentley WJ, Rice RE (2015) UC IPM pest management guidelines: Nectarine UC IPM Program, Kearney Agricultural Center, Parlier, USA. http://ipm.ucanr.edu/PMG/r540300411.html. Accessed 2 Nov 2018

BioNET-EAFRINET (2011) White grubs. Keys and fact sheets. http://keys.lucidcentral.org

Blackman RL, Eastop VF (2000) Aphids on the world's crops: an identification and information guide, 2nd edn. Wiley, New York

Butler CD, Trumble JT (2012) The potato psyllid, *Bactericera cockerelli* (Sulc) (*Hemiptera: Triozidae*): life history, relationship to plant diseases, and management strategies. Terr Arthropod Rev 5:87–111. https://doi.org/10.1163/187498312X634266

CABI (2012) *Liriomyza huidobrensis* (Blanchard). Invasive species compendium: datasheets, maps, images, abstracts and full text on invasive species of the world. http://www.cabi.org/isc/datasheet/30956. Accessed 2 Feb 2018

CABI (2016) *Bactericera cockerelli* (tomato/potato psyllid). http://www.cabi.org/isc/datasheet/45643. Accessed 20 Feb 2018

CABI (2017a) *Leptinotarsa decemlineata* (Colorado potato beetle) data sheet. In: Invasive species compendium. CAB International, Wallingford, UK. http://www.cabi.org/isc/datasheet/30380. Accessed 25 Feb 2018

CABI (2017b) *Ostrinia nubilalis* (European maize borer). In: Invasive species compendium. CAB International, Wallingford, UK. http://eol.org/pages/464566/details. Accessed 4 Feb 2018

CABI (2017c) *Myzus persicae* (green peach aphid). In: Invasive species compendium. CAB International, Wallingford, UK. http://www.cabi.org/isc/datasheet/35642. Accessed 20 Feb 2018

CABI (2017d) *Macrosiphum euphorbiae* (potato aphid). In: Invasive species compendium. CAB International, Wallingford, UK. http://www.cabi.org/isc/datasheet/32154. Accessed 5 Feb 2018

CABI (2017e) *Bemisia tabaci* [original text by Andrew Cuthbertson]. In: Invasive species compendium. CAB International, Wallingford, UK. http://www.cabi.org/isc/datasheet/8927. Accessed 6 Feb 2018

CABI (2017f) *Trialeurodes vaporariorum*. In: Invasive species compendium. CAB International, Wallingford, UK. http://www.cabi.org/isc/datasheet/54660. Accessed 7 Feb 2018

CABI (2017g) *Tetranychus urticae* (two-spotted spider mite) data sheet. In: Invasive species compendium. CAB International, Wallingford, UK. http://www.cabi.org/isc/datasheet/53366. Accessed 8 Feb 2018

CABI (2017h) *Polyphagotarsonemus latus* data sheet. In: Invasive species compendium. CAB International, Wallingford, UK. http://www.cabi.org/isc/datasheet/26876. Accessed 9 Feb 2018

CABI (2017i) *Frankliniella occidentalis* (western flower thrips). In: Invasive species compendium. CAB International, Wallingford, UK. https://www.cabi.org/ISC/datasheet/24426. Accessed 5 Feb 2018

CABI (2017j) *Trips tabaci* (onion thrips). In: Invasive species compendium. CAB International, Wallingford, UK. http://www.cabi.org/isc/datasheet/53746. Accessed 11 Feb 2018

CABI (2017k) *Tuta absoluta* datasheet. In: Invasive species compendium. CAB International, Wallingford, UK. http://www.cabi.org/isc/datasheet/49260. Accessed 12 Feb 2018

CABI (2018a) Agrotis ipsilon (black cutworm) datasheet. In: Invasive species compendium. Wallingford, UK: CAB International. https://www.cabi.org/isc/datasheet/3801. Accessed 10 Dec 2018

CABI (2018b) *Spodoptera frugiperda* (fall armyworm) datasheet. In: Invasive species compendium. CAB International, Wallingford, UK. https://www.cabi.org/isc/datasheet/29810. Accessed 10 Dec 2018

CABI (2018c) *Spodoptera eridania* (southern armyworm). In: Invasive species compendium. CAB International, Wallingford, UK. https://www.cabi.org/ISC/datasheet/44518. Accessed 10 Dec 2018

CABI (2018d) *Mamestra configurata* (bertha armyworm). In: Invasive species compendium. CAB International, Wallingford, UK. https://www.cabi.org/isc/datasheet/8492. Accessed 10 Dec 2018

Cabrera G (2003) Host range and reproductive traits of *Diabrotica speciosa* (Germar) and *Diabrotica viridula* (F.) (*Coleoptera: Chrysomelidae*), two species of South American pest rootworms, with notes on other species of *Diabroticina*. Environ Entomol 32(2):276–285. https://doi.org/10.1603/0046-225X-32.2.276

Canedo, V, Carhuapoma P, López E, Kroschel J (2016a) *Copidosoma koehleri* (Blanchard 1940). In: Kroschel J, Mujica N, Carhuapoma P, Sporleder M (eds) Pest distribution and risk atlas for Africa. Potential global and regional distribution and abundance of agricultural and horticultural pests and associated biocontrol agents under current and future climates. International Potato Center (CIP), Lima (Peru), pp. 208–219. https://doi.org/10.4160/9789290604761-16. ISBN 978-92-9060-476-1

Canedo V, Carhuapoma P, Dávila W, Kroschel J (2016b) *Apanteles subandinus* (Blanchard 1947). In: Kroschel J, Mujica N, Carhuapoma P, Sporleder M (eds) Pest distribution and risk atlas for Africa. Potential global and regional distribution and abundance of agricultural and horticultural pests and associated biocontrol agents under current and future climates. International Potato Center (CIP), Lima (Peru), pp 220–231. https://doi.org/10.4160/9789290604761-17

Canedo V, Carhuapoma P, Bartra C, Kroschel J (2016c) *Orgilus lepidus* (Muesebeck 1967). In: Kroschel J, Mujica N, Carhuapoma P, Sporleder M (eds) Pest distribution and risk atlas for Africa. Potential global and regional distribution and abundance of agricultural and horticultural pests and associated biocontrol agents under current and future climates. International Potato Center (CIP), Lima (Peru), pp 232–244. https://doi.org/10.4160/9789290604761-18

Capinera JL (2001) Handbook of vegetable pests. Academic Press, San Diego, p 729

Castillo JR (2010) *Prodiplosis longifila* in Peru. https://conference.ifas.ufl.edu/TSTAR/presentations/Tuesday/pm/4%2020pm%20J%20Castillo.pdf. Accessed 20 Feb 2018

CIP (1996) Major potato diseases, insects, and nematodes. International Potato Center, Lima, Peru, p 61

Cisneros F, Mujica N (1999a) Impacto del cambio climático en la agricultura: Efectos del fenómeno El Nino en los cultivos de la costa central. Consejo Nacional del Ambiente (CONAM), Lima (Peru), pp 115–133

Cisneros F, Mujica N (1999b) The leafminer fly in potato: plant reaction and natural enemies as natural mortality factors. International Potato Center, Lima, Peru, pp 129–140

Cisneros F, Mujica N (1999c) Biological and selective control of the sweet potato whitefly, *Bemisia tabaci* (*Gennadius*) (Hom.: *Aleyrodidae*). International Potato Center, Lima, Peru. Program Report 1997-98, pp 255–264

Cook KA, Ratcliffe ST, Gray ME, Steffey KL (2004) Potato leafhopper (*Empoasca fabae* (Harris)). Insect Fact Sheet, University of Illinois, Department of Crop Sciences. https://ipm.illinois.edu/fieldcrops/insects/potato_leafhopper.pdf

Dillon ME, Wang G, Huey RB (2010) Global metabolic impacts of recent climate warming. Nature 467:704–706

EPPO (2005a) *Tecia solanivora*. EPPO Bull 35:399–401

EPPO (2005b) *Tuta absoluta*. Data sheets on quarantine pests. Bull OEPP/EPPO Bull 35:434–435

EPPO (2014) PQR database. European and Mediterranean Plant Protection Organization, Paris, France. http://www.eppo.int/DATABASES/pqr/pqr.htm

EPPO (2015) *Prodiplosis longifila* (*Diptera: Cecidomyiidae*). In: EPPO technical document No. 1068, EPPO study on pest risks associated with the import of tomato fruit. EPPO Paris, pp 133–134. https://www.eppo.int/QUARANTINE/DT_1068_Tomato_study_MAIN_TEXT_and_ANNEXES_2015-01-26.pdf

European Parliament and the Council of the European Union (2009) Directive2009/128/EC of the European Parliament and of the Council of 21 October 2009 establishing a framework for Community action to achieve the sustainable use of pesticides. Official Journal of the European Union, L309/71. First published 2009. http://eur-lex.europa.eu/legal-content/EN/ALL/?uri=celex%3A32009L0128

Eyre D, Giltrap N (2013) *Epitrix* flea beetles: new threats to potato production in Europe. Pest Manage Sci 69(1):3–6

FAO (2018). http://www.fao.org/agriculture/crops/core-themes/theme/pests/ipm

Ferro DN (2000) Success and failure of Bt products: Colorado potato beetle—a case study. In: Kennedy GG, Sutton TB (eds) Emerging technologies for integrated pest management: concepts, research, and implementation. APS Press, St. Paul, pp 177–189

Forgash AG (1985) Insecticide resistance in the Colorado potato beetle. In: Ferro DN, Voss RH (eds) Proceedings, Symposium on the Colorado potato beetle. XVIIth International Congress of Entomology. Res. Bull. 704, Mass Agric Exp Stn Circ 347, pp 33–52

Franco-Lara L, Rodríguez D, Guzmán-Barney M (2013) Prevalence of potato yellow vein virus (PYVV) in *Solanum tuberosum* group *Phureja* fields in three states of Colombia. Am J Potato Res 90(4):324–330

Furth DG, Savini V, Chaboo CS (2015) Beetles (Coleoptera) of Peru: a survey of the families. Chrysomelidae: Alticinae (flea beetles). J Kans Entomol Soc 88(3):368–374

Gagné RJ (1986) Revision of *Prodiplosis* (*Diptera: Cecidomyiidae*) with descriptions of three new species. Ann Entomol Soc Am 79(1):235–245

Gamarra H, Mujica N, Carhuapoma P, Kreuze J, Kroschel J (2016a) Sweetpotato white fly, *Bemisia tabaci* (Gennadius 1989) (Biotype B). In: Kroschel J, Mujica N, Carhuapoma P, Sporleder M (eds) Pest distribution and risk atlas for Africa—potential global and regional distribution and abundance of agricultural and horticultural pests and associated biocontrol agents under current and future climates. International Potato Center (CIP), Lima, Peru, pp 85–99. https://doi.org/10.4160/9789290604761 (online publication: http://cipotato.org/riskatlasforafrica/)

Gamarra H, Carhuapoma P, Mujica N, Kreuze J, Kroschel J (2016b) Greenhouse whitefly, *Trialeurodes vaporariorum* (Westwood 1956). In: Kroschel J, Mujica N, Carhuapoma P, Sporleder M (eds) Pest distribution and risk atlas for Africa—potential global and regional distribution and abundance of agricultural and horticultural pests and associated biocontrol agents under current and future climates. International Potato Center (CIP), Lima, Peru, pp 154–168. https://doi.org/10.4160/9789290604761 (online publication: http://cipotato.org/riskatlasforafrica/)

Garzón TJA (2002) Asociación de *Paratrioza cockerelli* Sulc. con enfermedades en papa (*Solanum tuberosum*) y tomate (*Lycopersicum lycopersicum* Mil. Ex. Fawnl) en México. In: Memoria del Taller sobre 11 *Paratrioza cockerelli* (Sulc.) como plaga y vector de fitoplasmas en hortalizas. Culiacán, Sinaloa, México, pp 79–87

Ghosh SK, Chakraborty G (2012) Integrated field management of *Henosepilachna vigintiocto-punctata* (Fabr.) on potato using botanical and microbial pesticides. J Biopest 5(Suppl):151–154

Goftishu M, Dejene M, Kassaye A, Belay T (2016) Red Spider Mite, *Tetranychus urticae* Koch (Arachnida: Acari-Tetranychidae): a threatening pest to potato (*Solanum tuberosum*). Pest Manage J Ethiopia 19:53–59

Gonzales-Bustamante L (1996) *Prodiplosis longifila* infestando vainas del pallar en Lima. Rev Peru Entomol 39:118

Gupta S, Gavkare O (2014) White grub, *Brahmina coriacea*, a potential threat to potato. J Ind Pollut Control 30(2):357–359

Hazzard R (2008) New England vegetable management guide. New England extension systems, University of Massachusetts, Amherst, MA. https://ag.umass.edu/vegetable/fact-sheets/two-spotted-spider-mite. Accessed 2 Nov 2018

Hazzard R (2013) New England vegetable management guide. New England extension systems, University of Massachusetts, Amherst, MA. https://ag.umass.edu/vegetable/fact-sheets/two-spotted-spider-mite. Accessed 23 Sep 2019

Henne DC, Paetzold L, Workneh F, Rush CM (2010) Evaluation of potato psyllid cold tolerance, overwintering survival, sticky trap sampling, and effects of liberibacter on potato psyllid alternate host plants. In: Workneh F, Rush CM (eds), Proceedings of the 10th Annual Zebra Chip Reporting Session, pp. 149–153. Dallas, TX (November 7–10, 2010)

Hernandez LM (2014) Caracterización del daño y distribución geográfica de Cecidomyiidae (Diptera) y sus parasitoides asociados a Solanáceas y limón Tahití en Colombia. Tesis de Maestría en Ciencias Agrarias. Línea Protección de Cultivos. Universidad Nacional de Colombia. Palmira, 93 p

Hernandez LM, Guzman YC, Martínez-Arias A, Manzano MR, Selvaraj JJ (2015) The bud midge *Prodiplosis longifila*: damage characteristics, potential distribution and presence on a new crop host in Colombia. SpringerPlus 4:205. https://doi.org/10.1186/s40064-015-0987-6

Horn P, Page J (2008) Integrated pest management dealing with potato tuber moth and all other pests in Australian potato crops. In: Kroschel J, Lacey L (eds) Integrated pest management for the potato tuber moth, *Phthorimaea operculella* (Zeller)—a potato pest of global importance. Tropical Agriculture 20, Advances in Crop Research 10. Margraf Publishers, Weikersheim, Germany, pp 111–117

Jackson G (2016) Potato ladybird beetle (255). Pacific pests and pathogens fact sheet. PestNet, Queensland, Australia

Jacques RL, Fasulo TR (2015) Colorado potato beetle *Leptinotarsa decemlineata*. DPI Entomology Circular 271. Florida Department of Agriculture and Consumer Services, Division of Plant Industry and University of Florida. http://entnemdept.ufl.edu/creatures/veg/leaf/potato_beetles.htm

Jeger M, Bragard C, Caffier D, Candresse T, Chatzivassiliou E, Dehnen-Schmutz K, Gilioli G, Gregoire J-C, Jaques Miret JA, Navajas Navarro M, Niere B, Parnell S, Potting R, Rafoss T, Rossi V, Urek G, Van Bruggen A, Van der Werf W, West J, Winter S, Gardi C, Bergeretti F, MacLeod A (2018) Scientific opinion on the pest categorisation of *Tecia solanivora*. EFSA J 16(1):5102,. pp 25. https://doi.org/10.2903/j.efsa.2018.5102

Jeyasankar A, Premalatha S, Elumalai K (2014) Antifeedant and insecticidal activities of selected plant extracts against Epilachna beetle, *Henosepilachna vigintioctopunctata* (Coleoptera: Coccinellidae). Adv Entomol 2(1):14–19

Kabaluk JT, Goettel M, Erlandson M, Ericsson J, Duke G, Vernon R (2005) *Metarhizium anisopliae* as a biological control for wireworms and a report of some other naturally-occurring parasites. IOBC/WPRS Bull 28(2):109–115

Katakura H (1980) Classification and evolution of the phytophagous ladybirds belonging to *Henosepilachna vigintioctomaculata* complex (Coleoptera, Coccinellidae). PhD Dissertation, Hokkaido University, Japan

Kaya H, Alcazar J, Parsa S, Kroschel J (2009) Microbial control of the Andean potato weevil complex. Fruit Veg Cereal Sci Biotechnol 3(Special Issue 1):39–45

Keller S (2003) Integrated pest management of the potato tuber moth in cropping systems of different agroecological zones. In: Kroschel J (ed) Tropical Agriculture 11, Advances in Crop Research 1. Margraf Publishers, Weikersheim, Germany, p 153

Kennedy GG (1983) Effects of European corn borer (Lepidoptera: Pyralidae) damage on yields of spring-grown potatoes. J Econ Entomol 76(2):316–322

Kim JJ, Jeong G, Han JH, Lee S (2013) Biological control of aphid using fungal culture and culture filtrates of *Beauveria bassiana*. Mycobiology 41(4):221–224. https://doi.org/10.5941/MYCO.2013.41.4.221

Kogan M (1998) Integrated pest management: historical perspectives and contemporary developments. Annu Rev Entomol 43:243–270

Kroschel J (1994) Population dynamics of the potato tuber moth, *Phthorimaea operculella*, in Yemen and its effects on yield. In: Proceedings of the Brighton crop protection conference, pests and diseases, vol I. Brighton, UK, pp 241–246

Kroschel J (1995) Integrated Pest Management in potato production in the Republic of Yemen with special reference to the integrated biological control of the potato tuber moth (*Phthorimaea operculella* Zeller). Tropical Agriculture 8. Margraf Verlag, Weikersheim, Germany, p 227

Kroschel J, Koch W (1996) Studies on the use of chemicals, botanicals and *Bacillus thuringiensis* in the management of the potato tuber moth in potato stores. Crop Prot 15(2):197–203

Kroschel J, Schaub B (2013) Biology and ecology of potato tuber moths as major pests of potato. In: Giordanengo P, Vincent C, Alyokhin A (eds) Insect pests of potato: global perspectives on biology and management. Academic Press, Oxford, pp 165–192

Kroschel J, Sporleder M (2006) Ecological approaches to integrated pest management of the potato tuber moth, *Phthorimaea operculella* Zeller (Lepidoptera, Gelechidae). In: Proceedings of the 45th Annual Washington State Potato Conference, February 7–9, 2006, Moses Lake, Washington, pp 85–94

Kroschel J, Zegarra O (2010) Attract-and-kill: a new strategy for the management of the potato tuber moths *Phthorimaea operculella* (Zeller) and *Symmetrischema tangolias* (Gyen) in potato—Laboratory experiments towards optimizing pheromone and insecticide concentration. Pest Manage Sci 66:490–496

Kroschel J, Zegarra O (2013) Attract-and-kill as a new strategy for the management of the potato tuber moths *Phthorimaea operculella* (Zeller) and *Symmetrischema tangolias* (Gyen) in potato: evaluation of its efficacy under potato field and storage conditions. Pest Manage Sci 69(11):1205–1215

Kroschel J, Kaack HJ, Fritsch E, Huber J (1996) Biological control of the potato tuber moth (*Phthorimaea operculella* Zeller) in the Republic of Yemen using granulosis virus: Propagation and efficacy of the virus in field trials. Biocontrol Sci Technol 6:217–226

Kroschel J, Alcazar J, Pomar P (2009) Potential of plastic barriers to control Andean potato weevil *Premnotrypes suturicallus* Kuschel. Crop Prot 28:466–476

Kroschel J, Cañedo V, Alcázar J (2011) Manual de capacitación para extensionistas: Manejo de plagas de la papa en la región Andina del Perú. International Potato Center, Lima, Peru, p 88

Kroschel J, Mujica N, Alcazar J, Canedo V, Zegarra O (2012) Developing integrated pest management for potato: experiences and lessons from two distinct potato production systems of Peru. In: He Z, Larkin RP, Honeycutt CW (eds) Sustainable potato production: global case studies. Springer, UK, pp 419–450

Kroschel J, Sporleder M, Tonnang HEZ, Juarez H, Carhuapoma P, Gonzales JC, Simon R (2013) Predicting climate change caused changes in global temperature on potato tuber moth *Phthorimaea operculella* (Zeller) distribution and abundance using phenology modeling and GIS mapping. Agric For Meteorol 170:228–241

Kroschel J, Mujica N, Carhuapoma P, Sporleder M (2016a) Pest Distribution and Risk Atlas for Africa—potential global and regional distribution and abundance of agricultural and horticultural pests and associated biocontrol agents under current and future climates. International Potato Center (CIP), Lima, Peru, p 650. https://doi.org/10.4160/9789290604761. (online publication: http://cipotato.org/riskatlasforafrica/)

Kroschel J, Sporleder M, Carhuapoma P (2016b) Potato tuber moth, *Phthorimaea operculella* (Zeller). In: Kroschel J, Mujica N, Carhuapoma P, Sporleder M (eds) Pest distribution and risk atlas for Africa—potential global and regional distribution and abundance of agricultural and horticultural pests and associated biocontrol agents under current and future climates. International Potato Center, Lima, Peru, pp 7–23. https://doi.org/10.4160/9789290604761. (online publication: http://cipotato.org/riskatlasforafrica/)

Lacey L, Kroschel J (2009) Microbial control of the potato tuber moth (Lepidoptera: Gelechiidae). Fruit, Veg Cereal Sci Biotech 3:46–54

Lacey LA, Liu T-X, Buchman JL, Munyaneza JE, Goolsby JA, Horton DR (2011) Entomopathogenic fungi (Hypocreales) for control of potato psyllid, *Bactericera cockerelli* (Šulc) (Hemiptera: Triozidae) in an area endemic for zebra chip disease of potato. Biol Control 56 (3):271–278

Landis BJ, Davis EW (1947) Two-spotted spider mite damage to potatoes. J Econ Entomol 40(4):565–565

Lara FM, Scaranello AL, Baldin ELL, Boiça-Junior L, Lourenção AL (2004) Resistência de genótipos de batata a larvas e adultos de *Diabrotica speciosa*. Hortic Bras 22(4):761–765

Larrain P, Kalazich J, Carrillo R, Cisternas E (2003) Plagas de la papa y su manejo. Colección libros INIA- N° 9. Instituto de Investigaciones Agropecuarias, Ministerio de Agricultura. La Serena, Chile

Laznik Ž, Tóth T, Lakatos T, Vidrih M, Trdan S (2010) Control of the Colorado potato beetle (*Leptinotarsa decemlineata* [Say]) on potato under field conditions: a comparison of the efficacy of foliar application of two strains of *Steinernema feltiae* (Filipjev) and spraying with thiametoxam. J Plant Dis Prot 117(3):129–135

Learmonth S (2017) Thrips: potato pest in Indonesia and Western Australia. Department of Primary Industries and Regional Development. https://www.agric.wa.gov.au/potatoes/thrips-potato-pest-indonesia-and-western-australia

Li WD, Chen SX, Qin JG (2003) Identification of sympatric Asian and European corn borers. Entomol Knowl 40(1):31–35

Malumphy C, Everatt M, Eyre D, Giltrap N (2016) Potato flea beetle, *Epitrix* species. Plant pest factsheet. Department for Environment Food & Rural Affairs, United Kingdom, pp 1–6

Mamani D, Sporleder M, Kroschel J (2011) Efecto de materiales inertes de fórmulas bioinsecticidas en la protección de tubérculos almacenados contra las polillas de papa (Effect of inert materials of bioinsecticides formula to protect stored tubers of potato against the potato moths). Rev Peru Entomol 46:43–49

Martin NA (2016) Tomato potato psyllid—*Bactericera cockerelli*. Interesting insects and other invertebrates. New Zealand Arthropod Factsheet Series Number 60. http://nzacfactsheets.landcareresearch.co.nz/Index.html

Mauchline N, Stannard K (2013) Evaluation of selected entomopathogenic fungi and bioinsecticides against *Bactericera cockerelli* (Hemiptera). NZ Plant Prot 66:324-332. https://doi.org/10.30843/nzpp.2013.66.5707

Medeiros A, Tingey W, De Jong W (2004) Mechanisms of resistance to potato leafhopper, *Empoasca fabae* (Harris). Am J Potato Res 81(6):431–441

Misra SS (1995) White grub, *Holotrichia* (*Lachnosterna*) *coracea* (Hope)-A key pest of potatoes in Himachal Pradesh (India). J Entomol Res 9(2):181–182

Misra SS, Chandel RS (2003) Potato white grubs in India. Tech Bull 60

Muchemi SK, Ngichop FC, Fiaboe KKM (2014) Report of the establishmnet of the released hymenoptera *Phaedrotoma scabriventris* Nixon (Braconidae), Halticpotera arduine (Walker) (Pteromalidae) and Chrysocharis flacilla (Walker) (Eulophidae) in vegetables production systems of Kenya. Submitted to Kenyan Plant Health Inspectorate Service (KEPHIS). International Centre of Insect Physiology and Ecology (icipe), 13 pp.

Mujica N (1998) Uso de trampas amarillas en el control de adultos de mosca minadora y mosca blanca. International Potato Center, Lima, Peru. Hoja divulgativa N° 1

Mujica N (2016) Ecological approaches to manage the leafminer fly *Liriomyza huidobrensis* (Blanchard) in potato-based agroecosystems of Peru. In: Kroschel J (ed) Tropical agriculture 21–Advances in crop research, vol 11. Margraf Verlag, Weikersheim, Germany, p 255

Mujica N, Kroschel J (2013) Pest intensity-crop loss relationships for the leafminer fly *Liriomyza huidobrensis* (Blanchard) in different potato (*Solanum tuberosum* L.) varieties. Crop Prot 47:6–16

Mujica N, Kroschel J (2018) Ecological, economic and environmental assessments of Integrated Pest Management technologies: case study for potato lowland systems in the Cañete valley. Food Energy Secur 8:e00153. https://doi.org/10.1002/fes3.153

Mujica N, Alcázar J, Kroschel J (2013) Interacción del nematode entomopatogénico *Heterorhabditis indica* (RHABDITIDA: HETERORHABDITIDAE) y el ectoparasitoide *Diglyphus begini* (HYMENOPTERA: EULOPHIDAE) en el control de la mosca minadora *Liriomyza huidobrensis* (DIPTERA: AGROMYZIDAE). LV Convencion Nacional de Entomologia November 4–7, 2013, Lima-Peru

Mujica N, Carhuapoma P, Kroschel J (2016) Serpentine leafminer fly, *Liriomyza huidobrensis* (Blanchard 1926). In: Kroschel J, Mujica N, Carhuapoma P, Sporleder M (eds) Pest distribution and risk atlas for Africa—potential global and regional distribution and abundance of agricultural and horticultural pests and associated biocontrol agents under current and future climates. International Potato Center (CIP), Lima, Peru, pp 114–125. https://doi.org/10.4160/9789290604761 (online publication: http://cipotato.org/riskatlasforafrica/)

Mujica N, Sporleder M, Carhuapoma P, Kroschel J (2017) A temperature-dependent phenology model for the leafminer fly *Liriomyza huidobrensis* (Blanchard) (Diptera: Agromyzidae). J Econ Entomol 110(3):1333–1344

Mujica N, Fonseca C, Suarez F, Fabian F, Marchena M, Cisneros F (2000) Reducción del uso de insecticidas en el control de la mosca minadora, *Liriomyza huidobrensis* Blanchard, por medio del uso de técnicas etológicas. In: Arning G, A Lizárraga (eds.), Control Etológico. Uso de feromonas, trampas de colores y luz para el control de plagas en la agricultura sostenible. Lima, Perú

Munyaneza JE, Crosslin JM, Upton JE (2007) Association of *Bactericera cockerelli* (Homoptera: Psyllidae) with "zebra chip", a new potato disease in southwestern United State and Mexico. J Econ Entomol 100(3):656–663

Munyaneza JE, Sengoda VG, Crosslin JM, De la Rosa-Lozano G, Sánchez A (2009) First report of "*Candidatus Liberibacter psyllaurous*" in potato tubers with zebra chip disease in Mexico. Plant Dis 93(5):552

Nault BA, Kennedy GG (1996) Timing insecticide applications for managing European corn borer (Lepidoptera: Pyralidae) infestations in potato. Crop Prot 15(5):465–471

Naz F, Inayatullah M, Rafi MA, Ashfaque M, Ali A (2012) *Henosepilachna vigintioctopunctata* (Fab.) (Epilachninae; Coccinellidae); its taxonomy, distribution and host plants in Pakistan. Sarhad J Agric 28(3):421–427

Niederwieser F (2017) Guidelines for the control of tomato leafminer (*Tuta absoluta*) on potato. CHIPS 31(1):30–32

Niño L (2004) Revisión sobre la polilla de la papa *Tecia solanivora* en Centro y Suramerica. Suplemento de la Revista Latinoamericana de la papa, pp 4–22

Noetzel DM, Cutkomp LK, Harein PK (1985) Estimated annual losses due to insects in Minnesota 1981–1983. University of Minnesota Agricultural Experiment Station Bulletin 2541

Notz AP (1992) Distribution of eggs and larvae of *Scrobipalpula absoluta* in potato plants. Rev Facul Agron (Maracay) 18:425–432

Oerke EC, Dehne HW, Schönbeck F, Weber A (1994) Crop production and crop protection. Estimated losses in major food and cash crops. Elsevier Science, Amsterdam, p 808

OIRSA (2015a) El psílido de la papa y tomate *Bactericera* (*Paratrioza*) *cockerelli* (Sulc) (*Hemiptera: Triozidae*): ciclo biológico; la relación con las enfermedades de las plantas y la estrategia del manejo integrado de plagas en la región del OIRSA. https://www.oirsa.org/contenido/Manual%20Bactericera%20Cockerelli%20version%201.3.pdf

OIRSA (2015b) Manual de procedimientos para la vigilancia, prevención y control de la polilla del tomate *Tuta absoluta* (*Lepidóptera: Gelechiidae*) en la región del OIRSA. https://www.oirsa.org/contenido/2018/Sanidad_Vegetal/Manuales%20OIRSA%202015-2018/Manual%20T%C3%A9cnico%20de%20Tuta%20absoluta%20marzo%2010,%202017.pdf. Accessed 10 Oct 2017

OIRSA (2016) El psílido de la papa y tomate *Bactericera (Paratrioza) cockerelli* (Sulc) *(Hemiptera: Triozidae):* ciclo biológico; la relación con las enfermedades de las plantas y la estrategia del manejo integrado de plagas en la región del OIRSA. Editorial Tauro, Mexico. p. 48

Orlandini S, Magarey RD, Woo Park E, Sporleder M, Kroschel J (2017) Methods of agroclimatology: modeling approaches for pests and diseases. In: Hatfield JL, MVK S, Prueger JH (eds) Agroclimatology. Linking agriculture to climate. Agron Monogr 60. ASA, CSSA, and SSSA, Madison, WI. https://doi.org/10.2134/agronmonogr60.2016.0027

Orozco FA, Cole DC, Forbes G, Kroschel J, Wanigaratne S, Arica D (2009) Monitoring adherence to the International Code of Conduct—highly hazardous pesticides in central Andean agriculture and farmers' rights to health. Int J Occup Environ Health 15(3):255–268

Ortiz O, Alcázar J, Catalán W, Cerna V (1996) Economic impact of IPM practices on the Andean potato weevil in Peru. In: Walker TS, Crissman CC (eds) Case studies of the economic impact of CIP related technologies. International Potato Center (CIP), Lima, Peru, pp 95–113

Pacheco DM (2015) Patogenicidad de dos nematodos entomopatógenos (*Heterorhabditis* sp y *Steinernema* sp) sobre larvas y pupas de *Prodiplosis longifila* en cultivos de tomate (*Lycopersicum esculentum* Miller). Dissertation, Universidad Nacional Pedro Ruiz Gallo

Parker WE, Howard JJ (2001) The biology and management of wireworms (*Agriotes* spp.) on potato with particular reference to the UK. Agric For Entomol 3(2):85–98

Pathania M, Chandel RS (2016) Life history strategy and behaviour of white grub, *Brahmina coriacea* (Hope) (Coleoptera: Scarabaeidae: Melolonthinae) an invasive pest of potato and apple agro-ecosystem in northwestern India. Oriental Insects 51(1):46–69

Pell JK, Eilenberg J, Hajek AE, Steinkraus DS (2001) Biology, ecology and pest management potential of entomophthorales. In: Butt TM, Jackson C, Magan N (eds) Fungi as biocontrol agents: progress, problems and potential. CAB International, Wallingford, UK, pp 71–154

Pollet A, Barragan A, Zeddam JL, Laery X (2003) *Tecia solanivora*, a serious biological invasion of potato cultures in South America. Int Pest Control 45(3):139–144

Powell W, Pell JK (2017) Biological control. In: Van Emden HF, Harrington R (eds) Aphids as crop pests, 2nd edn. CABI Publishing, Wallingford, pp 469–514

Pundt L (2007) Biological control of two-spotted spider mites. College of Agriculture, Health and Natural Resources, UConn Extension, The University of Connecticut, USA. http://ipm.uconn.edu/documents/raw2/html/664.php?display=print. Accessed 2 Nov 2018

Puttarudriah M, Krishnamurti B (1954) Problem of *Epilachna* control in Mysore: Insecticidal control found inadvisbale when natural incidence of parasites is high. Indian J Entomol 16:137–141

Quiroz R, David AR, Kroschel J, Andrade-Piedra J, Barreda C, Condori B, Mares V, Monneveux P, Perez W (2018) Impact of climate change on the potato crop and biodiversity in its center of origin. Open Agric 3:273–283. https://doi.org/10.1515/opag-2018-0029

Radcliffe EB, Johnson KB (1994) Biology and Management of Leafhoppers on Potato. In: Zehnder GW, Powelson ML, Jansson RK, Raman KV (eds) Advances in Potato Pest Biology and Management, APS Press, Saint Paul, MN, pp 71–82

Reátegui F, Wilson J, Ayquipa G, Neyra S (2009) Actividad entomopatógena de cuatro especies de hongos sobre (Diptera: Cecidomyiidae). REBIOL 29(1):1–6

Rehman M, Melgar J, Rivera C, Urbina N, Idris AM, Brown JK (2010) First report of *Candidatus Liberibacter psyllaurous* or Ca. *Liberibacter* solanacearum associated with severe foliar chlorosis, curling, and necrosis and tuber discoloration of potato plants in Honduras. Plant Dis 94:376

Salazar L (1995) Los virus de la papa y su control. Centro Internacional de la Papa (CIP), Lima, Peru

Sanchez G, Vergara C (2002) Plagas del cultivo de papa. Departamento de Entomologia y Fitopatologia, UNALM, Lima, Peru

Sarmiento J (1997) Manejo de *Prodiplosis longifila* Gagné. Perú, p 1–4

Schaub B, Kroschel J (2017) Developing a biocontrol strategy to protect stored potato tubers from infestation with potato tuber moth species. J Appl Entomol 142(1–2):78–88. https://doi.org/10.1111/jen.12426

Schaub B, Carhuapoma P, Kroschel J (2016) Guatemalan potato tuber moth, *Tecia solanivora* (Povolny 1973). In: Kroschel J, Mujica N, Carhuapoma P, Sporleder M (eds) Pest distribution and risk atlas for Africa—potential global and regional distribution and abundance of agricultural and horticultural pests and associated biocontrol agents under current and future climates. International Potato Center (CIP), Lima, Peru, pp 24–38. https://doi.org/10.4160/9789290604761. (online publication: http://cipotato.org/riskatlasforafrica/)

SENASICA (2016) Palomilla del tomate *Tuta absoluta Meyrick*. Ficha Técnica 28:71

Song GH, Wu WW, Zhao QL (2008) Infection law and the control of 28-spot ladybird. Jilin Veg 1:50

Spencer KA (1973) Agromyzidae (Diptera) of Economic Importance. The Netherlands: Junk, The Hague

Spencer KA (1990) Host specialization in the world Agromyzidae Diptera. The Netherlands: Kluwer Academic Publisher

Sporleder M (2003) The granulovirus of the potato tuber moth *Phthorimaea operculella* Zeller – characterisation and prospects for effective mass production and pest control (Dissertation). Tropical Agriculture 13, Advances in Crop Research 3. Margraf Publishers, Weikersheim, Germany, 196 pp.

Sporleder M, Kroschel J (2008) The potato tuber moth granulovirus (PoGV): use, limitations and possibilities for field applications. In: Kroschel J, Lacey L (eds) Integrated pest management of the potato tuber moth, *Phthorimaea operculella* Zeller—a potato pest of global. Tropical Agriculture 20, Advances in Crop Research importance 10. Margraf Publishers, Weikersheim, Germany, pp 49–71

Sporleder M, Lacey L (2013) Biopesticides. In: Giordanengo P, Vincent C, Alyokhin A (eds) Insect pests of potato: global perspectives on biology and management. Academic Press, Oxford, pp 463–497

Sporleder M, Kroschel J, Gutierrez Quispe MR, Lagnaoui A (2004) A temperature-based simulation model for the potato tuber worm, *Phthorimaea operculella* Zeller (Lepidoptera; Gelechiidae). Environl Entomol 33(3):477–486

Sporleder M, Tonnang HEZ, Carhuapoma P, Gonzales JC, Juarez H, Kroschel J (2013) Insect Life Cycle Modeling (ILCYM) software—a new tool for regional and global insect pest risk assessments under current and future climate change scenarios. CABI Book Publication, pp 412–427

Sporleder M (2003) The granulovirus of the potato tuber moth *Phthorimaea operculella* Zeller – characterisation and prospects for effective mass production and pest control (Dissertation). Tropical Agriculture 13, Advances in Crop Research 3. Margraf Publishers, Weikersheim, Germany, 196 pp

Sporleder M, Carhuapoma P, Kroschel J (2016) Andean potato tuber moth, *Symmetrischema tangolias* (Gyen 1913). In: Kroschel J, Mujica N, Carhuapoma P, Sporleder M (eds) Pest distribution and risk atlas for Africa—potential global and regional distribution and abundance of agricultural and horticultural pests and associated biocontrol agents under current and future climates. International Potato Center (CIP), Lima, Peru, pp 39–53. https://doi.org/10.4160/9789290604761. (online publication: http://cipotato.org/riskatlasforafrica/)

Sporleder M, Schaub B, Aldana G, Kroschel J (2017) Temperature-dependent phenology and growth potential of the Andean potato tuber moth, *Symmetrischema tangolias* (Gyen) (Lep., Gelechiidae). J Appl Entomol 141:202–218. https://doi.org/10.1111/jen.12321

Steele C (2011) Wireworm Factsheet. Growers' advice. Potato Council. https://potatoes.ahdb.org.uk/sites/default/files/publication_upload/Growers%20Advice%20-%20Wireworm%20Factsheet.pdf

Stenberg JA (2017) A conceptual framework for integrated pest management. Trends Plant Sci 22(9):759–769. https://doi.org/10.1016/j.tplants.2017.06.010

Strand L (2006) Integrated pest management for potatoes in the western United States, vol 3316. UCANR Publications

Tolman JH, McLeod DGR, Harris CR (1986) Yield losses in potatoes, onions and rutabagas in Southwestern Ontario, Canada—the case for pest control. Crop Prot 5 (4):227–237

Venkatesha MG (2006) Seasonal occurrence of *Henosepilachna vigintioctopunctata* (F.) (Coleoptera: Coccinellidae) and its parasitoid on ashwagandha in India. J Asia-Pacific Entomol 9(3):265–268

Vereijssen J, Smith GR, Weintraub PG (2018) *Bactericera cockerelli* (Hemiptera: Triozidae) and Candidatus *Liberibacter solanacearum* in Potatoes in New Zealand: Biology, Transmission, and Implications for Management. J Integr Pest Manag 9(1):13. https://doi.org/10.1093/jipm/pmy007

Verma KD, Chandla VK, Trivedi TP (1994) Present status and management of potato pests in India. Annals of Entomology 12(2):73–79

Vincent C, Alyokhin A, Giordanengo P (2013) Potatoes and their pests—setting the stage. In: Giordanengo P, Vincent C, Alyokhin A (eds) Insect pests of potato: global perspectives on biology and management. Academic Press, Oxford, pp 3–8

Vishwakarma R, Prasad PH, Ghatak SS, Mondal S (2011) Bio-efficacy of plant extracts and entomopathogenic fungi against epilachna beetle, *Henosepilachna vigintioctopunctata* (Fabricius) infesting bottle gourd. J Insect Sci 24(1):65–70

Visser D (2012) Beetles and their grubs that damage potato tubers. Technical news. ARC-Roodeplaat. http://www.potatoes.co.za/SiteResources/documents/Beetles%20and%20their%20grubs%20that%20damaged%20potato%20tubers%202012.pdf

Visser D, Stals R (2012) *Temnorhynchus coronatus* (Fabricius) (Scarabaeidae: Dynastinae), potentially a pest of sweet potato in South Africa. Afr Entomol 20(2):402–407

Weintraub PG (2001) Effects of cyromazine and abamectin on the pea leafminer *Liriomyza huidobrensis* (Diptera: *Agromyzidae*) and its parasitoid *Diglyphus isaea* (Hymenoptera: *Eulophidae*) in potatoes. Crop Prot 20:207–213

Weintraub PG, Scheffer SJ, Visser D, Valladares G, Soares Correa A, Shepard BM, Rauf A, Murphy ST, Mujica N, MacVean C, Kroschel J, Kishinevsky M, Joshi RC, Johansen NS, Hallett RH, Civelek HS, Chen B, Blanco Metzler H (2017) The invasive *Liriomyza huidobrensis* (Diptera: Agromyzidae): understanding its pest status and management globally. J Insect Sci 17(1):1–27. https://doi.org/10.1093/jisesa/iew121

Workneh F, Henne DC, Childers AC, Paetzold L, Rush CM (2012) Assessments of the Edge Effect in Intensity of Potato Zebra Chip Disease. Plant Dis 96 (7):943–947

Xu R, Zhou J, Chen Z, Jia Y, Xu K (2013) Phototactic Behavior of *Henosepilachna vigintioctomaculata* Motschulsky (Coleoptera: Coccinellidae). The Coleopterists Bulletin 69(4):806–812

Yao XL, Li CQ, Gao ZL, Lu HM, Lan L, Liu H (1992) Preliminary observations on biology of potato ladybird. J Shaanxi Agric Sci 5(1):28–29

Zhou J, Xu R, Chen Z, Jia Y, Xu K (2015) Phototactic behavior of *Henosepilachna vigintioctomaculata* Motschulsky (Coleoptera: Coccinellidae). Coleopt Bull 69(4):806–812

Zuckerman L (1999) The potato: how the humble spud rescued the western world. North Point Press, New York

Chapter 9
Fungal, Oomycete, and Plasmodiophorid Diseases of Potato

Birgit Adolf, Jorge Andrade-Piedra, Francisco Bittara Molina, Jaroslaw Przetakiewicz, Hans Hausladen, Peter Kromann, Alison Lees, Hannele Lindqvist-Kreuze, Willmer Perez, and Gary A. Secor

Abstract This chapter discusses the major potato diseases worldwide: late blight, early blight, wart, and powdery scab. **Late blight**, caused by the oomycete *Phytophthora infestans*, continues to be the main biotic constraint of potato production. Annual losses have been estimated to be about €6.1 billion, with major consequences to food security, especially in developing countries. Symptoms of the disease can be seen in leaves (water-soaked light to dark brown spots), stems (brown spots), and tubers (slightly depressed areas with reddish-brown color). High humidity and mild temperatures are essential for disease development and, under optimal conditions, the disease can destroy a field in a few days. *Phytophthora infestans* evolves continuously, mainly through recombination and migration from other

B. Adolf · H. Hausladen
Technische Universität München, Freising, Germany
e-mail: b.adolf@lrz.tum.de; h.hausladen@lrz.tum.de

J. Andrade-Piedra · H. Lindqvist-Kreuze (✉) · W. Perez
International Potato Center, Lima, Peru
e-mail: j.andrade@cgiar.org; h.lindqvist-kreuze@cgiar.org

F. Bittara Molina
J. R. Simplot Company, Plant Sciences, Boise, ID, USA
e-mail: Francisco.BittaraMolina@simplot.com

J. Przetakiewicz
Plant Breeding and Acclimatization Institute, Radzikow, Blonie, Poland
e-mail: j.przetakiewicz@ihar.edu.pl

P. Kromann
International Potato Center, Nairobi, Kenya
e-mail: p.kromann@cgiar.org

A. Lees
The James Hutton Institute, Invergowrie, Dundee, UK
e-mail: alison.lees@hutton.ac.uk

G. A. Secor
Department of Plant Pathology, North Dakota State University, Fargo, ND, USA
e-mail: gary.secor@ndsu.edu

H. Campos, O. Ortiz (eds.), *The Potato Crop*,
https://doi.org/10.1007/978-3-030-28683-5_9

areas. Thus, monitoring of *P. infestans* populations is critical for the design of effective management strategies. Fungicides remain as the most common tactic for late blight management, but environmental considerations are increasing the pressure to use host resistance, sanitation, and other measures. New solutions being developed to manage late blight include, among others, smart phone-based decision support systems linked to portable molecular diagnostics kits that can disseminate disease information rapidly to a large number of farmers. Emerging research topics on *P. infestans* include the role of the pathogen–microbiota interaction in promotion or suppression of the disease, as well as the metabolism of *P. infestans*.

The fungus *Alternaria solani* is the main pathogen causing **early blight** on potatoes. Early blight can be found in most potato-growing countries. Typical symptoms on the leaves are dark brown to black spots with concentric rings (target spot). In susceptible potato cultivars in particular, as well as in locations (especially in warmer areas) with increased occurrence of *A. solani*, the disease can cause considerable yield losses. Integrated pest management to control early blight requires the implementation of several approaches. The disease is primarily controlled by the use of cultural practices (to reduce the soil born inoculum), less susceptible cultivars and the use of pesticides. But there is a loss in sensitivity toward two groups of fungicides described. The loss in sensitivity towards succinate dehydrogenase inhibitor (SDHI) and Quinone outside inhibitor (QoI) fungicides are caused by different point mutations. In many countries the occurrence of SDHI and QoI mutants is reported. Therefore, the control of early blight will be a considerable challenge in the future. The increasing importance of early blight in potatoes is due to a number of factors.

Synchytrium endobioticum is a soil-borne biotrophic fungus causing **potato wart** disease of cultivated potato. The fungus originates from the Andean zones of South America, from where it spread first to Europe in the late of the nineteenth century. Presently, the geographical distribution of this pathogen includes almost all European and Mediterranean Plant Protection Organization (EPPO) countries, Asia, North and South America as well as Oceania (New Zealand). The typical symptoms of cauliflower-like galls could develop on all meristematic tissues of potato except roots. *S. endobioticum* produces summer sporangia with mobile zoospores that can move in the soil. Winter (resting) sporangia are the dormant structures by which the fungus disperses to establish new infections. They can survive more than 40 years without plant hosts. The pathogen does not produce hyphae. Its long persistence in soil and the severe losses it inflicts to potato crops have prompted its inclusion into the A2 quarantine list of EPPO. Since the discovery of pathotype 2(G1) in Germany, more than 40 pathotypes were reported in Europe. In Europe, pathotypes 1(D1), 2(G1), 6(O1), and 18(T1) are the most relevant. Other pathotypes occur mainly in the rainy mountainous areas of central and eastern Europe. *S. endobioticum* is a still serious problem for crop production in countries with moderate climates. The strategies to confine the disease are strict quarantine and phytosanitary measures, and the cultivation of resistance cultivars of potato.

Spongospora subterranea causes root galling and tuber **powdery scab** leading to quality and yield losses in seed and ware crops worldwide and is also important as the natural vector of potato mop-top virus (PMTV), an economically important tuber blemish disease of potato. *S. subterranea* spreads by movement of infected seed tubers and soil and can survive long periods in soils and some asymptomatic hosts. Powdery scab is particularly favored by cool, damp conditions and is an intractable disease. Avoidance is the best control for powdery scab, but once soil is infested with *S. subterranea*, cultural practices and chemical treatments are ineffective control methods, and host resistance appears to be the most promising mechanism for long-term management. Although sources of host resistance have been identified, they are not widely deployed in practice. *S. subterranea* is an unculturable biotroph, making research difficult. Recent progress in understanding the biology of *S. subterranea* as a result of the application of basic molecular techniques, and future opportunities to further advance knowledge of this understudied pathogen, and the virus that it vectors, are included.

9.1 Late Blight

Hannele Lindqvist-Kreuze, Willmer Perez, Peter Kromann and Jorge Andrade-Piedra

9.1.1 Causal Organism

Late blight is caused by *Phytophthora infestans* Mont de Bary. It was previously classified as a fungus due to the superficial resemblance to filamentous fungi but is now classified as oomycete in the kingdom of stramenopiles (Kamoun et al. 2014). The vegetative stage of the mycelium in *P. infestans* is diploid, while in true fungi it is haploid. However, recent research has shown that in the modern-day lineages the progenies from sexual *P. infestans* populations are diploid, but the most important pandemic clonal lineages are triploid (Li et al. 2017). Virulence of oomycetes depends on large, rapidly evolving protein families including extracellular toxins, hydrolytic enzymes, and cell entering effectors that help the pathogen suppress the host plant defenses and gain nutrition from the host (Jiang and Tyler 2012). Elicitins are an example of structurally conserved extracellular proteins of *P. infestans* that have a function in the sequestration of sterols from the host plant, but can also act as pathogen-associated molecular patterns (PAMPs), and as such can activate PAMP triggered immunity (PTI) (Du et al. 2015). *P. infestans* secretes large numbers of effectors: apoplastic effectors, such as EPIC1, interact with the host cell wall, host proteases and other defense-related molecules in the host extracellular space, while cytoplasmic effectors, the RxLR proteins, and CRNs (crinkling and necrosis-induced proteins) function inside the plant cells (reviewed in Whisson et al. 2016). RxLR effectors act as activators of plant immunity, resulting in effector triggered immunity (ETI) (Oh et al. 2009; Wang et al. 2017a), while the apoplastic effectors, similarly to elicitins, act as activators of the PTI (Domazakis et al. 2017). Recent research has also shown that some *P. infestans* effectors can target host proteins whose activity enhances susceptibility possibly through the inhibition of positive regulators of immunity or promote the activity of susceptibility (S), that in turn can negatively regulate immunity (Boevink et al. 2016). The effector genes locate mostly in the gene sparse regions of the genome, that are rich in repetitive sequences and are rapidly evolving, probably enabling the evolutionary arms race between *P. infestans* and the host plant (Haas et al. 2009; Dong et al. 2015).

9.1.2 Symptoms

The asexual, aerially dispersed sporangia (Fig. 9.1) are responsible for most of the devastating epidemics on potato. When the sporangia lands on a plant surface it can germinate directly or first form zoospores, which encyst, germinate, and penetrate the host tissue (reviewed by Fry et al. 2015). This stage of infection is unnoticeable to the naked eye, but inside the plant cell a repertoire of molecular interactions takes place. After penetration and adhesion, the pathogen forms haustoria inside the plant cells, from where it secretes effector proteins (reviewed in Nowicki et al. 2012; Whisson et al. 2016; Wang et al. 2017a). At this biotrophic stage *P. infestans* requires living cells to obtain nutrients.

The first visible symptoms appear within 2–3 days when the pathogen switches to the necrotrophic stage. In leaves, lesions are light to dark brown in color, water-soaked, irregularly shaped, sometimes surrounded by a yellow halo and not limited by leaf veins. Symptoms typically begin to develop where water accumulates near the leaf edges or tips (Fig. 9.2) and in stems (Fig. 9.3) near petioles. Affected tubers show irregular, slightly depressed areas with brown color. In a cross-section, finger-like extensions can be seen from the external surface to the tuber medulla (Fig. 9.4) (Perez and Forbes 2010).

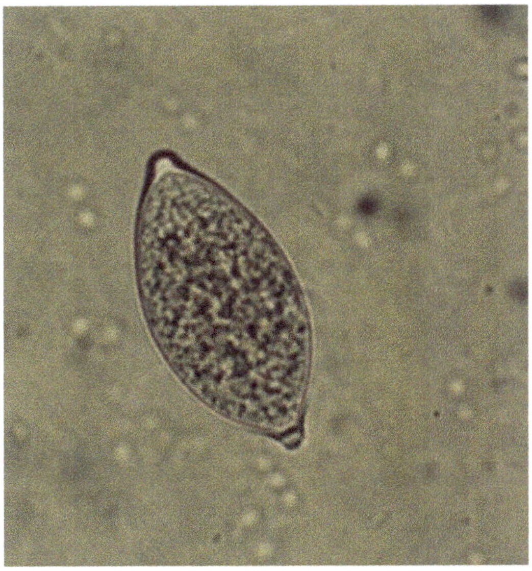

Fig. 9.1 Lemon-shaped sporangium of *Phytophthora infestans*

Fig. 9.2 Symptoms of late blight on leaves

Fig. 9.3 Symptoms of late blight on stems

Fig. 9.4 External and internal symptoms of late blight in tubers

The dying, necrotized cells serve as nutrient for the pathogen and under high humidity, a white mildew growth is formed on the underside of the leaves, which represents the sporangiophores and sporangia that emerge through the stomata (Nowicki et al. 2012). During spore formation and germination, large numbers of genes involved in pathogenesis, calcium signaling, and metabolism are upregulated or transcribed in waves (Ah-Fong et al. 2017), whereas genes of the fatty acid biosynthesis pathway are downregulated (Rodenburg et al. 2018). Identification of the proteins and enzymes essential for pathogen growth and development, and linked to the different symptomatic reactions in the plant, can lead to the discovery of potential targets for crop protection chemicals.

High level of moisture is essential for the lesion development, and under optimal conditions the disease can advance very fast and destroy the plant in matter of days (Perez and Forbes 2010).

9.1.3 Impact

The potential economic and social impact of potato late blight is best illustrated by the well-publicized role it played in the Irish Famine in the middle of the nineteenth century. Because of the famine, millions of Irish died or emigrated (Bourke 1993). Other devastating late blight outbreaks have been reported around the world, causing food insecurity, hunger (International Potato Center 2007), and oftentimes crippling the local potato industry. Haverkort et al. (2009) estimated that the global costs and losses due to late blight may take 16% of all global potato production. At 100 €/t the world potato production represents a value of €38 billion today. The 16% loss then represents an annual financial loss of €6.1 billion per annum today, considering that the increase in global potato production in the last decade has mainly been in developing countries, which suffer low yields and the vast majority of these estimated losses to late blight compared to developed countries.

9.1.4 Resistance to Late Blight

After the discovery of the Mexican wild species *Solanum demissum* as an excellent source of resistance, eleven major genes were introduced in cultivated tetraploid potato breeding lines (Black et al. 1953; Malcolmson and Black 1966). Although some of these genes can be considered defeated, others, for example R8, are still effective against current pathogen populations (Vossen et al. 2016). Over 50 R genes have been identified from wild Solanum species as detailed by Rodewald and Trognitz (2013), and the research field remains active with a growing list of genes available for potato breeding programs (Jo et al. 2015; Vossen et al. 2016; Witek et al. 2016; Yang et al. 2017). However, due to crossing barriers and linkage drag, there are only few successful cases where R genes have been introduced into improved tetraploid breeding lines by classical breeding (Bethke et al. 2017). Introduction of a single R gene from wild germplasm is a lengthy procedure as demonstrated by the examples of commercial varieties Bionica and Toluca that contain Rpi-blb2 originating from *S. bulbocastanum*, and were released almost 50 years after the first crosses were made (Haverkort et al. 2016). Genetic engineering bears promise and varieties containing stacked or single R genes are in the process of being released in the markets that accept this technology (Haverkort et al. 2016; Schiek et al. 2016; Pacifico and Paris 2016; Ghislain et al. 2018). As opposed to the major R genes, quantitative resistance was generally expected to be governed by many minor genes. However, recently it was shown that R genes can also have quantitative effects. The potato cultivar Sarpo Mira contains at least four R genes that confer complete resistance against incompatible isolates and a quantitative R gene, Rpi-Smira1 that confers broad-spectrum field resistance (Rietman et al. 2012). A biparental cross using a haploidized resistant clone from the CIP B3 population was used to locate a strong QTL in chromosome 9 (Li et al. 2012a). Subsequent association mapping confirmed the importance of the same genome region for late blight resistance in the tetraploid B3 breeding population (Lindqvist-Kreuze et al. 2014), and recently the R8 gene was identified in the QTL by dRenseq (Jiang et al. 2018).

Identification of new resistance sources and functional resistance or susceptibility genes has been recently greatly accelerated by modern techniques, such as effectoromics and resistance gene enrichment sequencing technologies. To date, all effector proteins identified that are recognized by the plant resistance (R) proteins belong to the RXLR category. Therefore, RXLR effectors cloned into expression vectors have been used to successfully identify functional new R genes from potato germplasm using agroinfection (Vleeshouwers et al. 2011; Vleeshouwers and Oliver 2014). Apoplastic effectors are recognized by pathogen recognition receptors (PRR) and can be used in a similar manner to identify resistance germplasm (Domazakis et al. 2017). In the resistant germplasm, the NB-LRR (nucleotide binding-site leucine-rich repeat) resistance genes can be rapidly identified and cloned using gene-targeted, resistance gene enrichment and sequencing method (Jupe et al. 2013; Witek et al. 2016).

Durability of quantitative resistance will, however, continue to depend on the size of the cultivation area of a variety as well as the dynamics of the pathogen population.

9.1.5 Phytophthora infestans *Populations*

Knowledge on the local pathogen population structure is important for the design of impactful disease management actions (Fry et al. 2015). In recent years, the initiatives EuroBlight (http://euroblight.net/), USABlight (http://www.usablight.org/), and TizonLatino (https://tizonlatino.github.io/) have been carrying out is monitoring of *P. infestans* populations. This work has confirmed that *P. infestans* populations are constantly evolving and novel usually more aggressive genotypes appear periodically replacing the previously dominating genotypes. New genotypes can emerge through divergence from other genotypes, through recombination, or migration from other areas (Knaus et al. 2016). The main mode of reproduction of *P. infestans* is asexual and variable numbers of clonal lineages exist in different countries and regions. Several studies have confirmed that appearance of new genotypes can often be attributed to migration (Fry et al. 2015; Knaus et al. 2016; Saville et al. 2016). Until recently the mating type A1 was dominating worldwide, except in the presumed center of origin, Mexico, where both mating types were found in similar frequencies (Goodwin et al. 1992). This situation has changed dramatically, and A2 has now been reported in Scandinavia and Estonia (Hermansen et al. 2000; Runno-Paurson et al. 2016; Montes et al. 2016), Central Europe (Flier et al. 2007; Li et al. 2012b; Mariette et al. 2016), China (Zhu et al. 2015), Bolivia, Argentina, Uruguay and Brazil (Plata 1998; Deahl et al. 2003; Forbes et al. 1998; Casa-Coila et al. 2017), the USA (Rojas and Kirk 2016), Tunisia (Harbaoui et al. 2014), Algeria (Rekad et al. 2017), India (Chowdappa et al. 2015), and Canada (Danies et al. 2014). However, even though both mating types are present in most cases, no evidence of frequent sexual reproduction has been found, suggesting that the sexual populations are ephemeral (Fry et al. 2015). There are notable exceptions however, such as the Nordic countries, where it was shown that the sexual reproduction in the field is frequent and the oospores surviving in the field over winter in plant debris has led to earlier onset of epidemics (e.g., Widmark et al. 2007). The diversity of *P. infestans* in South and Central America is a particularly interesting question, because these regions are extremely rich in biodiversity of Solanaceous species that are potential alternative hosts of this pathogen and thus can harbor divergent genotypes. Furthermore, the centers of origin of the economically most important hosts, potato and tomato, are there. Interestingly, in South America, no sexual reproduction of *P. infestans* has been reported, and populations maintain strictly clonal structures. In Colombia, Chile, Ecuador, and Peru the A1 mating type has been found mostly (Acuna et al. 2012; Perez et al. 2001; Forbes et al. 1997; Cardenas et al. 2011). In Mexico, in contrast, recombination is frequent and the population is extremely divergent with subdivisions associated with geographic regions (Wang et al. 2017b). Mexico was also shown to be the origin of the current genotypes found in South America and continues to play an important role as the source population of the newly emerged aggressive genotypes in the USA (Saville et al. 2016; Goss et al. 2014). Although *P. infestans* is generally heterothallic requiring two different mating types to form sexual oospores (Fig. 9.5), some isolates are homothallic. Recent studies have shown that these self-fertile isolates are found more frequently, constituting a new threat to potato and tomato crops because of their increased genotypic variability, better fitness, and greater aggressiveness (Zhu et al. 2016; Casa-Coila et al. 2017).

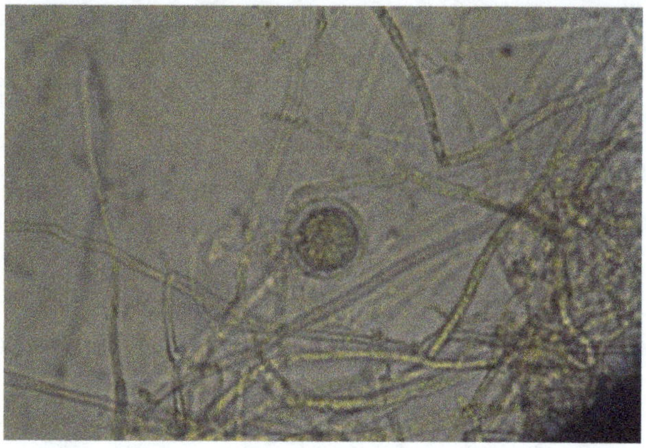

Fig. 9.5 Oospore of *Phytophthora infestans*

9.1.6 Management

Late blight of potatoes can be suppressed by a combination of approaches. As a polycyclic disease that explodes under favorable conditions, integrated strategies are crucial. These include sanitation measures that eliminate or reduce initial sources of the disease (e.g. infected seed, cull piles, infected neighboring plots, and volunteers), prophylactic fungicide sprays before the appearance of symptoms, curative fungicide sprays and use of resistant cultivars to reduce the rate of disease development, use of early-maturing cultivars to reduce the duration of the epidemic, or planting the crop in seasons or areas where the environment is not favorable for the pathogen.

The application of chemical fungicides continues to be the most common strategy for late blight control, making late blight one of the top drivers for pesticide use in the world. The demand for weekly applications generates a billion-dollar business globally every year (Haverkort et al. 2009). To optimize the use of fungicides, it is important to know the efficacy and type of activity of the active ingredients. The frequency and timing of fungicide applications may depend on the foliar resistance of the cultivar, fungicide characteristics, rate of growth of new foliage, weather conditions, irrigation, and incidence of blight in the region (Cooke et al. 2011). The range of fungicide types available to farmers vary depending on the numbers of products registered and commercialized in their area. In Europe the number of registered products is being reduced due to health and environmental concerns and in some European countries farmers have access to less than ten fungicide products for late blight control (http://www.endure-network.eu). In other countries the process of registration of pesticides is less restricted. For example, in Ecuador, hundreds of fungicide products are registered for late blight control based on more than 30 active ingredients of which most contain old generic substances like mancozeb, cymoxanil, and carbendazim (http://www.agrocalidad.gob.ec), yet in other countries only a couple or so products are readily available to farmers due to trade limitations.

The most efficient and arguably the most elegant strategy to control late blight is the use of host resistance. Today, it is well known that with the use of genetic resistance late blight can be controlled with less fungicide either by lowering the fungicide dose or using longer application intervals (Kirk et al. 2005; Nærstad et al. 2007; Cooke et al. 2011; Liljeroth et al. 2016; Haverkort et al. 2016). The use of resistant varieties could sharply reduce losses from late blight, especially in developing countries, where disease is less well managed for many reasons (e.g. high disease pressure, problems of access to fungicides, and inadequate farmer knowledge of disease dynamics). Nevertheless, the use of resistant varieties continues to be an uncommon disease management approach as susceptible varieties are promoted and required by many wholesalers and processing industries, leaving farmers with little option but to grow susceptible varieties (Forbes 2012). Integrating genetic resistance and chemical control helps in reducing the use of fungicides, decreases production costs, and reduces damage to human health and the environment (Perez and Forbes 2010; Cooke et al. 2011). One way of achieving better design and integration of management elements is through the use of simulation models (reviewed by Forbes et al. 2008) and especially decision support systems (DSS). These typically integrate and organize all available information on the life cycle of *P. infestans*, monitoring of inoculum, weather (historical and forecast), plant growth, fungicide characteristics, cultivar resistance, and thereby predict disease pressure and action thresholds that can guide decision-making. Based on the information provided by the DSS farmers can make informed disease management decisions. DSS can deliver general or very site-specific information to the users via extension officers, telephone, SMS, e-mail, and websites (Cooke et al. 2011). In the case of smallholders in developing countries, basic information to understand the disease is critical to improve management (Nelson et al. 2001; Andrade-Piedra et al. 2009; Ortiz et al. 2019).

Cultural control involves all the activities carried out during agronomic management which alter the microclimate, host condition, and pathogen behavior in such a way that they avoid or reduce pathogen activity (survival, dispersal, and reproduction) (Garrett and Dendy 2001). Among them are the elimination of volunteers and cull piles and associated debris, use of clean seed potatoes preferably certified seed, use of resistant varieties, adequate space between rows and plants, rotation with other crops not susceptible to late blight, adequate hilling, harvest in dry conditions and when the tubes are mature (Garrett and Dendy 2001; Perez and Forbes 2010). Mixtures of potato varieties (resistant and susceptible) offer partial improvement on disease suppression (Phillips et al. 2005; Pilet et al. 2006). Under temperate climate conditions and where tuber infections are a concern, potato vines are typically killed by applying chemical desiccants 2–3 weeks before harvest (Perez and Forbes 2010). Biological control consists of reducing disease through the interaction of one or more live organisms with the disease-causing pathogen or use of extract of plants. Some findings report the use of *Trichoderma* isolates (Yao et al. 2016), *Chaetomium globosum* (Shanthiyaa et al. 2013), *Trichoderma viride* and *Pennicillium viridicatum* (Gupta 2016) and bacteria from the genera *Bacillus, Pseudomonas*, *Rahnella,* and *Serratia* (Daayf et al. 2003) as biocontrol agents in the management of late blight disease in potato. Garlic has, for example, been suggested as a potential intercropping plant for the management of potato late blight disease under Ethiopian condition (Kassa and Sommartya 2006). Still, few biological control measures are used by nonorganic growers due to low efficacy and farmers' lack of knowledge about these options and access to the most efficient products.

9.1.7 Looking Forward

Phytophthora infestans has proven capable of overcoming fungicides and resistant varieties through decades. Potato late blight, thus, continues to be the main potato constraint worldwide despite huge investments in its management. The growing intercontinental trade in potato is also increasing the risk of worldwide dissemination of dominant *P. infestans* strains. The threat from the disease will without doubt continue in the future. Fortunately, encouraging solutions to improve its management are arising from new advances in molecular, sensor, computational, and smartphone technologies. Efficient inoculum monitoring tools are becoming more accessible that can indicate whether *P. infestans* is in an area for the guidance of the initial fungicide spray and *P. infestans* population's movements (Fall et al. 2015). These could even be connected to molecular diagnostics that can predict phenotypic traits of the pathogen population, such as fungicide sensitivity and R-gene interaction, and perhaps even predictions of aggressiveness and fitness in a matter of hours. Spore traps could eventually be mounted on tractors or drones allowing for real-time spatial monitoring (Fall et al. 2015; Fry 2016). New high-throughput methods for monitoring and assessing plant populations are being developed, such as remote sensing, image processing, and web-based farmer-extension service networks. Such indirect disease detection approaches are improving disease surveillance significantly with high potential in remote and low-input production systems. Smartphone-based extension systems linked to rapid and portable molecular diagnostics kits (e.g. Loop-mediated isothermal amplification—LAMP—and lateral flow immunodiagnostics tools) can support immediate dissemination of disease information to a large number of farmers. If made widely available such smartphone tools, handheld portable diagnostics kits and novel weather sensors could lead to precision management and significant impacts on production capacity. Farmer-extension service networks based on internet connectivity are already important options that improve disease surveillance and supplying fast access to appropriate and up-to-date knowledge on pathogen distribution and management (Fry 2016). Capacity building (for farmers and extension services) to improve disease management skills, coupled to early-maturing and resistant cultivars, novel diagnostic tools, improved DSSs, and low-toxicity fungicides have the potential to reduce crop loss, management costs, and environmental impact, and even more so as biological options become accessible to farmers.

Current research opportunities in late blight management focus on *P. infestans* studies, fungicide testing, dissemination of resistant cultivars and validation of DSSs and training. In addition to continued research and extension efforts, alliances with the agro-chemical industry seem to be necessary to fully achieve integrated pest management strategies (Pacilly et al. 2016). This is being promoted through local and regional research networks that via collaboration strengthen institutional capacity in research and extension related to late blight (see http://euroblight.net/). Emerging research topics on *P. infestans* include the role of the pathogen–microbiota interaction in promotion or suppression of the disease

(Larousse and Galiana 2017) as well as the metabolism of *P. infestans* (Judelson 2017), its effector repertoire on the plant, and, *in fine*, how it promotes or suppresses the disease. At the same time, *P. infestans* as an infectious entity is no longer only considered at the species or lineage level but also at the level of a resident microbiota or part thereof (Larousse and Galiana 2017).

9.2 Early Blight

Hans Hausladen and Birgit Adolf

9.2.1 Symptoms

Foliar symptoms of early blight (EB) are dark brown to black necrosis. First foliar symptoms become visible on the lowest and therefore oldest leaves just a few weeks after emergence. Initially, dark brown dot-like blotches appear, a few millimeters in diameter. The necrotic area gradually increases and the leaf symptoms grow to take up the whole of the green leaf tissue (Fig. 9.6). Often, the lesions are restricted by leaf veins and take on an angular shape. The size of the necrosis can vary in width, from a few millimeters to 2 cm. Within larger lesions, a series of dark concentric rings are visible. This target pattern is typical of EB symptoms. Subsequently, the necrotic leaf tissue is often surrounded by a chlorotic border caused by fungal mycotoxins, which turn the leaf tissue yellow. The chlorosis can extend to the whole infected leaf. During EB progression the infected areas enlarge, and the whole leaf becomes necrotic and falls off. In Europe, a heavy increase of EB infestation occurs from mid-July onwards, especially during hot and dry weather or when the potato crop is under stress, and on physiologically older plants. EB then starts spreading from the lower leaves to the middle and finally upper leaf levels.

Alternaria conidia which are washed of the leaves can also infect tubers. The symptoms of EB on tubers are dark, slightly sunken lesions (Fig. 9.7). The dry or

Fig. 9.6 Initial (left) and advanced (right) symptoms of potato early blight in foliage

Fig. 9.7 Symptoms of potato early blight in tubers

hard rot of tubers causes storage losses, reduces quality of table potatoes, and reduces germination capacity of seed potatoes.

It isn't possible to distinguish between the different *Alternaria* spp. causing EB based on the symptoms even though sometimes symptoms referred to as brown leaf spot (small, irregular to circular, dark brown spots ranging in size from a pinpoint to 4 mm) are attributed to *A. alternata* (Fairchild et al. 2013).

9.2.2 Causal Organism

The main causal agent of early blight on solanaceous crops is generally considered to be *Alternaria solani* Sorauer (Gannibal et al. 2014). However, there are reports of other large-spored *Alternaria* spp. involved in EB of potato. Rodrigues et al. (2010) found that *A. grandis* Simmons, but not *A. solani*, was the causal agent of EB affecting potato plants in several growing regions in Brazil, and Duarte et al. (2014) confirmed that this species can cause EB on potato in field trials using artificial inoculation with *A. grandis*. *Alternaria protenta* has been detected as the causal *Alternaria* spp. for EB in Algeria (Ayad et al. 2017) and was, together with *A. grandis* and *A. solani*, found to be part of the complex of *Alternaria* spp. detected on EB lesions in Belgium (Landschoot et al. 2017a). Of the small-spored *Alternaria* spp., *A. alternata* occurs on EB lesions on a regular basis, but is considered as a secondary invader (Leiminger et al. 2014; Stammler et al. 2014; Rotem 1994).

Alternaria solani overwinters as mycelium, chlamydiospores or conidia in the soil and infested plant debris (see disease cycle in Fig. 9.8). In spring, the primary infection occurs through inoculum (conidia) carried to the lower leaves by rain splashes. The pathogen is able to penetrate the leaf tissue directly through the intact epidermis or through stomata and wounds. First symptoms of EB on leaves become visible 2–4 weeks after the emergence of the potato crop. Initially, the older leaves closer to the ground are infested. The fungus is restricted to the lower leaf level for several weeks and seems initially to be of no concern. Conidia formation occurs on

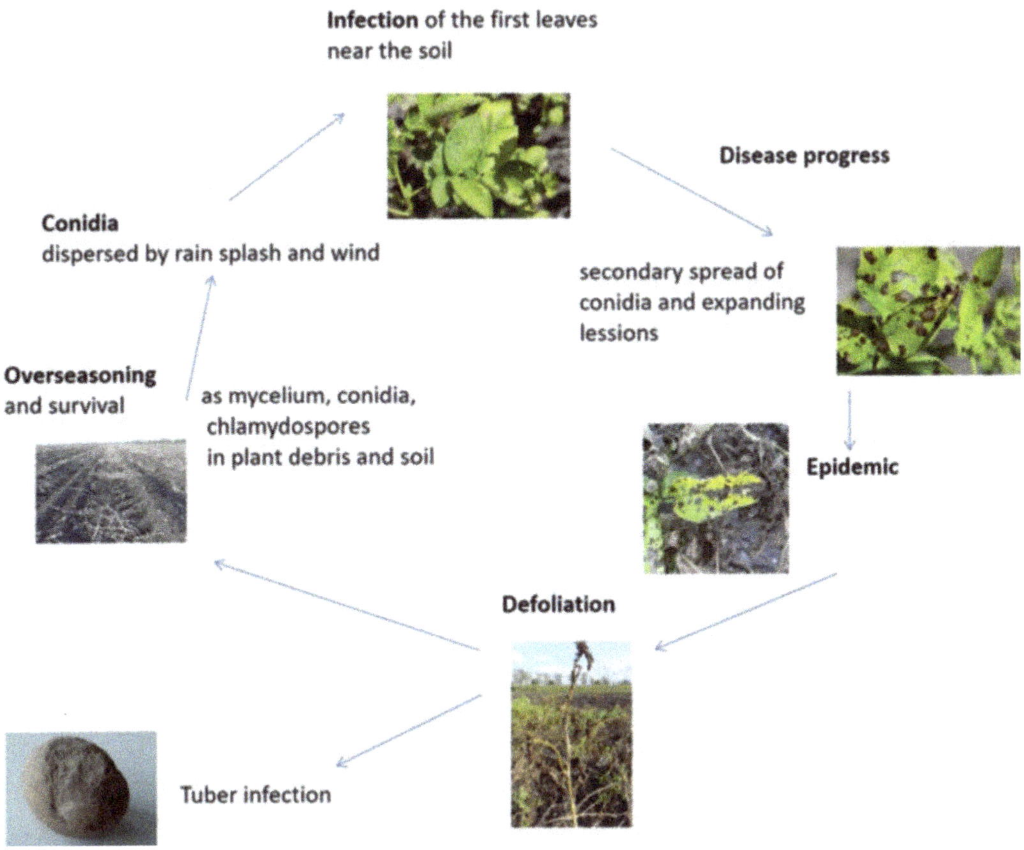

Fig. 9.8 Disease cycle of potato early blight

the necrotic leaf tissue at temperatures between 5 and 30 °C (optimum 20 °C). This secondary inoculum is disseminated by wind and causes infections of the surrounding leaves and stems. The latency period averages 3–7 days. During favorable infection conditions, and once the plant has got to a certain age, *A. solani* very rapidly colonizes the middle and upper leaf layers. In the field, a cascade-like progression of the fungus from the lower, via the middle, to the upper leaves is visible. Heavily infested leaves fall off and remain as inoculum source on and in the soil.

The disease progression of EB depends on weather parameters, plant age/crop growth stage, maturity group, susceptibility of the cultivar and inoculum concentration, the latter being influenced by short crop rotations.

Favorable weather conditions for infection with *A. solani* are temperatures above 22 °C and leaf wetness for more than 8 h. Furthermore, alternating wet and dry periods favor disease development.

An important factor for disease progression is the crop growth stage. In addition to the cascade-like spread within plants upwards from older to younger leaves, older plants are significantly more susceptible due to the earlier planting emergence time.

Tuber infestation is caused by conidia that are washed into the soil. The fungus can enter the tuber tissue through lenticels or mechanical injuries of the skin during harvest.

EB can also occur on other host plants apart from potatoes. It has been observed on hosts such as tomato (*Solanum lycopersicum* L.), eggplants (*S. melongena* L.), hairy nightshade (*S. sarrachoides* Sendt), black nightshade (*S. nigrum* L), horse nettle (*S. carolinense* L.), pepper (*Capsicum* spp.), and non-solanaceous weeds (Jones et al. 1993; Pscheidt 1985; Hausladen and Aselmeyer 2017).

Differentiation of *Alternaria* species within the large-spored or the small-spored group based on morphological traits is time-consuming and requires experience. The molecular approach to delineate large-spored isolates fast and precise is a multi-locus analysis based on different partial gene regions, like the internal transcribed spacer regions 1 and 2 and intervening 5.8S nrDNA (ITS), glyceraldehyde-3-phosphate dehydrogenase (GAPDH), RNA polymerase second largest subunit (RPB2), translation elongation factor 1-alpha (TEF1), the Alternaria major allergen gene (Alt a 1) (Woudenberg et al. 2014). Based on the RPB2 and calmodulin gene sequence *A. solani* can be distinguished from A. grandis and A. protenta (Landschoot et al. 2017a). For the small-spored species the glyceraldehyde-3-phosphate dehydrogenase (gapdh), RNA polymerase second largest subunit (rpb2), translation elongation factor 1-alpha (tef1), Alternaria major allergen gene (Alt a 1), endopolygalacturonase (endoPG), an anonymous gene region (OPA10-2), and the histone H3 gene can be used for differentiation (Woudenberg et al. 2015; Landschoot et al. 2017a).

9.2.3 Impact

After potato late blight (*Phytophthora infestans*), EB represents one of the most important fungal diseases of potato today. It can be found in most potato-growing countries (Woudenberg et al. 2014), but *A. solani* is described as a very important pathogen especially in warmer areas due to the requirement for higher temperatures. In susceptible potato cultivars in particular, as well as in locations with increased occurrence of *A. solani*, the disease can cause considerable yield losses. The potential EB-induced damage for different countries is shown in Table 9.1.

The necrosis of leaf tissue and considerable defoliation caused by *Alternaria* infestation reduces the assimilation area of the potato plant, and therefore has a

Table 9.1 Yield loss due to early blight in different regions

Country	Yield loss
Australia	>20% (Horsfield et al. 2010)
Brazil	up to 58% (Campo Arana et al. 2007)
Germany	2– >40% (Leiminger and Hausladen 2014)
Poland	6–34% (Kapsa 2004)
Israel	up to 24% (Shtienberg et al. 1996)
United States	18–39% (Harrison and Venette 1970)
South Africa	20–50% (Van der Waals et al. 2001)

negative effect on tuber size and starch content. *Alternaria solani* is also able to infect tubers. The subsequent hard rot of the tubers leads to reduced quality for marketing as table and processing potatoes. In Europe, the occurrence of tuber infections is known only in exceptional years.

9.2.4 Pathogen Populations

The genetic diversity among *A. solani* isolates is relatively high for an asexually reproducing fungus. Furthermore, when analyzed by virulence assays (VC), random amplified microsatellites (RAMS), random amplified polymorphic DNA (RAPD), and AFLP marker techniques, there is no clear clustering of isolates according to geographical origin, year, or even deriving from the same field (Van der Waals et al. 2004; Leiminger et al. 2013; Odilbekov et al. 2016). Polymorphism exists even within the usually strongly conserved mitochondrial DNA. Concerning the cytb gene, Leiminger et al. (2014) detected two genotypes in German *A. solani* populations which differed in their intron–exon structure. The occurrence of the two genotypes was confirmed for *A. solani* populations in the USA (Bauske et al. 2018), Belgium (Landschoot et al. 2017b), and Sweden (Odilbekov et al. 2016).

The registration of *Alternaria*-specific fungicides with single site modes of action for potato changed the population structure of *A. solani* in many countries due to the occurrence of point mutations, leading to reduced sensitivity of mutated isolates. For Quinone outside inhibitor fungicides (QoIs) like Azoxystrobin, the change of amino acid phenylalanine to leucine at position 129 in the cytb gene (F129L) has been demonstrated to be the reason for reduced sensitivity compared to wild-type isolates (Pasche et al. 2005). The presence of the F129L mutation in *A. solani* populations has been shown in different countries such as the USA (Pasche et al. 2004), Germany (Leiminger et al. 2014), Sweden (Odilbekov et al. 2016), and Belgium (Landschoot et al. 2017b). The loss in sensitivity towards succinate dehydrogenase inhibitor fungicides (SDHIs) like Boscalid can be caused by different point mutations, as the SDH is composed of four subunits and the binding site of the fungicides is formed by three of them (subunits B, C, D). For *A. solani* populations, five possible mutations have been described so far: H278Y and H278R in subunit B, H134R and H134Q in subunit C, and D123R and H133R in subunit D (Mallik et al. 2014; Metz et al. 2019). Landschoot et al. (2017b) describe the presence of two different SDH mutations in one isolate. Frequently, isolates carry both the F129L mutation in the cytb gene and one of the SDH mutations (Landschoot et al. 2017b).

The sensitivity of baseline isolates to demethylation inhibiting fungicides (DMIs) can vary substantially, but nonbaseline isolates remain sensitive, whereas a distinct loss in sensitivity to anilino-pyrimidine (AP) fungicide pyrimethanil exists. The primary mode of action has not yet been discovered for AP chemistries; therefore the resistance mechanism is currently unknown for this established fungicide group (Fonseka and Gudmestad 2016).

Resistance against QoIs in *A. alternata* populations is caused by the G143A mutation in the cytb gene (Ma et al. 2003). Mutations conferring reduced sensitivity to SDHIs for this Alternaria species are the H277R/Y in SDH subunit B, H134R in subunit C, and D123R and H133R in subunit D (Avenot et al. 2008, 2009). Fairchild et al. (2013) detected isolates resistant to pyrimethanil and Fonseka and Gudmestad (2016) stated high DMI EC50 values in baseline isolates.

9.2.5 Management

Integrated pest management to control early blight requires the implementation of several approaches. The disease is primarily controlled by the use of cultural practices (to reduce the soil born inoculum), less susceptible cultivars and the use of pesticides.

Phytosanitary aspects One of the main components in this case is the crop rotation, which influences the occurrence of early blight. The fungus *A. solani* persists as mycelium or spores in plant debris or soil in the field from one potato-growing season to the next. Therefore crop rotation, including the control of host plants such as weeds (black shadow) in the nonhost crops, reduces the initial soil born inoculum. A short crop rotation with host crops (tomato, potato) results in an earlier and more severe early blight epidemic (Shtienberg and Fry 1990). In addition, the removal or burning of infected plant debris reduces the inoculum level.

A further option to reduce the primary inoculum in the soil is biofumigation. Biofumigation means the suppression of soilborne pathogens by isothiocyanates (ITCs), which derive from hydrolization of glucosinolates by myrosinase in disrupted plant cells. Biofumigant plants (e.g. white mustard, leaf radish) can reduce the early blight disease progression in the crop (Volz et al. 2013).

Tuber harvest and storage *Alternaria solani* also can infect the potato tuber. The fungus cannot infect through the intact periderm, and so the risk of tuber infection can be reduced by allowing tubers to fully mature before harvest. Avoiding wounding at harvest and providing storage conditions to promote wound healing can also reduce tuber infection (Venette and Harrison 1973).

Pathogen-free seed The use of disease- and virus-free seed potatoes is the basis for an economical potato production. Virus-infected potato plants are more susceptible to early blight than healthy plants.

Biotic and abiotic stress Potato plants stressed by biotic or abiotic factors are more susceptible to early blight disease compared to nonstressed plants. There are different types of abiotic stress for plants during the growing season, such as drought, high temperature, and overhead irrigation. Additionally, overhead irrigation can prolong the leaf wetness period, allowing successful fungal infection.

Biotic stress is driven by insects (e.g. aphids, Colorado beetle), which are also known to be virus vectors.

Plant nutrition A balanced nutrition for the potato plants during the growing period is the basis for optimal plant growth and potato yield. Ideal soil fertility and plant nutrition can decrease the severity of *A. solani* (Lambert et al. 2005; MacDonald et al. 2007). Under specific conditions, such as drought when plants are unable to get enough nutrients from the soil through the roots, a foliar fertilizer can reduce the nutrient deficiency and reduce plant susceptibility to early blight.

The fertilizer form can also influence the disease progression of *A. solani*. The use of calcium cyanamide results in a delay of early blight disease, as the fungicidal side effects of degradation products of calcium cyanamide can reduce the initial inoculum in the soil (Volz et al. 2013).

Resistant cultivars Cultivars with reduced susceptibility to early blight are available; however, no completely resistant genotypes have been found so far. The observed field resistance of varieties to foliage infection is associated with plant maturity. Early maturing cultivars are in general more susceptible, and late maturing cultivars are more resistant to *A. solani* (Johanson and Thurston 1990; Abuley et al. 2017). There is no correlation between maturity group and the occurrence of the first early blight symptoms on the leaves, but a strong correlation between maturity group and disease progression. The epidemic in early maturing cultivars starts earlier. Interestingly, there are varieties within a maturity group which are more resistant to early blight (Johanson and Thurston 1990; Leiminger and Hausladen 2014). Overall, there is a possibility of influencing disease progression by planting more resistant cultivars.

Use of fungicides The most common method for controlling early blight in potatoes is the use of chemical pesticides. Some fungicides which are used for the control of late blight (Phytophthora infestans) also have some effect on Alternaria solani (e.g. maneb, mancozeb, chlorothalonil, triphenyl tin hydroxide). The most effective fungicides for control of early blight contain active ingredients from the strobilurins group, and azols.

Strobilurins, also known as Quinone outside inhibitors (QoIs), are an important class of fungicides in agriculture because they have a broad-spectrum activity. They inhibit mitochondrial respiration in fungi by binding to the Qo site of the cytochrome b (cytb) complex, blocking electron transfer and inhibiting ATP synthesis (Bartlett et al. 2002).

Carboxamides (SDHI) inhibit the enzyme succinate dehydrogenase (Sdh), a component of complex II in the mitochondrial electron transport chain (Kuhn 1984). Despite the two groups of fungicides having a similar mode of action, SDHI and QoI fungicides show no cross-resistance.

The triazol group belongs to the DMI fungicides, which inhibit one specific enzyme, C14-demethylase. This enzyme plays an important role in ergosterol biosynthesis. Ergosterol is important for functional membrane structure and for the development of functional cell walls.

Due to the single site mode of action of strobilurins (QoI) and carboxamides (SDHI), these fungicides have a high risk of development of resistance. In several potato-growing areas, mutants are found which show a reduced sensitivity in in vitro and in vivo trials and also a reduced efficacy in the field (see Pathogen populations).

Most of these fungicides have a very limited curative activity and should be used preventively. EB control is mainly achieved by multiple and frequent application of protectant fungicides. The optimization of fungicide usage for the control of early blight is a considerable challenge. The fungus produces huge amounts of secondary inoculum during the growing season. Therefore different DSS (decision support systems) are available to optimize the use of fungicide applications for the management of early blight in potato. One possibility is to use threshold values based on the disease progress (Leiminger and Hausladen 2012). In some countries, disease management is based on interactive computer-based systems dealing with forecasting favorable weather conditions for infection by Alternaria solani or temperature degree-day thresholds. Alternatively, the recommendation for the use of fungicides is based on plant development (plant size, plant age, onset of potato flowering).

Overall, the combination of plant age and host resistance, disease progress, and calculated weather-based infection risk is used as the basis for an integrated pest management.

9.2.6 Looking Forward

The control of early blight will be a considerable challenge in the future. The increasing importance of early blight in potatoes is due to a number of factors. Climatic change and global warming will result in more conducive conditions for the infection, growth, and disease progress of the fungus in several potato-growing areas. Increasing temperatures during the growing season result in favorable conditions for the pathogen, due to its requirement for higher temperatures. More abiotic stress (drought) increases the susceptibility of the plant, and will therefore also promote disease progress.

In addition, integrated pest management based on continuous pesticide treatments will become more and more ineffective. The occurrence of mutants towards different modes of action will result in reduced efficiency in the control of EB by fungicides. Therefore in the future, successful control of EB requires the implementation of cultural practices, the use of pesticides with a focus on fungicide resistance development, and the cultivation of less susceptible cultivars.

9.3 Wart

Jaroslaw Przetakiewicz

9.3.1 Causal Organism

The pathogen causing potato wart disease, *Synchytrium endobioticum* (Schilb.) Perc., was first discovered by Schilberszky in Hungary. In fact, the pathogen was known firstly in Europe. In 1876, potato wart disease was found for the first time in the UK (Hampson 1993; Flath et al. 2014). In the older classification this species has been included to Protista Kingdom. Nowadays, *Synchytrium endobioticum* (Schilb.) Perc., belongs to Fungi Kingdom, phylum Chytridiomycota, order Chytridiales, family Synchytriaceae, genus *Synchytrium*, and species *endobioticum*. The genus *Synchytrium* included about 200 species which are endobiotic halocarpic organisms that have inoperculate sporangia. All species of the genus *Synchytrium* are parasites but the most important economically and phytosanitary is *S. endobioticum* the causal agent of potato wart disease. The disease is also known by various common names like black wart, cauliflower disease, warty disease, potato tumor, potato cancer, black cancer, or black scab. *S. endobioticum* is an obligate soil-borne biotrophic fungus which is considered to be the most important worldwide quarantine plant pathogen of cultivated potato. Cultivated potato (*Solanum tuberosum*) is the primary host, but the fungus, under experimental conditions, can also infect wild species in genera *Capsicastrum*, *Duboisia*, *Hyoscyamus*, *Lycium*, *Nicotiana*, *Nicandria*, and *Physalis* (Obidiegwu et al. 2014). The pathogen is a primitive fungus which stimulates its host to produce hypertrophic outgrowths on young potato organs, such as eyes, sprouts, young tubers, stolons, stems, leaves, and even flowers but never roots. The fungus does not form hyphae but forms sporangia that produce about 200–300 motile zoospores (Obidiegwu et al. 2014). After infection, *S. endobioticum* produces two different kinds of sporangia in the galls. Summer sporangia (Fig. 9.9) have a thin cell wall and form haploid zoospores which are emerging and steady reinfection of the host tissue like sprouts, tubers, eye tubers, stolons, and roots (only in tomato) (Przetakiewicz 2014a). In appropriate conditions after isogamy of haploid zoospores to diploid zygotes which are able to infect host cells and form winter sporangia which are embedded deeper into the host tissue than the sori (always on the surface). Winter (resting) sporangia (Fig. 9.9) are the dormant structures by which the fungus disperses to establish new infections. They are usually spherical to ovoid in shape and 24–75 μm in diameter with thick-walled (triple wall) structure, which is ornamented with irregularly shaped wing-like protrusions. The spores can survive for a long time without plant hosts. After 43 years, in favorable conditions, disease may develop even from single spores of *S. endobioticum* (Przetakiewicz 2015b, 2016).

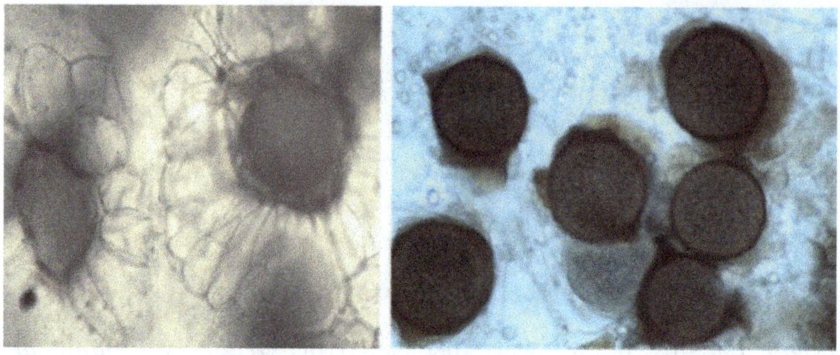

Fig. 9.9 Summer (left) and winter (resting) sporangia (right) of *Synchytrium endobioticum*. Thin-walled summer sporangia of pathotype 3(M1) are enclosed within one membrane forming sorus on surface of wart tissue of cultivar Asche Sämling. Thick-walled winter sporangia of pathotype 8(F1) in wart tissue of cultivar Sonda

Fig. 9.10 Symptoms of potato wart disease above ground (left) and underground (right). The galls on stem and foliage above ground are colored green, while subterranean warts on stem base, stolons, and tubers are colored white. Pathotype 2(Ch1) on cultivar Cykada (left). Pathotype 18(T1) on cultivar Deodara (right)

9.3.2 Symptoms

Usually, the symptoms of potato wart disease are not visible on plants, although there may be a reduction in plant vigor. Only in very suitable conditions small greenish warts (Fig. 9.10) might be visible on the top of plants: stem, foliage and in extremely conditions on inflorescences. (Obidiegwu et al. 2014). In the most cases the symptoms are visible on underground parts of potato (Fig. 9.10) on stolons, stems, bulbs, eye of matured tubers. In soil with the high content of winter sporangia (above 500/g) may lead to infection all eyes of seed potato and develop only warts without any emergences of potato (Przetakiewicz 2014a). The typical symptoms of the disease on tubers are the proliferating warts which may vary markedly in form but are primarily spherical to irregular. The warts are usually whitish or green if exposed to light, but gradually darken eventually rot and disintegrate. Rarely, the warts can be yellow or purple to brown. The color is dependent on potato variety. If variety produces purple sprouts, then the warts will be purple too. The warts are similar to cauliflower and they usually have the same color but sometimes they are compared to walnut kernel (Przetakiewicz 2014a). The warts are proliferated to gall. Galls vary markedly in form but are primarily spheroid. They are primarily parenchymatous and phytoteratological and not phytooncological. Although, the disease is described as potato tumor or cancer as well as black cancer they should not be referred to as tumors (Hampson 1993). Warts differ in size from pea-sized nodules to the size of a fist. The warts maintain in lab condition (for inoculum production) can reach 220 g from one eye/sprout. At maturity the galls become colored black and lead to total tuber decay (Przetakiewicz 2014a).

Simultaneous germination of all buds in one eye results in wart-like outgrowths very similar to those caused by *S. endobioticum*. However, these pseudo-warts consist of abundant pointed shoots compacted together. No winter sporangia are present in the affected tissue. Symptoms of powdery scab caused by *Spongospora subterranea* f.sp. *subterranea* or potato smut caused by *Thecaphora solani* can be mistaken for wart occurrence. A view under the microscope reveals that spore balls look different from winter sporangia of *S. endobioticum*. The size of warts depends not only on environmental conditions (cool summer, wet soil, etc.) but on potato variety. On extremely susceptible variety specified pathotype of *S. endobioticum* can differ markedly to very big size producing winter sporangia in the last stage of growing. On the slightly susceptible variety specified pathotype of *S. endobioticum* is able to produce very small warts or only to influence on weakly proliferation of host tissue but producing winter sporangia in the first stage of infection. The spores were visible after 13 days after inoculation of slightly susceptible varieties. In uncomfortable conditions (high temperature or dry soil) warts stop growing and begin to produce winter spores (Przetakiewicz 2014a).

9.3.3 *Impact*

Potato wart disease is so important that, for some 65 years, quarantine and domestic legislations have been in force throughout the world to prevent its spread (Anon 2015). The economic impact of disease caused by this pathogen is not only from disease losses but from loss of international trade markets, long-term quarantines, and regulatory restrictions placed on infested areas and the buffer zones (Przetakiewicz 2014a). Chemical control of *S. endobioticum* is not possible. The only strategies to confine the disease are strict quarantine and phytosanitary measures as well as cultivation of resistant cultivars (Obidiegwu et al. 2014). The availability of resistance cultivars allowed governments to issue regulations prohibiting the cultivation of susceptible cultivars (Baayen et al. 2006). For example, since 1955 only cultivars resistant to pathotype 1(D1) could be registered and grown on the Polish territory. As stated in the Food and Veterinary Office Mission Report, no potato plants or tubers with symptoms of *S. endobioticum* pathotype 1(D1) have been detected in Poland since the 1950s–early 1960s. This is a result of growing only resistant cultivars to mentioned pathotype of *S. endobioticum* (Przetakiewicz 2008). Worldwide prevention is based on the control of disease spread and pathogen exclusion via regulatory action. *S. endobioticum* has a very limited capacity for natural spread, which is principally why it has been possible to control it so effectively by statutory means. Nevertheless, *S. endobioticum* is a classic example of the distribution of plant pathogens by man. Regulatory action has largely restricted the spread of the disease within potato-growing regions, as the seriousness of the disease was quickly recognized (Hampson 1993). Once *S. endobioticum* has been introduced into a field, the whole crop may be rendered unmarketable and moreover the fungus is so persistent that potatoes cannot be grown again safely for many years, nor can the land be used for any plants intendent for export. Based on the results obtained by Przetakiewicz (2015b), winter sporangia of *S. endobioticum* are very persistent and capable of retaining viability for as long as 46 years. The single spores from the inter-host period were still infective after 43 years. These results should be taken into consideration when de-scheduling previously infested plots even after 40 years or longer, especially in the mountainous areas. *S. endobioticum* is still of great economic importance in cool areas and wet mountainous regions. The detection of potato wart disease on Prince Edward Island during the 2000 growing season resulted in an estimated $30 million loss to the island's economy in the first year.

9.3.4 Pathogen Populations and Distribution

Potato wart disease appears to have arisen in the potato-growing area of Andean South America (Hampson 1993). Wart-like outgrowths on early Peruvian tuber-shaped pottery were interpreted as potato wart disease. The disease is likely to have arrived in Europe from South America because the Great Potato famine of the 1840s in Ireland induced European growers to import potato germplasm from South America. The introduction of the pathogen to Europe was possible by diseased tubers, infested soil or contaminated bags along with shipments of guano (Hampson 1993). *S. endobioticum* spread at the end of the nineteenth century from the center of origin in the Andes first to Europe and North America and subsequently across whole potato-growing areas of Asia, Africa and Oceania. Historic account has it that potato wart disease entered England in 1876 or 1878 while another view upholds that the disease has been present in the Liverpool province of England in 1876 or 1878 (Obidiegwu et al. 2014). In 1901, the disease was officially recognized in the UK. Potato wart disease and its causal agent were described by Schilberszky who received in 1888 warted tubers of cultivar Maercker-Zwiebel that had been grown locally from seed tubers imported from England. It spread widely in Europe, but statutory measures finally restricted its distribution and it has spread only to a limited extent to other parts of the world. According to EPPO Pest Quarantine Database

(Anon 2015), *S. endobioticum* occurs locally in almost all EPPO countries. The distributions are fragmentary as a result of statutory control. According to national reports, it has been found but is not established in Belgium, France, Luxembourg as well as in Lebanon. Found in the past but eradicated in Portugal (unconfirmed). In Asia countries *S. endobioticum* occurs in Armenia, Bhutan, China, India, Georgia, Nepal, Turkey. In Africa: South Africa, Tunisia, Algeria, Zimbabwe, and Egypt (absent, unreliable record). North America: Canada, Mexico (absent formerly present), USA (eradicated). South America: Bolivia, Chile (eradicated), Ecuador, Falkland Islands, Peru, Uruguay (absent, confirmed by survey). Oceania: New Zealand (South Island). Numerous pathotypes of the fungus occur and are defined by their virulence on differential potato cultivars. In Europe, more than 40 pathotypes of *S. endobioticum* have already been identified (Obidiegwu et al. 2014; Przetakiewicz 2014b, 2015a). A pathotype is characterized by its pattern of virulence or avirulence to a series of differential cultivars of potato. Pathotypes 1(D_1), 2(G_1), 6(O_1), 8(F_1), and 18(T_1) are the most relevant in Europe. Other pathotypes occur mainly in the rainy mountainous areas of central and eastern Europe (Alps, Carpathians). They persist mainly in small garden potato plots and not in commercial potato crops (Przetakiewicz 2014a). Mitochondrial genomic variation shows that *S. endobioticum* has been introduced into Europe multiple times, that several pathotypes emerged multiple times, and that isolates represent communities of different genotypes (van de Vossenberg et al. 2018a).

9.3.5 Resistance

The biggest discovery was finding resistant cultivars of potato to *S. endobioticum* among cultivated ones. Systematic studies on resistance of potato cultivars to *S. endobioticum* started in England in 1909. Resistance sources to *S. endobioticum* were found in old cultivars such as Snowdrop and Flourball, which facilitated resistance breeding. Conventional breeding programs were successful in controlling potato wart disease through the development of resistant cultivars early in the twentieth century (Obidiegwu et al. 2014). Breeding for resistance was successful, thanks to the arability of a dominant gene that blocked development and reproduction of originally introduced pathotype 1(D1) of *S. endobioticum*. Unfortunately, in Europe wart development on resistant potato cultivars was first discovered in 1941 in Germany [pathotype 2(G1)] and former Czechoslovakia [pathotype 3(S1)]. The new pathotypes have been proved to be difficult to control and eradicate then the original pathotype 1(D1) (Baayen et al. 2006). Molecular mapping studies provide evidence that wart resistance to pathotype 1(D1) of *S. endobioticum* can be conferred by a single locus from different sources. Hehl et al. (1999) mapped the single dominant gene *Sen 1* for resistance to pathotype 1(D1) in diploid mapping population. The gene *Sen 1* is located on potato chromosome IX. Brugmans et al. (2006) also used a diploid potato linkage map to locate *Sen1-4*, a second dominant gene for resistance to pathotype 1(D1). This gene is located on the long arm of chromosome IV. In these two mentioned populations, the resistance segregated as a monogenic trait. Ballvora et al. (2011) discovered the first loci for virulent pathotypes 2(G1), 6(O1), and 18(T1) in two tetraploid half-sib families, in which the resistance to pathotype 1(D1) also segregated. In contrary to earlier studies in diploid populations (Hehl et al. 1999; Brugmans et al. 2006), the phenotypic distribution of wart resistance appeared quantitative in the two mapping populations analyzed (Ballvora et al. 2011). The quantitative resistance locus (QRL) *Sen2/6/18* on chromosome I expressed resistance to pathotypes 2(G1), 6(O1), and 18(T1). The QRL *Sen18* on chromosome IX expressed resistance to pathotype 18(T1). And the third QRL Sen1 on chromosome XI expressed resistance mainly to pathotype 1(D1) (Ballvora et al. 2011). Groth et al. (2013) mapped quantitative trait loci (QTL) for resistance to pathotype 1(D1), 2(G1), 6(O1), and 18(T1). The QRL for all four pathotypes were located on chromosomes II, VI, VIII, and IX, and QRL for pathotypes 2(G1), 6(O1), and 18(T1) on chromosomes VII and X. The QRL detected in this study were different from the ones in Ballvora et al. (2011). The cultivar Panda used in this study (Groth et al. 2013) as a resistant parent to pathotype 1(D1), 2(G1), 6(O1) and 18(T1), is in fact slightly susceptible to all four pathotypes. These results were confirmed in CORNET project (acronym SynTest) (Przetakiewicz, unpublished results 2014) Consequently, breeding is hampered by a lack of dominant major genes for resistance to virulent pathotypes. The wart resistance has several sources and diverse backgrounds which resulted in a few potato cultivars resistant to virulent pathotypes of *S. endobioticum*. Nevertheless, recent results indicated for identify a newly locus *Sen2* located on chromosome XI which provides resistance to at least seven various virulent pathotypes of *S. endobioticum* (Plich et al. 2018).

9.3.6 Management

The first acts were issued in 1908 in Ireland, and a few years later in Scotland, England, and Germany. The first legislation related primarily to prohibit the import from other countries of potatoes infected by *S. endobioticum*. In that time the acts did not take into account the possibility of the spread of the disease within the country. It was a reason of increasing number of outbreaks in such a short time in many countries. In consequence of the wide spread of potato wart disease was the development of research on the biology controlling of the fungus. EPPO includes *S. endobioticum* into the A2 quarantine list because of its long persistence in soil and the severe losses it inflicts to potato crops. The United State Department of Agricultural (USDA) has the fungus on its official list of selected agents and toxin. The European Union issued a specific requirement in the Council Directive 69/29/EC of 8 December 1969 on control of Potato wart disease and the Council Directive 2000/29/EC of 8 May 2000 on protective measures against the introduction into the Community of organisms harmful to plant or plant products and against their spread within the Community (Obidiegwu et al. 2014).

9.3.7 Looking Forward

Although *S. endobioticum* seems to be in remission in Europe, new foci have appeared in countries with warmer continental climate (Turkey, Georgia, Bulgaria, and Greece) (Anon 2015; Gorgiladze et al. 2014; Vloutoglou pers comm. 2015). This may suggest adaptation of *S. endobioticum* to warmer and/or dryer climate as well as the lack of adequate controls in countries where the disease has not been present before. Moreover, new pathotypes have been discovered recently (Çakir et al. 2009; Przetakiewicz 2015a). It can be expected that new pathotypes of *S. endobioticum* might appear, since there are many reasons for new pathotypes to arise (Melnik 1998). The newest Pathotype 39(P_1) was detected in the rainy mountainous area, in small garden potato plots, when the old traditional cultivars of potato are cultivated without any crop rotations. It seems therefore that where climatic conditions are suitable for *S. endobioticum* to take place and the growing of slightly susceptible cultivars is possible, the development of new pathotype is favored (Przetakiewicz 2015a). Although, local pathotypes persist mainly in small garden potato plots in economically unimportant potato-growing regions, they are still very important for quarantine and phytosanitary measures as spores can spread in rain water (Przetakiewicz 2015b). Phenotypic assessment of resistance to *S. endobioticum* is laborious and, time-consuming. Diagnostic DNA-based markers closely linked with or, even better, located within wart resistance genes would greatly facilitate the early detection and combination of different resistance sources and are therefore highly desirable (Obidiegwu et al. 2014). Natural DNA variation in wild and cultivated potato germplasm provides an excellent platform for the discovery of diagnostic tools for marker-assisted selection and resistance gene cloning (Obidiegwu et al. 2014). A new dominant gene *Sen2* on chromosome XI provides extreme resistance to pathotypes 1(D1), 2(G1), 2(Ch1), 3(M1), 6(O1), 8(F1), 18(T1), and 39(P1). In the future, this gene will offer potentials for the efficient selection of new commercial cultivars that are resistant to multiple *S. endobioticum* pathotypes. EPPO Standard PM 7/28 (2) (Anon 2017) recommends various biotests using differential potato cultivars for the identification only 4 pathotypes [1(D1), 2(G1), 6(O1) and 18(T1)]. Pathotype determination is labor-intensive and time-consuming too, especially for the identification of local pathotypes which require more differential cultivars (Przetakiewicz 2017; Przetakiewicz and Plich 2017). Molecular diagnostic tools (TaqMan PCR method) are currently available for the identification of pathotype 1(D1) and its discrimination from non-1(D1) pathotypes (Bonants et al. 2015; van de Vossenberg et al. 2018b). However, this method requires costly probes and cannot be used for the identification of virulent pathotypes. The recent report revealed no sequence polymorphisms between the five *S. endobioticum* pathotypes, indicating that *S. endobioticum* pathotypes may be exceptionally similar to each other. One reason for this may be that the development of new *S. endobioticum* pathotypes is caused by very limited changes in avirulence factors rather than extensive genetic recombination between divergent genotypes (Busse et al. 2017). Recently, polymorphic microsatellite markers were used to assess the genetic diversity of potato wart at the intraspecific level for the first time and will certainly contribute to a better understanding of the evolutionary history of this pathogen in the years to come (Gagnon et al. 2016). The molecular methods for *S. endobioticum* detection based on the RealTime PCR (Smith et al. 2014; Bonants et al. 2015) or PNA-based hybridization assay (Duy et al. 2015) have several advantages over the traditional method of microscopic examination in routine diagnostic testing. Molecular methods can significantly reduce the time to disease diagnosis and prevent the spread of *S. endobioticum* to other locations.

9.4 Powdery Scab

Alison Lees, Francisco Bittara, and Gary A. Secor

9.4.1 Introduction and Future Perspectives

Spongospora subterranea causes root hyperplasia (root galling) and tuber powdery scab, leading to losses in seed and ware crops of potato worldwide and is also important as the natural vector of Potato mop-top virus (PMTV), an economically important tuber blemish disease of potato found in some regions. Powdery scab is particularly favored by cool, damp conditions and is an intractable disease. Additionally, *S. subterranea* is an unculturable biotroph, making research difficult.

Progress in understanding many aspects of the biology of *S. subterranea* has been slow relative to other plant pathogens and even compared with other plasmodiophorids. However, the development of quantitative molecular assays and the increasing availability of sequence information have allowed recent progress to be made in understanding various aspects of the epidemiology of powdery scab and root galling, and biology of the pathogen.

The pathogen can survive for many years and very low levels of inoculum can cause relatively severe disease outbreaks, making control elusive: it is generally accepted that an integrated approach to disease control will prove most effective as no single control method is totally effective. Additional knowledge of inoculum-based risk, pathogen variation and infection conditions will all contribute towards such a risk-based integrated control system.

The tripartite potato/*S. subterranea* /PMTV interaction is a difficult system to manipulate and there has been little progress in understanding the basis of host specificity and molecular mechanisms of virus transmission. In recent years some progress has been made in developing molecular markers and genomic sequence information for the plasmodiophorids. In addition, host genomics has advanced with the discovery of genetic markers and identification of novel resistance. The availability of in vivo root culture systems for *S. subterranea* propagation, high-throughput nucleotide sequencing, comparative genomics and advanced imaging technologies will no doubt be applied in due course in order to advance knowledge of this intractable and understudied disease and the virus that it vectors. It is envisaged that availability of plasmodiophorid genomes will, in the future, lead to molecular interaction studies between *S. subterranea* and compatible host plants, and will allow the genes or molecular mechanisms involved in host recognition, infection of host cells, multiplication of *S. subterranea* within roots and the development of sporangial or sporogenic phases to be identified.

9.4.2 Causal Organism

The taxonomy of the plasmodiophorids is complicated and has previously been described in detail (Braselton 1995; Dick 2001; Down et al. 2002). The genus *Spongospora*, once ascribed to the fungi, is now considered to be a member of the family *Plasmodiophoridae* within the Super-group Rhizaria, Phylum Cercozoa, under the Class Phytomyxea, which includes members parasitic to higher plants and Stramenopila and which usually cause hypertrophy in the host cells (Bulman and Braselton 2014; Neuhauser et al. 2010). Traditionally, the species *S. subterranea* has been divided into two *formae speciales: Spongospora subterranea* (Wallr.) Lagerheim f.sp. *nasturtii* Tomlinson (the cause of crook root of watercress) and *Spongospora subterranea* (Wallr.) Lagerheim f.sp. *subterranea* (the cause of powdery scab and root galling on potato) (Neuhauser et al. 2010). However, host specificity characteristics, differences in sporangial states and habit as well as molecular data have provided additional evidence to support their placement into the species rank (i.e. *S. subterranea* and *S. nasturtii*; Bulman and Braselton 2014; Dick 2001; Gau et al. 2013; Neuhauser et al. 2010; Qu and Christ 2004) and they will be referred to as such hereafter. Due to the uncertain taxonomy of the group, the collective term "plasmodiophorids" is also commonly used (Braselton 1995). Key features of the plasmodiophorids are that they have zoospores with two anterior undilopodia ("flagella"), multinucleated protoplasts (plasmodia), and environmentally resistant resting spores, and that they are biotrophic parasites. Plasmodiophorids are a monophyletic group with cruciform nuclear division, obligate intracellular parasitism, biflagellated zoospores, and environmentally resistant resting spores (Bulman et al. 2001; Qu and Christ 2004).

A detailed description of life cycle of *S. subterranea* was given by Harrison et al. (1997). There are two major phases in the life cycle of *Spongospora*, each initiated by host cell infection through a single uninucleate plasmodium: in the sporogenic (spore-producing) phase, sporogenic plasmodia are located within infected plant tissue, either in tuber lesions or root, shoot or stolon galls. Following nuclear divisions within the plasmodium, thick-walled resting spores are produced, each being around 3.5–4.5 μm diameter (Jones 1978). These resting spores are aggregated together in sponge-like sporosori which vary in size from 19 to 85 μm. Falloon et al.

(2011) determined the mean numbers of resting spores in sporosori to be about 700. Resting spores may then persist in the soil, where they are able to survive in a dormant state in the absence of potato cultivation for many years, or on tubers. Under suitable conditions, the resting spores germinate and each releases a single biflagellate primary zoospore which can then infect host cells. The mechanisms of host infection are not well understood.

9.4.3 Symptoms

Spongospora subterranea can infect all underground organs of potato (i.e. stolons, tubers, and roots) where the pathogen stimulates the enlargement and division of host cells leading to the appearance of symptoms. Depending on environmental conditions initial tuber symptoms, which take 4–8 weeks to develop, are purplish brown lesions (1–2 mm diameter) that subsequently enlarge into raised mature lesions which burst, exposing large masses of sporosori resulting in characteristic symptoms of the disease (Figs. 9.11 and 9.12).

Sometimes infection of tuber buds can stimulate tubers to swell in that area, forming outgrowths or cankers which may also be infected through lenticels and be covered with scab lesions. Infection of roots or stolons can sometimes lead to development of galls (hyperplasia) (Fig. 9.13), where infected tissue is stimulated to grow and sporosori are formed inside the gall (Fig. 9.14). Galls burst when mature and release the sporosori into the soil. Root or stolon galls can be easily overlooked in the field and there is evidence that varieties with resistance to tuber infection can be susceptible to root or stolon gall production (Falloon 2008), resulting in an unseen build-up of inoculum in the soil.

Fig. 9.11 Characteristic lesions of powdery scab on cultivar Agria

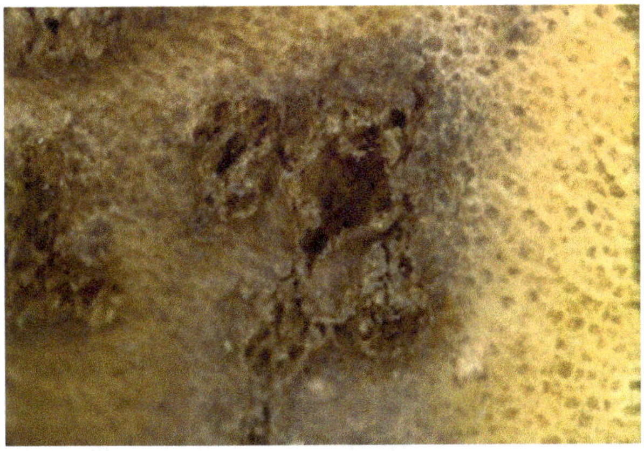

Fig. 9.12 A typical powdery scab lesion

Fig. 9.13 Root galling of potato caused by *Spongospora subterranea*

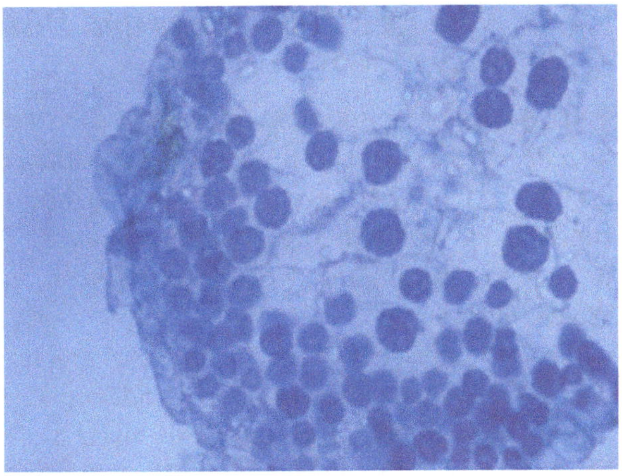

Fig. 9.14 Sporosori inside root gall

9.4.4 Impact

The main impact of powdery scab is cosmetic, due to a reduction in quality caused by lesions on the tuber surface, thus decreasing their value for either fresh or processing purposes. The disease also affects seed tuber certification, with the extent of the impact dependent on certification tolerances which vary from country to country (Falloon 2008; Wale 2000). Reports of losses due to unmarketable tubers can be as high as 50% in Australia (Hughes 1980) to 100% of the harvested product in Venezuela (Bittara et al. 2009). In addition, root infection by *S. subterranea* can reduce water absorption and nutrient intake (Shah et al. 2012). Some studies have reported a reduction in tuber yield and mean tuber weight due to disease caused by *S. subterranea* on either *Solanum tuberosum* spp. *andigena or S. tuberosum* spp. *tuberosum* (Gilchrist et al. 2011; Shah et al. 2012). In contrast, field studies using the cultivars Shepody and Umatilla Russet (*S. tuberosum* spp. *tuberosum*) showed no detrimental effect on either tuber yield or mean tuber weight due to the development of symptoms on roots and tubers (Johnson and Cummings 2015). *Spongospora subterranea* is also the vector of *Potato mop-top virus* (PMTV), one of the causes of spraing, a tuber blemish disease (Jones and Harrison 1969). PMTV is carried within zoospores and *S. subterranea* can remain viruliferous for many years. Little is known, however, about the virus–vector interactions, particularly with respect to the conditions conducive to transmission of the virus and the differential symptom expression of one or other of the diseases where infection by both organisms has occurred.

9.4.5 Pathogen Populations and Distribution

Since the first description of *Spongospora subterranea* under the name *Erysibe subterranea* by F. W. Wallroth in 1841, the pathogen has been reported in most potato-growing regions of the world, including hot and dry countries where farming is conducted at high altitudes or under irrigation. The number of first reports continues to increase across the world (Harrison et al. 1997; Merz and Falloon 2009; Wale 2000).

Few attempts have been made to characterize variation in *S. subterranea*, an unculturable biotroph with poorly understood genetics. It is difficult to obtain sufficient "clean" DNA for commonly used molecular marker techniques, sequence information is limited and assumptions about what constitutes an isolate or "strain" (sporosorus, resting spore or zoospore) are made. It is not demonstrated that sporosori occur as a result of sexual recombination, although this is thought to be the case (Braselton 1995). In general, analyses have been made on a single sporosori, which may also comprise many genotypes in the form of individual resting spores.

Genetic variation in Internal Transcribed Spacer sequences revealed limited differences among collections from Europe and Australasia and those from South

America (Bulman and Marshall 1998). Three ribotypes (I, II, and III) of *S. subterranea* were identified, with ribotype I, II and III is found in South America, ribotype II in North America and Australasia and ribotypes I and II in Great Britain (Osorio-Giraldo et al. 2012; Qu and Christ 2004). Most recently, Gau et al. (2013) applied SSR markers and ITS and actin sequence data to approximately 700 samples of *S. subterranea* obtained from 19 countries, different resting spore sources of the pathogen (root galls or tuber lesions) and from different potato host (sub) species. They described an overall low genetic diversity in *S. subterranea*, comprising three sample clusters; two occurring in South America (root galls and tuber lesions) and the third group comprising samples from elsewhere, independent of the resting spore source, and suggested that an ecological adaption in the native region due to coevolutionary processes and/or competitive exclusion may have taken place. South American populations were more diverse than those from other regions but no relationship between host species and pathogen diversity was noted.

9.4.6 Host Resistance

Resistance to powdery scab exists in some cultivars worldwide (Torres et al. 1995; Genet et al. 1996; Christ 1987; Falloon et al. 2003; Nitzan et al. 2008; Merz et al. 2012; Bittara et al. 2016). Falloon et al. (2003) found that although cultivars ranged from very susceptible to resistant, all developed zoosporangia and root galls and therefore none possessed immunity. Susceptibility of individual cultivars to root and tuber infection is not always closely correlated (Hughes 1980; Falloon et al. 2003; Bittara et al. 2016), particularly with regards to the relationship between root and tuber infection and root galling (Falloon et al. 2003; van de Graaf et al. 2007). The use of certain cultivars with high tuber resistance may therefore still maintain populations of *S. subterranea* in the soil.

Despite the availability of disease resistant cultivars, genetic resistance to *S. subterranea* currently plays a minor role in disease control as cultivars are usually selected by growers for characteristics other than their ability to resist powdery scab.

Resistance to powdery scab has also been demonstrated in other Solanum species, for example, *S. acaule* was reported to show resistance to pathogen infection in roots (Mäkäräinen et al. 1994) and among potato genotypes reported to have resistance to the formation of root galls at the Columbia Basin (WA) resistant clones were derived from the introgression of *S. bulbocastanum* and the resistant cultivar Summit Russet appeared more than once in their parental background (Nitzan et al. 2008). Although the mechanisms involved in resistance to disease caused by *S. subterranea* are poorly studied, evidence supports the hypothesis that root and tuber resistance to the disease is under control of multiple genes (Bittara et al. 2016; Falloon et al. 2003; Merz et al. 2012). In addition, disease resistance is inherited in an additive manner (Wastie et al. 1988) and is stable across environments (Bittara et al. 2016; Nitzan et al. 2010).

9.4.7 Management

Management of powdery scab is particularly difficult, and no single strategy currently controls powdery scab or root gall formation effectively, making an integrated approach essential (Falloon 2008). Disease avoidance using uncontaminated seed in uninfested soil represents the best method of disease prevention. The relative importance of soil inoculum level in causing disease on tubers was conclusively demonstrated by Brierley et al. (2013) who showed that when arbitrary soil inoculum threshold values of 0, <10 and >10 sporosori/g soil were set, it was observed that the number of crops developing powdery scab increased with the level of inoculum quantified in the field soil preplanting. In field trials carried out to investigate the link between the amount of inoculum added to the soil and disease development, disease incidence and severity on progeny tubers was found to be significantly ($P < 0.01$) greater in plots with increasing levels of inoculum. This information allows disease risk assessments to be made by taking soil inoculum concentration, in combination with other biotic and abiotic factors, into account.

The use of host resistance represents the most effective, sustainable and cost-effective approach for disease management; in addition, genetic resistance is especially suitable on pathogen populations with low genetic diversity (Gau et al. 2013). In a study performed across five European countries, no evidence of genotype x pathogen interaction was observed as disease development as was similar for all locations (Merz et al. 2012). Nevertheless, the use of host resistance is affected by factors defined by the consumer market (Harrison et al. 1997).

Resting spores produced by the pathogen are highly resistant to environmental stresses and can remain viable for >10 years (Merz and Falloon 2009), therefore up to 7 years of crop rotation or more that excludes potatoes might be required to reduce the risk of disease development once the pathogen has stablished in the field (Sparrow et al. 2015). Crop rotation may also have an influence on *S. subterranea* soil infestation as detectable inoculum was shown to be greater following a potato/wheat rotation compared to a potato/pea rotation (Shah et al. 2014). Additionally, a study by Qu and Christ (2006), found four of 16 crop and weed species infected with *S. subterranea* subsequently produced root galls, with sporosori detected on three host species (yellow mustard, oats and tomato). Similarly, Shah et al. (2010) confirmed the presence of sporosori in root galls formed on the solanaceous weeds hairy and black nightshade and subsequently demonstrated pathogenicity on potato with sporosorus inoculum derived from hairy nightshade. It is therefore possible for *S. subterranea* to complete its lifecycle in the absence of potatoes.

It is broadly accepted that high soil water content encourages zoospore release and that soil in which most pore spaces are filled with water facilitates movement of zoospores towards the host, and subsequent infection and disease development. This is supported by associations between seasons of high (or low) rainfall or irrigation treatments with high (or low) levels of powdery scab.

The chemical management of powdery scab is difficult and, in some cases, economically unsuitable. The use of seed and soil-applied chemicals to control powdery scab was previously reviewed in some detail by Merz and Falloon (2009).

Acknowledgements The section on late blight was undertaken as part of, and funded by, the CGIAR Research Program on Roots, Tubers and Bananas (RTB) and supported by CGIAR Trust Fund contributors. The section on wart was founded by the project CORNET (CORNET/2/13/2013) Acronym: SynTest. For the section on powdery scab, the authors thank Dr. Simon Bulman and Professor Richard Falloon, The New Zealand Institute for Plant & food Research for useful discussions regarding the taxonomy of *S. subterranea*. Alison Lees is funded by The Rural and Environment Science and Analytical Services Division of The Scottish Government. Thanks also to Professor Neil Gudmestad, North Dakota State University for helpful discussions.

References

Abuley IK, Nielsen BJ (2017) Evaluation of models to control potato early blight (*Alternaria solani*) in Denmark. Crop Prot 102:118–128

Acuna I, Sagredo B, Gutierrez M, Sandoval C, Fahrenkrog A, Secor G, Rivera V, Mancilla S (2012) Characterization of *Phytophthora infestans* population in Chile. In: Proceedings of the thirteenth EuroBlight workshop, St Petersburg, Russia, 9–12 October 2011. Praktijkonderzoek Plant & Omgeving, PPO, pp 145–150

Ah-Fong AM, Kim KS, Judelson HS (2017) RNA-seq of life stages of the oomycete *Phytophthora infestans* reveals dynamic changes in metabolic, signal transduction, and pathogenesis genes and a major role for calcium signaling in development. BMC Genomics 18(1):198

Andrade-Piedra JL, Cáceres PA, Pumisacho M, Forbes GA (2009) Humans: the neglected corner of the disease tetrahedron—developing a training guide for resource-poor farmers to control potato late blight. Acta Hortic (834):111–122

Anon (2015) CABI/EPPO *Synchytrium endobioticum*. Distribution maps of plant disease, No 10. Wallingford, UK: CABI, Map 1 (Edition 8) http://www.cabi.org/isc/abstract/20153399804

Anon (2017) EPPO Standard PM 7/28 (2) *Synchytrium endobioticum*. Bull OEPP/EPPO Bull 47:420–440

Avenot HF, Sellam A, Karaoglanidis G, Michailides TJ (2008) Characterization of mutations in the iron-sulphur subunit of succinate dehydrogenase correlating with boscalid resistance in *Alternaria alternata* from California pistachio. Phytopathology 98(6):736–742

Avenot H, Sellam A, Michailides T (2009) Characterization of mutations in the membrane-anchored subunits AaSDHC and AaSDHD of succinate dehydrogenase from *Alternaria alternata* isolates conferring field resistance to the fungicide boscalid. Plant Pathol 58(6):1134–1143

Ayad D, Leclerc S, Hamon B, Kedad A, Bouznad Z, Simoneau P (2017) First report of early blight caused by *Alternaria protenta* on potato in Algeria. Plant Dis 101(5):836

Baayen R, Cochius G, Hendriks H, Meffert J, Bakker J, Bekker M, van den Boogert P, Stachewicz H, van Leeuwen G (2006) History of potato wart disease in Europe—a proposal for harmonisation in defining pathotypes. Eur J Plant Pathol 116:21–23

Ballvora A, Flath K, Lübeck J, Strahwald J, Tacke E, Hofferbert H-R, Gebhardt C (2011) Multiple alleles for resistance and susceptibility modulate the defense response in the interaction of tetraploid potato (*Solanum tuberosum*) with *Synchytrium endobioticum* pathotypes 1, 2, 6 and 18. Theor Appl Genet 123:1281–1292

Bartlett DW, Clough JM, Godwin JR, Hall AA, Hamer M, Parr-Dobrzanski B (2002) The strobilurin fungicides. Pest Manag Sci 58:649–652

Bauske MJ, Mallik I, Yellareddygari SKR, Gudmestad NC (2018) Spatial and temporal distribution of mutations conferring QoI and SDHI resistance in *Alternaria solani* across the United States. Plant Dis 102(2):349–358

Bethke PC, Halterman DA, Jansky S (2017) Are we getting better at using wild potato species in light of new tools? Crop Sci 57:1241–1258

Bittara F, Rodriguez D, Sanabria M, Monroy J, Rodriguez J (2009) Fungicides and plant products evaluation for control of powdery scab of potato. Interciencia 34:265–269

Bittara F, Thompson A, Gudmestad N, Secor G (2016) Field evaluation of potato genotypes for resistance to powdery scab on tubers and root gall formation caused by *Spongospora subterranea*. Am J Potato Res 93:497–508

Black W, Mastenbroek C, Mills WR, Peterson LC (1953) A proposal for an international nomenclature of races of *Phytophthora infestans* and of genes controlling immunity in *Solanum demissum* derivatives. Euphytica 2(3):173–179

Boevink P, Wang X, McLellan H, He Q, Naqvi S, Armstrong M, Wei Z, Hein I, Gilroy E, Tian Z, Zhang W, Hein I, Gilroy EM, Tian Z, Birch PRJ (2016) A *Phytophthora infestans* RXLR effector targets plant PP1c isoforms that promote late blight disease. Nat Commun 7:10311

Bonants PJM, van Gent-Pelzer MPE, van Leeuwen GCM, van der Lee TAJ (2015) A real-time TaqMan PCR assay to discriminate between pathotype 1 (D1) and non-pathotype 1 (D1) isolates of *Synchytrium endobioticum*. Eur J Plant Pathol 143:495–506

Bourke A (1993) The visitation of god? The potato and the great Irish famine. Lilliput Press, Ltd., Dublin, Ireland

Braselton J (1995) Current status of the plasmodiophorids. Crit Rev Microbiol 21:263–275

Brierley J, Sullivan L, Wale S, Hilton A, Kiezebrink D, Lees A (2013) Relationship between *Spongospora subterranea* f. sp. *subterranea* soil inoculum level, host resistance and powdery scab on potato tubers in the field. Plant Pathol 62:413–420

Brugmans B, Hutten RGB, Rookmaker N, Visser RGF, van Eck HJ (2006) Exploitation of a marker dense linkage map of potato for positional cloning of a wart disease resistance gene. Theor Appl Genet 112:269–277

Bulman S, Braselton J (2014) Rhizaria: Phytomyxea. In: McLaughlin D, Spatafora J (eds) The Mycota VII, systematics and evolution, Part A, 2nd edn. Springer, Berlin, pp 99–112

Bulman S, Marshall J (1998) Detection of *Spongospora subterranea* in potato tuber lesions using the polymerase chain reaction (PCR). Plant Pathol 47:759–766

Bulman S, Kühn S, Marshal J, Schnepf E (2001) A phylogenetic analysis of the SSU rRNA from members of the Plasmodiophorida and Phagomyxida. Protist 152:43–51

Busse F, Bartkiewicz A, Terefe-Ayana D, Niepold F, Schleusner Y, Flath K, Sommerfeldt-Impe N, Lübeck J, Strahwald J, Tacke E, Hofferbert H-R, Linde M, Przetakiewicz J, Debener T (2017) Genomic and transcriptomic resources for marker development in *Synchytrium endobioticum*, an elusive but severe potato pathogen. Phytopathology 108(3):322–328

Çakır E, Van Leeuwen GCM, Flath K, Meffert JP, Janssen WAP, Maden S (2009) Identification of pathotypes of *Synchytrium endobioticum* found in infested fields in Turkey. EPPO Bull 39(2):175–178

Campo Arana RO, Zambolim L, Costa LC (2007) Potato early blight epidemics and comparison of methods to determine its initial symptoms in a potato field. Rev Fac Nac Agron Medellin 60(2):3877–3890

Cardenas M, Grajales A, Sierra R, Rojas A, Gonzalez-Almario A, Vargas A, Marín M, Fermín G, Lagos LE, Grünwald NJ, Bernal A, Salazar C, Restrepo S (2011) Genetic diversity of *Phytophthora infestans* in the Northern Andean region. BMC Genet 12:23

Casa-Coila VH, Lehner MDS, Hora Júnior BT, Reis A, Nazareno NRX, Mizubuti ES, Gomes CB (2017) First report of *Phytophthora infestans* self-fertile genotypes in Southern Brazil. Plant Dis 101(9):1682–1682

Chowdappa P, Nirmal Kumar BJ, Madhura S, Mohan Kumar SP, Myers KL, Fry WE, Cooke DEL (2015) Severe outbreaks of late blight on potato and tomato in South India caused by recent changes in the *Phytophthora infestans* population. Plant Pathol 64:191–199

Christ BJ (1987) Incidence and severity of powdery scab on 5 potato cultivars. Phytopathology 77(6):985–986

Cooke L, Schepers H, Hermansen A, Bain R, Bradshaw N, Ritchie F, Shaw D, Evenhuis A, Kessel GJT, Wander JGN, Andersson B, Hansen JG, Hannukkala A, Nærstad R, Nielsen BJ (2011) Epidemiology and integrated control of potato late blight in Europe. Potato Res 54(2):183–222

Daayf F, Adam L, Fernando WGD (2003) Comparative screening of bacteria for biological control of potato late blight (strain US-8), using in-vitro, detached-leaves, and whole-plant testing systems. Can J Plant Pathol 25(3):276–284

Danies G, Myers K, Mideros MF, Restrepo S, Martin FN, Cooke DE, Smart CD, Ristaino JB, Seaman AJ, Gugino BK, Grünwald NJ, Fry WE (2014) An ephemeral sexual population of *Phytophthora infestans* in the northeastern United States and Canada. PLoS One 9(12):e116354

Deahl KL, Pagani MC, Vilaro FL, Perez FM, Moravec B, Cooke LR (2003) Characteristics of *Phytophthora infestans* isolates from Uruguay. Eur J Plant Pathol 109(3):277–281

Dick M (2001) Straminipilous fungi: systematics of the Peronosporomycetes including accounts of the marine straminipilous protists, the plasmodiophorids and similar organisms. Kluwer, Dordrecht. 660 p

Domazakis E, Lin X, Aguilera-Galvez C, Wouters D, Bijsterbosch G, Wolters PJ, Vleeshouwers VG (2017) Effectoromics-based identification of cell surface receptors in potato. Methods Mol Biol 2017:337–353

Dong S, Raffaele S, Kamoun S (2015) The two-speed genomes of filamentous pathogens: waltz with plants. Curr Opin Genet Dev 35:57–65

Down G, Grenville L, Clarkson J (2002) Phylogenetic analysis of *Spongospora* and implications for the taxonomic status of the plasmodiophorids. Mycol Res 10:1060–1065

Du J, Verzaux E, Chaparro-Garcia A, Bijsterbosch G, Keizer LP, Zhou J, Liebrand TW, Xie C, Govers F, Robatzek S, Van Der Vossen EA (2015) Elicitin recognition confers enhanced resistance to *Phytophthora infestans* in potato. Nat Plants 1:15034

Duarte HSS, Zambolim L, Rodrigues FA, Paul PA, Pádua JG, Ribeiro Júnior JI, Júnior AFN, Rosado AWC (2014) Field resistance of potato cultivars to foliar early blight and its relationship with foliage maturity and tuber skin types. Trop Plant Pathol 39(4):294–306

Duy J, Smith RL, Collins SD, Connell LB (2015) Rapid colorimetric detection of the fungal phytopathogen *Synchytrium endobioticum* using cyanine dye-indicated PNA hybridization. Am J Potato Res 92:398–409

Fairchild KL, Miles TD, Wharton PS (2013) Assessing fungicide resistance in populations of *Alternaria* in Idaho potato fields. Crop Prot 49:31–39

Fall ML, Tremblay DM, Gobeil-Richard M, Couillard J, Rocheleau H, Van der Heyden H, Lévesque CA, Beaulieu C, Carisse O (2015) Infection efficiency of four *Phytophthora infestans* clonal lineages and DNA-Based quantification of sporangia. PLoS One 10(8):e0136312

Falloon R (2008) Control of powdery scab of potato: towards integrated disease management. Am J Potato Res 85:253–260

Falloon R, Genet R, Wallace A, Butler R (2003) Susceptibility of potato (Solanum tuberosum) cultivars to powdery scab (caused by *Spongospora subterranea* f. sp. *subterranea*), and relationships between tuber and root infection. Australas Plant Pathol 32:377–385

Falloon R, Merz U, Lister R, Wallace A, Hayes S (2011) Morphological enumeration of resting spores in sporosori of the plant pathogen *Spongospora subterranea*. Acta Protozool 50:121–132

Flath K, Przetakiewicz J, van Rijswick PCJ, Ristau V, van Leeuwen GCM (2014) Interlaboratory tests for resistance to *Synchytrium endobioticum* in potato by the Glynne-Lemmerzahl method. OEPP/EPPO Bull 44:510–517

Flier WG, Kroon LPNM, Hermansen A, Van Raaij HMG, Speiser B, Tamm L, Fuchs JG, Lambion J, Razzaghian J, Andrivon D, Wilcockson S, Leifert C (2007) Genetic structure and pathogenicity of populations of *Phytophthora infestans* from organic potato crops in France, Norway, Switzerland and the United Kingdom. Plant Pathol 56:562–572

Fonseka DL, Gudmestad NC (2016) Spatial and temporal sensitivity of *Alternaria* species associated with potato foliar diseases to demethylation inhibiting and anilino-pyrimidine fungicides. Plant Dis 100(9):1848–1857

Forbes GA (2012) Using host resistance to manage potato late blight with particular reference to developing countries. Potato Res 55(3–4):205–216

Forbes GA, Escobar XC, Ayala CC, Revelo J, Ordoñez ME, Fry BA, Doucett K, Fry WE (1997) Population genetic structure of *Phytophthora infestans* in Ecuador. Phytopathology 87:375–380

Forbes GA, Goodwin SB, Drenth A, Oyarzun P, Ordoñez ME, Fry WE (1998) A global marker database for Phytophthora infestans. Plant Dis 82:811–818

Forbes GA, Fry WE, Andrade-Piedra JL, Shtienberg D (2008) Simulation models for potato late blight management and ecology. In: Ciancio A, Mukerji K (eds) Integrated management of diseases caused by fungi, phytoplasma and bacteria. Integrated management of plant pests and diseases, vol 3. Springer, Dordrecht

Fry WE (2016) *Phytophthora infestans*: new tools (and old ones) lead to new understanding and precision management. Annu Rev Phytopathol 54:529–547

Fry WE, Birch PRJ, Judelson HS, Grünwald NJ, Danies G, Everts KL, Gevens BK, Gugino BK, Johnson DA, Johnson SB, McGrath MT, Myers KL, Ristaino JB, Roberts PD, Secor G, Smart CD (2015) Five reasons to consider *Phytophthora infestans* a reemerging pathogen. Phytopathology 105(7):966–981

Gagnon MC, vander Lee TAJ, Bonants PJM, Smith DS, Li X, Lévesque CA, Bilodeau GJ (2016) Development of polymorphic microsatellite loci for potato wart from next-generation sequence data. Phytopathology 106:636–644

Gannibal PB, Orina AS, Mironenko NV, Levitin MM (2014) Differentiation of the closely related species, *Alternaria solani* and *A. tomatophila*, by molecular and morphological features and aggressiveness. Eur J Plant Pathol 139:609–623

Garrett KA, Dendy SP (2001) Cultural practices in potato late blight management. In: Complementing resistance to late blight in the Andes, 13-16 February 2001. International Potato Center, Cochabamba, Bolivia

Gau R, Merz U, Falloon R, Brunner P (2013) Global genetics and invasion history of the potato powdery scab pathogen, *Spongospora subterranea* f.sp. *subterranea*. PLoS One 8(6):e67944. https://doi.org/10.1371/journal.pone.0067944

Genet R, Falloon R, Braithwaite M, Wallace A, Fletcher J, Nott H, Braam F (1996) Disease resistance and chemical for control of Powdery scab (*Spongospora subterranea*): progress toward integrated control in New Zealand. In: Abstracts of the 13th triennial conference of the European association for potato research, Veldhoven, The Netherlands, 14–19 July 1996

Ghislain M, Byarugaba AA, Magembe E, Njoroge AW, Rivera C, Roman ML, Tovar JC, Gamboa S, Forbes G, Kreuze JF, Barekye A, Kiggundu A (2018) Stacking three late blight resistance genes from wild species directly into African highland potato varieties confers complete field resistance to local blight races. Plant Biotechnol J. ISSN 1467-7652. Early view 22 Nov 2018. 11 p

Gilchrist E, Soler J, Merz U, Reynaldi S (2011) Powdery scab effect on the potato Solanum tuberosum ssp. andigena growth and yield. Trop Plant Pathol 36:350–355

Goodwin SB, Spielman LJ, Matuszak JM, Bergeron SN, Fry WE (1992) Clonal diversity and genetic differentiation of *Phytophthora infestans* populations in northern and central Mexico. Phytopathology 82:955–961

Gorgiladze L, Meparishvili G, Sikharulidze Z, Natsarishvili K, Meparishvili S (2014) First report of *Synchytrium endobioticum* causing potato wart in Georgia. New Dis Rep 30:4

Goss EM, Tabima JF, Cooke DEL, Restrepo S, Fry WE, Forbes GA, Fieland VJ, Cardenas M, Grünwald NJ (2014) The Irish potato famine pathogen Phytophthora infestans originated in central Mexico rather than the Andes. Proc Natl Acad Sci U S A 111:8791–8796. PMID: 24889615

Groth J, Song Y, Kellermann A, Schwarzfischer A (2013) Molecular characterisation of resistance against potato wart races 1, 2, 6 and 18 in a tetraploid population of potato (*Solanum tuberosum* subsp. *tuberosum*). J Appl Genet 54:169–178

Gupta J (2016) Efficacy of biocontrol agents against *Phytophthora infestans* on potato. Int J Eng Sci Comput 6(9):2249–2251

Haas BJ, Kamoun S, Zody MC, Jiang RHY, Handsaker RE, Cano LM et al (2009) Genome sequence and analysis of the Irish potato famine pathogen *Phytophthora infestans*. Nature 461:393–398

Hampson MC (1993) History, biology and control of potato wart disease in Canada. Can J Plant Pathol 15:223–244

Harbaoui K, Hamada W, Li Y, Vleeshouwers VG, van der Lee T (2014) Increased difficulties to control late blight in Tunisia are caused by a genetically diverse *Phytophthora infestans* population next to the clonal lineage NA-01. Plant Dis 98(7):898–908

Harrison MD, Venette JR (1970) Chemical control of potato early blight and its effect on potato yield. Am J Potato Res 47(3):81–86

Harrison J, Searle R, Williams N (1997) Powdery scab disease of potato—a review. Plant Pathol 46:1–25

Hausladen H, Aselmeyer A (2017) Studies about infection of different *Alternaria solani* isolates on *Solanum tuberosum, Lycopersicon esculentum* and *Solanum nigrum*. PPO-Special Report no 18: 201

Haverkort AJ, Struik PC, Visser RGF, Jacobsen E (2009) Applied biotechnology to combat late blight in potato caused by *Phytophthora infestans*. Potato Res 52(3):249–264

Haverkort AJ, Boonekamp PM, Hutten R, Jacobsen E, Lotz LAP, Kessel GJT, Vossen JH, Visser RGF (2016) Durable late blight resistance in potato through dynamic varieties obtained by cisgenesis: scientific and societal advances in the DuRPh project. Potato Res 59(1):35–66

Hehl R, Faurie E, Hesselbach J, Salamani F, Whitham S, Baker B, Gebhardt C (1999) TMV resistance gene *N* homologues are linked to *Synchytrium endobioticum* resistance in potato. Theor Appl Genet 98:379–386

Hermansen A, Hannukkala A, Hafskjold Nærstad R, Brurberg MB (2000) Variation in populations of *Phytophthora infestans* in Finland and Norway: mating type, metalaxyl resistance and virulence phenotype. Plant Pathol 49:11–22

Horsfield A, Wicks T, Davies K, Wilson D, Paton S (2010) Effect of fungicide use strategies on the control of early blight (*Alternaria solani*) and potato yield. Australas Plant Pathol 39(4):368–375

Hughes I (1980) Powdery scab (*Spongospora subterranea*) of potatoes in Queensland: occurrence, cultivar susceptibility, time of infection, effect of soil pH, chemical control. Aust J Exp Agric Animl Husb 20:625–632

International Potato Center (2007) Annual report 2007: root and tubers the overlooked opportunity. International Potato Center

Jiang RH, Tyler BM (2012) Mechanisms and evolution of virulence in oomycetes. Annu Rev Phytopathol 50:295–318

Jiang R, Li J, Tian Z, Du J, Armstrong M, Baker K, Tze-Yin Lim J, Vossen JH, He H, Portal L, Zhou J, Bonierbale M, Hein I, Lindqvist-Kreuze H, Xie C (2018) Potato late blight field resistance from QTL dPI09c is conferred by the NB-LRR gene R8. J Exp Bot 69(7):1545–1555

Jo KR, Visser RG, Jacobsen E, Vossen JH (2015) Characterization of the late blight resistance in potato differential MaR9 reveals a qualitative resistance gene, R9a, residing in a cluster of Tm-2 2 homologs on chromosome IX. Theor Appl Genet 128(5):931–941

Johanson A, Thurston HD (1990) The effect of cultivar maturity on the resistance of potatoes to early blight caused by *Alternaria solani*. Am J Potato Res 67(9):615–623

Johnson D, Cummings T (2015) Effect of powdery scab root galls on yield of potato in the Columbia Basin. Potato Prog 15:1–12

Jones D (1978) Scanning electron microscopy of cystosori of *Spongospora subterranea*. Trans Br Mycol Soc 70:292–293

Jones R, Harrison B (1969) The behavior pf potato mop-top virus in soil, and evidence for its transmission by *Spongospora subterranea* (Wallr.) Lagerh. Ann Appl Biol 63:1–17

Jones JB, Jones JP, Stall RE, Zitter TA (1993) Compendium of tomato diseases. American Phytopathological Society, St. Paul

Judelson HS (2017) Metabolic diversity and novelties in the oomycetes. Annu Rev Microbiol 71:21–39

Jupe F, Witek K, Verweij W, Śliwka J, Pritchard L, Etherington GJ, Maclean D, Cock PJ, Leggett RM, Bryan GJ, Cardle L (2013) Resistance gene enrichment sequencing (RenSeq) enables reannotation of the NB-LRR gene family from sequenced plant genomes and rapid mapping of resistance loci in segregating populations. Plant J 76(3):530–544

Kamoun S, Furzer O, Jones JD, Judelson HS, Ali GS, Dalio RJ, Roy SG, Schena L, Zambounis A, Panabieres F, Cahill D, Ruocco M, Figueiredo A, Chen XR, Hulvey J, Stam R, Lamour K, Gijzen M, Tyler BM, Grünwald NJ, Mukhtar MS, Tome D, Tor M, Van den Ackerveken G, McDowell J, Daay F, Fry WE, Lindqvist-Kreuze H, Meijer HJ, Petre B, Ristaino J, Yoshida K, Birch PR, Gover F (2014) The top 10 oomycete pathogens in molecular plant pathology. Mol Plant Pathol 16(4):413–434. https://doi.org/10.1111/mpp.12190

Kapsa J (2004) Early blight (*Alternaria spp.*) in potato crops in Poland and results of chemical protection. J Plant Protect Res 44(3):231–238

Kassa B, Sommartya T (2006) Effect of intercropping on potato late blight, *Phytophthora infestans* (Mont.) de Bary development and potato tuber yield in Ethiopia. Kasetsart J (Nat Sci) 40:914–924

Kirk WW, Abu-El Samen FM, Muhinyuza JB, Hammerschmidt R, Douches DS, Thill CA, Groza H, Thompson AL (2005) Evaluation of potato late blight management utilizing host plant resistance and reduced rates and frequencies of fungicide applications. Crop Prot 24(11):961–970

Knaus BJ, Tabima JF, Davis CE, Judelson HS, Grünwald NJ (2016) Genomic analyses of dominant US clonal lineages of *Phytophthora infestans* reveals a shared common ancestry for clonal lineages US11 and US18 and a lack of recently shared ancestry among all other US lineages. Phytopathology 106(11):1393–1403

Kuhn PJ (1984) Mode of action of carboxamides. In: Trinci APJ, Ryley JF (eds) Mode of action of antifungal agents. Cambridge University Press, Cambridge, pp 155–183

Lambert DH, Powelson ML, Stevenson WR (2005) Nutritional interactions influencing diseases of potato. Am J Potato Res 82(4):309–319

Landschoot S, Carrette J, Vandecasteele M, De Baets B, Höfte M, Audenaert K, Haesaert G (2017a) Identification of *A. arborescens*, *A. grandis*, and *A. protenta* as new members of the European *Alternaria* population on potato. Fungal Biol 121:172–188

Landschoot S, Vandecasteele M, Carrette J, De Baets B, Höfte M, Audenaert K, Haesaert G (2017b) Assessing the Belgian potato *Alternaria* population for sensitivity to fungicides with diverse modes of action. Eur J Plant Pathol 148(3):657–672

Larousse M, Galiana E (2017) Microbial partnerships of pathogenic oomycetes. PLoS Pathog 13(1):e1006028. https://doi.org/10.1371/journal.ppat.1006028

Leiminger JH, Hausladen H (2012) Early blight control in potato using disease-orientated threshold values. Plant Dis 96(1):124–130

Leiminger JH, Hausladen H (2014) Untersuchungen zur Befallsentwicklung und Ertragswirkung der Dürrfleckenkrankheit (*Alternaria* spp.) in Kartoffelsorten unterschiedlicher Reifegruppe. Gesunde Pflanzen 66:29–36

Leiminger JH, Auinger H-J, Wenig M, Bahnweg G (2013) Genetic variability among *Alternaria solani* isolates from potatoes in Southern Germany based on RAPD-profiles. J Plant Dis Protect 120(4):164–172

Leiminger JH, Adolf B, Hausladen H (2014) Occurrence of the F129L mutation in *Alternaria solani* populations in Germany in response to QoI application, and its effect on sensitivity. Plant Pathol 63:640–650

Li J, Lindqvist-Kreuze H, Tian Z, Liu J, Song B, Landeo J, Portal L, Gastelo M, Frisancho J, Sanches L, Meijer D, Xie C, Bonierbale M (2012a) Conditional QTL underlying resistance to late blight in a diploid potato population. Theor Appl Genet 124(7):1339–1350

Li Y, van der Lee TAJ, Evenhuis A, van den Bosch GBM, van Bekkum PJ, Förch MG, van Gent-Pelzer MPE, van Raaij HMG, Jacobsen E, Huang SW, Govers F, Vleeshouwers VGAA, Kessel GJT (2012b) Population dynamics of *Phytophthora infestans* in the Netherlands reveals expansion and spread of dominant clonal lineages and virulence in sexual offspring. G3 (Bethesda, Md.) 2(12):1529–1540

Li Y, Shen H, Zhou Q, Qian K, van der Lee T, Huang S (2017) Changing ploidy as a strategy: The Irish Potato Famine pathogen shifts ploidy in relation to its sexuality. Mol Plant-Microbe Interact 30(1):45–52

Liljeroth E, Lankinen Å, Wiik L, Burra DD, Alexandersson E, Andreasson E (2016) Potassium phosphite combined with reduced doses of fungicides provides efficient protection against potato late blight in large-scale field trials. Crop Prot 86:42–55

Lindqvist-Kreuze H, Gastelo M, Perez W, Forbes GA, de Koeyer D, Bonierbale M (2014) Phenotypic stability and genome-wide association study of late blight resistance in potato genotypes adapted to the tropical highlands. Phytopathology 104(6):624–633

Ma Z, Felts D, Michailides TJ (2003) Resistance to azoxystrobin in *Alternaria* isolates from pistachio in California. Pestic Biochem Physiol 77:66–74

MacDonald W, Peters R, Coffin R, Lacroix C (2007) Effect of strobilurin fungicides on control of early blight (*Alternaria solani*) and yield of potatoes grown under two N fertility regimes. Phytoprotection 88(1):9–15

Mäkäräinen E, Rita H, Teperi E, Valkonen J (1994) Resistance to *Spongospora subterranea* in tuber-bearing and non tuber-bearing Solanum spp. Potato Res 37:123–127

Malcolmson JF, Black W (1966) New R genes in *Solanum demissum* Lindl. and their complementary races of *Phytophthora infestans* (Mont.) de Bary. Euphytica 15(2):199–203

Mallik I, Arabiat S, Pasche JS, Bolton MD, Patel JS, Gudmestad NC (2014) Molecular characterization and detection of mutations associated with resistance to succinate dehydrogenase-inhibiting fungicides in *Alternaria solani*. Phytopathology 104(1):40–49

Mariette N, Mabon R, Corbière R, Boulard F, Glais I, Marquer B, Pasco C, Montarry J, Andrivon D (2016) Phenotypic and genotypic changes in French populations of *Phytophthora infestans*: are invasive clones the most aggressive? Plant Pathol 65:577–586

Melnik PA (1998) Wart disease of potato, Synchytrium endobioticum (Schilbersky) Percival. EPPO Technical documents 1032, Paris

Merz U, Falloon R (2009) Review: powdery scab of potato—increased knowledge of pathogen biology and disease epidemiology for effective disease management. Potato Res 52:17–37

Merz U, Lees A, Sullivan L, Schwärzel R, Hebeisen T, Kirk H, Bouchek-Mechiche K, Hofferbert H (2012) Powdery scab resistance in *Solanum tuberosum*: an assessment of cultivar x environment effect. Plant Pathol 61:29–36

Metz N, Adolf B, Chaluppa N, Hückelhoven R, Hausladen H (2019) Occurrence of sdh mutations in German Alternaria solani isolates and potential impact on boscalid sensitivity in vitro, in the greenhouse and in the field. Plant Dis. https://doi.org/10.1094/PDIS-03-19-0617-RE

Montes MS, Nielsen BJ, Schmidt SG, Bødker L, Kjøller R, Rosendahl S (2016) Population genetics of *Phytophthora infestans* in Denmark reveals dominantly clonal populations and specific alleles linked to metalaxyl-M resistance. Plant Pathol 65:744–753

Nærstad R, Hermansen A, Bjor T (2007) Exploiting host resistance to reduce the use of fungicides to control potato late blight. Plant Pathol 56(1):156–166

Nelson R, Orrego R, Ortiz O, Tenorio J, Mundt C, Fredrix M, Vien NV (2001) Working with resource-poor farmers to manage plant diseases. Plant Dis 85:684–695

Neuhauser S, Bulman S, Kirchmair M (2010) Plasmodiophorids: the challenge to understand soil-borne, obligate biotrophs with a multiphasic life cycle. In: Gherbawy Y, Voigt K (eds) Molecular identification of fungi. Springer, Berlin, pp 51–78

Nitzan N, Cummings T, Johnson D, Miller J, Batchelor D, Olsen C, Quick R, Brown C (2008) Resistance to root galling caused by the powdery scab pathogen *Spongospora subterranea* in potato. Plant Dis 92:1643–1649

Nitzan N, Haynes K, Miller J, Johnson D, Cummings T, Batchelor D, Olsen C, Brown C (2010) Genetic stability in potato germplasm for resistance to root galling caused by the pathogen *Spongospora subterranea*. Am J Potato Res 87:497–501

Nowicki M, Foolad MR, Nowakowska M, Kozik EU (2012) Potato and tomato late blight caused by *Phytophthora infestans*: an overview of pathology and resistance breeding. Plant Dis 96:4–17

Obidiegwu JE, Flath K, Gebhardt C (2014) Managing potato wart: a review of present research status and future perspective. Theor Appl Genet 127:763–780

Odilbekov F, Edin E, Garkava-Gustavsson L, Hovmalm HP, Liljeroth E (2016) Genetic diversity and occurrence of the F129L substitutions among isolates of *Alternaria solani* in south-eastern Sweden. Hereditas 153:10

Oh SK, Young C, Lee M, Oliva R, Bozkurt TO, Cano LM, Win J, Bos JIB, Liu HY, Van Damme M, Morgan W, Choi D, Van der Vossen EAG, Vleeshouwers VGAA, Kamoun S (2009) In planta expression screens of *Phytophthora infestans* RXLR effectors reveal diverse phenotypes, including activation of the *Solanum bulbocastanum* disease resistance *protein* Rpi-blb2. Plant Cell 21:2928–2947

Ortiz O, Nelson R, Olanya M, Thiele G, Orrego R, Pradel W, Kakuhenzire R, Woldegiorgis G, Gabriel J, Vallejo J, Xie K (2019) Human and technical dimensions of potato integrated pest management using farmer field schools: International Potato Center and partners' experience with potato late blight management. J Integr Pest Manag 10:1–8

Osorio-Giraldo I, Orozco-Valencia M, Gutiérrez-Sánchez P, González-Jaimes P, Marin-Montoya M (2012) Variabilidad genética de *Spongospora subterranea* f. sp. *subterranea* en Colombia. Bioagro 24:151–162

Pacifico D, Paris R (2016) Effect of organic potato farming on human and environmental health and benefits from new plant breeding techniques. Is it only a matter of public acceptance? Sustainability 8(10):1054

Pacilly FCA, Groot JCJ, Hofstede GJ, Schaap BF, Lammerts van Bueren ET (2016) Analyzing potato late blight control as a social-ecological system using fuzzy cognitive mapping. Agron Sustain Dev 36:35

Pasche JS, Wharam CM, Gudmestad NC (2004) Shift in sensitivity of *Alternaria solani* in response to QoI fungicides. Plant Dis 88(2):181–187

Pasche JS, Piche LM, Gudmestad NC (2005) Effect of the F129L mutation in *Alternaria solani* on fungicides affecting mitochondrial respiration. Plant Dis 89(3):269–278

Perez W, Forbes G (2010) Potato late blight: technical manual. International Potato Center (CIP), Lima. http://www.cipotato.org/publications/pdf/005446.pdf

Perez WG, Gamboa JS, Falcon YV, Coca M, Raymundo RM, Nelson RJ (2001) Genetic structure of Peruvian populations of *Phytophthora infestans*. Phytopathology 91(10):956–965

Phillips SL, Shaw MW, Wolfe MS (2005) The effect of potato variety mixtures on epidemics of late blight in relation to plot size and level of resistance. Ann Appl Biol 147:245–252

Pilet F, Chacón G, Forbes GA, Andrivon D (2006) Protection of susceptible potato cultivars against late blight in mixtures increases with decreasing disease pressure. Phytopathology 96(7):777–783

Plata G (1998) Fenotipos de virulencia en Morochata y tipo sexual de apareamiento en Bolivia de *Phytophthora infestans* que afectan al cultivo de la en la zona de Morochata y determinación del tipo sexual de apareamiento de *Phytophthora infestans* en Bolivia. Tesis Ing. Agr., Facultad de Ciencias Agrícolas Pecuarias y Forestales "Martín Cárdenas" Universidad Mayor de San Simón. Cochabamba, Bolivia

Plich J, Przetakiewicz J, Śliwka J, Flis B, Wasilewicz-Flis I, Zimnoch-Guzowska E (2018) Novel gene Sen2 conferring broad-spectrum resistance to *Synchytrium endobioticum* mapped to potato chromosome XI. Theor Appl Genet 131:2321–2331. In: New locus *Sen2* conferring potato resistance against seven virulent pathotypes of *Synchytrium endobioticum* European Association for Potato Research (EAPR), 20th EAPR Triennial Conference

Przetakiewicz J (2008) Assessment of the resistance of potato cultivars to *Synchytrium endobioticum* (Schilb.) Per. in Poland. EPPO Bull 38:211–215

Przetakiewicz J (2014a) Manual of security sensitive microbes and toxins: *Synchytrium endobioticum*. Liu D (ed). CRC Press, Boca Raton, pp 823–829

Przetakiewicz J (2014b) First report of *Synchytrium endobioticum* (potato wart disease) pathotype 18(T1) in Poland. Plant Dis 98:688

Przetakiewicz J (2015a) First report of new pathotype 39(P1) of *Synchytrium endobioticum* causing potato wart disease in Poland. Plant Dis 99(2):285

Przetakiewicz J (2015b) The viability of winter sporangia of *Synchytrium endobioticum* (Schilb.) Perc. from Poland. Am J Potato Res 92(6):704–708

Przetakiewicz J (2016) A modification of the Potoček's tube test for diagnostic of *Synchytrium endobioticum* (Schilb.) Perc. a causal agent of potato wart disease. Indian Phytopathol 69(4s):260–265

Przetakiewicz J (2017) Sampling, maintenance and pathotype identification of *Synchytrium endobioticum* (schilb.) Perc. Plant Breed Seed Sci 76:29–36

Przetakiewicz J, Plich J (2017) Assessment of potato resistance to *Synchytrium endobioticum*. Plant Breed Seed Sci 76:37–43

Pscheidt JW (1985) Epidemiology and control of potato early blight, caused by *Alternaria solani*. Ph.D. Dissertation. University of Wisconsin, Madison, WI, USA

Qu X, Christ B (2004) Genetic variation and phylogeny of *Spongospora subterranea* f.sp. *subterranean* based on ribosomal DNA sequence analysis. Am J Potato Res 81:385–394

Qu X, Christ B (2006) The host range of *Spongospora subterranea* f.sp. *subterranea* in the United States. Am J Potato Res 83:343–348

Rekad FZ, Cooke DEL, Puglisi I, Randall E, Guenaoui Y, Bouznad Z, Evoli M, Pane A, Schena L, Magnano di San Lio G, Cacciola SO (2017) Characterization of *Phytophthora infestans* populations in northwestern Algeria during 2008-2014. Fungal Biol 121(5):467–477

Rietman H, Bijsterbosch G, Cano LM, Lee HR, Vossen JH, Jacobsen E, Visser RG, Kamoun S, Vleeshouwers VG (2012) Qualitative and quantitative late blight resistance in the potato cultivar Sarpo Mira is determined by the perception of five distinct RXLR effectors. Mol Plant-Microbe Interact 25(7):910–919

Rodenburg SY, Seidl MF, de Ridder D, Govers F (2018) Genome-wide characterization of *Phytophthora infestans* metabolism: a systems biology approach. Mol Plant Pathol 19(6):1403–1413

Rodewald J, Trognitz B (2013) *Solanum* resistance genes against *Phytophthora infestans* and their corresponding avirulence genes. Mol Plant Pathol 14(7):740–757

Rodrigues TTMS, Berbee ML, Simmons EG, Cardoso CR, Reis A, Maffia LA, Mizubuti ESG (2010) First report of *Alternaria tomatophila* and *A. grandis* causing early blight on tomato and potato in Brazil. New Dis Rep 22:28–28

Rojas A, Kirk WW (2016) Phenotypic and genotypic variation in Michigan populations of *Phytophthora infestans* from 2008 to 2010. Plant Pathol 65:1022–1033

Rotem J (1994) The genus *Alternaria*, biology and pathogenicity. APS Press, Minnesota

Runno-Paurson E, Kiiker R, Joutsjoki T, Hannukkala A (2016) High genotypic diversity found among population of *Phytophthora infestans* collected in Estonia. Fungal Biol 120(3):385–392

Saville AC, Martin MD, Ristaino JB (2016) Historic late blight outbreaks caused by a widespread dominant lineage of *Phytophthora infestans* (Mont.) de Bary. PLoS One 11(12):e0168381

Schiek B, Hareau G, Baguma Y, Medakker A, Douches D, Shotkoski F, Ghislain M (2016) Demystification of GM crop costs: releasing late blight resistant potato varieties as public goods in developing countries. Int J Biotechnol 14(2):112–131

Shah F, Falloon R, Bulman S (2010) Nightshade weeds (*Solanum* spp.) confirmed as hosts of the potato pathogens *Meloidogyne fallax* and *Spongospora subterranea* f. sp. *subterranea*. Australas Plant Pathol 39:492–498

Shah F, Falloon R, Butler R, Lister R (2012) Low amounts of *Spongospora subterranea* sporosorus inoculum cause severe powdery scab, root galling and reduced water use in potato (*Solanum tuberosum*). Australas Plant Pathol 41:219–228

Shah F, Falloon R, Butler R, Lister R, Thomas S, Curtin D (2014) Agronomic factors affect powdery scab of potato and amounts of *Spongospora subterranea* DNA in soil. Australas Plant Pathol 43:679–689

Shanthiyaa V, Saravanakumar D, Rajendran L, Karthikeyan G, Prabakar K, Raguchander T (2013) Use of *Chaetomium globosum* for biocontrol of potato late blight disease. Crop Prot 52:33–38

Shtienberg D, Fry WE (1990) Influence of host resistance and crop rotation on initial appearance of potato early blight. Plant Dis 74(11):849–852

Shtienberg D, Blachinsky D, Ben-Hador G, Dinoor A (1996) Effects of growing season and fungicide type on the development of *Alternaria solani* and on potato yield. Plant Dis 80(9):994–998

Smith DC, Rocheleau H, Chapados JT, Abbott C, Ribero S, Redhead SA, Lévesque CA, De Boer SH (2014) Phylogeny of the genus *Synchytrium* and the development of TaqMan PCR assay for sensitive detection of *Synchytrium endobioticum* in soil. Phytopathology 104(4):422–432

Sparrow L, Rettke M, Corkrey S (2015) Eight years of annual monitoring of DNA of soil-borne potato pathogens in farm soils in south eastern Australia. Australas Plant Pathol 44:191–203

Stammler G, Böhme F, Philippi J, Miessner S, Tegge V (2014) Pathogenicity of *Alternaria*-species on potatoes and tomatoes. PPO special report 16:85–96

Torres H, Pacheco M, French E (1995) Resistance of potato to powdery scab (*Spongospora subterranea*) under Andean field conditions. Am Potato J 72:355–363

van de Graaf P, Wale S, Lees A (2007) Factors affecting the incidence and severity of *Spongospora subterranea* infection and galling in potato roots. Plant Pathol 56:1005–1013

van de Vossenberg BTLH, Brankovics B, Nguyen HDT, van Gent-Pelzer MPE, Smith D, Dadej K, Przetakiewicz J, Kreuze J, Boerma M, van Leeuwen G, Levesque CA, van der Lee TAJ (2018a) The linear mitochondrial genome of the quarantine chytrid *Synchytrium endobioticum*; insights into the evolution and recent history of an obligate biotrophic plant pathogen. BMC Evol Biol 18:136

van de Vossenberg BTLH, Westenberg M, Adams I, Afanasenko O, Beniusis A, Besheva A, Boerma M, Bonants P, Choiseul J, Dekker T, Flath K, Heungens K, Karelov A, Przetakiewicz J, Schlenzig A, Yakovleva V, van Leeuwen G (2018b) EUPHRESCO Sendo: an international laboratory comparison study of molecular tests for *Synchytrium endobioticum* detection and identification. Eur J Plant Pathol 151:757–766

Van der Waals JE, Korsten L, Aveling TAS (2001) A review of early blight of potato. Afr Plant Protect 7(2):91–102

Van der Waals JE, Korsten L, Slippers B (2004) Genetic diversity among *Alternaria solani* isolates from potatoes in South Africa. Plant Dis 88:959–964

Venette JR, Harrison MD (1973) Factors affecting infection of potato tubers by *Alternaria solani* in Colorado. Am J Potato Res 50(8):283–292

Vleeshouwers VGAA, Oliver RP (2014) Effectors as tools in disease resistance breeding against biotrophic, hemibiotrophic, and necrotrophic plant pathogens. Mol Plant-Microbe Interact 27:196–206

Vleeshouwers VG, Raffaele S, Vossen JH, Champouret N, Oliva R, Segretin ME, Rietman H, Cano LM, Lokossou A, Kessel G, Pel MA (2011) Understanding and exploiting late blight resistance in the age of effectors. Annu Rev Phytopathol 49:507–531

Volz A, Tongle H, Hausladen H (2013) An integrated concept for early blight control in potatoes. PPO special report 14:12–15

Vossen JH, van Arkel G, Bergervoet M, Jo KR, Jacobsen E, Visser RG (2016) The *Solanum demissum* R8 late blight resistance gene is an Sw-5 homologue that has been deployed worldwide in late blight resistant varieties. Theor Appl Genet 129(9):1785–1796

Wale S (2000) Summary of the session on national potato production and the powdery scab situation. In: Merz U, Lees A (eds) Proc of the 1st European powdery scab workshop. Aberdeen, Scotland, pp 3–9

Wang H, Ren Y, Zhou J, Du J, Hou J, Jiang R, Wang H, Tian Z, Xie C (2017a) The cell death triggered by the nuclear localized RxLR effector PITG_22798 from *Phytophthora infestans* is suppressed by the effector AVR3b. Int J Mol Sci 18(2):409

Wang J, Fernández-Pavía SP, Larsen MM, Garay-Serrano E, Gregorio-Cipriano R, Rodríguez-Alvarado G, Grünwald NJ, Goss EM (2017b) High levels of diversity and population structure in the potato late blight pathogen at the Mexico centre of origin. Mol Ecol 26:1091–1107

Wastie R, Caligari P, Wale (1988) Assessing the resistance of potatoes to powdery scab (*Spongospora subterranea* (Wallr.) Lagerh.). Potato Res 31:167–171

Whisson SC, Boevink PC, Wang S, Birch PRJ (2016) The cell biology of late blight disease. Curr Opin Microbiol 34:127–135

Widmark AK, Andersson B, Cassel-Lundhagen A, Sandström M, Yuen J (2007) *Phytophthora infestans* in a single field in southwest Sweden early in spring: symptoms spatial distribution and genotypic variation. Plant Pathol 56:473–579

Witek K, Jupe F, Witek AI, Baker D, Clark MD, Jones JD (2016) Accelerated cloning of a potato late blight-resistance gene using RenSeq and SMRT sequencing. Nat Biotechnol 34(6):656–660

Woudenberg JHC, Truter M, Groenewald JZ, Crous PW (2014) Large-spored *Alternaria* pathogens in section Porri disentangled. Stud Mycol 79:1–47

Woudenberg JHC, Seidl MF, Groenewald JZ, de Vries M, Stielow JB, Thomma BPHJ, Crous PW (2015) *Alternaria* section *Alternaria*: species, formae speciales or pathotypes? Stud Mycol 82:1–21

Yang L, Wang D, Xu Y, Zhao H, Wang L, Cao X, Chen Y, Chen Q (2017) A new resistance gene against potato late blight originating from *Solanum pinnatisectum* located on potato chromosome 7. Front Plant Sci 8:1729

Yao Y, Li Y, Chen Z, Zheng B, Zhang L, Niu B, Meng J, Li A, Zhang J, Wang Q (2016) Biological control of potato late blight using isolates of *Trichoderma*. Am J Potato Res 93:33–42

Zhu W, Yang LN, Wu EJ, Qin CF, Shang LP, Wang ZH, Zhan J (2015) Limited sexual reproduction and quick turnover in the population genetic structure of *Phytophthora infestans* in Fujian, China. Sci Rep 5:srep10094

Zhu W, Shen L, Fang Z, Yang L, Zhang J, Sun D, Zhan J (2016) Increased frequency of self-fertile isolates in *Phytophthora infestans* may attribute to their higher fitness relative to the A1 isolates. Sci Rep 6, Article number: 29428. https://doi.org/10.1038/srep29428

Chapter 10
Bacterial Diseases of Potato

Amy Charkowski, Kalpana Sharma, Monica L. Parker, Gary A. Secor, and John Elphinstone

Abstract Bacterial diseases are one of the most important biotic constraints of potato production, especially in tropical and subtropical regions, and in some warm temperate regions of the world. About seven bacterial diseases affect potato worldwide and cause severe damages especially on tubers, the economically most important part of the plant. Bacterial wilt and back leg are considered the most important diseases, whereas potato ring rot, pink eye, and common scab are the minor. Knowledge about zebra chip is extremely rare, as it occurs in a very isolated area and is an emerging disease in New Zealand, Europe, the USA and Mexico. Potato crop losses due to bacterial diseases could be direct and indirect; and they have several dimensions, some with short-term consequences such as yield loss and unmarketability of the produce and others with long-term consequences such as economic, environmental, and social. Some of them are of national and international importance and are the major constraints to clean seed potato production, with considerable indirect effects on trade. This review focuses on *Clavibacter* spp., *Ralstonia* spp., *Pectobacterium* spp., *Dickeya* spp., *Streptomyces* spp., and *Liberibacter* spp. pathogenic to potato, and looks at the respective pathogen in terms of their taxonomy and nomenclature, host range, geographical distribution, symptoms, epidemiology, pathogenicity and resistance, significance and economic losses, and management strategies. Nevertheless, the information collected here deal more with diseases known in developed and developing countries which cause severe economic losses on potato value chain.

A. Charkowski
Colorado State University, Fort Collins, CO, USA
e-mail: amy.charkowski@colostate.edu

K. Sharma (✉) · M. L. Parker
International Potato Center, Nairobi, Kenya
e-mail: kalpanasharma@cgiar.org; m.parker@cgiar.org

G. A. Secor
Department of Plant Pathology, North Dakota State University, Fargo, ND, USA
e-mail: gary.secor@ndsu.edu

J. Elphinstone
Food and Environment Research Agency (Fera) Ltd., Sand Hutton, York, UK
e-mail: john.elphinstone@fera.co.uk

H. Campos, O. Ortiz (eds.), *The Potato Crop*,
https://doi.org/10.1007/978-3-030-28683-5_10

10.1 Brown Rot and Bacterial Wilt of Potato Caused by *Ralstonia solanacearum*

10.1.1 Taxonomy and Nomenclature

The genus *Ralstonia* is classified in the ß-Proteobacteria within the family Burkholderiaceae. The species complex of *Ralstonia solanacearum* has long been recognized as a group of phenotypically diverse strains, originally characterized as pathogenic races and biovars (Buddenhagen 1962; Hayward 1964). More recently, Fegan and Prior (2005) described four phylotypes in the species complex, each comprising multiple phylogenetic variants (sequevars) according to sequence diversity within barcoding genes (including 16S rRNA, *hrpB*, *mutS*, and *egl*). Recently, the complex has been reclassified on the basis of whole genome comparisons into three distinct species (Safni et al. 2014; Prior et al. 2016): *R. solanacearum* (Phylotype II), *R. pseudosolanacearum* (Phylotypes I and III), and *R. syzygii* (Phylotype IV) (Fig. 10.1).

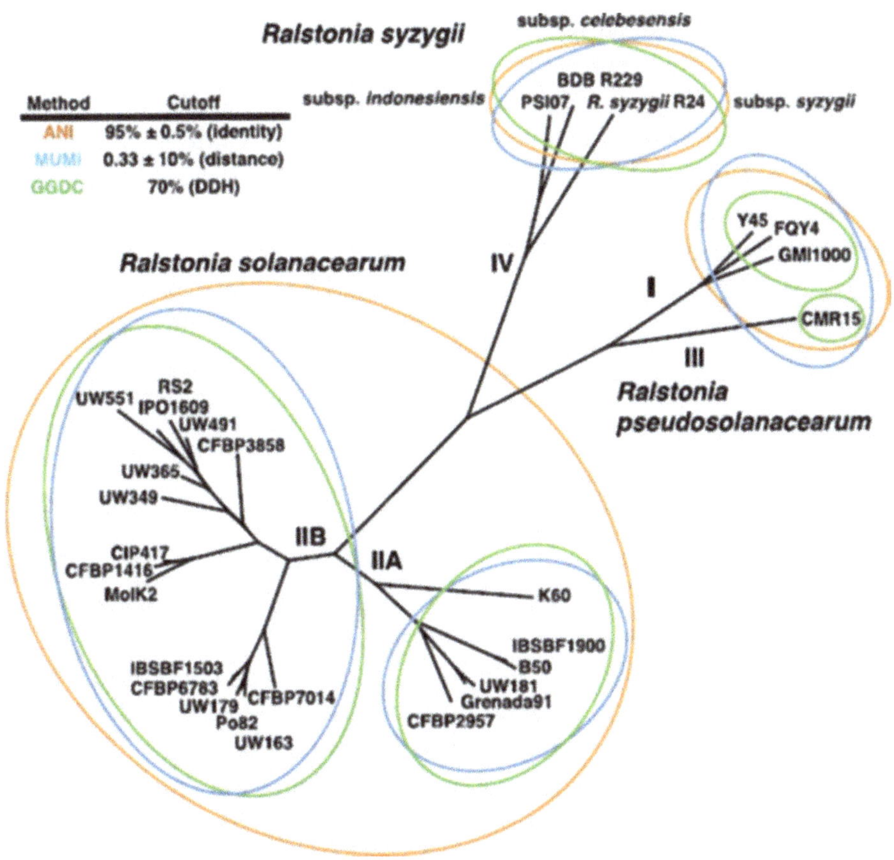

Fig. 10.1 Phylogenetic relatedness of strains within the *Ralstonia solanacearum* species complex, from Prior et al. (2016). Concatenated tree of distance matrices generated from average nucleotide identity (ANI), maximum unique matches index (MUMi), and genome-to-genome distance calculator (GGDC) to compare DNA:DNA homology (DDH)

10.1.2 Host Range

In addition to potato (*Solanum tuberosum* and *S. phureja*), the large range of economically important hosts includes banana and plantain, cucurbits, eggplant, *Eucalyptus,* ginger, groundnut, mulberry, tobacco, tomato, and many ornamental plants. *R. solanacearum*, *R. pseudosolanacearum*, and *R. syzygii* each comprise strains that were originally designated as race 1 and which occur in tropical areas all over the world, attacking a very wide host range of over 250 hosts in 54 botanical families. Some *R. solanacearum* genotypes within Phylotype II (sequevars IIA-6, IIA-24, IIA-41, IIA-53, IIB-3, IIB-4, and IIB-25), originally described as race 2, cause Moko disease of *Musa* spp. (banana and plantain) and *Heliconia*. *R. syzygii* comprises three subspecies: subsp. *syzygii* found on clove, subsp. *celebesensis* the cause of banana blood disease and subsp. *indonesiensis* found on solanaceaous crops (potato, tomato, and chilli pepper) as well as clove.

There are strains within each of the four phylotypes that can cause bacterial wilt and brown rot of potato; however, a single strain of *R. solanacearum* (sequevar 1) within Phylotype IIB (PIIB1), formerly known as race 3/biovar 2, is most widely associated with potato. This genotype has a lower *in planta* temperature optimum (27 °C) than most other genotypes (35 °C), often occurring in latent (symptomless) infections at high altitudes in the tropics and in subtropical and temperate potato-growing areas. This strain can also cause bacterial wilt of tomato and can survive in perennial nightshades, which act as secondary hosts. For example, the PIIB1 strain has overwintered in infected underground stolons of *Solanum dulcamara* (woody nightshade), growing along some European rivers, spreading to potato crops when the bacteria were transmitted in contaminated river water which was used for irrigation (Janse et al. 1998). The same strain has also been spread internationally on geranium cuttings produced in Africa and Central America (Williamson et al. 2002).

10.1.3 Geographical Distribution

Phylotype I strains are regarded to be of Asian origin, Phylotype II strains are thought to be of South American origin, whereas Phylotype III appears to have evolved in Africa and Phylotype IV in Indonesia. The *R. solanacearum* species complex is widely designated as a quarantine organism in many countries in an effort to prevent its movement across geographical borders. Nevertheless, the PIIB1 strain has spread from its origin to many potato-growing areas worldwide, presumably with movement in trade of infected seed tubers (Elphinstone 2005). However, this strain has never been reported on potato in the USA, despite it having been introduced on infected geraniums (Williamson et al. 2002) and findings of other Phylotype II strains on potato and other hosts in the southern states. The PIIB1 strain has in fact been designated as a select agent in the USA because of its perceived potential to pose a severe threat to agriculture. Moko disease-causing Phylotype II strains mainly occur in South and Central America and the Carribean, but also appear to have spread to the Philippines where the same strains have been found on cooking banana (ABB and BBB genotypes), causing the so-called bugtok disease.

10.1.4 Symptoms

Wilting is a common symptom of infections of most hosts with all phylotypes. The youngest leaves usually wilt first, appearing at the warmest time of day. Wilting may be visible in only one stem, on one side of a plant or even sectoral in part of a leaf, depending where vascular infections are restricted in sectors of stems and leaf petioles. Leaves may become bronzed or chlorotic and epinasty may occur. Wilting of the whole plant may follow rapidly if environmental conditions are favorable for pathogen growth. As the disease develops, a brown discoloration of the xylem vessels in the stem may be observed above the soil line and adventitious roots may develop. A creamy, slimy mass of bacteria exudes from vascular bundles when the stem is cut. Wilting and collapse of whole plants can lead to rapid death.

Symptoms on infected potato tubers may or may not be visible, depending on the state of development of the disease in relation to the prevailing temperature (Fig. 10.2). Cutting a diseased tuber will reveal browning and necrosis of the vascular ring and in adjacent tissues. A creamy fluid exudate usually appears spontaneously from the vascular ring at the cut surface. Bacterial ooze can emerge from the eyes and stem-end attachment of whole tubers, to which soil adheres. If cut stem or tuber vascular tissue is placed in water, threads of bacterial ooze exude.

Fig. 10.2 Symptoms of potato brown rot with bacteria oozing from cut vascular tissues (**a**) and eyes (**b**) (UK Crown Copyright—Courtesy of Fera Science Ltd.), and wilted plant in the field (**c**) (Courtesy of International Potato Center)

10.1.5 Epidemiology

Although often described as soilborne pathogens, survival is usually short lived at low temperature in bare soil but is significant in alternative wild host plants (especially perennial nightshade species growing in waterlogged conditions or overwintering volunteers from susceptible crops). The bacteria have been shown to survive in a viable but nonculturable (VBNC) form under stress conditions in soil and water (Kong et al. 2014), but the epidemiological relevance of this is unclear. Disease is usually most severe at temperatures of 24–35 °C, although the PIIB1 strain is more cold tolerant than other strains. High soil moisture or periods of wet weather or rainy seasons are associated with high disease incidence. Entry into plants is usually through root injuries from where the bacteria move by colonization of the xylem where they adhere by polar attraction to the vessel walls, or invade the lumen. Blocking of the vessels by bacterial extracellular polysaccharide (EPS) is considered to be the major cause of wilting. The bacteria can also be transmitted mechanically during pruning operations or when cuttings are taken for propagation.

Long distance movement of vegetative propagating material (e.g. seed potatoes, rhizomes of ginger and turmeric, and banana suckers) can carry latent infections. Natural infection of true seed has only been established for groundnut in Indonesia and China. There have been findings of contaminated seed of other hosts (including tomato, Capsicum, eggplant, and soyabean) although seed infection and transmission has not been substantiated. At present, transmission through water or soil and movement of infected vegetative plant parts are considered to be more important for most host plants than transmission via true seed. In contrast, some strains of *R. solanacearum* and *R. syzygii*, which cause Moko disease and blood disease of banana and Sumatra disease of clove are transmitted by insects (including pollinating flies, bees, wasps, and thrips on banana and xylem-feeding spittlebugs of *Hindola* spp. on clove) with potential for rapid spread over several kilometers.

10.1.6 Pathogenicity Determinants and Resistance

Factors determining pathogenicity, virulence, and avirulence in the bacteria have been recently reviewed (Genin and Denny 2012; Meng 2013). After invasion into root intercellular spaces, expression of *hrpB* is induced, in response to plant signals, activating other *hrp* (hypersensitive response and pathogenicity) genes in construction (*hrpK* and *hrpY*) and regulation (*hrpB* and *hrpG*) of a type 3 secretion system (T3SS), a molecular syringe which is essential for pathogenicity. The bacteria proliferate in intercellular spaces with the aid of a variety of effector proteins, secreted through the T3SS, which suppress plant defenses by interfering with host signal pathways. *HrpB* positively regulates expression not only of *hrp* genes but also of genes encoding a number of plant cell wall-degrading exoproteins secreted through a type II secretion system (T2SS). These include polygalacturonases (PehA, PehB, and PehC), endoglucanase (egl), pectin methylesterase (Pme), and cellobiohydrolase (cbhA), which contribute not only to invasion of xylem vessels, leading to systemic infection, but also to quantitative control of virulence. When cell densities reach a threshold, *PhcA* is activated by build-up of a volatile quorum sensing signal, 3-hydroxy palmitic acid methyl ester (3-OH PAME), inducing genes (*epsABCDEF & P*) controlling biosynthesis of exopolysaccharide (EPS), a major virulence factor. Transcriptome analysis (Jacobs et al. 2012) has also confirmed expression of these and other genes encoding various virulence traits at high cell densities during host infection, including genes imparting stress tolerance (*bcp*, *acrA*, *acrB*, and *dps*) and motility and attachment structures (*pilA* and *fliC*).

Available field resistance to the *R. solanacearum* species complex is limited and tends to be unstable under different environmental conditions and/or strain variability. Traditional breeding has not yet yielded new resistant varieties because of the difficulty in transferring multiple unknown genes from wild germplasm with polygenic resistance into cultivars without cotransferring undesirable linked traits. Furthermore, high-level resistance to host colonization as well as to disease development is needed to avoid the risk of spreading the pathogens in symptomless latent infections. Most studies on the genetic basis of resistance to bacterial wilt have been conducted in the model plants *Arabidopsis thaliana* and *Medicago truncatula* (Huet 2014). Quantitative trait loci (QTL) have been identified that include R genes that encode proteins that recognize bacterial effector avirulence (AVR) proteins, triggering resistance to the bacterium. Transfer of selected R genes from *A. thaliana* into tomato conferred immunity to the *Ralstonia* strain with the corresponding AVR gene (Narusaka et al. 2013). Both broad-spectrum and strain-specific quantitative trait loci (QTLs) have been identified in tomato (Wang et al. 2013), tobacco (Qian et al. 2013) and eggplant (Salgon et al. 2017). Discovery of possible resistance/avirulence (R/Avr) gene for gene resistance mechanisms is particularly interesting since known bacterial effectors can be used to screen for homologous resistance genes in related crops, including potato.

10.1.7 Significance and Economic Loss

Recently ranked by international phytobacteriologists as the second most important of all plant pathogenic bacteria after *Pseudomonas syringae* (Mansfield et al. 2012), the plant pathogenic *Ralstonia* spp. have an extremely wide geographic distribution and host range. On potato alone, it is thought to be responsible for approximately US$1 billion in losses each year, affecting some 3 m farm families over 1.5 m ha in around 80 countries (Elphinstone 2005). Moko disease has affected banana and plantain over thousands of square miles in Central and South America, particularly affecting small susbsistence farmers. In many countries in which the organism has quarantine status, important losses occur as a result of regulatory eradication measures and restrictions introduced on further production on contaminated land.

10.1.8 Management

Disease management remains limited and is hampered by the ability of the pathogens to survive in wet environments on plant debris or in asymptomatic weed hosts, which act as inoculum reservoirs. In the absence of any curative chemical control methods, prevention of bacterial wilt largely relies on the availability of pathogen-free planting material and effective surveillance and monitoring to protect areas free from the bacteria. For potato, effective disease management has mainly resulted from the use of limited generation seed multiplication from pathogen-free nuclear stocks with zero tolerances for the disease in official seed certification programs. Regular post-harvest testing of seed potato tubers is usually also necessary to avoid distribution of latent infections. Similarly, for other vegetatively propagated crops, there is a need to ensure planting material has been tested free of infection and that there are restrictions on the movement of planting material from affected to disease-free areas. Disinfection of pruning and harvesting tools is important in preventing spread of disease e.g. in banana and plantain production. In areas where the pathogen could be spread in contaminated irrigation water, prohibition of irrigation with surface water has been an effective control measure. For hydroponic glasshouse production systems, disinfection of recirculating water (e.g. using chlorine dioxide) can prevent spread of the bacterium. This effectively halted international spread of *R. solanacearum* PIIB1 in geranium cuttings produced in Central America and East Africa following export to the USA and Europe.

10.2 Bacterial Blackleg and Tuber Soft Rot Symptoms Caused by *Pectobacterium*

10.2.1 Taxonomy and Nomenclature

The genera *Pectobacterium* is a member of the ß-Proteobacteria in the family *Pectobacteriaceae* within the order *Enterobacterales*. The *Pectobacteriaceae* family also contains the genera *Brenneria*, *Dickeya*, *Lonsdalea*, and *Sodalis* (Adeolu et al. 2016). *Pectobacterium* originally belonged to the genus *Erwinia* (Winslow et al. 1917) with the name *Pectobacterium* being proposed by Waldee (1942). However, the name *Erwinia* persisted until Hauben et al. (1998), using 16S rDNA analysis, re-proposed the name *Pectobacterium*, which has been used since. There are currently 12 species of *Pectobacterium* including *P. aroidearum*, *P. atrosepticum*, *P. betavasculorum*, *P. brasiliense*, *P. cacticida*, *P. carotovorum*, *P. odoriferum*, *P. parmentieri*, *P. peruviense*, *P. polaris*, *P. punjabense*, and *P. wasabiae* (Dees et al. 2017a, b; Khayi et al. 2016; Nabhan et al. 2013; Sarfraz et al. 2018; Waleron et al. 2018; Zhang et al. 2016a, b).

10.2.2 Host Range

Pectobacterium species have a wide range of host plants with approximately a third of these overlapping with hosts for *Dickeya* species (Charkowski 2018; Ma et al. 2007a, b). For *Pectobacterium* species, hosts have been identified in at least 20 dicot families in 13 orders and 12 monocot families in 6 orders, often with only a single isolate being associated with a particular family or order. This may be due to lack of reporting rather than a clear difference in the abilities of these two genera to infect (Charkowski 2018; Ma et al. 2007a, b). However, some important specialization may exist since *Pectobacterium* appears to be found more frequently than *Dickeya* on cabbage, cotton, and mango, and *Dickeya* but not *Pectobacterium* on rice and maize. Some species such as *P. atrosepticum*, *P. betavascularum*, and *P. parmentieri* appear to have a very narrow host range, and *P. aroidearum* appears to be more virulent on monocots than other *Pectobacterium* species (Nabhan et al. 2013). The *Pectobacterium* species most commonly found on potato include *P. atrosepticum*, *P. brasiliense*, *P. carotovorum*, *P. odoriferum*, *P. parmentieri*, *P. peruviense*, *P. polaris*, and *P. punjabense*.

10.2.3 Geographical Distribution

Pectobacterium species are found on all continents where potato is grown, and are likely to be present as saprophytes in the soil, water, and are also regular inhabitants of plant roots when not causing disease. While there are likely to be some regional differences in the species distribution, some appear to be ubiquitous. For example, *P. atrosepticum, P. brasiliense, P. parmentieri*, and *P. carotovorum* are found on multiple continents (De Boer et al. 2012; Duarte et al. 2004; Kim et al. 2009; Ngadze et al. 2012; Pérombelon and Kelman 1987; Pitman et al. 2008, 2010; She et al. 2017; van der Merwe et al. 2010; van der Wolf et al. 2017; Wang et al. 2017a, b, c).

In Europe, *P. atrosepticum* has been the predominant species responsible for blackleg disease on potato, with *P. carotovorum* responsible a minority of blackleg disease incidents but often associated with soft rot in storage. Recently, at least some of these *P. carotovorum* strains were been reclassified as *P. wasabiae* and subsequently as *P. parmentieri* (Khayi et al. 2016; Nykyri et al. 2012). *P. brasiliense*, which was originally identified as causing disease on potato in Brazil (Duarte et al. 2004), has been common in the United States since at least 2001, as has *P. parmentieri* (Kim et al. 2009; Yap et al. 2004). *P. brasiliense* was not known to cause disease on potato in Europe prior to 2012–2013 but has since increased greatly in its incidence in many European countries (de Werra et al. 2015) and is now recognized as an important pathogen in Africa as well (van der Merwe et al. 2010).

10.2.4 Symptoms

Pectobacterium causes blackleg, which is a stem necrosis that originates from the planted seed tuber (Pérombelon 2002) (Fig. 10.3). Necrotic symptoms often extend several centimeters up the stem and necrotic vascular tissue is typically present inside the stem several centimeters beyond where general stem necrosis occurs. The pith of the stem is often decayed. Plant leaves may turn bright yellow and the plant will eventually wilt and die. Infected plants produce few or no tubers. *Pectobacterium* can enter daughter tubers through the xylem or through wounds caused by insects, frost damage, or harvest equipment. Once inside a tuber, it will decay the inside of the tuber, but not the tuber periderm, causing bacterial soft rot. The bacteria will also decay stems damaged by cultivation equipment or severe weather, causing aerial stem rot. In all cases, it is common to find multiple *Pectobacterium* species or *Pectobacterium* and *Dickeya* together when blackleg, aerial stem rot, or soft rot symptoms are present (Kim et al. 2009; Yap et al. 2004). In the United States, *P. partmentieri* is often found with other decay pathogens, such as *Clavibacter michiganensis* and potato rot nematode.

10.2.5 Epidemiology

The most common *Pectobacterium* strains in a region change from year to year and the strains and species present are also not consistent across a particular continent (Dees et al. 2017a, b). The species also differ in optimal and upper limits of growth temperatures. For example, *P. atrosepticum* and *P. parmentieri* die above 33 °C, but *P. carotovorum* and *P. brasiliense* can grow at temperatures up to 39 °C. Initial seed potato production relies on pathogen-free micropropagated plantlets. These plantlets are grown in greenhouses or screen houses to produce minitubers, which are used for field planting. *Pectobacterium* is sometimes found in or on minitubers, but it is more common on potato ones the tubers have been grown in the field. Each generation of potato multiplication tends to increase *Pectobacterium* incidence on potato tubers. Since *Pectobacterium* is common in the environment and can be found in soil, water, weeds, and insects, it is not feasible to produce potatoes free of this pathogen (Charkowski 2015). The bacteria may also be spread by insects (Kloepper and Schroth 1981), but the importance of insects compared to other routes of spread remains unknown. *Pectobacterium* appears to spread mainly at harvest. Bacterial numbers increase dramatically on senescing vines and the bacteria will contaminate harvest equipment and may become aerosolized during harvest. In tubers, the bacteria are found in lenticels and inside the stolon scare. Asymptomatic infestations are common, so it is not possible to visually assess seed potato lots for risk. Blackleg development is highly dependent on the environment and it is unpredictable, even when a seed lot is known to be contaminated with *Pectobacterium*. Detection protocols useful for studying *Pectobacterium* epidemiology were recently complied (Humphris et al. 2015).

Fig. 10.3 Bacterial blackleg and tuber soft rot symptoms caused by *Pectobacterium* on potato. (**a**) Plants with blackleg are shorter and have curled leaves (A), the stem is blackened on the outside (**b**), the pith inside is decayed and the xylem are brown (**c**). Brown or black decay may spread into leaves (**d**) or leaves may turn bright yellow (**e**). Tubers may have swollen lenticels and sunken lesions (**f**). The soft rot bacteria may enter the tuber through the stolen and decay the center of the tuber (**g**) (Courtesy of Amy O. Charkowski, Colorado State University)

10.2.6 Pathogenicity Determinants and Resistance

Pectobacterium pathogenicity depends upon secreted plant cell wall-degrading enzymes, although several other factors also contribute to virulence (Charkowski et al. 2012). The genetic basis for the observed host range limitation in some *Pectobacterium* species and differences in ability to grow at temperatures above 33 °C remain unknown. There are no examples of gene-for-gene resistance with this necrotrophic pathogen. The antimicrobial peptide Snakin-1 enhances resistance when overexpressed in potato 329–338 and some wild potato species exhibit resistance to *Pectobacterium* (Rietman et al. 2014), but the basis for resistance in wild potato species is poorly understood. There are no resistant commercial potato varieties, but varieties differ in tolerance. There is a large environmental component to disease development for blackleg and little effort has been made to correlate laboratory assays for tolerance with results observed on grower farms. The recent identification of numerous new *Pectobacterium* species suggests that additional novel and potentially high virulent species remain to be discovered and also that this high level of diversity will hinder development of tolerant potato varieties.

10.2.7 Significance and Economic Loss

Pectobacterium has served as a model pathogen for phytobacteriology research for longer than almost any other bacterial pathogen, except *Erwinia amylovora*, and *Pectobacterium* research has resulted in some notable firsts, such as the first demonstration of the role of quorum sensing in bacterial pathogenicity (Pirhonen et al. 1993). It remains an economically significant disease worldwide. Farmers lose millions annually to blackleg, aerial stem rot, and tuber soft rot. Of these, tuber soft rot can be particularly devastating since it occurs after the farmer has invested a full season of inputs into growing the crop.

10.2.8 Management

Pectobacterium management relies mainly on cultural practices (Charkowski 2015; Czajkowski et al. 2011). Growers initiate potato production with micropropagated plantlets that are free of *Pectobacterium*, but tubers are quickly contaminated once they are planted in fields. To reduce the risk of disease at planting, growers should fully suberize seed if they are using cut seed and they should not plant cold seed into wet ground. During the growing season, they should irrigate with ground water if possible and not overfertilize with nitrogen. Rouging infected plants is likely to spread the disease, so this is not recommended. At harvest, the bacteria will multiply on the vines as they senesce, so quickly killing potato vines may aid in reducing disease incidence the following year. Tubers should be allowed to heal before cooling storages. Good airflow and high humidity in potato warehouses will also aid in reducing soft rot in storage. High levels of carbon dioxide in warehouses will promote soft rot development.

10.3 Blackleg and Soft Rot of Potato Caused by *Dickeya*

10.3.1 Taxonomy and Nomenclature

The genus *Dickeya* is a member of the ß-Proteobacteria in the family *Pectobacteriaceae* within the order *Enterobacterales*. The *Pectobacteriaceae* family also contains the genera *Brenneria*, *Lonsdalea Pectobacterium*, and *Sodalis* (Adeolu et al. 2016). Members of the *Dickeya* genus originally belonged to the genus *Erwinia* represented by strains within species *E. chrysanthemi* (Burkholder et al. 1953). Later this species was reclassified as *Pectobacterium chrysanthemi* (Hauben et al. 1998), until Samson et al. (2005) elevated the species to the genus *Dickeya* with six species. There have since been some changes and additions to these species, which currently include *D. aquatica*, *D. chrysanthemi*, *D. dadantii*, *D. dianthicola*, *D. fangzhongdai*, *D. paradisiaca*, *D. solani*, and *D. zeae* (Brady et al. 2012; Parkinson et al. 2014; Samson et al. 2005; Tian et al. 2016).

10.3.2 Host Range

Dickeya has a broad host range and can infect plant species in at least 12 dicot families in 10 orders and 10 monocot families in 5 orders, and include ornamentals such as chrysanthemum, carnation, dahlia, and calla lily as well as important crops including carrot, tomato, and, the most economically important, potato (Charkowski 2018; Ma et al. 2007a, b; Samson et al. 2005). While all *Dickeya* species, with the exception of *D. paradisiaca*, have been found on ornamentals in Europe, only *D. dianthicola* and *D. solani* have caused significant economic losses on potato (Toth et al. 2011). In both cases, the lack of genetic diversity between isolates on potato and ornamental hosts suggests the organisms may have spread to potato from such a host (Parkinson et al. 2009; Slawiak et al. 2009). Only *D. aquatica*, which was isolated from waterways in the UK (Parkinson et al. 2014) and Maine (J. Hao, personal communication), has not yet been associated with a plant disease.

10.3.3 Geographical Distribution

As with *Pectobacterium*, *Dickeya* species have been reported on a wide range of hosts in different countries around the world (Samson et al. 2005). While *D. zeae, D. solani,* and *D. dianthicola* have wide geographic distributions, *D. paradisiaca* appears to be restricted to Colombia (Samson et al. 2005; Toth et al. 2011).

D. dianthicola was the first *Dickeya* species to be associated with plant disease in Europe, occurring on *Dianthus* in the Netherlands, Denmark, and the UK and later spreading to other nations (Hellmers 1958). It was later associated with other ornamentals and crops in a number of European countries, including potato. In some cases, *D. dianthicola* replaced *P. atrosepticum* as the dominant blackleg pathogen (Parkinson et al. 2009; Toth et al. 2011). *D. solani* was recognized independently as a new *Dickeya* pathogen on potato by several groups from 2004 through 2010 (Laurila et al. 2008; Parkinson et al. 2009; Slawiak et al. 2009). Isolates of both *D. dianthicola* and *D. solani* show little genetic diversity compared to isolates from ornamentals, and within these species there is a high degree of genetic similarity. Therefore, it seems likely that these pathogens have independently jumped host from an ornamental onto potato (Toth et al. 2011).

10.3.4 Symptoms

Although *Dickeya* can cause tuber soft rot, it primarily causes blackleg on potato. Blackleg symptoms include necrosis of the potato stem, originating from the mother tuber and spreading several centimeters above ground (Fig. 10.4). Plant leaves will wilt and curl as the disease develops and the plant vascular system will become necrotic. The pith of the stem is often decayed. *D. dianthicola* can also cause severe seed decay and lack of plant emergence in severe cases. Infected plants produce few or no tubers and any tubers produced may decay prior to harvest. Both *Dickeya* and *Pectobacterium* may be present together in diseased plants. In the United States, *P. parmentieri* is the most common species found together with *Dickeya.*

10.3.5 Epidemiology

Initial seed potato production relies on pathogen-free micropropagated plantlets. These plantlets are grown in greenhouses or screenhouses to produce minitubers, which are used for field planting (Frost et al. 2013). *Dickeya* will kill micropropagated plants within a few days and is not typically found in greenhouses or screenhouses. It appears to contaminate potatoes after they have been grown for at least one generation in the field, with the risk of contamination increasing with each

Fig. 10.4 Foliar symptoms of *Dickeya dianthicola* on potato. Initial symptoms are either a lack of emergence or leaf curling (**a**). The base of the stem turns dark brown or black and this necrosis can extend several centimeters from the soil line (**b**). The pith inside symptomatic stems is often decayed and the xylem are necrotic for several centimeters above the external stem necrosis and the pith decay (**c**). Disease symptoms may only develop on one stem of a multi-stem plant (**d**) (Courtesy of Amy O. Charkowski, Colorado State University)

generation in the field. *Dickeya* does not appear to survive in soil, but it can contaminate waterways and survive for long periods in surface water (Toth et al. 2011). It may also survive in weeds (Fikowicz-Krosko and Czajkowski 2017) or volunteer potatoes and spread by insects (Rossmann et al. 2018). Like *Pectobacterium*, *Dickeya* appears to spread mainly at harvest, where it can spread from infected vines and tubers to previously uncontaminated tubers. The bacteria are mainly found on tuber lenticels, but may also be present in the tuber stolon scar. Asymptomatic infestations are common, so it is not possible to visually assess seed potato lots for risk.

Blackleg development is highly dependent on the environment and it is unpredictable, even when a seed lot is known to be contaminated with *Dickeya*. Plants grown from infested seed lots planted in warm, humid areas tend to develop disease, while plants grown from the same infested seed lot planted in cooler, drier climates may remain healthy. Temperatures above 30 °C during the growing season appear to be particularly conducive to disease development. Co-contamination with *Pectobacterium* and *Dickeya* appears to lead to disease development more frequently than when only *Dickeya* is present.

10.3.6 Pathogenicity Determinants and Resistance

Dickeya pathogenicity relies mainly on pectate lyases and other plant cell wall-degrading enzymes secreted by the bacterial cell, although several other virulence genes are known (Charkowski et al. 2012). Although both *Pectobacterium* and *Dickeya* use plant cell wall-degrading enzymes, there are some important differences in enzyme genes and gene regulation between the genera that may account for some of the differences in disease symptoms. There are no examples of gene-for-gene resistance with *Dickeya* and the basis for resistance to *Dickeya* in wild potato species or for host range is poorly understood. There are no resistant commercial potato varieties, but varieties do differ in tolerance.

10.3.7 Significance and Economic Loss

The relative importance of *Dickeya* as a potato pathogen appears to be increasing (Toth et al. 2011). *D. solani* caused severe losses in the early 2000s in multiple countries and in 2015 *D. dianthicola* was in up to 20% of seed potato lots in some states in the US. Recent development of species-specific PCR assays for *Dickeya* will likely reveal that it is widespread in potato. As with *Pectobacterium*, farmers lose millions annually to blackleg caused by *Dickeya*.

10.3.8 Management

Cultural practices are important for *Dickeya* management and the recommendations are essentially the same as for *Pectobacterium* (Czajkowski et al. 2011, 2013). Growers initiate potato production with micropropagated plantlets that are free of *Dickeya*, but tubers may become contaminated once they are planted in fields. To reduce the risk of disease spread, growers should sanitize equipment thoroughly between seed fields, especially if blackleg is present. At planting, growers should fully suberize seed if they are using cut seed, and they should not plant seed that is too cold or into saturated ground. During the growing season, they should irrigate with ground water if possible and not overfertilize with nitrogen. Rouging infected plants is likely to spread the pathogen if diseased plants are present. At harvest, the *Dickeya* may multiply on the vines as they senesce, so quickly killing potato vines may aid in reducing disease incidence the following year. Tubers should be allowed to heal before cooling storages. Good airflow and high humidity in potato warehouses will also aid in reducing soft rot in storage. High levels of carbon dioxide in warehouses will promote soft rot development. Seed potatoes may be tested for *Dickeya* prior to planting (Czajkowski et al. 2015; Humphris et al. 2015) and growers should avoid planting contaminated seed lots in areas where growing conditions are conducive to blackleg.

10.4 Potato Ring Rot Caused by *Clavibacter michiganensis* Subsp. *sepedonicus*

10.4.1 Taxonomy and Nomenclature

Clavibacter michiganensis subsp. *sepedonicus* is a Gram positive, coryneform, aerobic, non-spore-forming bacterium in the Microbacteriaceae family of the Actinobacteria. *C. michiganensis* is the only species currently recognized within the genus; all six of its subspecies (subspp. *Insidiosus*, *michiganensis*, *nebraskensis*, *phaseoli*, *sepedonicus*, and *tesselarius*) are plant pathogens. *C. michiganensis* subsp. *sepedonicus* (*Cms*) was formerly known under the synonyms *Corynebacterium sepedonicum*, *Corynebacterium michiganense* pv. *sepedonicum*, and *Corynebacterium michiganense* subsp. *sepedonicum*.

10.4.2 Host Range

The only economically important host is potato (*Solanum tuberosum*), although natural infection was recently reported for the first time on tomato (van Vaerenbergh et al. 2016). Many members of the Solanaceae, including tomato and eggplant, are susceptible after artifical inoculation. Some solanaceous weeds, e.g. hairy nightshade (*Solanum sarrachoides*) and buffalobur (*S. rostratum*), may harbor the bacterium following potato crops with ring rot (van der Wolf et al. 2005a).

10.4.3 Geographical Distribution

First reported after an outbreak in Germany in 1905 (Appel 1906), it is one of the few major plant pathogens that is not present in the area where the crop evolved, i.e. Andean South America. In North America it was first reported in Quebec (Canada) in 1931 and by 1940 it had spread to all important potato-producing districts in Canada and the USA due to movement in trade of infected seed potato tubers. Subsequent zero tolerances imposed in quarantine and seed certification controls in most potato-growing areas have effectively limited the numbers of findings, although total eradication is difficult. Currently, it tends to occur sporadically in cool, northern latitudes of North America (Northern USA and Canada) with only a single report in Mexico. Strict regulation in Europe has also reduced findings in annual surveys, especially in certified seed crops, with only occasional recent findings in some countries (Bulgaria, Czech Republic, Estonia, Finland, Germany, Greece, Hungary, Latvia, Lithuania, Netherlands, Norway, Slovakia, Sweden, and Turkey). Isolated former outbreaks have been declared eradicated in Austria, Belgium, Cyprus, Denmark, France, Spain, and UK (England and Wales). However, it remains prevalent in areas where formal seed certification is absent (parts of northern, western, and central Russia, Ukraine, Poland, and Romania). It is also reported in Asia (several provinces of China, Japan, Kazakhstan, Korea, Nepal, Pakistan, and Uzbekistan) although its distribution is not clearly defined. Ring rot has never been confirmed in Africa, Australasia, or South America.

In the field, foliar symptoms are not always observed or may occur only at the end of the season when they are difficult to distinguish in the senescing plant and are easily missed during crop inspections. Unlike bacterial wilt, caused by *Ralstonia,* wilting due to the ring rot bacterium is usually slow and initially limited to the leaf margins (Fig. 10.5). Young infected leaves expand more slowly in the infected zones and become distorted. Leaves affected by xylem blockages further down the stem often develop chlorotic, yellow to orange, interveinal areas. Infected leaflets, leaves, and even stems may eventually die. Leaves and tubers may simply be reduced in size and occasionally whole plants can be stunted.

10.4.4 Epidemiology

Factors affecting development and spread of potato ring rot were reviewed in detail by van der Wolf et al. (2005b). Seed potato tubers infected or contaminated with *Cms* are the primary source of infection. Inadvertent dissemination of the bacterium to new places of production occurs with the movement and planting of latently infected seed tubers. The bacterium also spreads from infected tubers through direct contact and via contaminated surfaces of equipment used in potato production, such as seed cutters, planters, harvesters, graders, and transport vehicles as well as in contaminated stores and containers. Plant-to-plant spread in the field is usually low but there is some experimental evidence that insects can transmit the disease (Christie et al. 1991) although the full significance of this is not understood. *Cms* survives for extended periods of many months to years in a dry and cool environment. Its persistence on farm equipment, in stores, and on transport vehicles is an important means by which the bacterium is maintained and spread within farm units and disseminated to other production units.

The bacteria migrate systemically from seed tubers to the stems via the vascular tissue, and subsequently into progeny tubers through the stolons. The pathogen population density increases during the growing season but sometimes can be detected in stems within 3–4 weeks after planting infected seed. Survival of *Cms* in soil is not thought to contribute greatly to ring rot epidemiology although it can overwinter in the field in volunteer tubers (groundkeepers) and in potato tissue

Fig. 10.5. Symptoms of potato ring rot: initial tuber symptoms (water-soaked vascular ring, bacterial exudate, and start of vascular necrosis) (**a**), necrosis around vascular ring (**b**), advanced necrosis and secondary rotting (**c**), interveinal chlorosis and wilting/epinasty at leaf margin (**d**, **e**), leaf distortion (**f**). (UK Crown Copyright—Courtesy of Fera Science Ltd.)

debris. Survival is longest in cold dry conditions. The bacterium survives particularly well when dried in smears of decayed tuber tissue on equipment, machinery, potato sacks, and storage containers and can remain infectious in the dried state for at least 18 months at temperatures from 5 to −40 °C. *Cms* has been reported to be associated with solanaceous weeds, but any role of these potential inoculum sources in the epidemiology of the ring rot disease of potato is unclear. *Cms* has a low optimum growth temperature (21–23 °C) and is confined mainly to cooler potato-growing regions.

10.4.5 Pathogenicity Determinants and Resistance

The genome of *Cms* was first sequenced by Bentley et al. (2008). In addition to the 3.26 Mb chromosome, which is highly similar amongst all *Clavibacter* subspecies, a 50-kb circular plasmid (pCS1) and a 90-kb linear plasmid (pCSL1) are carried by all *Cms* strains. The plasmid pCS1 is essential for symptom development, but genes required for host recognition, efficient colonization, infection, and evasion or suppression of plant defense are located on the chromosome (Eichenlaub and Gartemann 2011). Two proteins, CelA and Chp-7, have been shown to be required for full virulence (Laine et al. 2000; Nissinen et al. 2001). The gene *celA* is located on the plasmid pCS1 and encodes the cellulase endo-β-1,4-glucanase. Located on the chromosome, *chp-7* encodes a serine protease effector that directly elicits a hypersensitive response in nonhost tobacco plants (Lu et al. 2015). Gene expression studies in *Cms* cells growing in either potato tissue or rich media (Holtsmark et al. 2008) have identified other putative virulence genes. In addition to *celA*, a homologous gene *celB*, two serine proteases and a xylanase was also upregulated in the plant tissue. Three other serine protease genes, including the *chp-7*, were downregulated. Unlike Gram-negative bacterial pathogens of potato, the gram-positive *Clavibacter* pathogens do not have a type 3 secretion system (T3SS) to translocate effectors into host plant cells.

Eichenlaub and Gartemann (2011) have described the likely infection process by *Clavibacter*. During infection the bacteria enter the xylem vessels of host plants, and subsequently spread systemically to colonize the whole plant. *Clavibacter* can be considered as a biotrophic phytopathogen that recruits nutrients (carboxylic acids and sugars) from the xylem fluid. Following growth and colonization, the cell walls of the xylem vessels and surrounding parenchymatic cells are then hydolyzed by expression of cellulases and other extracellular enzymes, potentially including polygalacturonase, pectic lyase, xylanases, and other endoglucanases, leading to symptom development.

There are no currently available potato cultivars with immunity or useful resistance to ring rot. The concept of cultivar tolerance to ring rot is not yet understood and little is known of the status of most commonly grown cultivars with respect to their susceptibility to infection and colonization under varying environmental conditions. Although potato cultivars vary in their propensities to express ring rot symptoms, less variation between cultivars in their susceptibilities to latent infections is observed (De Boer and McCann 1990). Since tolerant cultivars, which tend not to develop symptoms, can act as symptomless carriers of *Cms*, they have been removed from seed certification schemes in North America (Manzer and McKenzie 1988). Laurila et al. (2003) demonstrated that an accession (PI472655) of the wild potato species *Solanum acaule* was susceptible to latent infection by *Cms* at 15 °C but appeared immune to infection at 25 °C.

10.4.6 Significance and Economic Loss

Direct losses due to wilting and tuber rotting in field and store are usually moderate, especially where modern seed certification systems are in place. Nevertheless, ring rot constitutes a constraint on seed potato production, with considerable indirect effects on trade. These result from statutory measures taken against ring rot outbreaks, which include loss of certification, restrictions on further cropping, purchase of new seed stocks, costs of disinfection, and disposal of infected and associated crops and subsequent effects on reputation and export trade.

10.4.7 Management

In the absence of effective chemical or biological control measures, or potato cultivars with adequate levels of resistance, management of potato ring rot must rely on the production and safe distribution of seed potatoes that are free from infection. Control is achieved primarily through strict application of quarantine and seed certification regulations, which involve a zero tolerance for the disease during seed and import inspections and for the pathogen during regular testing of consignments. By laboratory testing for latent infections, infected lots can be detected early and eliminated from seed programs before further spread of the pathogen occurs. Phytosanitary measures must be aimed at the entire potato production system on account of the insidious nature of the disease. Seed potatoes should be imported only from countries which can show, by regular surveys and tests, that they operate a seed-potato production and distribution system free from ring rot.

Implementation of crop rotation, disinfection, and other sanitation practices is most important whenever the disease has occurred to prevent recurrence of the disease and spread of the pathogen. Ring rot infected crops, and any adjacent crops that may have become contaminated should be eliminated from the production system and new certified seed should be acquired for any future production. Disinfectants effective against *Cms* include quaternary ammonia-, chlorine-, or iodine-containing compounds. These should be applied, after cleaning of equipment and other contaminated surfaces, to ensure a minimum of 10 min contact under low organic load. Control of potato volunteers and solanaceous weeds is also important. The use of whole rather than cut seed helps to reduce any potential spread of the disease.

10.5 Common Scab of Potato Caused by *Streptomyces* Species

10.5.1 Taxonomy and Nomenclature

Streptomyces is a gram positive, aerobic, filamentous, spore-forming bacterium in the Streptomycetaceae family of the Actinobacteria. The filamentous mycelia have few or no cross walls. Spores are formed in spiral chains at the tips of hyphae. *Streptomyces* is the largest genus in the Actinobacteria and nearly 600 species are recognized. Most *Streptomyces* are soil-dwelling saprophytes and some species have a beneficial symbiosis with eukaryotes, including plants. At least 12 *Streptomyces* species cause common scab, netted scab, and/or pitted scab on potato. The names of several pathogenic *Streptomyces* species, such as *S. scabies*, were grammatically incorrect when they were first named and the scientific community has only recently begun using corrected names, such as *S. scabiei*.

Common scab is usually caused by *S. scabiei* (Thaxter 1892; Lambert and Loria 1989b), *S. acidiscabiei* (Lambert and Loria 1989a), or *S. turgidiscabiei* (Miyajima et al. 1998). Other species that cause scab symptoms on potato include the pitted scab pathogen *S. caviscabiei* (Goyer et al. 1996), three species first reported in France, including *S. europaeiscabiei*, *S. reticuliscabiei*, and *S. stelliscabiei* (Bouchek-Mechiche et al. 2000), three species first reported in Korea, including *S. luridiscabiei*, *S. niveiscabiei*, and *S. puniciscabiei* (Park et al. 2003), and one species reported in Japan, *S. cheloniumii* (Oniki et al. 1986a, b). The species *S. reticuliscabiei* is genomically the same as *S. turgidiscabiei*, but causes netted scab symptoms rather than typical common scab lesions (Bouchek-Mechiche et al. 2000, 2006). *S. diastatochromogenes* was recently reported as a common scab pathogen of potato, but there is no information available on its relative importance and the species identification was based solely on 16S rDNA sequence (Yang et al. 2017). A related species, *S. ipomeae*, causes root rot of sweet potato. Additional *Streptomyces* species capable of causing common scab have been isolated, but not yet described as species (see Table 1 in Bignell et al. 2014) and additional pathogenic species certainly remain to be discovered. Nonpathogenic strains exist within the pathogenic species, and none of the nonpathogenic strains appear to encode the phytotoxin thaxtomin (Wanner 2006, 2007, 2009).

10.5.2 Host Range

Potato is the most economically important host of plant pathogenic *Streptomyces* species. Plant pathogenic species are also able to cause disease on root crops, such as carrot, beet, parsnip, radish, sweet potato, and turnip (Goyer and Beaulieu 1997), and on peanut pods (Kritzman et al. 1996), but the economic impact of *Streptomyces* on these crops is less important than other diseases that infect these root crops.

10.5.3 Geographical Distribution

Pathogenic *Streptomyces* are present in soils wherever potato is grown and, as the name denotes, the disease it causes is one of the most common and most important potato diseases worldwide. Multiple species are present in individual fields and tubers (Wanner 2009; Lehtonen et al. 2004; Dees et al. 2013). Some species have only been reported from limited geographical regions, but no comprehensive global surveys have been done, so the distribution of pathogenic *Streptomyces* species remains mostly unexplored.

The spread of this pathogen is managed mainly through quality regulations which prohibit planting or shipping of severely affected seed, so there are essentially no limits on the spread of pathogenic *Streptomyces* through seed potatoes. Establishment of new *Streptomyces* strains in field soil is dependent on numerous complex factors, including soil chemistry and resident soil microbes, making establishment of pathogenic *Streptomyces* strains transported on seed potatoes unpredictable.

10.5.4 Symptoms

Streptomyces can cause necrosis on all underground parts of a potato (Fig. 10.6), including roots, stolons, and stems, and it can reduce growth of roots from seed tubers (Han et al. 2008). This pathogen can also cause necrosis on and kill potato seedlings grown from true potato seed. It does not directly cause foliar symptoms, although plant vigor may be reduced due to root necrosis caused by *Streptomyces*.

There is a wide variation in tuber symptoms caused by *Streptomyces*, including pitted scab, erumpent scab, and mild netted scab and symptom type depends, at least in part, on which toxins the infecting strain produces and the potato genotype. The pathogen colonizes tubers as they initiate, often entering the tube through lenticels. Whitish-grey bacterial mycelia and spores are sometimes visible in pitted scab lesions at harvest. The disease does not progress in storage, although tubers with severe pitted scab lesions will dehydrate and will not sprout the following season.

Fig. 10.6 Common scab symptoms on potato (Courtesy of AHDB Potatoes Sutton Bridge Crop Storage Research)

10.5.5 Epidemiology

Streptomyces has a relatively complex life cycle compared to many bacterial pathogens. It grows vegetatively as filamentous mycelia-like cells. When resources are depleted, the vegetative cells undergo programed cell death, nutrients are transferred to aerial reproductive hyphae, and spores are formed. These hyphae are sometimes visible without magnification inside scab lesions. Pathogenic *Streptomyces* grow best in soils with a pH between 5.2 and 8.0, and temperature of 20–22 °C, which are conditions that also favor potato growth.

Streptomyces survives and disperses mainly through cylindrical spores formed at hyphal tips. The spores can disperse in water, on soil-dwelling invertebrates, and on seed tubers. *Streptomyces* spores can survive in soil for 20 or more years and the spores are heat resistant. The pathogen spores germinate and enter the plant through natural openings, such as lenticels, or through wounds. Tubers are most susceptible to *Streptomyces* colonization during the first month of development. *Streptomyces* cannot cause lesions on mature tubers and lesion size and severity does not progress during storage, although tubers with severe pit scab may become dehydrated and will not sprout the following season.

Because multiple *Streptomyces* species are present in field soil and on diseased plants, epidemiological studies now rely on molecular detection of the species present in order to understand the impacts of management methods, soil characteristics, or biocontrol strains. PCR assays capable of distinguishing *Streptomyces* species are available (Wanner 2009). PCR assays designed to detect genes encoding thaxtomin are also used in epidemiological studies because detection of thaxtomin DNA is correlated with ability of an isolate to cause common scab (Wanner 2006, 2007, 2009; Flores-González et al. 2008) and with development of common scab symptoms in field soils (Qu et al. 2008).

Soils that suppress common scab exist and ongoing work is aimed at identifying the communities that lead to suppressiveness. Soils that suppress common scab have high *Streptomyces* populations. These saprophytic streptomycetes produce antibiotics that inhibit pathogenic *Streptomyces* or that compete with pathogenic *Streptomyces* for resources, thereby reducing common scab (for a comprehensive review, see Schlatter et al. 2017).

10.5.6 Pathogenicity Determinants and Resistance

Bacteria in this genus have unusually large linear genomes of 10–12 Mb and they produce diverse secondary metabolites. In the plant pathogenic *Streptomyces*, large pathogenicity islands encompassing several hundred genes encode virulence genes required for production of secondary metabolites, such as toxins, cytokinin, nitric oxide, and secreted proteins (Bignell et al. 2010; Joshi and Loria 2007). At least two of these pathogenicity islands are mobile (Bukhalid et al. 2002) and one of them can mobilize at least one otherwise nonmobile pathogenicity island (Zhang and Loria 2017). As a result, pathogenicity can be transferred to previously nonpathogenic species (Zhang and Loria 2017).

Phytotoxins are the main *Streptomyces* pathogenicity determinants and the toxin thaxtomin appears to be required for pathogenicity (for a recent review, see Bignell et al. 2014). Thaxtomins, which are nitrated dipeptides (tryptophan and phenylalanine), are required for the development of common scab symptoms (King et al. 1989, 1991; Kinkel et al. 1998). Thaxtomin appears to weaken plant cell walls and cause plant cell hypertrophy through inhibition of cellulose synthesis and cell wall acidification (Fry and Loria 2002; Bischoff et al. 2009). This toxin can be used in potato breeding since seedling tolerance to thaxtomin is correlated with tolerance to common scab in the field (Hiltunen et al. 2011).

The other types of toxins produced by pathogenic *Streptomyces*, including coronatine-like toxins (Fyans et al. 2015), concanamycin (Natsume et al. 2017), borrelidin (Cao et al. 2012), and FD-891 (Natsume et al. 2005), are not necessarily produced by all pathogenic strains and production of these toxins may affect whether an individual strain produces pitted, net, or erumpent common scab symptoms. For example, concanamycin, a type of toxin produced by *S. scabies*, but not by some other *Streptomyces* species, may be required for formation of pitted scab lesions and appears to be synergistic with thaxtomin (Natsume et al. 2017).

Enzymes may also play a role in *Streptomyces* pathogenicity. *Streptomyces* lesions typically do not autofluoresce, suggesting that suberin formation is either inhibited or digested. Two genes that encode potential suberinases are present in the *S. scabies* genome and biochemical evidence supports that suberin is degraded (Beaulieu et al. 2016; Komeil et al. 2013). Degradation of suberin also appears to increase expression of the numerous cellulases produced by *S. scabies* (Padilla-Reynaud et al. 2015). *Streptomyces* toxin production is induced by plant-derived molecules, including the disaccharide cellobiose, a breakdown product of cellulose.

Little is known about the genetic basis of resistance to common scab. Suggested mechanisms include phellum layer thickness (Thangavel et al. 2016), phellum suberization (Thangavel et al. 2016; Khatri et al. 2011), detoxification of thaxtomin (Acuna et al. 2001), or sustained expression of disease defense genes (Merete Wiken Dees et al. 2016). Differences in ability of potato varieties to support growth of nonpathogenic *Streptomyces* species may also affect susceptibility to common scab (Wanner 2007).

10.5.7 Significance and Economic Loss

Common scab can cause complete loss, although this is usually associated with mismanagement of the crop, such as adding too much lime to a field, insufficient irrigation, or highly susceptible varieties planted in fields with high disease pressure.

Direct losses occur annually, however, worldwide, and common scab is often listed among the most important potato diseases (for example, Hill and Lazarovits 2005).

10.5.8 Management

The best option is disease tolerance or resistance, but currently there are limited options for potato varieties with high tolerance to common scab. Common scab symptom development is affected by soil moisture and chemistry, the soil microbial community, crop rotation, and host genetics in a complex manner that has made predicting common scab severity and managing this disease difficult. A comprehensive review of these challenges was published by Dees and Wanner (2012). Recommendations for management of common scab usually include adequate irrigation during tuber formation, and low soil pH (<5.2). Typically, sulfur fertilizers are used to reduce soil pH and this can reduce disease severity (Pavlista 2005). However, these methods sometimes fail to provide adequate management and can lead to other production problems. For example, over-irrigation during tuber formation can lead to development of powdery scab and several other potato diseases, and low soil pH limits farmer options for crop rotations and selects for *S. acidiscabies*.

Chemical treatments can work for a season, but are often expensive and damaging to the soil, making this the least sustainable disease management option. Some commonly used effective chemicals include fludioxonil as a seed piece treatment, chloropicrin as a soil fumigant, and pentachloronitrobenzene as an in-furrow treatment (Al-Mughrabi et al. 2016; Powelson and Rowe 2008). Fluazinam may also provide some control of common scab (Santos-Cervantes et al. 2017).

Crop rotation choices can also reduce common scab severity (Powelson and Rowe 2008; Larkin and Halloran 2014; Larkin et al. 2011; Larkin and Griffin 2007). These crop rotations tend to include brassica crops as a biofumigant and commonly planted green manures that are allelopathic and that help control multiple soil-borne potato diseases. Soil amendments, such as rice bran, chelated iron, or peat can decrease common scab, likely by increasing the population of nonpathogenic streptomycetes (Tomihama et al. 2016; Sarikhani et al. 2017). Some soil amendments, such as manure, which increases soil pH, will increase common scab severity.

Biocontrol with nonpathogenic *Streptomyces* strains also shows promise and the mechanism of biocontrol is likely similar to that seen in suppressive soils, which is thought to be due to both resource completion and antibiotic production (Schlatter et al. 2017). Suppressive soils develop through repeated monoculture of potato, but this practice results in accumulation of other soil-borne pathogens. However, repeated inoculations of soils with a single antagonistic *Streptomyces* strain can result in common scab suppression in as little as 3 years, and suppressive lasted for 2 years beyond the last inoculation (Hiltunen et al. 2017).

10.6 Zebra Chip of Potato Caused by *Liberibacter*

10.6.1 Taxonomy and Nomenclature

The genus "*Candidatus* Liberibacter" is a gram-negative bacterium in the Rhizobeaceae family. At least seven *Ca. Liberibacter* species exist. Of these, "*Candidatus* Liberibacter solanacearum" (Lso), which is a phloem-limited pathogen, is the only known potato-infecting species. There are at least five Lso haplotypes, with haplotypes A and B causing disease on potato and the remaining three haplotypes infecting carrots and celery (Nelson et al. 2011; Teresani et al. 2014). At 1.26 Mbp, the circular Lso genome is relatively small (Hong Lin et al. 2011) and there are relatively few genomic differences among Lso haplotypes (Wang et al. 2017a, b, c). Compared to related free-living bacteria, such as *Agrobacterium*, Lso has a low G + C content and lacks many genes involved in metabolism.

10.6.2 Host Range

Potato is the most economically important host of Lso haplotypes A and B, but Lso can also infect other solanaceous crops and weeds. All Ca. Liberibacter are spread by *Bactericera* species and Lso also infects its vector and can reduce vector fitness (Yao et al. 2016). Although Lso is only spread in potato by *B. cockerelli*, but it can also be found in other *Bactericera* species, suggesting that vector feeding preferences limit the species of vectors important for zebra chip and not Lso-vector interactions (Borges et al. 2017).

10.6.3 Geographical Distribution

Potato psyllids are native to North and Central America, and it recently invaded New Zealand (Teulon et al. 2009). The bacterial pathogen has spread with its vector and can be found wherever potato psyllids are found. The highest disease incidence is typically found in the Southwestern United States, Mexico, and Central America.

10.6.4 Symptoms

Zebra chips symptoms are severe on both the foliage and the tubers. The upper parts of infected plants have leaf curling, chlorosis, shortened internodes, aerial tubers, and early necrosis and death (Buchman et al. 2012) (Fig. 10.7). The tubers appear to have glassy or brown streaks that darken when they are fried, giving the

Fig. 10.7 Typical Lso symptoms on potato. Leaf curl symptoms on foliage (**a**); Necrotic specks in tuber (**b**); Blackened splotches on a fried chip (healthy on left, and diseased on right) (**d**) (Courtesy of Gary A. Secor, North Dakota State University)

disease its name, zebra chip. Tuber development slows or ceases in symptomatic plants, resulting in yield losses. Lso appears to reduce protease inhibitor levels in tubers, and as a result, tubers from infected plants have less protein (Kumar et al. 2015). Infected tubers either do not sprout or have only hair sprouts (Rashed et al. 2015). If plants emerge from infected tubers, they die shortly after emergence. Storage temperature affects symptom development, with cooler storage (3 °C) resulting more tuber symptoms than warmer storage temperatures (6 or 9 °C) (Wallis et al. 2017).

10.6.5 Epidemiology

Lso is transmitted solely by *B. cockerelli*, which feeds on phloem with its piercing-sucking mouthparts. The pathogen is transmitted in a persistent, propagative, and circulative fashion and it is also transmitted transovarially in the psyllid (Cicero et al. 2016; Hansen et al. 2008). A 2-week latent period occurs between psyllid acquisition of Lso and ability to transmit the pathogen (Sengoda et al. 2014). In most regions in North America where the pathogen and vector are present, potatoes are infected late in the season. Surprisingly, Lso reduces the fitness of its insect vector, with haplotype B resulting in more insect mortality than haplotype A (Yao et al. 2016).

Once transmitted into a leaf, the pathogen does not cause symptoms for at least 3 weeks. It is also not evenly distributed in plants, which makes it difficult to detect prior to symptom development and this has hampered epidemiological studies. Lso is not culturable, which also makes epidemiological studies more challenging and as a result, researchers rely mainly on PCR assays for pathogen detection (Ananthakrishnan et al. 2013; Secor et al. 2009). Based on symptom development, Lso appears to be sensitive to temperatures above 32 °C and to thrive at 27–32 °C. Solanaceous weeds serve as important reservoirs for Lso and can provide a green bridge between potato crops (Thinakaran et al. 2015).

Infected tubers rarely sprout and when they do, they tend to develop hair sprouts. As a result, this disease is poorly transmitted through seed potatoes and insect transmission remains the most important mode of spread. For this reason, zebra chip is not currently regulated through seed potato certification in North America. It has, however, impacted export of potatoes from North America.

10.6.6 Pathogenicity Determinants and Resistance

Liberibacter pathogenicity determinants were recently thoroughly reviewed (Wang et al. 2017a, b, c). There are no resistant potato varieties, although timing and severity of symptoms differ among varieties (Lévy et al. 2015). Tolerant lines still support Lso levels similar to those found in susceptible varieties, but the Lso has less impact on plant physiology and symptom development in tubers in tolerant lines (Rashidi et al. 2017; Wallis et al. 2015). Recent results also suggest that psyllids are not able to transmit the pathogen with equal efficiency into all potato lines (Rashidi et al. 2017).

10.6.7 Significance and Economic Loss

Zebra chip has caused millions in losses in North America, and although seed tubers are not a major source of inoculum, it has affected the potato export market. The spread of this pathogen and its vector to New Zealand has also caused significant losses there. In addition to losses in yield and quality, the high cost of vector management has added to financial losses caused by *L. solanacearum*.

10.6.8 Management

Insecticides are the main management method used for control of zebra chip. Growers in North America monitor psyllids and determine when psyllids appear and the percentage of Lso-infected psyllids present. They may spray insecticides a dozen or more times during the growing season to protect the potato crop, with imidocloprid and spirotetramat among the most commonly used (Guenthner et al.

2012). Since the insects tend to be present on the underside of leaves, effectively covering the underside of the leaves is essential. These sprays are expensive and the potential for insecticide resistance and loss of natural enemies due to frequent sprays makes this approach unsustainable in the long term.

10.7 Concluding Remarks

Bacterial diseases of potato have remained an economically significant disease worldwide. Farmers lose millions of dollars annually due to bacterial diseases. Bacterial wilt, soft rot, and ring rot have got international attention as they constitute a huge constraint on seed potato production, with considerable indirect effects on trade. Rigorous seed certification and testing programs in developed countries have limited the impact of these diseases within their value chains, while developing countries commonly lack these safeguards. Lack of certified disease-free planting material in many developing country contexts contributes to further distribution of these pathogens via latently infected tubers, as well as tuber quality and yield degeneration caused by farmers replanting diseased seed year to year. Bacterial disease management efforts in developing countries should follow the systems approach that incorporates specific operational practices to reduce the likelihood of incursion, establishment, and growth of these pathogens in potato crops. This includes training farmers in proper production practices, on-farm management tools, using healthy seed tubers, and planting in clean soils. Additional factors to consider in controlling these diseases can include sanitation, cultural practices, crop rotation with nonhost plants, and the use of tolerant or resistant varieties.

References

Acuna IA, Strobel GA, Jacobsen BJ, Corsini DL (2001) Glucosylation as a mechanism of resistance to thaxtomin A in potatoes. Plant Sci 161(1):77–88

Adeolu M, Alnajar S, Naushad S, Gupta RS (2016) Genome-based phylogeny and taxonomy of the 'Enterobacteriales': proposal for Enterobacterales ord. nov. divided into the families Enterobacteriaceae, Erwiniaceae fam. nov., Pectobacteriaceae fam. nov., Yersiniaceae fam. nov., Hafniaceae fam. nov., Morganellaceae fam. nov., and Budviciaceae fam. nov. Int J Syst Evol Microbiol 66(12):5575–5599. https://doi.org/10.1099/ijsem.0.001485

Al-Mughrabi KI, Vikram A, Poirier R, Jayasuriya K, Moreau G (2016) Management of common scab of potato in the field using biopesticides, fungicides, soil additives, or soil fumigants. Biocontrol Sci Tech 26(1):125–135

Ananthakrishnan G, Choudhary N, Roy A, Sengoda VG, Postnikova E, Hartung JS, Stone AL, Damsteegt VD, Schneider WL, Munyaneza JE (2013) Development of primers and probes for genus and species specific detection of 'Candidatus Liberibacter species' by real-time PCR. Plant Dis 97(9):1235–1243

Appel O (1906) Neuere Untersuchungen Über Kartoffel-Und Tomaten-Erkrankungen. Borntraeger, Stuttgart

Beaulieu C, Sidibé A, Jabloune R, Simao-Beaunoir A-M, Lerat S, Monga E, Bernards MA (2016) Physical, chemical and proteomic evidence of potato suberin degradation by the plant pathogenic bacterium Streptomyces scabiei. Microbes Environ 31(4):427–434

Bentley SD, Corton C, Brown SE, Barron A, Clark L, Doggett J, Harris B, Ormond D, Quail MA, May G (2008) Genome of the actinomycete plant pathogen Clavibacter michiganensis subsp. sepedonicus suggests recent niche adaptation. J Bacteriol 190(6):2150–2160

Bignell DRD, Huguet-Tapia JC, Joshi MV, Pettis GS, Loria R (2010) What does it take to be a plant pathogen: genomic insights from Streptomyces species. Antonie Van Leeuwenhoek 98(2):179–194

Bignell DRD, Fyans JK, Cheng Z (2014) Phytotoxins produced by plant pathogenic Streptomyces species. J Appl Microbiol 116(2):223–235

Bischoff V, Cookson SJ, Wu S, Scheible W-R (2009) Thaxtomin A affects CESA-complex density, expression of cell wall genes, cell wall composition, and causes ectopic lignification in Arabidopsis thaliana seedlings. J Exp Bot 60(3):955–965

Borges KM, Rodney Cooper W, Garczynski SF, Thinakaran J, Jensen AS, Horton DR, Munyaneza JE, Cueva I, Barcenas NM (2017) 'Candidatus Liberibacter solanacearum' associated with the psyllid, Bactericera maculipennis (Hemiptera: Triozidae). Environ Entomol 46(2):210–216

Bouchek-Mechiche K, Gardan L, Normand P, Jouan B (2000) DNA relatedness among strains of Streptomyces pathogenic to potato in France: description of three new species, S. europaeiscabiei sp. nov. and S. stelliscabiei sp. nov. associated with common scab, and S. reticuliscabiei sp. nov. associated with netted scab. Int J Syst Evol Microbiol 50(1):91–99

Bouchek-Mechiche K, Gardan L, Andrivon D, Normand P (2006) Streptomyces turgidiscabies and Streptomyces reticuliscabiei: one genomic species, two pathogenic groups. Int J Syst Evol Microbiol 56(12):2771–2776

Brady CL, Cleenwerck I, Denman S, Venter SN, Rodríguez-Palenzuela P, Coutinho TA, De Vos P (2012) Proposal to reclassify Brenneria quercina (Hildebrand and Schroth 1967) Hauben et Al. 1999 into a new genus, Lonsdalea gen. nov., as Lonsdalea quercina comb. nov., descriptions of Lonsdalea quercina subsp. quercina comb. nov., Lonsdalea quercina subsp. iberica subsp. nov. and Lonsdalea quercina subsp. britannica subsp. nov., emendation of the description of the genus Brenneria, reclassification of Dickeya dieffenbachiae as Dickeya dadantii subsp. Dieffenbachiae comb. nov., and emendation of the description of Dickeya dadantii. Int J Syst Evol Microbiol 62(7):1592–1602. https://doi.org/10.1099/ijs.0.035055-0

Buchman JL, Fisher TW, Sengoda VG, Munyaneza JE (2012) Zebra chip progression: from inoculation of potato plants with Liberibacter to development of disease symptoms in tubers. Am J Potato Res 89(2):159–168

Buddenhagen IW (1962) Designations of races in pseudomonas Solanacearum. Phytopathology 52:726

Bukhalid RA, Takeuchi T, Labeda D, Loria R (2002) Horizontal transfer of the plant virulence gene, Nec1, and flanking sequences among genetically distinct Streptomyces strains in the Diastatochromogenes cluster. Appl Environ Microbiol 68(2):738–744

Burkholder WR, Mcfadden LA, Dimock EW (1953) A bacterial blight of chrysanthemums. Phytopathology 43(9). https://www.cabdirect.org/cabdirect/abstract/19541101119

Cao Z, Khodakaramian G, Arakawa K, Kinashi H (2012) Isolation of Borrelidin as a phytotoxic compound from a potato pathogenic Streptomyces strain. Biosci Biotechnol Biochem 76(2):353–357

Charkowski AO (2015) Biology and control of Pectobacterium in potato. Am J Potato Res 92(2):223–229

Charkowski AO (2018) The changing face of bacterial soft-rot diseases. Annu Rev Phytopathol 56(1):269–288. https://doi.org/10.1146/annurev-phyto-080417-045906

Charkowski A, Blanco C, Condemine G, Expert D, Franza T, Hayes C, Hugouvieux-Cotte-Pattat N et al (2012) The role of secretion systems and small molecules in soft-rot Enterobacteriaceae pathogenicity. Annu Rev Phytopathol 50(1):425–449. https://doi.org/10.1146/annurev-phyto-081211-173013

Christie RD, Sumalde AC, Schulz JT, Gudmestad NC (1991) Insect transmission of the bacterial ring rot pathogen. Am Potato J 68(6):363–372

Cicero JM, Fisher TW, Qureshi JA, Stansly PA, Brown JK (2016) Colonization and intrusive invasion of potato Psyllid by 'Candidatus Liberibacter solanacearum'. Phytopathology 107(1):36–49

Czajkowski R, Pérombelon MCM, van Veen JA, van der Wolf JM (2011) Control of blackleg and tuber soft rot of potato caused by Pectobacterium and Dickeya species: a review. Plant Pathol 60(6):999–1013. https://doi.org/10.1111/j.1365-3059.2011.02470.x

Czajkowski R, de Boer WJ, van der Wolf JM (2013) Chemical disinfectants can reduce potato blackleg caused by 'Dickeya solani'. Eur J Plant Pathol 136(2):419–432. https://doi.org/10.1007/s10658-013-0177-8

Czajkowski R, Pérombelon MCM, Jafra S, Lojkowska E, Potrykus M, van der Wolf JM, Sledz W (2015) Detection, identification and differentiation of Pectobacterium and Dickeya species causing potato blackleg and tuber soft rot: a review. Ann Appl Biol 166(1):18–38. https://doi.org/10.1111/aab.12166

De Boer SH, McCann M (1990) Detection of Corynebacterium sepedonicum in potato cultivars with different propensities to express ring rot symptoms. Am Potato J 67(10):685

De Boer SH, Li X, Ward LJ (2012) Pectobacterium spp. associated with bacterial stem rot syndrome of potato in Canada. Phytopathology 102(10):937–947

de Werra P, Bussereau F, Keiser A, Ziegler D (2015) First report of potato blackleg caused by Pectobacterium carotovorum subsp. Brasiliense in Switzerland. Plant Dis 99(4):551

Dees MW, Wanner LA (2012) In search of better management of potato common scab. Potato Res 55(3–4):249–268

Dees MW, Sletten A, Hermansen A (2013) Isolation and characterization of Streptomyces species from potato common scab lesions in Norway. Plant Pathol 62(1):217–225

Dees MW, Lysøe E, Alsheikh M, Davik J, Brurberg MB (2016) Resistance to Streptomyces turgidiscabies in potato involves an early and sustained transcriptional reprogramming at initial stages of tuber formation. Mol Plant Pathol 17(5):703–713

Dees MW, Lebecka R, Perminow JIS, Czajkowski R, Motyka A, Zoledowska S, Śliwka J, Lojkowska E, Brurberg MB (2017a) Characterization of Dickeya and Pectobacterium strains obtained from diseased potato plants in different climatic conditions of Norway and Poland. Eur J Plant Pathol 148(4):839–851

Dees MW, Lysøe E, Rossmann S, Perminow J, Brurberg MB (2017b) Pectobacterium polaris sp. nov., isolated from potato (Solanum tuberosum). Int J Syst Evol Microbiol 67(12):5222–5229

Duarte V, De Boer SH, Ward LJ, De Oliveira AMR (2004) Characterization of atypical Erwinia carotovora strains causing blackleg of potato in Brazil. J Appl Microbiol 96(3):535–545

Eichenlaub R, Gartemann K-H (2011) The Clavibacter michiganensis subspecies: molecular investigation of gram-positive bacterial plant pathogens. Annu Rev Phytopathol 49:445–464

Elphinstone JG (2005) The current bacterial wilt situation: a global overview. In: Allen C, Piror P, Hayward AC (eds) Bacterial wilt disease and the Ralstonia solanacearum species complex. APS Press, St Paul, MN, pp 9–28

Fegan M, Prior P (2005) How complex is the Ralstonia solanacearum species complex. APS Press, St. Paul

Fikowicz-Krosko J, Czajkowski R (2017) Systemic colonization and expression of disease symptoms on bittersweet nightshade (Solanum dulcamara) infected with a GFP-tagged Dickeya solani IPO2222 (IPO2254). Plant Dis 102(3):619–627. https://doi.org/10.1094/PDIS-08-17-1147-RE

Flores-González R, Velasco I, Montes F (2008) Detection and characterization of Streptomyces causing potato common scab in Western Europe. Plant Pathol 57(1):162–169

Frost KE, Groves RL, Charkowski AO (2013) Integrated control of potato pathogens through seed potato certification and provision of clean seed potatoes. Plant Dis 97(10):1268–1280. https://doi.org/10.1094/PDIS-05-13-0477-FE

Fry BA, Loria R (2002) Thaxtomin A: evidence for a plant cell wall target. Physiol Mol Plant Pathol 60(1):1–8

Fyans JK, Altowairish MS, Li Y, Bignell DRD (2015) Characterization of the coronatine-like phytotoxins produced by the common scab pathogen Streptomyces scabies. Mol Plant-Microbe Interact 28(4):443–454

Genin S, Denny TP (2012) Pathogenomics of the Ralstonia solanacearum species complex. Annu Rev Phytopathol 50:67–89

Goyer C, Beaulieu C (1997) Host range of Streptomycete strains causing common scab. Plant Dis 81(8):901–904

Goyer C, Faucher E, Beaulieu C (1996) Streptomyces Caviscabies sp. nov., from deep-pitted lesions in potatoes in Québec, Canada. Int J Syst Evol Microbiol 46(3):635–639

Guenthner J, Goolsby J, Greenway G (2012) Use and cost of insecticides to control potato psyllids and zebra chip on potatoes. Southwest Entomol 37(3):263–270

Han L, Dutilleul P, Prasher SO, Beaulieu C, Smith DL (2008) Assessment of common scab-inducing pathogen effects on potato underground organs via computed tomography scanning. Phytopathology 98(10):1118–1125

Hansen AK, Trumble JT, Stouthamer R, Paine TD (2008) A new Huanglongbing species,'Candidatus Liberibacter Psyllaurous,' found to infect tomato and potato, is vectored by the Psyllid Bactericera Cockerelli (Sulc). Appl Environ Microbiol 74(18):5862–5865

Hauben L, Moore ERB, Vauterin L, Steenackers M, Mergaert J, Verdonck L, Swings J (1998) Phylogenetic position of phytopathogens within the Enterobacteriaceae. Syst Appl Microbiol 21(3):384–397. https://doi.org/10.1016/S0723-2020(98)80048-9

Hayward AC (1964) Characteristics of Pseudomonas solanacearum. J Appl Bacteriol 27(2):265–277

Hellmers E (1958) Four wilt diseases of perpetual-flowering Carnations in Denmark. Dansk botanisk Arkiv 18(2):1–100. https://www.cabdirect.org/cabdirect/abstract/19581101855

Hill J, Lazarovits G (2005) A mail survey of growers to estimate potato common scab prevalence and economic loss in Canada. Can J Plant Pathol 27(1):46–52

Hiltunen LH, Alanen M, Laakso I, Kangas A, Virtanen E, Valkonen JPT (2011) Elimination of common scab sensitive progeny from a potato breeding population using Thaxtomin A as a selective agent. Plant Pathol 60(3):426–435

Hiltunen LH, Kelloniemi J, Valkonen JPT (2017) Repeated applications of a nonpathogenic Streptomyces strain enhance development of suppressiveness to potato common scab. Plant Dis 101(1):224–232

Holtsmark I, Takle GW, Brurberg MB (2008) Expression of putative virulence factors in the potato pathogen Clavibacter michiganensis subsp. sepedonicus during infection. Arch Microbiol 189(2):131–139

Huet G (2014) Breeding for resistances to Ralstonia Solanacearum. Front Plant Sci 5:715

Humphris SN, Cahill G, Elphinstone JG, Kelly R, Parkinson NM, Pritchard L, Toth IK, Saddler GS (2015) Detection of the bacterial potato pathogens Pectobacterium and Dickeya Spp. using conventional and real-time PCR. In: Lacomme C (ed) Plant pathology: techniques and protocols, Methods in molecular biology. Springer, New York, pp 1–16. https://doi.org/10.1007/978-1-4939-2620-6_1

Jacobs JM, Babujee L, Meng F, Milling A, Allen C (2012) The in planta transcriptome of Ralstonia solanacearum: conserved physiological and virulence strategies during bacterial wilt of tomato. MBio 3(4):e00114–e00112

Janse JD, Araluppan FAX, Schans J, Wenneker M, Westerhuis W (1998) Experiences with bacterial brown rot Ralstonia solanacearum biovar 2, race 3 in the Netherlands, Bacterial wilt disease. Springer, Berlin, pp 146–152

Joshi MV, Loria R (2007) Streptomyces turgidiscabies possesses a functional cytokinin biosynthetic pathway and produces leafy galls. Mol Plant-Microbe Interact 20(7):751–758

Khatri BB, Tegg RS, Brown PH, Wilson CR (2011) Temporal association of potato tuber development with susceptibility to common scab and Streptomyces scabiei-induced responses in the potato periderm. Plant Pathol 60(4):776–786

Khayi S, Cigna J, Chong TM, Quêtu-Laurent A, Chan K-G, Hélias V, Faure D (2016) Transfer of the potato plant isolates of Pectobacterium wasabiae to Pectobacterium parmentieri sp. nov. Int J Syst Evol Microbiol 66(12):5379–5383

Kim H-S, Ma B, Perna NT, Charkowski AO (2009) Phylogeny and virulence of naturally occurring type III secretion system-deficient Pectobacterium strains. Appl Environ Microbiol 75(13):4539–4549

King RR, Harold Lawrence C, Clark MC, Calhoun LA (1989) Isolation and characterization of phytotoxins associated with Streptomyces scabies. J Chem Soc Chem Commun 13:849–850

King RR, Harold Lawrence C, Clark MC (1991) Correlation of phytotoxin production with pathogenicity of streptomyces scabies isolates from scab infected potato tubers. Am Potato J 68(10):675–680

Kinkel LL, Bowers JH, Shimizu K, Neeno-Eckwall EC, Schottel JL (1998) Quantitative relationships among Thaxtomin A production, potato scab severity, and fatty acid composition in streptomyces. Can J Microbiol 44(8):768–776

Kloepper JW, Schroth MN (1981) Relationship of in vitro antibiosis of plant growth-promoting rhizobacteria to plant growth and the displacement of root microflora. Phytopathology 71: 1020–1024

Komeil D, Simao-Beaunoir A-M, Beaulieu C (2013) Detection of potential suberinase-encoding genes in Streptomyces scabiei strains and other actinobacteria. Can J Microbiol 59(5):294–303

Kong HG, Bae JY, Lee HJ, Joo HJ, Jung EJ, Chung E, Lee S-W (2014) Induction of the viable but nonculturable state of Ralstonia solanacearum by low temperature in the soil microcosm and its resuscitation by catalase. PLoS One 9(10):e109792

Kritzman G, Shani-Cahani A, Kirshner B, Riven Y, Bar Z, Katan J, Grinstein A (1996) Pod wart disease of peanuts. Phytoparasitica 24(4):293–304

Kumar GNM, Knowles LO, Richard Knowles N (2015) Zebra chip disease decreases tuber (Solanum tuberosum L.) protein content by attenuating protease inhibitor levels and increasing protease activities. Planta 242(5):1153–1166

Laine MJ, Haapalainen M, Wahlroos T, Kankare K, Nissinen R, Kassuwi S, Metzler MC (2000) The cellulase encoded by the native plasmid of Clavibacter michiganensis ssp. sepedonicus plays a role in virulence and contains an expansin-like domain. Physiol Mol Plant Pathol 57(5):221–233

Lambert DH, Loria R (1989a) Streptomyces scabies sp. nov., nom. rev. Int J Syst Evol Microbiol 39:387–392

Lambert DH, Loria R (1989b) Streptomyces acidiscabies sp. nov. Int J Syst Evol Microbiol 39:393–396

Larkin RP, Griffin TS (2007) Control of Soilborne potato diseases using Brassica green manures. Crop Prot 26(7):1067–1077

Larkin RP, Halloran JM (2014) Management effects of disease-suppressive rotation crops on potato yield and soilborne disease and their economic implications in potato production. Am J Potato Res 91(5):429–439

Larkin RP, Wayne Honeycutt C, Griffin TS, Modesto Olanya O, Halloran JM, He Z (2011) Effects of different potato cropping system approaches and water management on soilborne diseases and soil microbial communities. Phytopathology 101(1):58–67

Laurila J, Metzler MC, Ishimaru CA, Rokka V-M (2003) Infection of plant material derived from Solanum acaule with Clavibacter michiganensis ssp. sepedonicus: temperature as a determining factor in immunity of S. acaule to bacterial ring rot. Plant Pathol 52(4):496–504

Laurila J, Ahola V, Lehtinen A, Joutsjoki T, Hannukkala A, Rahkonen A, Pirhonen M (2008) Characterization of Dickeya strains isolated from potato and river water samples in Finland. Eur J Plant Pathol 122(2):213–225. https://doi.org/10.1007/s10658-008-9274-5

Lehtonen MJ, Rantala H, Kreuze JF, Bång H, Kuisma L, Koski P, Virtanen E, Vihlman K, Valkonen JPT (2004) Occurrence and survival of potato scab pathogens (Streptomyces species) on tuber lesions: quick diagnosis based on a PCR-based assay. Plant Pathol 53(3):280–287

Lévy JG, Scheuring DC, Koym JW, Henne DC, Tamborindeguy C, Pierson E, Creighton Miller J (2015) Investigations on putative zebra chip tolerant potato selections. Am J Potato Res 92(3):417–425

Lin H, Lou B, Glynn JM, Doddapaneni H, Civerolo EL, Chen C, Duan Y, Zhou L, Vahling CM (2011) The complete genome sequence of 'Candidatus Liberibacter solanacearum', the bacterium associated with potato zebra chip disease. PLoS One 6(4):e19135

Lu Y, Hatsugai N, Katagiri F, Ishimaru CA, Glazebrook J (2015) Putative serine protease effectors of Clavibacter michiganensis induce a hypersensitive response in the apoplast of Nicotiana species. Mol Plant-Microbe Interact 28(11):1216–1226

Ma B, Hibbing ME, Kim HS, Reedy RM, Yedidia I, Breuer J, Breuer J, Glasner JD, Perna NT, Kelman A, Charkowski AO (2007a) Host range and molecular phylogenies of the soft rot enterobacterial genera Pectobacterium and Dickeya. Phytopathology 97(9):1150–1163. https://apsjournals.apsnet.org/doi/abs/10.1094/PHYTO-97-9-1150

Ma B, Hibbing ME, Kim H-S, Reedy RM, Yedidia I, Breuer J, Breuer J, Glasner JD, Perna NT, Kelman A (2007b) Host range and molecular phylogenies of the soft rot enterobacterial genera Pectobacterium and Dickeya. Phytopathology 97(9):1150–1163

Mansfield J, Genin S, Magori S, Citovsky V, Sriariyanum M, Ronald P, Dow MAX, Verdier V, Beer SV, Machado MA (2012) Top 10 plant pathogenic bacteria in molecular plant pathology. Mol Plant Pathol 13(6):614–629

Manzer FE, McKenzie AR (1988) Cultivar response to bacterial ring rot infection in Maine. Am Potato J 65(6):333–339

Meng F (2013) The virulence factors of the bacterial wilt pathogen Ralstonia solanacearum. J Plant Pathol Microbiol 4(168):10–4172

Miyajima K, Tanaka F, Takeuchi T, Kuninaga S (1998) Streptomyces turgidiscabies sp. nov. Int J Syst Evol Microbiol 48(2):495–502

Nabhan S, De Boer SH, Maiss E, Wydra K (2013) Pectobacterium aroidearum sp. nov., a soft rot pathogen with preference for monocotyledonous plants. Int J Syst Evol Microbiol 63(7):2520–2525

Narusaka M, Kubo Y, Hatakeyama K, Imamura J, Ezura H, Nanasato Y, Tabei Y, Takano Y, Shirasu K, Narusaka Y (2013) Interfamily transfer of dual NB-LRR genes confers resistance to multiple pathogens. PLoS One 8(2):e55954

Natsume M, Komiya M, Koyanagi F, Tashiro N, Kawaide H, Abe H (2005) Phytotoxin produced by Streptomyces sp. causing potato russet scab in Japan. J Gen Plant Pathol 71(5):364–369

Natsume M, Tashiro N, Doi A, Nishi Y, Kawaide H (2017) Effects of concanamycins produced by Streptomyces scabies on lesion type of common scab of potato. J Gen Plant Pathol 83(2):78–82

Nelson WR, Fisher TW, Munyaneza JE (2011) Haplotypes of 'Candidatus Liberibacter solanacearum' suggest long-standing separation. Eur J Plant Pathol 130(1):5–12

Ngadze E, Brady CL, Coutinho TA, Van der Waals JE (2012) Pectinolytic bacteria associated with potato soft rot and blackleg in South Africa and Zimbabwe. Eur J Plant Pathol 134(3):533–549

Nissinen R, Kassuwi S, Peltola R, Metzler MC (2001) In planta-complementation of Clavibacter michiganensis subsp. sepedonicus strains deficient in cellulase production or HR induction restores virulence. Eur J Plant Pathol 107(2):175–182

Nykyri J, Niemi O, Koskinen P, Nokso-Koivisto J, Pasanen M, Broberg M, Plyusnin I, Törönen P, Holm L, Pirhonen M (2012) Revised phylogeny and novel horizontally acquired virulence determinants of the model soft rot phytopathogen Pectobacterium wasabiae SCC3193. PLoS Pathog 8(11):e1003013

Oniki M, Suzui T, Araki T, Sonoda RI, Chiba T, Takeda T (1986a) Causal agent of russet scab of potato. Bull Nat Inst of Agro Environ Sci 2:45–60

Oniki M, Suzui T, Araki T, Sonoda R, Chiba T, Takeda T (1986b) Causal agent of russet scab of potato [Streptomyces sp.]. Bull Nat Inst Agro-Environ Sci. http://agris.fao.org/agris-search/search.do?recordID=JP870464388

Padilla-Reynaud R, Simao-Beaunoir A-M, Lerat S, Bernards MA, Beaulieu C (2015) Suberin regulates the production of cellulolytic enzymes in Streptomyces scabiei, the causal agent of potato common scab. Microbes Environ 30(3):245–253

Park DH, Kim JS, Kwon SW, Wilson C, Yu YM, Hur JH, Lim CK (2003) Streptomyces luridiscabiei sp. nov., Streptomyces puniciscabiei sp. nov. and Streptomyces niveiscabiei sp. nov., which cause potato common scab disease in Korea. Int J Syst Evol Microbiol 53(6):2049–2054

Parkinson N, Stead D, Bew J, Heeney J, Tsror (Lahkim) L, Elphinstone J (2009) Dickeya species relatedness and clade structure determined by comparison of RecA sequences. Int J Syst Evol Microbiol 59(10):2388–2393. https://doi.org/10.1099/ijs.0.009258-0

Parkinson N, DeVos P, Pirhonen M, Elphinstone J (2014) Dickeya aquatica sp. nov., isolated from waterways. Int J Syst Evol Microbiol 64(7):2264–2266. https://doi.org/10.1099/ijs.0.058693-0

Pavlista AD (2005) Early-season applications of sulfur fertilizers increase potato yield and reduce tuber defects. Agron J 97(2):599–603

Pérombelon MCM (2002) Potato diseases caused by soft rot erwinias: an overview of pathogenesis. Plant Pathol 51(1):1–12

Pérombelon MCM, Kelman A (1987) Blackleg and other potato diseases caused by soft rot erwinias: proposal for revision of terminology. Plant Dis 71(3):283–285

Pirhonen M, Flego D, Heikinheimo R, Tapio Palva E (1993) A small diffusible signal molecule is responsible for the global control of virulence and exoenzyme production in the plant pathogen Erwinia carotovora. EMBO J 12(6):2467–2476

Pitman AR, Wright PJ, Galbraith MD, Harrow SA (2008) Biochemical and genetic diversity of pectolytic enterobacteria causing soft rot disease of potatoes in New Zealand. Australas Plant Pathol 37(6):559–568

Pitman AR, Harrow SA, Visnovsky SB (2010) Genetic characterisation of Pectobacterium wasabiae causing soft rot disease of potato in New Zealand. Eur J Plant Pathol 126(3):423–435

Powelson ML, Rowe RC (2008) Managing diseases caused by seedborne and soilborne fungi and fungus-like pathogens. Potato Health Manage:2183–2195

Prior P, Ailloud F, Dalsing BL, Remenant B, Sanchez B, Allen C (2016) Genomic and proteomic evidence supporting the division of the plant pathogen Ralstonia solanacearum into three species. BMC Genomics 17(1):90

Qian Y-l, Wang X-s, Wang D-z, Zhang L-n, Chao-long Z, Gao Z-l, Zhang H-j, Wang Z-y, Sun X-y, Yao D-n (2013) The detection of QTLs controlling bacterial wilt resistance in tobacco (N. Tabacum L.). Euphytica 192(2):259–266

Qu X, Wanner LA, Christ BJ (2008) Using the TxtAB operon to quantify pathogenic Streptomyces in potato tubers and soil. Phytopathology 98(4):405–412

Rashed A, Workneh F, Paetzold L, Rush CM (2015) Emergence of 'Candidatus Liberibacter solanacearum'-infected seed potato in relation to the time of infection. Plant Dis 99(2):274–280

Rashidi M, Novy RG, Wallis CM, Rashed A (2017) Characterization of host plant resistance to zebra chip disease from species-derived potato genotypes and the identification of new sources of zebra chip resistance. PLoS One 12(8):e0183283

Rietman H, Finkers R, Evers L, van der Zouwen PS, van der Wolf JM, Visser RGF (2014) A stringent and broad screen of Solanum spp. tolerance against Erwinia bacteria using a petiole test. Am J Potato Res 91(2):204–214

Rossmann S, Dees MW, Perminow J, Meadow R, Brurberg MB (2018) Soft rot Enterobacteriaceae are carried by a large range of insect species in potato fields. Appl Environ Microbiol 84(12):e00281–e00218. https://doi.org/10.1128/AEM.00281-18

Safni I, Cleenwerck I, De Vos P, Fegan M, Sly L, Kappler U (2014) Polyphasic taxonomic revision of the Ralstonia solanacearum species complex: proposal to emend the descriptions of *Ralstonia solanacearum* and *Ralstonia syzygii* and reclassify current *R. syzygii* strains as Ralstonia syzygii subsp. syzygii subsp. nov., R. solanacearum phylotype IV strains as *Ralstonia syzygii* subsp. Indonesiensis subsp. nov., Banana blood disease bacterium strains as *Ralstonia syzygii* subsp. Celebesensis subsp. nov. and *R. solanacearum* phylotype I and III strains as *Ralstonia pseudosolanacearum* sp. nov. Int J Syst Evol Microbiol 64(9):3087–3103

Salgon S, Jourda C, Sauvage C, Daunay M-C, Reynaud B, Wicker E, Dintinger J (2017) Eggplant resistance to the Ralstonia solanacearum species complex involves both broad-spectrum and strain-specific quantitative trait loci. Front Plant Sci 8:828

Samson R, Legendre JB, Christen R, Saux MF-L, Achouak W, Gardan L (2005) Transfer of Pectobacterium chrysanthemi (Burkholder et Al. 1953) Brenner et Al. 1973 and Brenneria paradisiaca to the genus Dickeya gen. nov. as Dickeya chrysanthemi comb. nov. and Dickeya paradisiaca comb. nov. and delineation of four novel species, Dickeya dadantii sp. nov., Dickeya dianthicola sp. nov., Dickeya dieffenbachiae sp. nov. and Dickeya zeae sp. nov. Int J Syst Evol Microbiol 55(Pt 4):1415–1427. https://doi.org/10.1099/ijs.0.02791-0.

Santos-Cervantes ME, Felix-Gastelum R, Herrera-Rodríguez G, Espinoza-Mancillas MG, Mora-Romero AG, Leyva-López NE (2017) Characterization, pathogenicity and chemical control of Streptomyces acidiscabies associated to potato common scab. Am J Potato Res 94(1):14–25

Sarfraz S, Riaz K, Oulghazi S, Cigna J, Sahi ST, Khan SH, Faure D (2018) Pectobacterium punjabense sp. nov., isolated from blackleg symptoms of potato plants in Pakistan. Int J Syst Evol Microbiol 68(11):3551–3556

Sarikhani E, Sagova-Mareckova M, Omelka M, Kopecky J (2017) The effect of peat and iron supplements on the severity of potato common scab and bacterial community in tuberosphere soil. FEMS Microbiol Ecol 93(1)

Schlatter D, Kinkel L, Thomashow L, Weller D, Paulitz T (2017) Disease suppressive soils: new insights from the soil microbiome. Phytopathology 107(11):1284–1297

Secor GA, Rivera VV, Abad JA, Lee I-M, Clover GRG, Liefting LW, Li X, De Boer SH (2009) Association of 'Candidatus Liberibacter solanacearum'with zebra chip disease of potato established by graft and psyllid transmission, electron microscopy, and PCR. Plant Dis 93(6):574–583

Sengoda VG, Rodney Cooper W, Swisher KD, Henne DC, Munyaneza JE (2014) Latent period and transmission of 'Candidatus Liberibacter solanacearum' by the potato Psyllid Bactericera cockerelli (Hemiptera: Triozidae). PLoS One 9(3):e93475

She XM, Lan GB, Tang YF, He ZF (2017) Pectrobacterium caratovorum subsp. Brasiliense causing pepper black spot disease in China. J Plant Pathol 99(3):769–772

Sławiak M, van Beckhoven JRCM, Speksnijder AGCL, Czajkowski R, Grabe G, van der Wolf JM (2009) Biochemical and genetical analysis reveal a new clade of biovar 3 Dickeya spp. strains isolated from potato in Europe. Eur J Plant Pathol 125(2):245–261. https://doi.org/10.1007/s10658-009-9479-2

Teresani GR, Bertolini E, Alfaro-Fernández A, Martínez C, Tanaka FAO, Kitajima EW, Roselló M, Sanjuán S, Ferrándiz JC, López MM (2014) Association of 'Candidatus Liberibacter solanacearum'with a vegetative disorder of celery in Spain and development of a real-time PCR method for its detection. Phytopathology 104(8):804–811

Teulon DAJ, Workman PJ, Nielsen MC, Thomas KL, Anderson DP (2009) Bactericera cockerelli: incursion, dispersal and current distribution on vegetable crops in New Zealand. New Zealand Plant Prot 62:136–144

Thangavel T, Tegg RS, Wilson CR (2016) Toughing it out—disease-resistant potato mutants have enhanced tuber skin defenses. Phytopathology 106(5):474–483

Thaxter R (1892) Potato scab. Potato Scab:153–160. https://www.cabdirect.org/cabdirect/abstract/20057000599

Thinakaran J, Pierson E, Kunta M, Munyaneza JE, Rush CM, Henne DC (2015) Silverleaf nightshade (Solanum Elaeagnifolium), a reservoir host for 'Candidatus Liberibacter solanacearum', the putative causal agent of zebra chip disease of potato. Plant Dis 99(7):910–915

Tian Y, Zhao Y, Yuan X, Yi J, Fan J, Zhigang X, Baishi H, De Boer SH, Li X (2016) Dickeyafangzhongdai sp. nov., a plant-pathogenic bacterium isolated from pear trees (Pyrus Pyrifolia). Int J Syst Evol Microbiol 66(8):2831–2835. https://doi.org/10.1099/ijsem.0.001060

Tomihama T, Nishi Y, Mori K, Shirao T, Iida T, Uzuhashi S, Ohkuma M, Ikeda S (2016) Rice bran amendment suppresses potato common scab by increasing antagonistic bacterial community levels in the rhizosphere. Phytopathology 106(7):719–728

Toth IK, van der Wolf JM, Saddler G, Lojkowska E, Hélias V, Pirhonen M, Tsror (Lahkim) L, Elphinstone JG (2011) Dickeya species: an emerging problem for potato production in Europe. Plant Pathol 60(3):385–399. https://doi.org/10.1111/j.1365-3059.2011.02427.x

van der Merwe JJ, Coutinho TA, Korsten L, van der Waals JE (2010) Pectobacterium carotovorum subsp. brasiliensis causing blackleg on potatoes in South Africa. Eur J Plant Pathol 126(2):175–185

Van der Wolf JM, Elphinstone JG, Stead DE, Metzler M, Müller P, Hukkanen A, Karjalainen R (2005a) Epidemiology of Clavibacter michiganensis subsp. sepedonicus in relation to control of bacterial ring rot. PRI Bioscience (Report/Plant Research International 95)—38

Van der Wolf JM, Van Beckhoven JRCM, Hukkanen A, Karjalainen R, Müller P (2005b) Fate of *Clavibacter michiganensis* ssp. sepedonicus, the causal organism of bacterial ring rot of potato, in weeds and field crops. J Phytopathol 153(6):358–365

Van der Wolf JM, de Haan EG, Kastelein P, Krijger M, de Haas BH, Velvis H, Mendes O, Kooman-Gersmann M, van der Zouwen PS (2017) Virulence of Pectobacterium carotovorum subsp. brasiliense on potato compared with that of other Pectobacterium and Dickeya species under climatic conditions prevailing in the Netherlands. Plant Pathol 66(4):571–583

Van Vaerenbergh J, De Paepe B, Hoedekie A, Van Malderghem C, Zaluga J, De Vos P, Maes M (2016) Natural infection of *Clavibacter michiganensis* subsp. sepedonicus in tomato (Solanum Lycopersicum). New Dis Rep 33:7–7

Waldee EL (1942) Comparative studies of some peritrichous phytopathogenic bacteria. Retrospective Theses and Dissertations. p 14150. https://lib.dr.iastate.edu/rtd/14150

Waleron M, Misztak A, Waleron M, Franczuk M, Wielgomas B, Waleron K (2018) Transfer of Pectobacterium carotovorum subsp. carotovorum strains isolated from potatoes grown at high altitudes to *Pectobacterium peruviense* sp. nov. Syst Appl Microbiol 41(2):85–93

Wallis CM, Munyaneza JE, Chen J, Novy R, Bester G, Buchman JL, Nordgaard J, Hest PV (2015) 'Candidatus Liberibacter solanacearum' titers in and infection fffects on potato tuber chemistry of promising germplasm exhibiting tolerance to zebra chip disease. Phytopathology 105(12):1573–1584

Wallis CM, Rashed A, Workneh F, Paetzold L, Rush CM (2017) Effects of holding temperatures on the development of zebra chip symptoms,'Candidatus Liberibacter solanacearum' titers, and phenolic levels in 'Red La Soda' and 'Russet Norkotah' tubers. Am J Potato Res 94(4):334–341

Wang J-F, Ho F-I, Truong HTH, Huang S-M, Balatero CH, Dittapongpitch V, Hidayati N (2013) Identification of major QTLs associated with stable resistance of tomato cultivar 'Hawaii 7996'to Ralstonia solanacearum. Euphytica 190(2):241–252

Wang J, Haapalainen M, Schott T, Thompson SM, Smith GR, Nissinen AI, Pirhonen M (2017a) Genomic sequence of 'Candidatus Liberibacter solanacearum' haplotype C and its comparison with haplotype A and B genomes. PLoS One 12(2):e0171531

Wang N, Pierson EA, Setubal JC, Xu J, Levy JG, Zhang Y, Li J, Rangel LT, Jr JM (2017b) The Candidatus Liberibacter–host interface: insights into pathogenesis mechanisms and disease control. Annu Rev Phytopathol 55:451–482

Wang J, Wang YH, Dai PG, Chen DX, Zhao TC, Li XL, Huang Q (2017c) First report of tobacco bacterial leaf blight caused by Pectobacterium carotovorum subsp. Brasiliense in China. Plant Dis 101(5):830–830

Wanner LA (2006) A survey of genetic variation in Streptomyces isolates causing potato common scab in the United States. Phytopathology 96(12):1363–1371

Wanner LA (2007) High proportions of nonpathogenic Streptomyces are associated with common scab-resistant potato lines and less severe disease. Can J Microbiol 53(9):1062–1075

Wanner LA (2009) A patchwork of Streptomyces species isolated from potato common scab lesions in North America. Am J Potato Res 86(4):247–264

Williamson L, Nakaho K, Hudelson B, Allen C (2002) Ralstonia solanacearum race 3, biovar 2 strains isolated from geranium are pathogenic on potato. Plant Dis 86(9):987–991

Winslow C-EA, Jean B, Buchanan RE, Krumwiede C Jr, Rogers LA, Smith GH (1917) The families and genera of the bacteria: preliminary report of the Committee of the Society of American Bacteriologists on characterization and classification of bacterial types. J Bacteriol 2(5):505

Yang FY, Yang DJ, Zhao WQ, Liu DQ, Yu XM (2017) First report of Streptomyces diastatochromogenes causing potato common scab in China. Plant Dis 101(1):243–243

Yao J, Saenkham P, Levy J, Ibanez F, Noroy C, Mendoza A, Huot O, Meyer DF, Tamborindeguy C (2016) Interactions 'Candidatus Liberibacter solanacearum'—Bactericera cockerelli: haplotype effect on vector fitness and gene expression analyses. Front Cell Infect Microbiol 6:62

Yap M-N, Barak JD, Charkowski AO (2004) Genomic diversity of Erwinia carotovora subsp. carotovora and its correlation with virulence. Appl Environ Microbiol 70(5):3013–3023

Zhang Y, Loria R (2017) Emergence of novel pathogenic Streptomyces species by site-specific accretion and Cis-mobilization of Pathogenicity Islands. Mol Plant-Microbe Interact 30(1):72–82

Zhang Y, Bignell DRD, Zuo R, Fan Q, Huguet-Tapia JC, Ding Y, Loria R (2016a) Promiscuous pathogenicity islands and phylogeny of pathogenic Streptomyces spp. Mol Plant-Microbe Interact 29(8):640–650

Zhang Y, Fan Q, Loria R (2016b) A re-evaluation of the taxonomy of phytopathogenic genera Dickeya and Pectobacterium using whole-genome sequencing data. Syst Appl Microbiol 39(4):252–259

Chapter 11
Viral Diseases in Potato

J. F. Kreuze, J. A. C. Souza-Dias, A. Jeevalatha, A. R. Figueira, J. P. T. Valkonen, and R. A. C. Jones

Abstract Viruses are among the most significant biotic constraints in potato production. In the century since the discovery of the first potato viruses we have learned more and more about these pathogens, and this has accelerated over the last decade with the advent of high-throughput sequencing in the study of plant virology. Most reviews of potato viruses have focused on temperate potato production systems of Europe and North America. However, potato production is rapidly expanding in tropical and subtropical agro-ecologies of the world in Asia and Africa, which present a unique set of problems for the crop and affect the way viruses can be managed. In this chapter we review the latest discoveries in potato virology as well as the changes in virus populations that have occurred over the last 50 years, with a particular focus on countries in the (sub-)tropics. We also review the different management approaches including use of resistance, seed systems, and cultural approaches that have been employed in different countries and reflect on what can be learnt from past research on potato viruses, and what can be expected in the future facing climate change.

J. F. Kreuze (✉)
International Potato Center, Lima, Peru
e-mail: j.kreuze@cgiar.org

J. A. C. Souza-Dias
Instituto Agronômico de Campinas (IAC)/APTA/SAA-SP, Campinas, São Paulo, Brazil
e-mail: jcaram@iac.sp.gov.br

A. Jeevalatha
ICAR-Central Potato Research Institute, Shimla, Himachal Pradesh, India

A. R. Figueira
Department of Plant Pathology, Lavras Federal University, Lavras, Minas Gerais, Brazil
e-mail: antonia@ufla.br

J. P. T. Valkonen
Department of Agricultural Sciences, University of Helsinki, Helsinki, Finland
e-mail: jari.valkonen@helsinki.fi

R. A. C. Jones
Institute of Agriculture, University of Western Australia, Crawley, Perth, Australia
e-mail: roger.jones@uwa.edu.au

H. Campos, O. Ortiz (eds.), *The Potato Crop*,
https://doi.org/10.1007/978-3-030-28683-5_11

11.1 Introduction

In terms of human consumption, potato (*Solanum tuberosum*) is currently the third most important food crop globally after rice and wheat, and over half of its production currently occurs in developing countries (Devaux et al. 2014). Worldwide, over the last few decades, potato production has increased at a much higher rate compared to other major staple crops. This increase has occurred principally in developing countries located in largely tropical and subtropical regions. Due to its ability to produce high amounts of digestible energy per unit time and unit area for home consumption, but at the same time provide income as a cash crop, planting potato tends to be popular with farmers wherever they are able to grow them. In the future, development of new heat tolerant and early maturing potato cultivars will likely lead to further expansion of its production into warmer areas of the tropics. However, as temperatures increase virus vectors often become more abundant and the incidence of virus epidemics increases. This increase in insect vectors and virus disease incidence, combined with the fact that virus-tested seed systems are weak or entirely absent, explains why potato virus diseases are of particular importance in the developing world and estimated to account for 50% or more of the potential total yield being lost (Harahagazwe et al. 2018). In addition, the presence of year-round potato cultivation in some tropical regions and the lack of cool upland areas where insect vector pressure is low enough to produce high-quality seed potatoes, both exacerbate potato virus disease problems in these regions. Several global or regional reviews and intercontinentally focused research papers on potato virus diseases have been written during the last decade, devoted to many different aspect or particular viruses, including economic losses, detection methods, molecular variability, resistance genes, and evolution (Valkonen 2007; Gray et al. 2010; Karasev and Gray 2013; Jones 2014; Gibbs et al. 2017; Lacomme et al. 2017; Santillan et al. 2018) and we refer to them for details on those specific aspects. In this chapter we will review potato viruses with a focus on developing world regions and changes that have occurred over the past 50 years. The reason for this is that they have traditionally received less attention in the literature, but also are generally located in places with warmer climates, and with global temperatures rising, may be representative for what the future holds also for the currently more temperate regions. We will start with a general overview of potato infecting viruses of global importance, the damages they cause to potato production, and where they occur. Next, we describe the viruses found in the center of origin of wild and cultivated potatoes, the Andes. Then we describe the situation in two emerging economies in (sub-)tropics with functional seed systems, one, Brazil which is largely based on imported basic seed tubers and one, India, which largely produces its own potato cultivars and seed. This is followed by a brief review of the situation in Africa, and a description of the situation in two contrasting developed economies, Australia and Europe. Finally, we consider the main control methods for potato viruses, how they are being applied in the different agro-ecologies, and how they might be affected by changing climates.

11.2 Viruses of Potato

Whereas more than 50 different viruses and one viroid have been reported infecting potatoes worldwide (Table 11.1, Fig. 11.1), only a handful of them cause major losses globally. However, some are locally and/or temporarily relevant, while others

Table 11.1 List of viruses of potato worldwide

Virus[a]	Genus, family	Transmission	Distribution
Potato virus Y (PVY)	*Potyvirus, Potyviridae*	Aphids	Worldwide
Potato virus A (PVA)			Worldwide
Potato virus V (PVV)			Europe, South America
Wild potato mosaic virus (WPMV)			Andes, only reported in wild potatoes
Potato virus X (PVX)	*Potexvirus, Alphaflexiviridae*	Contact	Worldwide
Potato aucuba mosaic virus (PAMV)			Worldwide, very rare
Potato leaf roll virus (PLRV)	*Polerovirus*	Aphids	Worldwide
Potato virus S (PVS)	*Carlavirus, Betaflexiviridae*	Contact, aphids	Worldwide
Potato latent virus (PotLV)		Aphids	North America, rare
Potato virus M (PVM)		Aphids	Worldwide
Potato virus H (PVH)		unknown	China
Potato virus P (PVP syn. Potato rought dwarf virus: PRDV)		unknown	Brazil & Argentina
Potato virus T (PVT)	*Tepovirus, Betaflexiviridae*	Contact, seed	Southern Andean region
Andean potato mottle virus (APMoV)	*Comovirus, Secoviridae*	Beetles	Andean region, Brazil
Potato black ringspot virus (PBRSV = TRSV-Ca)	*Nepovirus, Secoviridae*	true seed, nematodes	Peru
Potato virus U (PVU)		nematodes	Peru, only reported once
Potato virus B (PVB)		nematodes?	Peru, recently reported, relatively common
Cherry leaf roll virus (CLRV)		Nematodes, TPS, pollen?	Europe, North & South America
Lucerne Australian latent virus (LALV)		Unknown	Australia and New Zealand, rare in potato
Tomato black ring virus (TBRV)		Nematodes	Europe, rare
Cherry rasp leaf virus (CRLV)	*Cheravirus, Secoviridae*		North America, only reported once
Arracacha virus B (AVB)		TPS, pollen	Andes

(continued)

Table 11.1 (continued)

Virus[a]	Genus, family	Transmission	Distribution
Tomato spotted wilt virus (TSWV)	*Tospovirus, Bunyaviridae*	Thrips	Worldwide
Tomato chlorotic spot virus (TCSV)			South America
Groundnut bud necrosis virus (GBNV)			India
Groundnut ringspot virus (GRSV)			Americas
"Tomato yellow fruit ring virus" (TYFRV)			Reported from potato in Iran
Impatiens necrotic spot virus (INSV)			Worldwide, reported in greenhouse grown potatoes in USA
Andean potato latent virus (APLV)	*Tymovirus, Tymoviridae*	Beetles	Andean region
Andean potato mild mottle virus (APMMV)			Andean region
Potato yellow vein virus (PYVV)	*Crinivirus, Closteroviridae*	Whiteflies	Northern Andean region, Panama
Tomato chlorosis virus (ToCV)			Brazil, Spain, India
"Potato yellowing virus" (PYV)	*Ilarvirus, Bromoviridae*	Unknown	Andean region
Tobacco streak virus (TSV)		Pollen, thrips	Worldwide, reported in potato in Brazil
Cucumber mosaic virus (CMV)	*Cucumovirus, Bromoviridae*	Aphids	Worldwide, sporadic in potato
Alfalfa mosaic virus (AlMV)	*Alfamovirus, Bromoviridae*	Aphids	Worldwide, sporadic in potato
Tomato leaf curl New Delhi virus (ToLCNDV)	*Begomovirus, Geminiviridae*	Whiteflies	India
Tomato severe rugose virus (ToSRV)			Brazil
Tomato yellow vein streak virus (ToYVSV=PDMV)			Brazil, Argentina
Tomato mottle Taino virus (ToMoTV)			Cuba
"Solanum apical leaf curl virus" (SALCV)			Peru, only reported once
Potato yellow mosaic virus (PYMV)			Carribean
Beet curly top virus (BCtV)	*Curtovirus, Geminiviridae*	Leaf hopper	Americas, Europe, Asia under dry conditions

(continued)

Table 11.1 (continued)

Virus[a]	Genus, family	Transmission	Distribution
Potato mop-top virus (PMTV)	*Pomovirus, Virgaviridae*	Spongospora	Americas, Europe, Asia in cool and humid environments
"Colombian potato soil-borne virus" (CPSbV)		Spongospora?	Colombia, isolated from potato soils; CPSbV could infect potatoes symptomless
"Soil-borne virus 2" (SbV2)		Spongospora?	
Tobacco rattle virus (TRV)	*Tobravirus, Virgaviridae*	Nematodes	Worldwide, common in cool climates, or Australia
Tobacco mosaic virus (TMV)	*Tobamovirus, Virgaviridae*	Contact	Worldwide, rare on potato
Tomato mosaic virus (ToMV)			Europe, Andes, rare on potato
Tobacco necrosis virus (TNV)	*Necrovirus, Tombusviridae*	Fungus	Worldwide, rare on potato
Sowbane mosaic virus (SbMV)	*Sobemovirus*, unassigned	Contact	Worldwide, rare on potato
SB26/29 "potato rugose stunting virus"	Torradovirus-like, *Secoviridae*	Psyllids	Southern Peru
Potato yellow dwarf virus (PYDV)	*Nucleorhabdovirus, Rhabdoviridae*	Leafhoppers	North America, has become rare
Eggplant mottle dwarf virus (EMDV)		Aphids	Europe, Africa, Asia, occasionally infects potatoes
Cauliflower mosaic virus (CaMV)	*Caulimovirus, Caulimoviridae*	Aphids	Intercepted once in potato from South America
Potato spindle tuber viroid (PSTVd)	*Pospiviroid, Pospiviroidae*	Contact, aphids (when co infecting with PLRV)	Worldwide
"Potato stunt virus" (PStV)	?	?	Europe

[a] Officially accepted virus species names italicized, whereas unofficial names are between quotation marks and not in italics

are currently only of minor importance anywhere in the world. PVY (see Table 11.1 for virus acronyms) and PLRV are now the most damaging viruses of potato worldwide, with PVY having overtaken PLRV as the most important. Tuber yield losses are caused by either of them in single infections and can reach more than 80% in combination with other viruses. PVX occurs commonly worldwide and causes losses of 10–40% in single infections and is particularly damaging in combination with PVY or PVA. This is due to its synergism with both potyviruses leading to tuber yield losses of up to 80%. PVS also occurs commonly worldwide but generally causes only minor tuber yield losses unless severe strains are present or it occurs in mixed infection with PVX. PVA can cause yield losses of up to 40% by itself but is far less prevalent than PVY, PVS, or PLRV. PVM is relatively uncommon in most

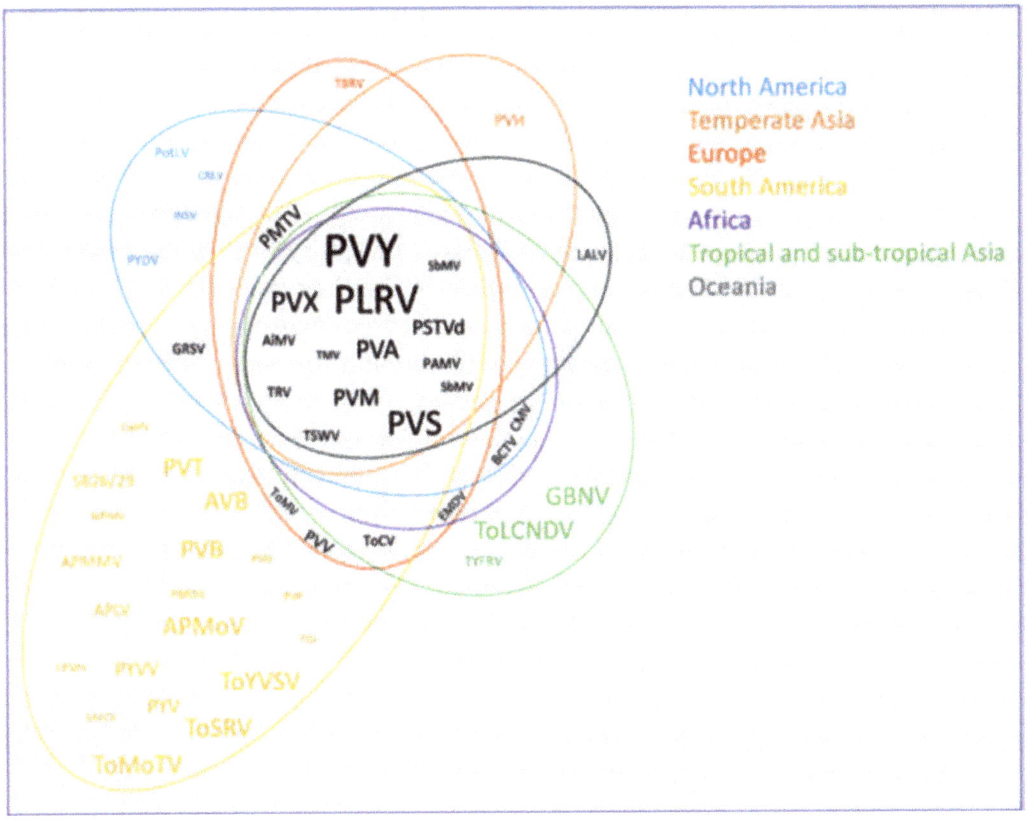

Fig. 11.1 VENN diagram of geographic occurrence of commoner potato viruses worldwide. Their relative importance/incidence globally, or regionally in case they are geographically restricted, is illustrated by the size of the virus acronym. See Table 11.1 for virus acronyms

countries and, like PVS, mostly causes only minor tuber yield losses, except in mixed infection with PVX or other viruses.

Besides yield reduction, several viruses cause economic losses by affecting potato quality, particularly by inducing internal and surface tuber necrosis. PLRV sometimes causes necrosis of the tuber vascular system known as "net necrosis." Tuber necrosis, consisting of necrotic rings or arcs in the flesh, sometimes develop with the thrips-transmitted virus TSWV, and with soil-borne viruses like nematode vectored TRV (Sahi et al. 2016) and protist-transmitted PMTV (Abbas and Madadi 2016). TSWV generally infects potato in warmer regions but TRV and PMTV both occur globally in cooler regions where their vectors are established. Certain phylogenetically defined recombinant strains of PVY cause similar necrotic symptoms known as "potato tuber ringspot disease." Over the last three decades, these have caused particularly heavy economic losses to potato industries in Europe and North America as well as in many developing countries in Asia and South America but have not yet reached all parts of the world, e.g. south-west Australia (Kehoe and Jones 2016) or Peru (Fuentes et al. 2019a). Therefore, PVY "strains" have been heavily studied worldwide over the past two decades revealing an exceptional amount of variation and a plethora of genotypes, many of them recombinants. PVY "strains" separate into at least 13 different subgroups defined either biologically or by phylogenetics (Karasev and Gray 2013; Kehoe and Jones 2016; Glais et al. 2017; Gibbs et al. 2017).

Biological strains of PVY are differentiated by the phenotypes they develop when different strain-specific hypersensitive HR resistance genes are present in potato cultivar differentials and whether they introduce necrotic symptoms in tobacco. Strain groups PVY^C, PVY^O and PVY^Z elicit HR phenotypes with hypersensitivity genes *Nc, Ny*, or *Nz,* respectively. Strain groups PVY^N and PVY^E overcome all three hypersensitivity genes, but differ in the phenotypes they induce in tobacco, only PVY^N eliciting veinal necrosis (Chikh-Ali et al. 2014; Karasev and Gray 2013; Rowley et al. 2015; Kehoe and Jones 2016; Jones and Vincent 2018). Such biological strains do not necessarily coincide with the phylogenetic lineages named after them. For example, previously potato biological strain groups PVY^C and PVY^O were thought to coincide with major lineages PVY^C and PVY^O, respectively. However, this proved incorrect as isolates within biological strain group PVY^D fitted within phylogroup PVY^C and those in PVY^Z within phylogroup PVY^O (Kehoe and Jones 2016; Jones and Vincent 2018). As the number of complete PVY genome sequences from different world regions grows, phylogenetic nomenclature based on biological, geographical, and sequence names is becoming increasingly unsustainable so substituting Latinised numerals for current PVY subgroup names was suggested (Jones 2014; Kehoe and Jones 2016; Jones and Kehoe 2016). Biological strains of PVA and PVV are also differentiated by the phenotypes they develop with strain-specific hypersensitive resistance genes present in potato cultivar differentials, but their phylogenetics is little studied. PLRV strains are differentiated biologically based on the severity of symptom expression in potato, but are phylogenetically very homogenous with limited sequence variation between isolates worldwide.

Potato spindle tuber disease caused by a viroid, PSTVd also impairs tuber quality in addition to direct yield loss. Although it led to several disease outbreaks in potato in different parts of the world in the past, through implementation molecular detection and eradication programs its presence in potato has now been significantly diminished in North America and Europe. By contrast, PSTVd is still prevalent in Central-Asia (CIP, unpublished) and China (Qiu et al. 2016). Although PSTVd presence in potato has declined recently globally, the opposite is the case for tomato where outbreaks have been increasing due to its worldwide spread in tomato seed via the international seed trade (Constable et al. 2019). This is of concern for global potato production as tomato PSTVd can cause severe yield and tuber quality losses in potato (Mackie et al. 2019). Additional collateral damage can be caused by virus infections as was demonstrated for PVY infection, which compromises plant defense responses rendering them more vulnerable to Colorado potato beetle (*Leptinotarsa decemlineata*) attack (Petek et al. 2014). Similarly, Lin et al. (2014) found that PVS infection rendered late blight (*Phytophpora infestans*) resistant cultivars more susceptible to late blight.

Among the most important viruses PVY, PLRV, PVA, PVS, and PVM are all aphid-transmitted. All of these except PLRV are transmitted nonpersistently by aphids, whereas PLRV is persistently transmitted. Insecticides have long been known to be effective only against persistently transmitted viruses, which likely explains the decline in prevalence observed in PLRV over the last 50 years in developed and emerging economies. Thrips- and whitefly-transmitted viruses continue to cause outbreaks in potato in warmer climatic regions. These outbreaks are usually only occasional, but TSWV is found commonly infecting potatoes in some countries, e.g. Australia and Argentina such that it is among the common viruses tested for in seed potato production schemes. Such outbreaks of thrips and whitefly-transmitted viruses are becoming steadily more frequent due to a warming climate, and at least one of these viruses, ToLCNDV has recently become a major potato pathogen in India (Jeevalatha et al. 2017a).

Recent phylogenetic studies, that use dating programs to compare the complete genomic sequences of common potato virus isolates, obtained at different times, are providing new insights into their evolution. So far, this has only been done with PVY and PVS (Gibbs et al. 2017; Santillan et al. 2018). Gibbs et al. (2017) inferred the phylogeny of the genomic sequences of 240 PVY isolates collected since 1938, 42% of which were nonrecombinants; sequences from the Andean region were lacking. The nonrecombinants all fitted into major lineages C, O, and N, and recombinants all into lineages R-1 and R-2. The main parents of R-1 and R-2 were PVY^N or PVY^O, respectively, and vice versa for their minor parents. The minor phylogroups within these major lineages [roman numerals in parentheses are from PVY classification system of Kehoe and Jones 2016] were: C with C1(II) and C2(III); O with O (=I) and O5 (=X); N with Eu-N (=IV), XIII and NA-N (=IX); R-1 with NTN-NW + SYR-I (XII), NTN-B (VI), NTN-NW + SYR-II (XI), N-Wi (VII), and N:O (VIII); and R-2 with NTN-A (V). Analysis of the nonrecombinant genomes found the estimated "time to most recent common ancestor" (TMCRA) for PVY to be around 1000 CE which corresponds with the end of the Tiahuanaco and start of the Inca civilizations in the Andes. A more comprehensive study including Andean PVY sequences recently found that PVY-N (=III) could be divided into three phylogroups (N1, N2 and N3), and two of them were unique to the Andes, suggesting PVY-N originated from the Andean region in contrast to PVY-O for which no such evidence could be found (Fuentes et al. 2019a).

Santillan et al. (2018) studied the phylogenetics of PVS genomic sequences collected since 1976, including Andean region sequences. The nonrecombinant genomes belonged to three major PVS lineages, two evenly branched and predominantly South American and a non-South American one with a long basal branch and many distal subdivisions. The South American lineages contained isolate sequences from three cultivated potato species, pepino (*Solanum muricatum*) and arracacha (*Arracacia xanthorrhiza*), whereas only isolates from a single cultivated potato species (*Solanum tuberosum*) were present in the other lineage. The two nodes of the basal PVS trifurcation were dated at 1079 and 1055 CE corresponding with the end of the Tiahuanaco and well before the start of the Inca civilizations, and the basal node to the non-SA lineage at 1837 CE corresponding roughly with the start of the European potato famine caused by late blight (*Phythophora infestans*). The PRDV/PVP cluster diverged from PVS 5–7000 years ago. This suggests a potato-infecting proto-PVS/PRDV/PVP emerged in South America, and spread into a range of local *Solanum* and other species, one early lineage spreading worldwide in potato.

Kutnjak et al. (2014) studied the phylogeny of PVX genomic sequences from the Andes. What they found was similar to the PVS situation with three major lineages, two of which were South American and one non-South American. However, an earlier study with PVX coat protein gene sequences had found several European and North American sequences in the single major South American lineage known at that time, and their presence was confirmed by Kutnjak et al. (2014).

At least 37 of the known potato viruses are found in South America and this number is set to increase further with the application of high-throughput sequencing (HTS) techniques to screen for virus infections (Kreuze 2014; Fuentes et al. 2019b; CIP, http://potpathodiv.org/). Only the above-mentioned viruses and PAMV have so far established themselves globally as potato infecting viruses, whereas most of the other global viruses may be the result of generalist viruses that have managed to become established in potatoes, or more recent newcomers from related crops that have achieved a foothold in potato due to increasing vector populations, principally whiteflies (crini- and begomoviruses) and thrips (tospoviruses). At least 20 potato viruses remain restricted to South America and most of these represent viruses that evolved together with wild and/or cultivated potatoes in the Andean region.

11.2.1 Viruses in the Andean Region

Cultivated potatoes were first domesticated in the Andean region of South America where they show the highest level of genetic diversity including four cultivated potato species with various ploidy levels and many native cultivar groups and wild potato relatives. Their viruses evolved with them and it is therefore not surprising that more viruses are found infecting potatoes in this region than elsewhere in the world (Fig. 11.1). Besides the usual viruses associated with potato production throughout the world and which were distributed globally through infected tubers, the Andean region hosts several unique viruses that do not seem to have established themselves beyond their geographical region of origin. These include the nepoviruses PBRSV, PVU, and PVB, the tymoviruses APLV, APMMV, the Ilarvirus PYV, the crinivirus PYVV, the cheravirus AVB, and the tepovirus PVT. HTS-based approaches have recently also detected the presence of new viruses corresponding to at least 14 different genera (Fuentes et al. 2019b; CIP, http://potpathodiv.org/). Many of the Andean potato viruses have also been reported infecting other root and tuber crops that are grown in the same environments as potatoes, such as Ulluco (PVS, PVT, PLRV, AVA, APLV/APMMV), Oca (PBRSV, PVT, AVB), Mashua (PVT), Aracacha (PVS, AVA, AVB, PBRSV) & Maca (APLV/APMMV), and in the solanaceous bush fruit pepino (PVS). These are crops from such diverse species that it is likely many wild hosts may also be infected constituting a continual environmental reservoir for these viruses present in the absence of cultivated plant hosts. In addition, the viruses commonly found in other parts of the world often show much higher level of variability in the Andean region (Gil et al. 2016a, b; Kalyandurg et al. 2017; Santillan et al. 2018; Gutierrez et al. 2013; Gibbs et al. 2017; Henao-Díaz et al. 2013).

In the pre-ELISA era, surveys to establish the occurrence of potato viruses in Andean countries mostly involved potato germplasm collections, but in the 1970s Peruvian potato crops were widely sampled. They were undertaken using older serological detection assays and inoculation to indicator hosts complemented by electron microscopy. The viruses found in that era included nine also found in Europe and North America (PLRV, PVX, PVY, PVV, PVS, PVA, PVM, PMTV, PAMV) and eight others only found in Andean countries (APLV, APMoV, PVT, AVB, PBRV, PVU, PYVV, WPMV) (Jones 1981, references therein). Since the 1980s, surveys have been conducted using ELISA to detect the most common potato viruses (PVY, PVX, PVS, PLRV, APMoV, APLV) in potatoes growing at higher altitude (>3000 m) in the Peruvian highlands. The most frequently detected viruses have consistently been contact-transmitted with PVX (30–82% incidence) and PVS (20–50%) being the commonest followed by APMoV (4–15%) and APLV (2–6%). Similar viruses and incidences were found in higher altitude plantings in Ecuador where PYVV was also found occurring at low frequency (0–3% incidence). In contrast, PLRV and PVY were usually only detected at 0–5% in these materials and, when included, PMTV was uncommon (Pérez Barrera et al. 2015). When similar surveys were undertaken at lower altitudes in the Andean region (<3000 m), the findings resembled those in other areas of the world, with PVY and PLRV dominating. The differences in PVY and PLRV incidences between potato crops growing at different altitudes likely reflects the greater abundance of their aphid vectors below 3000 m.

Whereas the potyviruses PVY, PVA, and, to a lesser extent, PVV are established worldwide, another potato potyirus, WPMV has never been reported infecting cultivated potato even in the Andean region. So far, it has only been reported from a wild potato species growing in an isolated Lomas ecosystem and the cultivated bush fruit crop pepino both in the coastal desert in Peru (Fribourg et al. 2019). In earlier studies, when Andean potato cultivars were inoculated with PVY isolates belonging to biological strain groups PVY^C, PVY^O, PVY^Z, and PVY^N, one developed HR phenotypes consistent with presence of genes *Nc* and *Ny*, one an HR phenotype consistent with gene *Nc* alone, and 1 with neither, so both genes were easily found in commercial cultivars in potato's original center of domestication. With European potato cultivars inoculated at the same time, the corresponding figures were six with *Nc*, six with *Ny* and one with neither. An HR phenotype consistent with *Nz* presence and an ER phenotype consistent with *Ry* presence developed in two cultivars each. There is still a need to determine how common resistance genes *Ry, Nc, Nz, Ny*, and putative *Nd* are among Andean potato cultivars and the degree of protection they provide against PVY in Andean potato crops. This applies not only to commercial plantings grown with or without access to healthy seed programs but also to Andean native potato landraces belonging to the four potato species that grow in Andean subsistence plantings (Jones 1981; Jones and Vincent 2018).

The potexvirus PVX consists of four biological strains differentiated by their phenotypic reactions when they infect potato cultivars with strain-specific hypersensitive resistance genes *Nx* and *Nb*. Group 1 strains fail to overcome either gene, group 2 strains overcome *Nx*, group 3 strains overcome *Nb*, and group 4 strains overcome both genes. All four strain groups occur in the Andean region, along with an additional strain (HB) that has not been reported elsewhere in the world, which overcomes not only these two genes but also extreme PVX resistance gene *Rx*. PVX^{HB} caused a mild or symptomless infection in eight native potato landraces, systemic necrotic symptoms in cultivar "Mi Peru," and bright yellow leaf markings in "Renacimiento." Phylogenetic analysis of coat protein gene sequences placed HB in the major PVX lineage that contained group 2 and 4 isolates from South America, North America, or Europe, whereas strain group 1, 3, and 4 sequences, none of which were from South America, were in the main lineage that lacked any South American sequences. Thus, only strain group 4 sequences were in both lineages (Kutnjak et al. 2014).

Three carlaviruses have been reported infecting potatoes in South America (PVS, PVM, and PRDV). Early PVS isolates from the Andean region caused systemic infection in *Chenopodium quinoa,* but isolates from elsewhere only infected its inoculated leaves. The former were therefore called Andean strain (PVS^A) and the latter ordinary strain (PVS^O). However, PVS^A was found subsequently in many countries (Jones 2014, references therein), and, more recently, PVS^O was found in the Andean region (Santillan et al. 2018). These biological defined strains are not coincident with phylogenetically defined PVS strains as PVS^O occurs in both South American and non-South American lineages (Santillan et al. 2018). A new strain of PVS was recently found infecting Arracacha (Santillan et al. 2018; de Souza et al. 2018). The carlavirus PVM has been reported from Bolivia, Chile, and Peru and in the Andean region of northern Argentina, but in recent surveys is conspicuously absent from Peru. The carlavirus PRDV on the other hand has been reported only from Argentina and Brazil (PVP isolate) and probably infected potato from indigenous hosts as they have not been reported from the Andean region, although in evolutionary terms PRDV and PVP are considered likely ancestral parents of PVS from this region (Santillan et al. 2018).

Several nepoviruses infect potatoes in the Andes all of which were originally isolated from potato plants showing calico symptoms (Fig. 11.2), although none of them have been linked to this syndrome by reproducing the symptoms in experimentally infected potato plants. The first was discovered simultaneously by two research groups who separately named it PBRV and the calico strain of TRSV. Subsequently, however, the virus was shown not to be TRSV (Souza Richards et al. 2013) so only the name PBRV remains in use. PVB was recently reported infecting potatoes in Peru (de Souza et al. 2017) where it is now known to be relatively common (Fuentes et al. 2019b; CIP, http://potpathodiv.org/). Although PVB was first identified from plants showing calico symptoms, these symptoms were not necessarily caused by it as other viruses, such as PVX, were also present. The virus has not yet been characterized biologically and is impact on tuber yield is unknown. Another subgroup C nepovirus, PVU, was isolated previously from potatoes with calico symptoms in central Peru, but sequence comparison distinguished it from PVB (de Souza et al. 2017).

Fig. 11.2 Field with typical calico and yellowing symptoms, typically associated with nepoviruses, in the highlands of Peru. Photo credits: CIP

The comovirus APMoV (family *Secoviridae*) was identified infecting potato in Peru, Argentina, and in Brazil where it was also found infecting eggplants. In Peru, it was relatively common and widespread in the past based on ELISA results, but in a recent survey using HTS, was not commonly encountered (Fuentes et al. 2019b; http://potpathodiv.org/); it is transmitted by beetles of the genus Diabrotica, as well as by seed and contact. It also occurs outside South America having been found infecting tabasco peppers in Honduras in Central America.

The tymoviruses APLV and APMMV (family *Tymoviridae*), which was recently separated from APLV (Kreuze et al. 2013; Koenig and Ziebell 2013), have been identified in potato germplasm from Colombia, Ecuador, Peru, and Bolivia. Nevertheless, they seem to be becoming less common in field grown potatoes in the Andean region than when they were first found in the 1970s.

PVT is the only known member of the genus *Tepovirus* (family *Betaflexiviridae*). It has been detected not only in Peruvian, Bolivian, and Chilean potato germlasm, but also ulluco (*Ullucus tuberosus*), oca (*Oxalis tuberosa*), and mashua (*Tropaeolum tuberosum*) plants growing in the field in these countries. It is transmitted through contact and potato true seed, but causes only mild mosaic or no symptoms in potato plants and seems relatively uncommon (Lizárraga et al. 2000).

Two ilarviruses (Family *Bromoviridae*), AlMV and PYV, both sometimes infect potatoes in the Andean region. Whereas AMV is found worldwide and normally causes calico symptoms, including in the highlands of Peru, PYV is largely symptomless and restricted to the Andean region where it has been identified in germplasm from Ecuador, Peru, Bolivia, and Chile.

PYVV (genus *Crinivirus*, family *Closteroviridae*) causes obvious veinal yellowing symptoms, which were first seen in an early Andean potato germplasm collection in Europe. It has been known in Colombia for many years (Jones 1981, references therein; Franco-Lara et al. 2013). PYVV is thought to have originated from the Andean region of Northern Ecuador and Central West Colombia and causes up to 50% yield losses. It is transmitted in a semi-persistent manner by the greenhouse whitefly (*Trialeurodes vaporariorum* Westwood; *Hemiptera*: *Aleyrodidae*) (Salazar et al. 2000), through tuber seed and underground stem-grafts. It also infects tomatoes and various weed species. The virus also occurs in potato-producing areas of Northern Peru, and in the Venezuelan Andes, and recently spread to Panama in Central America (CIP, unpublished). Its whitefly vector occurs globally so it could spread to other continents. Its prevalence in plantings at lower altitudes in the Andes reflects the restriction of its whitefly vector to warmer conditions (Jones 2016, references therein).

A begomovirus PALCV was reported infecting potatoes and wild solanaceaous hosts from the highland jungle region of central Peru in the late 1980s (Hooker and Salazar 1983). Although apparently relatively common at the time, the original isolates were lost and the virus was never found again in the same region. On the other hand, during the late 90s a novel virus coded SB-26/29 and transmitted by brown leafhoppers (*Russelliana solanicola*) was associated with a novel and rapidly spreading rugose stunting disease in Southern Peru. The disease caused yield reductions of 20–90% (Tenorio et al. 2003). The disease has now become rare, likely due to changes in cropping patterns that led to reduction in leafhopper populations. Partial sequence determination identified this virus as related to torradoviruses (CIP, unpublished), and recent surveys (Fuentes et al. 2019b; http://potpathodiv.org/) in Peru indicate it, and related viruses, can still be found with some frequency in potatoes. In Colombia, two new Pomoviruses related to PMTV were identified in soil samples from potato fields (Gil et al. 2016b). At least one of them (CPSbV) was shown to infect potatoes and transmitted through tubers, but the virus could only be detected in roots, and the plants were without symptoms.

11.2.2 Brazil

In Brazil, the potato is grown throughout the year, in three successive crops: rainy season, with harvest from December to March, with more than 50% of the total production; dry season, with harvest from April to August, representing about 30% of production, and winter season, with harvest from September to November, with lower production volume (IBGE 2017). The positive aspect of three harvesting seasons is a supply of fresh potatoes on the market throughout the year. However, this year-long field production also means a greater opportunity for uninterrupted spread of insect-transmitted viruses. This is a major factor why Brazil imports virus-tested seed-potato stocks from abroad annually, especially from Netherlands, Germany, France, Canada and Chile. This frequent importation has, historically, allowed the introduction of new pathogens into the country.

A first report of potato cultivation in Brazil dates from the 1920s in São Paulo, and since extended to other neighboring states of the South and Southeast, and later further expanded to the Northeast and Central-West regions. However, the largest production still occurs in the South and Southeast regions, with the State of Minas Gerais ranking as the first, with about 1.3 million tons/year (Agrianual 2016). Since the beginning of potato cultivation in Brazil, *Potato leafroll virus* (PLRV) had always been the most important viral agent associated with seed-potato tuber degeneration (Souza-Dias et al. 2013). The predominance of PLRV among seed-potato viruses in Brazil lasted until the mid-1990s, when two new strains of PVY were introduced, nearly simultaneously, through seeds imported from countries where their incidence was already known. After their introduction, this virus became a major cause of rapid seed-potato degeneration, overtaking the historical importance of PLRV as main cause for rejecting early field generations (G-1 or G-2), based on tolerance limits for viruses of the Federal Brazilian seed-potato tuber production-certification program.

As mentioned, PLRV has shown high incidences in Brazilian potato fields in the past. Until the mid 1990s it was normal for qualified and traditional seed-potato producers to face high incidences of PLRV (over 20%) in the very first (G-1) field multiplication of imported seed-potato stocks, which are officially considered, in Brazil, as G-0 (Souza-Dias et al. 2016a). This traditional and prevalent virus problem started to decline, coincidently in space and time with the introduction and fast outbreak of new PVY strains detailed below.

A possible reason for the shifting from PLRV to PVY was that seed-potato producers had over the years become more conscious regarding PLRV infected plants, learning to recognize its symptoms such as interveinal yellowing and rolled leaves, but did not recognize or understand the relevance of new symptoms such as mosaic, chlorosis and leaf deformation characteristic of PVY. As PVY expanded in association with the introduction of imported potato seeds, this virus soon became the major cause of seed-potato tuber degeneration (Barrocas et al. 2000). To counter this, a more intensive and efficient action toward controlling aphids, including the use of new insecticides such as neonicotinoids, took place. The vector transmission mechanism of PLRV (persistent circulatory, requiring significant time for acquisition and transmission) and PVY (nonpersistent, almost immediate transmission) would be more affected by the insecticidal effect, which could explain the noticeable reduction in the field incidence and spread of PLRV, whereas it is well know that insecticides have limited effect on nonpersistently transmitted viruses such as PVY. Nowadays, the detection of PLRV in seeds within official certification programs is extremely rare. In the State of Minas Gerais, as well as in most of the potato producing states, PLRV has not been associated with the rejection of seed lots since the beginning of the new millennium. However, imported seed potatoes are a big concern; Villela et al. (2017), testing national and imported potato seeds, found virus incidences as high as 10% in seed potatoes imported by Brazil.

The probable first report of PVY necrotic strain (PVY^N) in South America was in the beginning of 1940s (Nóbrega and Silberschmidt 1944). They studied some PVY isolates from Peru, inducing vein necrosis in leaves of tobacco. It took 20 years before a first official scientific report on PVY in Brazil was published in Sao Paulo State (Kitajima et al. 1962) where the authors describe the virus particle morphology and the histological symptoms. During the following years, although PVY was not considered a major problem for Brazilian potato producers, attempts to discover new methods to avoid PVY infection and to find new indicator plants were carried out in Brazil.

Surveying for PVY isolates in experimental fields, Andrade and Figueira (1992) detected five different strains, based on the reactions induced in the tobacco cultivar "Turkish NN." Although PVY^N was present, the PVY^O strain had a much higher incidence and was identified in almost all cultivars planted during 1980–1990. The investigation of PVY strains, done between 1983 and 1988, showed the presence of PVY^O in 80% of samples infected with PVY collected in experimental fields, planted with "Achat," "Baraka," "Baronesa," "Bintje," "Granola," and "Mona Lisa" cultivars. On the other hand, the incidence of PVY^N ranged from 0 to 12.4%. Even if present, PVY was never found in high incidences in fields of either ware or seed potatoes. Possibly the strains present in the country at that time were not as easily spread as the PVY strains introduced in Brazil in mid-1990s.

It is considered that the first significant introduction of a new PVY strain into Brazil came with seed potatoes of the cultivars "Achat" and "Baraka" imported from Germany, where it encountered optimal conditions for dissemination. This strain was later recognized as being the Wilga strain of PVY (PVY-N^{wi}; Galvino-Costa et al. 2012). All over the country, where seeds of these cultivars were used, large outbreaks of PVY were soon observed, reaching incidence rates above 70% in plants produced from infected seeds. A second introduction is suggested to have occurred through imported seed-potato of cv. Atlantic from Canada. It was later confirmed to be PVY^{NTN}, which causes necrotic rings in tubers of susceptible cultivars, such as "Monalisa". This strain also adapted itself very well to the Brazilian conditions, spreading rapidly to all potato growing regions (Galvino Costa et al. 2012). The introduction of these two strains brought a major problem to the potato growers who were used to plant cvs "Achat", "Baraka", and "Bintje", and easily recognized PLRV infected plants as showing, symptoms of leaf roll and yellowing of the lower leaves, but not chlorosis and mosaic that started appear as new PVY strains expanded. In 1995, in Minas Gerais, the seed-potato areas under certification were covered by more than 50% with cvs "Achat" and "Baraka," and a little less than 18% with "Bintje". Due to the increasing incidence of new PVY strains, these cultivars were abandoned, and only 3 years later, in 1998, little more than 15% of the potato seed production area was planted with those three potato cultivars. Conversely, where "Monalisa", a ware potato, used to occupy only around 5% of the certified area, it rose to about 29%. By 2000, "Monalisa" reached more than 50% of the production area, after which PVY^{NTN} began to spread rapidly, mainly associated to regions were "Atlantic" was grown, implicating it as the source of introduction of PVY^{NTN}. Due to a high susceptibility to PVY^{NTN} and sensitivity to the typical superficial tuber necrotic rings, "Monalisa" rapidly became unmarketable. Soon after in 2001, the Dutch cv "Agata" began to be tested and gained broad acceptance among the producers due to its high productivity, marketable phenotype, soil adaptability, and resistance to different PVY strains (Ramalho et al. 2012). The lack of expression of PVY^{NTN} symptoms in tubers of "Agata", in striking contrast to "Monalisa", contributed much to the fast replacement of "Monalisa" by "Agata". Thus, 4 years later, "Agata" was occupying over 30% of the area planted in Minas Gerais and nowadays is the main cultivar planted in Brazil (Silva et al. 2015).

Early characterization of PVY strains in Brazil was based on host symptoms, and the serological tests employed were DAS-ELISA which did not identify specific strains. The first serological tests using monoclonal antibodies and molecular studies started in the 1990s, confirming the presence of $PVY^{N\text{-}Wi}$ and PVY^{NTN} strains Brazil. Based on the reaction of PVY isolates to monoclonal antibodies, and on the symptoms shown by host plants, a great variability among them became evident (Galvino-Costa et al. 2012). More recent investigations (authors, unpublished data) have shown that there is a large abundance of PVY isolates with uncommon serology and that apparently N and O strains have disappeared from the Brazilian fields. Galvino-Costa et al. (2014) found that the incidence of $PVY^{N:O/N\text{-}Wi}$ was either equal to or greater than that of PVY^{NTN}, depending on the region in which the tubers have been produced and that mixed infections with both strains occur often, although this is sometimes only detectable with more sensitive techniques, such as RT-qPCR.

"Agata" has reached about 80% of the potato producing areas in Brazil and being a symptomless carrier of PVY, it acts as an efficient silent disseminator of the virus, particularly among "home saved" (not certified) seed-potato producers. This scenario has been a serious sanitary problem for potato production in Brazil, as it is strongly correlated with the successive increase of PVY reservoir, favoring spread of this virus into certified seed-potato fields.

Two other viruses that have been monitored in the field by the official certification programs are PVS and PVX. The incidence of these viruses in the field is sporadic and has never been associated with crop losses in Brazil. However, their monitoring is because some countries that export seed potatoes to Brazil have high incidences of these two viruses in the field (Souza-Dias et al. 2016a). If potato seeds with high PVS/PVX incidence reach Brazil, where high incidence of PVY usually occurs, the consequences could be disastrous. Recent surveys have shown incidences as high as 10% of PVX and 20% of PVS in imported potato seeds (Villela et al. 2017).

An increasing number of potato fields showing over 50% of typical PLRV-like symptoms, brought concerns about a possible PLRV outbreak, associated with whitefly instead of with aphids as virus vector. However, later on it was identified as ToCV (Souza-Dias et al. 2013; Lima 2016). The field symptoms of ToCV are characterized by internerval chlorosis and slight curling of the leaf edges, which begin in the apical leaves. ToCV can be transmitted by at least five species of whitefly (Orfanidou et al. 2016). In recent years an outbreak of the whitefly *Bemisia tabaci*, has been noticed in potato crops in Brazil (Moraes et al. 2017), favoring the spread of ToCV, probably from infected tomato plants in the vicinity. Despite of continued reports of whitefly infestation in Brazilian potato fields over the past 5 years, high occurrences of ToCV have been associated with tomato but not with potato crops (Orfanidou et al. 2016). Some outbreaks in potato have been reported in certain areas such as in the states of Goiás and São Paulo (Souza-Dias et al. 2013). However, at least for the time being, it does not appear to be a recurring problem for this crop. Thus, special care is being taken to monitor this disease in the field, as well as other whitefly-transmitted viruses, also reported in commercial Brazilian potato fields (see below).

The overlapping cropping cycles of tomato and potato, combined with favorable climatic conditions and the frequent proximity between the areas where they are planted, has caused other tomato viruses to occasionally migrate into potato crops. Two species of *Begomovirus* (family *Geminiviridae*) have been described in potato in Brazil: *Tomato yellow vein streak virus* (ToYVSV) and *Tomato severe rugose virus* (ToRSV). ToYVSV and ToRSV, the latter which has been prevailing in tomato crops, seems to be also prevalent in potato fields and both inducing similar deforming yellow mosaic symptoms (Souza-Dias et al. 2016a). The vector of both viruses is also the whitefly *Bemisia tabaci* (Pantoja et al. 2014). In contrast to what has been normal for tomato producing areas in Brazil, so far, there has not been any record of widespread begomovirus outbreaks in Brazilian potato producing areas. However, considering they are whitefly-transmitted viruses, and the high populations of whiteflies observed in potato crops in recent times (Moraes et al. 2017), careful monitoring of begomoviruses in potato should take place, as recommended for ToCV.

Tospoviruses, whose type member is *Tomato Spotted wilt virus* (TSWV), are transmitted by several species of thrips in a persistent manner. The viruses are acquired only at larvae stage, replicate in the insect vector and persist through the several stages of its life cycle (Rotenberg et al. 2015). They have always had a sporadic occurrence in potato crops in Brazil; but, in general, they are recorded as current season infection, and not as seed-tuber perpetuated virus. Therefore, they were never considered important as causing damage to this crop. However, from 2010 to 2015 there was a long period of drought in the Southeast of Brazil, with a significant increase in temperature, contributing to the increase of viral diseases in several crops, including potato, clearly associated to the same favorable conditions for increase in vector population. Field surveys by Souza-Dias (data not published) showed a high virus incidence of tospoviruses with a rare and isolated observation of virus perpetuation between 2010 and 2015. The more common species are *Tomato spotted wilt virus* (TSWV), *Groundnut ringspot virus* (GRSV) and *Tomato chlorotic spot virus* (TCSV) (Lima and Michereff 2016). Similar outbreaks have been described elsewhere, such as in Argentina (Salvalaggio et al. 2017) and United States of America (Abad et al. 2005).

As a norm, usually not all stems of a plant-hill show tospovirus symptoms: potato tubers are not only symptomless but also tospovirus-free, even when produced from infected plants. However, as a rare event for potato tospoviruses in Brazil, necrotic rings, both on the surface and penetrating the tuber flesh were observed in some of the tuber progeny of a tospovirus-infected plant (cv "Agata") (Souza-Dias et al. 2016b).

The plants that emerged from these tospovirus symptomatic tubers were however free of the virus. In other countries, there has been evidences of tospovirus species perpetuating via tubers produced by infected plants. These observations cause concern for seed-potato production (Abad et al. 2005; Salvalaggio et al. 2017). Therefore, a close monitoring of the incidence of tospoviruses in Brazilian potato fields is recommended in order to control not only its dissemination in the current season but also its perpetuation by tuber seed transmission.

11.2.3 India

PLRV, PVY, PVX, PVA, PVS, PVM, GBNV, and PAMV are known to occur in India. ToLCNDV-[potato], a begomovirus is reported to infect potato only in India. Mosaics and leafroll are the most common and severe symptoms in the subtropical and tropical climates of India. PLRV is important and occurs widely in almost all varieties. The mosaic causing viruses, PVY, PVA, and PVM as well as severe strains of PVX occur either singly and/or in different combinations. PVA and PVM are not common. PVY^{N} is almost not known in India, but recent study indicates the possible presence of PVY^{N} and PVY^{NTN}. ToLCNDV-[potato] has emerged as a serious threat to potato production during recent times. Its incidence is reported in almost all major potato growing states (Jeevalatha et al. 2017a). GBNV is reported in the early planted crop in the central and western parts of India (Jain et al. 2004). However, its occurrence in Pant nagar (Pundhir et al. 2012) and northwestern hills of India (Raigond et al. 2017) indicates the adaptation and spread of the virus to new areas. Recently, mixed infection of CMV with other potato viruses was reported (Sharma et al. 2016). It was found mostly in association with PVX, PVY^{n} followed by PVA, $PVY^{o/c}$, and PVM. Rarely, it was found associated with PAMV (Ghorai et al. 2017).

PVY is an important potato virus, which occurs widely in almost all the potato cultivars in India. Severe strains of PVY have the potential to reduce yield up to 80%. In India, PVY^{O} is most common and PVY^{C} strain has also been reported earlier based on the reactions on biological indicator host. Recently, based on host reactions, serology with monoclonal antibodies and complete genome sequence, the evidence of occurrence of a recombinant strain (N:O type) of PVY (isolate PVY-Del-66) was provided for the first time (Jailani et al. 2017). Isolate PVY-Del-66 shared closest sequences identity of 97.7–99.9% and a close phylogenetic relatedness with the N:O strains reported from USA and Germany. Del-66 isolate caused necrosis in tobacco and reacted positively with the MAb to common strain PVY^{O} but not with necrotic or chlorotic strains of PVY (Jailani et al. 2017). PCR analysis with strain-specific primers showed the possible presence of PVY^{N} and PVY^{NTN} in India and is being further confirmed by biological assays (CPRI, unpublished). PVA causes mild mosaic symptoms and not common in India. It reduces yield up to 30–40% and higher in combination with PVY or PVX.

PLRV is one of the most prevalent viral diseases of potato in India. All Indian potato varieties are susceptible to this virus. Yield loss normally ranges from 20 to

50% in India but in extreme cases may be as high as 50–80%, and infected plants produce only a few, small to medium tubers in severe secondary infections. At genome level, Indian isolates are closer to European and Canadian isolates than to an Australian isolate (Jeevalatha et al. 2013a).

PVX is one of the mosaic-causing viruses in almost all varieties of potato. In India, PVX infection may depress yield up to 10–30% and in the presence of PVA or PVY reduces yield up to 40% in potato. Indian PVX isolates were characterized for their biological properties, host range and transmission. Molecular analysis of complete and partial genomes of PVX found that all Indian isolates cluster in clade 1 with isolates from Europe and Asia, and none of them with clade II from south America (Jeevalatha et al. 2016c). Amino acid analysis suggested that these isolates cannot overcome *Rx1* gene or *Nx* gene mediated resistance (Jeevalatha et al. 2016c).

Like most parts of the world, PVS and PVM also infect potato in India. The Andean strain of PVS is reported in India (Garg and Hegde 2000). Complete genome sequence of one isolate of PVM, PVM-Del-144 has been sequenced. PVM isolates from northern plains showed considerable diversity in coat protein gene region (Jebasingh and Makeshkumar 2017).

Tomato leaf curl New Delhi virus-[potato] (ToLCNDV), a species of the genus begomovirus (family *Geminiviridae*) causes apical leaf curl disease of potato in India (Fig. 11.3). Infection leads to severe seed degeneration particularly in susceptible varieties. Primary symptoms appear as curling/crinkling of apical leaves with distinct mosaic symptoms and in case of secondary infection, the entire plant shows severe leaf curling and stunting symptoms (Jeevalatha et al. 2013b; Sohrab et al. 2013). The association of a geminivirus with potato apical leaf curl disease was first reported in northern India and the virus was named tentatively as Potato apical leaf curl virus. However, later it was confirmed that this virus is a strain of

Fig. 11.3 Symptoms of primary (**a**) and secondary (**b**) infection by ToLCNDV in potato. Photo credits: CIP

ToLCNDV. ToLCNDV-[potato] is a bipartite begomovirus with two genomic components referred as DNA-A and DNA-B. The DNA A components of the ToLCNDV-[potato] isolates shared more than 90.0% similarity to ToLCNDV isolates from vegetable crops such as tomato and okra, 89.0–90.0% to papaya isolates and 70.4–74.0% to other ToLCVs (Jeevalatha et al. 2017a).

Initially, sporadic incidence of the disease was reported in 1996 at Hisar in Haryana, later severe infections were observed in western Uthar Pradesh and other parts of northern India (Saha et al. 2014). It now occurs in almost all the major potato growing states in India and is reported in all cultivated varieties with varying severity levels (Jeevalatha et al. 2013b, 2017a). All the Indian potato varieties are susceptible to this disease except Kufri Bahar, which shows lowest seed degeneration and no/mild leaf curl symptoms even under favorable field (Kumar et al. 2015) and glass house conditions (Jeevalatha et al. 2017b). The virus is transmitted by whiteflies and the infection is more common in crops planted during October than in November because of the large whitefly population. Between 40 and 75% of incidence was recorded in the cultivars grown in Indo-Gangetic plains of India, up to 100% of incidence from the Hisar (Haryana) in susceptible varieties and recently, up to 40% incidence is reported from West Bengal (Saha et al. 2014). Infection results in significant decrease in size and number of tubers. Losses in marketable yield were reported to be as high as 50% in early planted susceptible cultivars. Currently, it is one of the most important viral diseases of potato in India. Repeated use of the same seed stock for 5 years led to 44.83–60.78% yield reduction in susceptible cultivar in Hisar and seed tubers of these cultivars cannot be reused profitably for more than 2 years. Since the virus spreads through seed tubers, it is critical to ensure quality of seed tubers through effective diagnostic tools. Diagnostic protocols like nucleic acid spot hybridization (NASH), polymerase chain reaction (Jeevalatha et al. 2013b), RCA-PCR (Jeevalatha et al. 2014), qPCR (Jeevalatha et al. 2016a) and LAMP assays (Jeevalatha et al. 2018) are available for the detection of ToLCNDV-[potato] in potato. PCR is being used to screen mother plants meant for tissue culture based seed production and also stage I plants in healthy potato seed production. So far, the infection of potato by ToLCNDV is known to occur only in India.

A tospovirus, GBNV causing severe stem/leaf necrosis disease in plains/plateaux of central/western India heavily infects the early crop of potato. It was first reported through morphological and serological studies by Jain et al. (2004). Stem necrosis incidence was recorded up to 90% in some parts of Madhya Pradesh and Rajasthan and up to 50% in Pant nagar. Its occurrence in northwestern hills of India despite of unfavorable conditions indicates possible adaption of the pathogen to new climatic conditions (Raigond et al. 2017).

In northwestern India leaf samples from potato plants with yellow mosaic or flecking symptoms showed positive reaction with PVX and CMV subgroup II in DAS-ELISA and the mixed infection with these viruses was further confirmed by PCR assay using specific primers and sequencing (Sharma et al. 2016). Ghorai et al. (2017) reported above 10% incidence of CMV in potato grown in Punjab. CMV infection in potato occurred mostly in association with PVX (60%), PVY^n (60%)

followed by PVA (40%), $PVY^{o/c}$ (30%), and PVM (30%). Rarely, it was found associated with PAMV (10%). Severe symptoms like malformation of leaves, blistering, stunting and reduced leaf size of potato were observed when CMV was present in potato in association with other potato viruses like PVX, PVY^{n}, $PVY^{o/c}$, PVA, PAMV, and PVM (Ghorai et al. 2017). Since the cropping pattern in Punjab corresponds to potato during October to February followed by cucurbits from February to May, the potato serves as an over wintering host of CMV when preferred host plants are not available and CMV is transmitted from potato to cucurbits through aphids (Ghorai et al. 2017).

11.2.4 Africa

Relatively little has been published regarding the viruses infecting potatoes in Africa and the few studies performed have focused on the globally common viruses using antisera. However the same viruses, most of them commonly found elsewhere in the world, have also been found throughout the continent where surveys have been performed. Thus, in Kenya Gildemacher and coworkers (2009) tested over 1000 tubers from 11 markets in seven districts for PLRV, PVY, PVA, and PVX and found average incidences of 71%, 57%, 75%, and 41%, respectively. Mixed infections were common and only 2.4% of tubers were free of any of these viruses. In Tanzania, Chiunga and Valkonen (2013) surveyed for the occurrence of the same viruses, but also PVM and PVS in plants from 16 fields in the south western highlands and found incidences of 55%, 39%, 14% and 5% for PVS, PLRV, PVX, and PVM whereas PVY and PVM were only detected in two locations. In a survey performed in South-West Uganda during 2014, PVX and PLRV were most frequent, followed by PVY and PVM, whereas PVA was not detected (CIP/IITA, unpublished). AlMV and *Beet curly top virus* (BCTV) were found to be locally frequent in Sudan around Kartoum (Baldo et al. 2010).

Although there have been no reports of whitefly-transmitted viruses, whiteflies can be abundant in potato crops in some locations during some seasons and because potatoes are often grown in close proximity to other vegetables there is a clear risk of transfer and possibly emergence of whitefly-transmitted viruses as has already been observed in India and Brazil.

11.2.5 Europe

There are some ten viruses infecting potatoes and causing significant yield losses in Europe (Table 11.1; Fig. 11.1). The main viral pathogens include PLRV, PVY, PVA, PLRV, PVM, and PVS which are all transmitted by aphids. PVV is also aphid-transmitted but occurs only in a few cultivars. They also include PSTVd and PVX which are solely contact-transmitted, and PMTV and TRV both of which are soil-borne and mainly cause problems in the northern more cooler countries of the continent. PLRV was formerly the most important potato virus but has been on the decline for many decades. Conversely, PVY has become the most important, especially its new necrogenic strains which often cause mild foliar symptoms that are often difficult to see in field inspections and tend to be more efficiently aphid transmitted. Their importance resides in the necrotic symptoms they induce in tubers.

One of the viruses with potential to become a serious problem in potato production in Europe could be *Tomato spotted wilt virus* (TSWV). This virus has an unusually large host range of over 1000 plant species, including potato (Bulajic et al. 2014). TSWV is transmitted by thrips preferring climates warmer than those typical for potato growing areas in Europe, which may be a reason why damage caused by TSWV in potato crops has remained mostly modest or negligible and the yield losses affect mainly greenhouse production. However, climate change is predicted to increase temperatures in Europe (Lamichhane et al. 2015), which would provide more favorable conditions for thrips to thrive outdoors in more diversified living environments.

Geographical location and climate seem to create the conditions where different potato viruses get established and spread over the years. Therefore, while the potato viruses transmitted persistently by their vectors can spread over long distances, it is not self-evident that the virus gets established in the new area. For example, PVY is found in potato crops in all potato production areas of Europe, whereas PLRV is rare in northern Europe, i.e., in Finland and the northern parts of Sweden and Norway, despite the fact that both viruses are transmitted by aphids, and transmission of PLRV in the vector aphids continues much longer than PVY. Winged viruliferous aphids carrying PLRV cross the Baltic Sea during warm weather and suitable wind, and transfer PLRV from the potato fields in northern Germany and Poland to southern Finland. Nevertheless, PLRV has not become a significant pathogen in Finland and the infection rate of PLRV in potato crops has remained negligible. Why the abundance of potato viruses is different in different parts in Europe is not fully understood, but the climatic conditions are anticipated to play a role.

Another example of differences in geographical distribution of potato viruses in Europe is PMTV. It is common in potato crops in Scotland, Northern Ireland, all Nordic countries including Denmark, and in Czech Republic and Austria, but rare in the other countries at the southern side of Baltic Sea. It was only recently detected in Poland (Santala et al. 2010). The likely means for spread of PMTV over long distances are seed potatoes produced in an area where soils are contaminated with PMTV and its vector. However, PMTV can spread over long distances also in the resting spores of *S. subterranea* adhering to tools, equipment or vehicles, and in traded materials containing soil, e.g., ornamental plants. Taking the seed potato trade between the European countries to consideration, it seems that factors which are not well-known limit establishment of PMTV and/or development of the necrotic symptoms in tubers in many areas in Europe. Learning more about those factors might also help to design means for control of PMTV.

The advanced seed potato producers in Europe base production on multiplication of pathogen-free in vitro plants of potato cultivars. In practice, the propagation material is tested only for selected viruses considered to be the most harmful and included in the phytosanitary regulations. There are new methods to ensure that the plants are free of those viruses that are not among those routinely tested. First, cryotherapy is an efficient approach to ensure that the promising potato cultivars and breeding lines are virus-free and free from phytoplasma before they are introduced to long-term maintenance in vitro (Wang and Valkonen 2009). Secondly, all known plant viruses and also related, unknown viruses can be detected by analyzing the small RNAs generated by RNA silencing, the main antiviral defense mechanism in plants. The small RNAs (21–24 nucleotides) are extracted from plant tissue, sequenced and used to assemble longer sequences (contigs) using methods of bioinformatics. Viruses in the sample are identified by comparing the contigs with sequences available in databases (Kreuze 2014). This new method called small RNA sequencing and assembly (sRSA) is as sensitive for virus detection as the widely used PCR-based methods (Santala and Valkonen 2018). The advantage of sRSA is that it detects all types of plant viruses in the same assay without need for virus-specific primers or probes.

11.2.6 Australia

The viruses so far found infecting potato in the Australian continent are PLRV, PVY, PVA, PVS, PVM, PAMV, AMV, CMV, TSWV, and *Lucerne Australian latent virus* (LALV), and the viroid PSTVd has also been found (Buchen-Osmond et al. 1988). *Beet western yellows* virus was reported infecting potato in Tasmania but later shown to be confused with PLRV. Of the viruses infecting the potato crop, the most prevalent are PLRV, PVY, PVX, PVS, and TSWV, and these five viruses are the ones tested for routinely in Australian seed potato production schemes. For many years, PVM, PAMV, CMV, and LALV have not been recorded infecting Australian potato crops, but AlMV infection typified by bright yellow calico symptoms still occurs sporadically. Soil-borne viruses, such as TRV and PMTV, that cause problems in other world regions have not yet been recorded infecting potato in Australia, although the PMTV vector *Spongospora subterranea* and TRV vectors *Trichodorus* and *Paratrichodorus* are present.

The most important potato viruses in Australia are PLRV and PVY. The PVY recombinant PVY^{NTN} has been found infecting potato crops in four eastern Australian states (Queensland, New South Wales, Victoria, South Australia) where it is causing similar problems in seed potato production to the ones it causes in Europe. However, it has not, as yet, been found infecting potatoes in Tasmania or Western Australia (Kehoe and Jones 2016). PLRV remains the most prevalent and important potato virus in south-west Australia. PVX and PVS are common contaminants detected during Australian seed potato production but their incidence is now much lower than in the past when roguing was focused on PLRV rather than viruses causing mild foliar symptoms and no routine virus testing was done. TSWV is another common contaminant detected during seed potato production and tuber necrosis due to TSWV infection (Wilson 2001) continues to be found in ware potatoes in some states due to the common occurrence of this virus in weed hosts growing in or near to potato fields and its spread to potato plants by its thrips vector. In the last decade (2000–2010) relaxation of seed potato regulations concerning isolation from commercial potato crops in two Australian states (Victoria, Tasmania) led to a temporary upsurge in the incidence of common potato viruses in high grade seed potatoes. More thorough seed production regulations had to be reintroduced to counteract this situation.

Recent studies on potato viruses in Australia have focused mainly on PVX, PVS, PLRV and PVY. Nyalugwe et al. (2012) inoculated PVX isolates belonging to two strain groups to 38 cultivars grown in Australia to identify phenotypic responses and presence or absence of different PVX resistance genes. They also found that infection with PVX and PVS increased the titer of PVS and enhanced expression of foliar symptoms in potato plants. In a similar study with PVY, Jones and Vincent (2018) inoculated PVY^{O} and PVY^{D} to 39 cultivars to identify phenotypic responses and presence of absence of different PVY resistance genes. Coutts and Jones (2015) investigated PVY^{O}'s contact transmissibility, stability on surfaces, and inactivation with disinfectants. It was contact-transmitted to potato foliage but not to tubers, remained infective for up 24 h on some surfaces, and both bleach and the less caustic nonfat milk were useful PVY disinfectants.

When Cox and Jones (2010a) studied the CP nucleotide sequences of 13 PVS isolates from mainland Australia, all isolates were in phylogroup PVS^{O}. None of them invaded *C. quinoa* systemically so they were all in biological strain PVS^{O}. However, when Lambert et al. (2012) studied 42 PVS isolates from the Island of Tasmania, based on abiity to invade *C. quinoa* systemically three of them belonged to PVS^{A}, while the others belonged to PVS^{O}. When their CP genes were sequenced, they all belonged to phylogroup PVS^{O}. Santillan et al. (2018) included two complete PVS genomes from Australia in their evolutionary study on PVS, and both belonged to the main non-South American grouping, i.e. PVS^{O}. When Cox and Jones (2010b) studied the CP nucleotide sequences of 11 PVX isolates from Australia, all 11 belonged to the main non-South American grouping, i.e. phylogroup I. There was no relationship between biological strain and phylogroup as phylogroup I contained PVX isolates in biological strain groups 1, 3 and 4, whereas minor phylogroups II-1 and II-2 both contained isolates in strain groups 2 and 4. Kehoe and Jones (2014) compared the biological and genomic properties of eight historical European (1943–1984) and five Australian (2003–2012) PVY isolates from potato. Based on eliciting hypersensitivity genes *Nc*, *Ny*, or *Nz*, the European isolates belonged to biological strain groups PVY^{C}, PVY^{O} or PVY^{Z}, whereas the Australian isolates belonged to PVY^{O}, PVY^{Z} or new strain group PVY^{D} which elicited putative hypersensitivity gene *Nd*. The Australian and historical European isolates all fitted in phylogroups Y^{O} or Y^{C}. Moreover, biologically defined PVY^{O} and PVY^{Z} isolates were both within phylogroup Y^{O} while biologically defined Y^{C} and Y^{D} isolates were both phylogroup Y^C revealing disagreement between the current biological and phylogenetic PVY nomenclature systems.

11.3 Control of Potato Viruses

Potato is clonally propagated by planting tubers, which increases the risk of virus accumulation in the next crop and tuber generations. Apart from semi-persistently or persistently vector-transmitted viruses, such as PLRV, for which insecticide application as seed tuber dressings or foliar sprays are effective during seed potato production, such treatments are generally ineffective at controlling nonpersistently vector-borne viruses like PVY (Jones 2014). Thus, most potato viruses are controlled by three principal methods: host plant resistance, clean seed systems and cultural practices. Nowadays, in developed countries potato viruses are by and large controlled through formal certified clean seed production systems and to some extent through virus resistance. On the other hand, despite many years of intensive investment, formal certified seed systems have had only very limited, if any, penetration in many developing countries, where farmers mostly obtain their seed from their previous crop or through informal trade involving low-quality planting material. High cost of seed production, lack of adequate infrastructure and economic resources of small scale family farms are some of the reasons contributing to this situation.

In the past, simple seed potato schemes that, for example, relied solely on visual inspection and roguing combined with flooding and livestock to remove any tubers left behind after harvest proved effective at removing PLRV and other viruses causing obvious foliar symptoms, but ineffective at removing viruses causing mild symptoms e.g. PVS and PVX. Formal certified seed systems are expensive to implement in most developing countries as they require rigorous visual inspections and diagnostic testing. Relying solely on visual inspections is cheaper but leads to selection of viral strains that show few foliar symptoms, as occurred with some strains of PVY. Diagnostic testing often requires laboratories. While well-established method such as ELISA (Enzyme Linked Immunosorption Assay) are relatively cheap, they may lack sensitivity. Various PCR (Polymerase Chain Reaction), reverse transcription PCR and real-time PCR, protocols have been developed and multiplexed (e.g. Raigond et al. 2013; Meena et al. 2017; Jeevalatha et al. 2016a) which can provide ultrasensitive detection of viruses in samples. Field diagnostics with viruses is also possible using lateral flow devices that are commercialized by several companies globally but suffer from similar sensitivity issues as regular ELISA and are not available for all viruses. On the other hand, Loop Mediated Isothermal Amplification (LAMP) has recently emerged as a technology that can provide highly sensitive in field detection of potato viruses, with assays developed for PVY (Treder et al. 2018), PLRV (Ahmadi et al. 2013; Almasi et al. 2013), PVX (Jeong et al. 2015), PSTVd (Learcic et al. 2013) and ToLCNDV (Jeevalatha et al. 2018). LAMP assays can rapidly be designed to detect newly identified viruses and can be multiplexed, making it a flexible technology. LAMP is also compatible with crude nucleic acid extractions, can achieve high sensitivity, and be combined with the availability of relatively cheap battery powered real-time devices, such as Bioranger or real-time Genie series of devices, so may soon see more routine use in determining virus infections.

11.4 Potato Viruses and Seed Systems

Because of the previously mentioned factors, implementation of healthy seed systems in tropical countries is challenging, especially if there is a lack of cool areas or growing seasons with low aphid vector pressure available to reduce rate of reinfection during seed production. Due to this, as well as lack of appropriate infrastructure, investment, and commercial opportunity for smallholders to recover their investment in expensive seed potatoes, the amount of certified seeds used is generally negligible in most developing countries (Thomas-Sharma et al. 2016). Nevertheless, emerging economies, such as India and Brazil, have implemented seed systems with a level of success.

In **India**, a conventional seed tuber production system based on the "seed plot technique (SPT)" has successfully been used for the last five decades. Since its introduction, the SPT revolutionized the indigenous quality seed production system in the subtropical plains of India by extending it from the hills to the plains. The principle of SPT is growing the seed potato crop using healthy seed during a period with low aphid prevalence from October to the first week of January, coupled with IPM, roguing and dehaulming the seed crop during January before aphids reach critical threshold numbers. Today, 90% of seeds are being produced in northern (Punjab), north central (Gwalior), northwestern (Modipuram), and eastern plains (Patna) of the country. This seed is being supplied to the north east, Deccan plateau, and southern parts of the country which are not suitable for quality seed production. The seed production system in India includes tuber indexing for all major viruses and clonal multiplication of virus free mother tubers in four cycles for breeders seed production. The breeder seed produced by ICAR-CPRI is supplied to various State Government Organizations for further multiplication in three more cycles, viz. Foundation Seed 1 (FS-1), Foundation Seed 2 (FS-2) and Certified Seed (CS) under strict health standards. However, the current situation of breeder seed multiplication by the State Governments is not following the desired seed multiplication chain and breeder seed supplied by ICAR-CPRI is often being multiplied only up to FS-1 stage. Therefore, there is a shortage of certified seed in the country (ICAR-CPRI, Shimla). Incorporation of hi-tech seed production systems coupled with advanced virus detection techniques is the only way out in fulfilling the very large demand of quality seed potatoes in the country.

Continuous monitoring of aphid vector dynamics revealed that aphids cross critical limits 1 week earlier in Punjab and 1–2 weeks earlier in western UP in the recent past. In general terms, vector pressure has increased many folds as compared to the

1980s which is a cause of concern. Therefore, "SPT" is being refined to cope with the changing climate and vector pressure. There is also an urgent need to explore possibilities of seed production in nontraditional areas using modern techniques (Singh et al. 2014).

The major problem for potato production in **Brazil** is related to the low availability of virus-free seed tubers. As mentioned earlier, Brazil has three potato growing seasons per year in a climate where high population density of virus vectors occurs. Therefore, one of the most important measures that must be taken in Brazil is the use of virus-free seed tubers, because if there is a source of virus inoculum in the field, there will be a rapid spread during the potato production cycle. As a consequence, the tubers produced suffer rapid degeneration during their multiplication in the field. To address this problem, Brazil has adopted and periodically revised a seed certification system that establishes norms for seed-potato production in diverse categories: (1) G-0, which is the first generation derived from in-vitro plants, although imported basic classes can also be considered as G-0; (2) the basic and certified seeds, usually going up to G-4, but having virus incidence thresholds mandatorily respected. This is currently regulated under the Federal MAPA IN-32 (as of 20/11/2012, http://www.agricultura.gov.br/assuntos/insumos-agropecuarios/insumos-agricolas/sementes-e-mudas/publicacoes-sementes-e-mudas/INN32de20denovembrode2012.pdf). Producers have a laboratory support system accredited by INMETRO and certified by the Ministry of Agriculture for analysis and diagnosis of viruses in seed material. Currently a small part of the employed seed originates from tissue culture and another part from the production of minitubers in the greenhouse.

An effective low cost alternative system to produce high grade, virus-free, minituber/seed-potato lots involves the sprout/seed-potato technology (Virmond et al. 2017; Souza Dias et al. 2018). This may be accomplished either by the seed-potato exporting or importing country, and the research has shown that sprouts are durable, easy to handle and economic, when compared with potato tubers. The sprouts have to be detached from virus-free seed potatoes when they reach at least 3–5 cm tall (0.5–0.8 g), making sure that they will not be removed by cutting or sectioning, but just by hand-removing at the stolon, and with primordial root formation on the base (Fig. 11.4). After removal, they can easily be packaged into sealed polyethylene bags (100 units) for transport. Storage, before or after export and upon arrival in the country has to be done in under proper environmental conditions, avoiding insect vectors, and having favorable conditions for sprout growth, such as a dark room, 20 °C and 70–80% RU. Upon arrival at the final destination, the sprouts can be directly planted under greenhouse conditions, using horticultural soil-substrate or in a hydro- or aeroponics system to produce minituber/seed-potatoes. Details about the methodology can be found in Virmond et al. (2017). The sprout/seed-potato technology is a novel method to increase the multiplication rate of high-grade (G-0, basic classes), national or imported seed-potato stocks. It has been applied by small and large-scale potato producers in Brazil to obtain minitubers/seed-potato, free of viruses and true-to-type. It has been officially accepted as of 20-11-2012 (MAPA IN 32) to use sprouts, originally from G-0 basic seed-potato lots (national laboratory-

Fig. 11.4 Producing potato seed from tuber sprouts in Brazil. Photo credits: J. A. C. Souza-Dias

greenhouse minituber/seed-potato, or the annually imported tuber/seed-potato basic classes), as propagating material to produce certified minituber/seed-potato stocks. If properly handled, under the same conditions as used for minituber production, they can have the same phytosanitary health status. Moreover, minitubers/seed-potatoes obtained from sprouts through successive generations (G-3) are safer when compared to tissue culture, because they do not run the risk of presenting the common mutations seen with that technique (Souza-Dias et al. 2017).

Due to the difficulty of potato seed production in the country, producers often import seeds from European countries, Canada and the United States. This exposes the country to the entry of new pathogens, as happened with strains of PVY, that, once in Brazil, underwent a rapid adaptation and changed the epidemiology of the virus in Brazilian potato crops.

Another problem that has occurred with the production of seed potatoes in Brazil was the start in 2011 of the Brazilian legislation for production of potato seeds for personal use. This released the farmer from testing the seed planted, allowing him to plant them without laboratory analysis. Based on visual evaluation only, the producer is at risk of planting seeds with high incidences of virus, especially when dealing with cv. Agata. Besides compromising the yield, it can act as symptomless PVY carrier, especially of PVY^{NTN} as tuber necrotic rings are rarely visible in cv Agata, thus providing a generous source of virus inoculum for potato and other Solanaceous crops nearby.

11.5 Resistance

Potato is clonally propagated by planting tubers, which increases the risk of accumulation of viruses in the next crop and tuber generations. Apart from semi-persistent or persistently transmitted viruses such as PLRV, viruses cannot be controlled readily with pesticides, so chemical control of virus vectors provides only partial protection at best or is ineffective. On the other hand, production of healthy seed tubers is an expensive process. Therefore, resistance to viruses in potato cultivars is the most efficient and cost-effective means to control virus diseases in potato when effective seed production systems are absent, as in most developing countries. In developed countries with sophisticated and effective seed tuber production schemes, virus resistance becomes less important than other cultivar characteristics, such as high yield, tuber quality, and adaptation to the local environment.

PVY is now the most widespread viral pathogen in potatoes in most countries. Fortunately, breeders have introduced resistance genes that control PVY to many potato cultivars. Many of them, however, recognize only certain PVY strains. These strain-specific resistance genes can act quickly upon recognition of PVY and kill most of the PVY-infected cells at an early stage of infection leading to localized necrotic lesions, although they are sometimes slower acting resulting in systemic movement followed by single shoot or complete plant death. Therefore, they are called "hypersensitivity resistance" (HR) genes and contrast to "extreme resistance" (ER) genes which do not lead to any visible lesions during the resistance reaction. Furthermore, plasmodesmata connecting the plant cells and used by viruses for movement from cell to cell are sometimes blocked, preventing further spread of the virus. However, mutations in the viral genome can overcome resistance. For example, a few mutations in the helper component protein (HCpro) overcome resistance to PVY^{O} conferred by the resistance gene *Ny* in potato (Tian and Valkonen 2013). Consequently, the mutant of PVY^{O} can multiply, spread, and cause leaf mosaic and growth reduction in the potato plants carrying *Ny*.

Jones and Vincent (2018) studied strain-specific HR and ER phenotypes elicited in potato plants by PVY isolates in strain groups PVY^{O} and PVY^{D}. These isolates were inoculated to 39 Australasian, European, or North American potato cultivars. HR elicited by infection with strain group PVY^{D} occurred in 34 of the 39 cultivars, including 2 released as early as 1893–1894 in North America. Since PVY^{D} elicits putative gene *Nd*, this had apparently been present but unrecognized since the earliest epoch of potato breeding in the second half of the nineteenth century. Systemic hypersensitive resistance (SHR) elicited by strain group PVY^{O} in presence of hypersensitivity gene *Ny* was present in 23 of the same 34 potato cultivars with putative *Nd*, occurring widely amongst cultivars released in each of the three world regions. Two European cultivars always developed ER following sap and graft inoculation so carried comprehensive PVY resistance gene *Ry*, but no Australasian or North American cultivars carried it. One Australasian and two European cultivars always developed susceptible phenotypes so lacked genes *Ry*, *Ny*, and putative *Nd*. When breeding new PVY-resistant potato cultivars for countries lacking healthy seed potato stocks, or where subsistence farmers cannot afford them, the next best option to gene *Ry* inclusion is incorporating as many strain-specific PVY resistance genes as possible (Jones and Vincent 2018).

The degree of PVY^O resistance conferred by *Ny* varies between potato cultivars depending on the extent of localized hypersensitive resistance (LHR) and/or severe SHR versus weak SHR that develops. With LHR the source of infection for further virus spread is removed. When SHR involves death of all systemically infected shoots or entire plant death, foci of PVY infection are eliminated from within the crop so they are unavailable to become infection sources for secondary spread. By contrast, weak SHR that allows PVY-infected plants to persist means they can act as virus sources for secondary spread (Jones and Vincent 2018). A recent example of its effectiveness against PVY^O in the field was provided by an investigation in a potato growing region of North America following widespread planting of cultivars with *Ny*. Over a 5-year period, incidence of PVY^O dropped from 63 to 7% of the PVY population (Funke et al. 2017). Another example comes from potato cv. Yukon Gold, which is grown in Australia, Canada, Europe and USA, and carries genes *Ny*, *Nz*, putative *Nd* (Rowley et al. 2015; Kehoe and Jones 2016; Jones and Vincent 2018). The quick elimination of PVY-infected plants by the SHR response is beneficial, since it occurs before any tubers of useful size have developed. For example, in Finland it has been possible to grow Yukon Gold over 10 years from farm-owned seed.

Although the SHR response to PVY or PVX infection is a frequently observed phenotypic reaction in breeding populations (of e.g. CIP), breeders have traditionally ignored them as this reaction usually kills the plants or severely stunts them. As the above example of Yukon Gold exemplifies, this phenotype is nevertheless effective in controlling virus infection under field conditions. Growing cultivars with *Ny* and *Nz* is likely to be most helpful in potato growing regions where the recombinant PVY strains that overcome it are still rare or absent.

Nyalugwe et al. (2012) studied strain-specific HR and ER phenotypes elicited in potato plants by isolates in PVX strain groups 1 and 3. They inoculated these isolates to 38 potato cultivars. Presence of extreme PVX resistance gene *Rx* was identified in four Australian, two European cultivars, and one North American cultivar. PVX hypersensitivity gene *Nx* was identified two Australian, four European, and one North American cultivar. PVX hypersensitivity gene *Nb* was identified in one Australian, five European, and one North American cultivars. When breeding new PVX-resistant cultivars potato cultivars for developing countries, incorporation of gene *Rx* is the best option. However, Andean PVX resistance breaking strain XHB not only overcomes *Rx*, but also overcomes *Nx* and *Nb*, so *Rx* is likely to be less effective in potato cultivars growing in the center of origin of the crop.

Breeders have during the past decades focused on the use of ER genes, that are usually strain unspecific and cause no or only microscopic HR reactions in the plants. Thus, the ER genes Ry_{adg}, Ry_{sto}, and Ry_{chc} have been used to introduce resistance for PVY and Rx_1 and Rx_2 for resistance to PVX. To facilitate introgression of PVY resistance, molecular markers have been developed and used, e.g. Bhardwaj et al. (2015) screened potato germplasms and varieties employing SCAR and SSR marker linked to Ry_{adg} and Ry_{sto} genes and identified some elite parental lines that can be exploited for transferring the virus resistance into new potato cultivars. On the other hand, triplex parental potato lines containing three copies of the Ry_{adg} gene have been developed in various breeding programs ensuring 96% of progeny contain at least one copy of the resistance gene (Kaushik et al. 2013; Kneib et al. 2017). High resolution melting markers developed for Ry_{sto} (Nie et al. 2016), Ry_{adg} (Del Rosario et al. 2018) and *Rx1* and *2* (Nie et al. 2018) can accurately predict allele dosage and significantly aide in developing such parental lines. Nevertheless, even ER genes may be sensitive to changes in environment, as exemplified in recent study showing reduced efficiency of resistance to PVY by Ry_{chc} in response to increasing temperatures observed in Japan (Ohki et al. 2018). In contrast to PVY and PVX, a good source of resistance to PLRV has long evaded breeders, but a dominant gene Rl_{adg} conferring high levels of resistance was identified about a decade ago in a potato accession LOP-868 and the subsequently developed SCAR marker (Mihovilovich et al. 2014) has enabled rapid introgression into elite germplasm (Carneiro et al. 2017). Markers have also been developed for another dominant resistance gene, Rlr_{etb} originating from the non-tuber bearing wild species *S. etuberosum* (Kuhl et al. 2016), but its introgression into advanced breeding populations may still take time due to linkage drag from its wild progenitor. Screening of germplasm lines for ToLCNDV-[potato] resistance in field or glass house conditions showed possible presence of resistance source (Kumar et al. 2015; Maan et al. 2017; Jeevalatha et al. 2016b).

Additional resistance genes to PVA, PVV, PVS and PVM have also been identified (Palukaitis 2012) and mapped in potatoes but have to date not been widely utilized due to the considered limited importance of these viruses. Naderpour and Sadeghi (2018) developed a multiplex PCR assay including markers for resistance to PVY, PVS, and PLRV to facilitate introgression of multiple resistances into new varieties. Nevertheless, due to the complex genetics of potato it has not been easy to combine virus resistance with the myriad of other necessary traits needed for a successful variety. Transgenic approaches can readily incorporate resistance to multiple viruses into specific potato varieties (Chung et al. 2013), but considering current controversies surrounding transgenic crops, such products will not likely be released for cultivation in the near future.

11.6 Cultural Approaches

Landraces that survive for many years tend to be ones that possess multiple virus resistances as evidenced by the frequent occurrence of virus resistance genes in Andean potato landraces in germplasm collections (Jones 1981, and references therein). However, cultural approaches (such as roguing out plants with obvious virus symptoms, removing volunteer potato plants or weeds likely to harbor potato viruses, deploying reflective mulches to deter insect vector landings, manipulating the planting date to avoid peak flights of insect vectors, and early haulm destruction to avoid late virus infections) are rarely used by developing country farmers unless they are seed producers. In fact, the common habit of small holder farmers of selling and or consuming large tubers and keeping the small ones as seed for a next crop probably maintains virus loads in the seed high, as virus-infected plants often are the ones producing the smallest tubers. Gildemacher et al. (2011)and Schulte-Geldermann et al. (2012) showed how positive selection of healthy looking mother plants to provide seed tubers could reduce virus incidences in subsequent crops by 35–40% and a corresponding yield increase of 30%. Modeling approaches have similarly indicated that the approach of selecting healthy plants for seed production can be as effective as certified seed in maintaining seed quality (Thomas-Sharma et al. 2017). However, this would only apply where viruses causing little or no symptoms are being discounted. Also, the penetration of positive selection techniques among regular farmers has until now been limited.

The seed plot technique as practiced in India (whereas it starts out with certified virus free seed) is largely based on cultural practices to keep tuber seed healthy, growing during seasons and areas with low vector pressure coupled with IPM (Integrated Pest Management), rouging (negative selection), and dehaulming the seed crop before vectors reach a critical threshold limit. The use of straw mulch (Kirchner et al. 2014), mineral oil sprays and intercropping has been shown to enable control of PVY infection, particularly when used in combination (Dupuis et al. 2017a, b), although the economics of it would only justify their application for seed potato production (Dupuis 2017). Insecticide application to prevent PLRV spread in seed potato crops is also routinely used where seed potato stocks are multiplied in more aphid vector prone areas, especially in developed countries.

A unique practice is performed in the Andean region (and also the Himalayas) where farmers traditionally grow their potatoes at higher altitudes to reinvigorate them after several years of cultivation at lower altitudes (De Haan and Thiele 2003). Bertschinger et al. (2017) recently demonstrated that growing potatoes at higher altitude significantly reduced the number of virus-infected tubers produced from infected mother plants. Together with the absence of insect vector populations that could reinfect healthy plants at high altitude, this helps explain how this practice can reduce virus infections in the crop and resulting in subsequent higher yields. The mechanism of this phenomenon (reduced infection of tubers of infected mother plants at higher altitude), however, still remains unknown and should be an interesting topic of further study. Possibly RNA silencing mechanisms as affected by environmental conditions may be involved and understanding them may lead to new approaches for cleaning virus-infected plants.

Other mechanisms may also play a role in reducing losses by virus infections in the Andean region. Anecdotal evidence suggest that healthy planting material repatriated to farming communities often rapidly succumbs to new and more severe virus infections than the original material. An explanation for this may be that farmers have over the generations selected for plants which are infected with mild strains of the viruses, causing only limited yield losses, but protecting from more severe virus strains. This protection is lost when plants are cleaned from viruses.

Whereas this has not been researched until now, it may be worthwhile to investigate this phenomenon as it could lead to identification of new methods to control yield losses by viruses in potato using mild strains.

11.7 Final Remarks

Viruses remain a problem for global potato production, even though, over the years, the importance of certain viruses has increased or decreased globally. These changes in relative importance result from a range of factors including not only increased global trade but also regional changes in cultivar usage, cropping patterns, implemented seed systems and diagnostic testing regimes, appearance and evolution of new viruses and virus strains, and vector populations. All of these factors interact with each other and are further affected by climate change, making it difficult to predict what the future will hold. The two examples of potato-infecting geminivirus and torradovirus in Peru represent interesting cases where viruses have rapidly emerged as a significant local problem, only to subsequently disappear again into the background. This may be something that, in the past, has frequently occurred unnoticed in the potato's center of domestication in the Andes, or elsewhere in the world (e.g. PYDV was a devastating virus in the US during the early twentieth century, but has now all but disappeared). Many factors may influence whether a virus eventually manages to get a foothold in a region and become a permanent threat to the potato crop, but it has obviously happened on several occasions during the past 500 years since potato was introduced worldwide, as evidenced by viruses infecting potato uniquely in geographical areas outside of the Andes. The latter represent viruses that the potato did not encounter until it was moved away from the Andean region. Whitefly- and thrips-transmitted viruses should form a particular concern as witnessed by recurrent outbreaks occurring in (sub-)tropical regions, the increasing geographical spread of crinivirus PYVV and the establishment of the begomovirus ToYLCNDV as a major potato pathogen in India. Besides in Brazil, ToCV has also been reported from potatoes in Spain (Fortes and Navas-Castillo 2012) and identified in India associated with leaf-roll disease (CIP, unpublished), and thus the virus does seem to have recurring opportunities to infect potatoes worldwide where conditions are appropriate and it may only be a matter of time until an adaptive mutation appears for it to establish as a significant potato pathogen.

With a warming climate, producing high-quality virus-free seed is set to become more difficult, as opportunities to move to cooler areas with low vector pressure have become fewer in warmer countries, especially those that lack mountainous regions. Some cooler countries have the opportunity to move their seed potato production towards more extreme latitudes, though this rarely applies to developing countries. Thus, there is an increased requirement for new technologies for rapid multiplication of healthy plants under controlled conditions to be able to supply high-quality seeds at an affordable level. Despite the availability of effective resistance genes to, for example, PVY, due to the complex genetics of potatoes, it is difficult to recombine these with other critical traits necessary for a successful cultivar, and as a result only a fraction of potatoes grown globally possess non-strain-specific host resistance. The use of more efficient molecular markers and the promise of diploid potato breeding (Taylor 2018) may change that in the future, but for now, a combined approach for degeneration control adjusted to the local socio-economical and climatic context, as suggested by (Thomas-Sharma et al. 2016), may be the best way to go in developing countries where sophisticated seed production schemes are not currently a viable option.

References

Abad JA, Moyer JW, Kennedy GG, Holmes GA, Cubeta MA (2005) Tomato spotted wilt virus on potato in eastern North Carolina. Am J Potato Res 82(3):255

Abbas A, Madadi M (2016) A review paper on Potato Mop-Top Virus (PMTV): occurrence, properties and management. World J Biology and Biotechnology 1(3):129–134

AGRIANUAL (2016) Anuário da Agricultura Brasileira. FNP Consultoria, São Paulo

Ahmadi S, Almasi MA, Fatehi F, Struik PC, Moradi A (2013) Visual detection of *potato leafroll virus* by one-step reverse transcription loop-mediated isothermal amplification of DNA with hydroxynaphthol blue dye. J Phytopathol 161(2):120–124

Almasi MA, Manesh ME, Jafary H, Dehabadi SMH (2013) Visual detection of *Potato Leafroll* virus by loop-mediated isothermal amplification of DNA with the GeneFinder™ dye. J Virol Methods 192(1–2):51–54

Andrade ER, Figueira AR (1992) Incidência e sintomatologia de estirpes do vírus Y(PVY) nas regiões produtoras de batata do sul de Minas Gerais. Ciência e Prática 16:371–376

Baldo NH, Elhassan SM, Elballa MM (2010) Occurrence of viruses affecting potato crops in Khartoum State-Sudan. Potato Res 53(1):61–67

Barrocas EN, Figueira AR, Morais FR, Santos RC (2000) PVY and PLRV occurrence in lots of potato seeds coming from different regions of Minas Gerais State-Brazil. Fitopatol Bras 25(S):437

Bertschinger L, Bühler L, Dupuis B, Duffy B, Gessler C, Forbes GA, Keller ER, Scheidegger UC, Struik PC (2017) Incomplete infection of secondarily infected potato plants–an environment dependent underestimated mechanism in plant virology. Front Plant Sci 8:74

Bhardwaj V, Sharma R, Dalamu SAK, Baswaraj R, Singh R, Singh BP (2015) Molecular characterization of potato virus Y resistance in potato (*Solanum tuberosum* L.). Ind J Genet 75(3):389–392. https://doi.org/10.5958/0975-6906.2015.00062.0

Buchen-Osmond C, Crabtree K, Gibbs A, McLean G (1988) Viruses of plants in Australia: descriptions and lists from the VIDE database (No. 632.8 V821v). Australian National University, Canberra

Bulajic AR, Stankovic IM, Vucurovic AB, Ristic DT, Milojevic KN, Ivanovic MS, Krstic BB (2014) *Tomato spotted wilt virus* - potato cultivar susceptibility and tuber transmission. Am J Potato Res 91:186–194

Bertschinger L, Bühler L, Dupuis B, Duffy B, Gessler C, Forbes GA, Keller ER, Scheidegger UC, Struik PC (2017) Incomplete infection of secondarily infected potato plants – an environment dependent underestimated mechanism in plant virology. Front Plant Sci (8):74. https://doi.org/10.3389/fpls.2017.00074

Carneiro OL, Ribeiro SR, Moreira CM, Guedes ML, Lyra DH, Pinto CA (2017) Introgression of the Rl adg allele of resistance to potato leafroll virus in Solanum tuberosum L. Crop Breed Appl Biotechnol 17(3):242–249

Chiunga E, Valkonen JPT (2013) First report on viruses infecting potato in Tanzania. Plant Dis 97:1260

Chikh-Ali M, Rowley JS, Kuhl J, Gray S, Karasev AV (2014) Evidence of a monogenic nature of the Nz gene against *Potato virus Y* strain Z (PVYZ) in potato. Am J Potato Res 91:649–654

Chung BN, Yoon JY, Palukaitis P (2013) Engineered resistance in potato against potato leafroll virus, potato virus A and potato virus Y. Virus Genes 47(1):86–92

Constable F, Chambers G, Penrose L, Daly A, Mackie J, Davis K, Rodoni B, Gibbs M (2019) Viroid-infected Tomato and Capsicum Seed Shipments to Australia. Viruses 11(2):98

Cox BA, Jones RAC (2010a) Genetic variability of the coat protein gene of potato virus S and distinguishing its biologically distinct strains. Arch Virol 155:1163–1169

Cox BA, Jones RAC (2010b) Genetic variability of the coat protein gene of potato virus X, and the current relationship between phylogenetic placement and resistance groupings. Arch Virol 155:1349–1356

Coutts BA, Jones RAC (2015) Potato virus Y: contact transmission, stability on surfaces, and inactivation with disinfectants. Plant Dis 99:387–394

De Haan S, Thiele G (2003) In situ conservation and potato seed systems in the Andes. In: Devra I, Sevilla-Panizo JR, Chávez-Servia JL, Hodgkin T (eds) Seed systems and crop genetic diversity on-farm. The International Plant Genetic Resources Institute (IPGRI), Pucallpa, p 126

De Souza J, Gamarra H, Muller G, Kreuze J (2018) First report of potato virus S naturally infecting Arracacha (Arracacia xanthorrhiza) in Peru. Plant Dis 102:1. https://doi.org/10.1094/PDIS-07-17-0945-PDN

De Souza J, Muller G, Perez W, Cuellar W, Kreuze JF (2017) Complete sequence and variability of a new subgroup B nepovirus infecting potato in central Peru. Arch Virol 62:885–889. https://doi.org/10.1007/s00705-016-3147-6

Del Rosario HM, Vidalon LJ, Montenegro JD, Riccio C, Guzman F, Bartolini I, Ghislain M (2018) Molecular and genetic characterization of the Ry adg locus on chromosome XI from Andigena potatoes conferring extreme resistance to potato virus Y. Theor Appl Genet 131(9):1925–1938

Devaux A, Kromann P, Ortiz O (2014) Potatoes for sustainable global food security. Potato Res 57:185–199

Dupuis B (2017) Development of a crop management method to control the spread of Potato Virus Y (PVY). Doctoral dissertation, UCL-Université Catholique de Louvain

Dupuis B, Bragard C, Carnegie S, Kerr J, Glais L, Singh M, Nolte P, Rolot JL, Demeulemeester K, Lacomme C (2017a) Potato virus Y: control, management and seed certification programmes, Potato virus Y: biodiversity, pathogenicity, epidemiology and management. Springer, Cham, pp 177–206

Dupuis B, Cadby J, Goy G, Tallant M, Derron J, Schwaerzel R, Steinger T (2017b) Control of potato virus Y (PVY) in seed potatoes by oil spraying, straw mulching and intercropping. Plant Pathol 66(6):960–969

Fortes IM, Navas-Castillo J (2012) Potato, an experimental and natural host of the crinivirus Tomato chlorosis virus. Eur J Plant Pathol 134:81–86

Franco-Lara L, Rodríguez D, Guzmán-Barney M (2013) Prevalence of *Potato yellow vein virus* (PYVV) in *Solanum tuberosum* group Phureja fields in three states of Colombia. Am J Potato Res 90:324–330. https://doi.org/10.1007/s12230-013-9308-1

Fribourg CE, Gibbs AJ, Adams IP, Boonham N, Jones RAC (2019) Biological and molecular properties of isolates from pepino. Plant Dis 103(7):1746–1756

Funke CN, Nikolaeva OV, Green KJ, Tran LT, Chikh-Ali M, Quintero-Ferrer A, Cating RA, Frost KE, Hamm PB, Olsen N, Pavek MJ, Gray SM, Crosslin JM, Karasev AV (2017) Strain-specific resistance to Potato virus Y (PVY) in potato and its effect on the relative abundance of PVY strains in commercial potato fields. Plant Dis 101:20–28

Fuentes S, Jones RAC, Matsuoka H, Ohshima K, Kreuze JF, Gibbs A (2019a) Potato virus Y; the Andean connection. Virus Evolution, in press https://doi.org/10.13140/RG.2.2.15998.33602

Fuentes S, Perez A, Kreuze JF (2019b) Dataset for: The peruvian potato virome. International Potato Center, V1. https://doi.org/10.21223/P3/YFHLQU

Galvino-Costa SBF, Figueira AR, Rabelo-Filho FAC, Moraes FHR, Nikolaeva OV, Karasev AV (2012) Molecular and serological typing of potato virus Y Isolates from Brazil reveals a diverse set of recombinant strains. Plant Dis 96:1451–1458. https://doi.org/10.1094/PDIS-02-12-0163-RE

Galvino-Costa SBF, Santos BA, Figueira AR, Geraldino-Duarte PS (2014) Detection of potato virus Y (PVY) strains in plants with single and mixed infection. Virus Rev Res 19:215–216

Garg ID, Hegde V (2000) Biological characterization, preservation and ultrastructural studies of Andean strain of Potato virus S. Ind Phytopathol 53:256–260

Ghorai AK, Kang SS, Sharma A, Sharma S (2017) Occurrence of *Cucumber mosaic virus* on potato and its transmission to muskmelon under potato-cucurbit cropping pattern followed in Punjab, India. Int J Curr Microbiol App Sci 6(11):2947–2965

Gibbs AJ, Ohshima K, Yasaka R, Mohammad M, Gibbs MJ, Jones RAC (2017) The phylogenetics of the global population of potato virus Y and its necrogenic recombinants. Virus Evol 3(1):vex002. https://doi.org/10.1093/ve/vex002

Gil JF, Adams I, Boonham N, Nielsen SL, Nicolaisen M (2016a) Molecular and biological characterisation of Potato mop-top virus (PMTV, Pomovirus) isolates from potato growing regions in Colombia. Plant Pathol 65:1210–1220

Gil JF, Adams I, Boonham N, Nielsen SL, Nicolaisen M (2016b) Molecular and biological characterisation of two novel pomo-like viruses associated with potato (*Solanum tuberosum*) fields in Colombia. Arch Virol 161(6):1601–1610

Gildemacher P, Demo P, Barker I, Kaguongo W, Woldegiorgis G, Wagoire W, Wakahiu M, Leeuwis C, Struik P (2009) A description of seed potato systems in Kenya, Uganda and Ethiopia. Am J Potato Res 86:373–382

Gildemacher PR, Schulte-Geldermann E, Borus D, Demo P, Kinyae P, Mundia P, Struik PC (2011) Seed potato quality improvement through positive selection by smallholder farmers in Kenya. Potato Res 54(3):253

Glais L, Bellstedt DU, Lacomme C (2017) Diversity, characterisation and classification of PVY, Potato virus Y: biodiversity, pathogenicity, epidemiology and management. Springer, Cham, pp 43–76

Gray S, De Boer S, Lorenzen J, Karasev A, Whitworth J, Nolte P, Singh R, Boucher A, Xu H (2010) Potato virus Y: an evolving concern for potato crops in the United States and Canada. Plant Dis 94(12):1384–1397

Gutierrez PA, Alzate JF, Marín-Montoya MA (2013) Complete genome sequence of a novel potato virus S strain infecting *Solanum phureja* in Colombia. Arch Virol 158:2205–2208. https://doi.org/10.1007/s00705-013-1730-7

Harahagazwe D, Condor B, Barreda C et al (2018) How big is the potato (Solanum tuberosum L.) yield gap in Sub-Saharan Africa and why? A participatory approach. Open Agriculture 3(1):180–189. https://doi.org/10.1515/opag-2018-0019

Henao-Díaz E, Gutiérrez-Sánchez P, Marín-Montoya M (2013) Phylogenetic analysis of potato virus y (PVY) isolates obtained in potato (*Solanum tuberosum*) and tamarillo (*Solanum betaceum*) crops from Colombia. Actual Biol 35(99):219–232

Hooker WJ, Salazar LF (1983) A new plant virus from the high jungle of the eastern Andes; Solanum apical leaf curling virus (SALCV). Ann Appl Biol 103:449–454

IBGE (Instituto Brasileiro de Geografia e Estatística) (2017) Levantamento Sistemático de Produção Agrícola. Grupo de Coordenação Estatística Agropecuárias/GCEA/DPE/COAGRO, Disponibleat. http:www.ibge.gov.br. Accessed 10 April 2018

Jailani AAK, Shilpi S, Mandal B (2017) Rapid demonstration of infectivity of a hybrid strain of potato virus Y occurring in India through overlapping extension PCR. Physiol Mol Plant Pathol 98:62–68

Jain RK, Khurana SMP, Bhat AI, Chaudhary V (2004) Nucleocapsid protein gene sequence studies confirm that potato stem necrosis is caused by a strain of groundnut bud necrosis virus. Ind Phytopath 57:169–173

Jebasingh T, Makeshkumar T (2017) Characterisation of carlaviruses occurring in India. In: Mandal B, Rao G, Baranwal V, Jain R (eds) A century of plant virology in India. Springer, Singapore

Jeevalatha A, Chakrabarti SK, Sharma S, Sagar V, Malik K, Raigond B, Singh BP (2017a) Diversity analysis of *Tomato leaf curl* New Delhi virus–[potato], causing apical leaf curl disease of potato in India. Phytoparasitica 45:33–43. https://doi.org/10.1007/s12600-017-0563-4

Jeevalatha A, Kaundal P, Kumar A, Guleria A, Sundaresha S, Pant RP, Sridhar J, Venkateswarlu V, Singh BP (2016a) SYBR green based duplex RT-qPCR detection of a begomovirus, *Tomato leaf curl* New Delhi virus-[potato] along with Potato virus X and Potato leaf roll virus in potato. Potato J 43:125–137

Jeevalatha A, Kaundal P, Kumar R, Raigond B, Kumar R, Sharma S, Chakrabarti SK (2018) Optimized loop-mediated isothermal amplification assay for *Tomato leaf curl* New Delhi virus-[potato] detection in potato leaves and tubers. Eur J Plant Pathol 150(3):565–573

Jeevalatha A, Kaundal P, Kumar R, Raigond B, Gupta M, Kumar A, Sharma S, Sagar V, Nagesh M, Singh BP (2016c) Analysis of the coat protein gene of Indian Potato virus X isolates for identification of strain groups and determination of the complete genome sequence of two isolates. Eur J Plant Pathol 145:447–458

Jeevalatha A, Kaundal P, Shandil RK, Sharma NN, Chakrabarti SK, Singh BP (2013a) Complete genome sequence of Potato leafroll virus isolates infecting potato in the different geographical areas of India shows low level genetic diversity. Ind J Virol 24(2):199–204. https://doi.org/10.1007/s13337-013-0138-z

Jeevalatha A, Kaundal P, Venkatasalam EP, Chakrabarti SK, Singh BP (2013b) Uniplex and duplex PCR detection of geminivirus associated with potato apical leaf curl disease in India. J Virol Methods 193:62–67

Jeevalatha A, Singh BP, Kaundal P, Kumar R, Raigond B (2014) RCA-PCR: a robust technique for the detection of *Tomato leaf curl* New Delhi virus-[potato] at ultra-low virus titre. Potato J 41:76–80

Jeevalatha A, Sundaresha S, Kumar A, Kaundal P, Guleria A, Sharma S, Nagesh M, Singh BP (2017b) An insight into differentially regulated genes in resistant and susceptible genotypes of potato in response to tomato leaf curl New Delhi virus-[potato] infection. Virus Res 232:22–33

Jeevalatha A, Vanishree G, Bhardwaj V, Singh R, Nagesh M, Singh BP (2016b) Screening of potato germplasms for apical leaf curl disease resistance. CPRI Newslett 66:1

Jeong J, Cho SY, Lee WH, Lee KJ, Ju HJ (2015) Development of a rapid detection method for Potato virus X by reverse transcription loop-mediated isothermal amplification. Plant Pathol J 31(3):219

Jones RAC (1981) The ecology of viruses infecting wild and cultivated potatoes in the Andean Region of South America. In: Thresh JM (ed) Pests pathogens and vegetation. Pitman, London, pp 89–107

Jones RAC (2014) Virus disease problems facing potato industries worldwide: viruses found, climate change implications, rationalizing virus strain nomenclature and addressing the Potato virus Y issue. The potato: botany, production and uses. CABI, Wallingford, pp 202–224

Jones RAC (2016) Future scenarios for plant virus pathogens as climate change progresses. Adv Virus Res 95:57–147

Jones RAC, Kehoe MA (2016) A proposal to rationalize within-species plant virus nomenclature: benefits and implications of inaction. Arch Virol 161:2051–2057

Jones RA, Vincent SJ (2018) Strain-specific hypersensitive and extreme resistance phenotypes elicited by potato virus Y among 39 potato cultivars released in three world regions over a 117-year period. Plant Dis 102(1):185–196

Kalyandurg P, Gil JF, Lukhovitskaya NI, Flores B, Müller G, Chuquillanqui C, Palomino L, Monjane A, Barker I, Kreuze J, Savenkov E (2017) Molecular and pathobiological characterization of 61 Potato mop-top virus full-length cDNAs reveals great variability of the virus in the centre of potato domestication, novel genotypes and evidence for recombination. Mol Plant Pathol. https://doi.org/10.1111/mpp.12552

Karasev AV, Gray SM (2013) Continuous and emerging challenges of Potato virus Y in potato. Annu Rev Phytopathol 51:571–586

Kaushik SK, Sharma R, Garg ID, Singh BP, Chakrabarti SK, Bhardwaj V, Pandey SK (2013) Development of a triplex (YYYy) parental potato line with extreme resistance to potato virus Y (PVY) using marker assisted selection (MAS). J Hortic Sci Biotechnol 88:580–584

Kehoe MA, Jones RAC (2016) Improving Potato virus Y strain nomenclature: lessons from comparing isolates obtained over a 73-year period. Plant Pathol 65:322–333

Kirchner SM, Hiltunen LH, Santala J et al (2014) Comparison of straw mulch, insecticides, mineral oil, and birch extract for control of transmission of potato virus Y in seed potato crops. Potato Res 57:59–75

Kitajima EW, Carvalho AMB, Costa AS (1962) Microscopia eletrônica de estirpes do vírus Y da batatinha que ocorrem em São Paulo. Bragantia 21:755–763

Kneib RB, Kneib RB, Pereira ADS, Castro CM (2017) Allele dosage of PVY resistance genes in potato clones using molecular markers. Crop Breed Appl Biotechnol 17(4):306–312

Koenig R, Ziebell H (2013) Sequence-modified primers for the differential RT-PCR detection of Andean potato latent and Andean potato mild mosaic viruses in quarantine tests. Arch Virol. https://doi.org/10.1007/s00705-013-1859-4

Kreuze J (2014) siRNA deep sequencing and assembly: piecing together viral infections, Detection and diagnostics of plant pathogens. Springer, Dordrecht, pp 21–38

Kreuze J, Koenig R, De Souza J, Vetten HJ, Müller G, Flores B, Ziebell H, Cuellar W (2013) The complete genome sequences of a Peruvian and a Colombian isolate of Andean potato latent virus and partial sequences of further isolates suggest the existence of two distinct potato-infecting tymovirus species. Virus Res 173:431–435

Kuhl JC, Novy RG, Whitworth JL, Dibble S, Schneider B, Hall D (2016) Development of molecular markers closely linked to the Potato leafroll virus resistance gene, Rlretb, for use in marker-assisted selection. Am J Potato Res 93(3):203–212

Kumar M, Gupta A, Singh J, Singh F (2015) Screening of germplasm lines against Potato apical leaf curl virus disease in potato crop. Int J Trop Agric 33(2):1283–1285

Kutnjak D, Silvestre R, Cuellar W, Perez W, Müller G, Ravnikar M, Kreuze JF (2014) Complete genome sequences of new divergent potato virus X isolates and discrimination between strains in a mixed infection using small RNAs sequencing approach. Virus Res 191:45

Lacomme C, Glais L, Bellstedt D, Dupuis B, Karasev A, Jacquot E (eds) (2017) Potato virus Y: biodiversity, pathogenicity, epidemiology and management. Springer, Cham

Lambert SJ, Scott JB, Pethybridge SJ, Hay FS (2012) Strain characterization of Potato virus S isolates from Tasmania, Australia. Plant Dis 96(6):813–819

Lamichhane JR, Barzman M, Booij K, Boonekamp P, Desneux N, Huber L, Kudsk P, Langrell SRH, Ratnadass A, Ricci P, Sarah JL, Messean A (2015) Robust cropping systems to tackle pests under climate change. A review. Agron Sustain Dev 35:443–459

Learcic R, Morisset D, Mehle N, Ravnikar M (2013) Fast real-time detection of Potato spindle tuber viroid by RT-LAMP. Plant Pathol 62(5):1147–1156

Lima MF (2016) Crinivírus e geminivírus transmitidos por mosca branca: ameaça à produção nacional de batata. Batata Show 44:8–10

Lima MF, Michereff FM (2016) Vira-cabeça: ameaça à bataticultura no Brasil. Batata Show 46:22–25

Lin YH, Johnson DA, Pappu HR (2014) Effect of Potato virus S infection on late blight resistance in potato. Am J Potato Res 91(6):642–648

Lizárraga C, Querci M, Santa-Cruz M, Bartolini I, Salazar LF (2000) Other natural hosts of potato virus T. Plant Dis 84:736–738

Maan DS, Bhatia AK, Rathee M (2017) Screening and evaluation of potato (*Solanum tuberosum*) genotypes to identify the sources of resistance to Potato apical leaf-Curl disease. Int J Pure App Biosci 5(3):53–61. https://doi.org/10.18782/2320-7051.2658

Mackie A, Barbetti MJ, Rodoni B, McKirdy S, Anthony R, Jones C (2019) Effects of a tomato strain on the symptoms, biomass and yields of classical indicator and currently grown potato and tomato cultivars. Plant Disease 3009–3017

Meena PN, Kumar R, Baswaraj R, Jeevalatha A (2017) Simultaneous detection of potato viruses A and M using CP gene specific primers in an optimized duplex RT-PCR. J Pharmacogn Phytochem 6(4):1635–1640

Mihovilovich E, Aponte M, Lindqvist-Kreuze H, Bonierbale M (2014) An RGA-derived SCAR marker linked to PLRV resistance from Solanum tuberosum ssp. andigena. Plant Mol Biol Report 32(1):117–128

Moraes LA, Marchi BR, Bello VH, Watanabe LFM, Yuki VA, Marubayashi JM, Pavan MA, Krause-Sakate R (2017) Uma nova ameaça no Brasil: bemisiatabaci espécie mediterranean (Biótipo Q). Batata show 47:20–21

Naderpour M, Sadeghi L (2018) Multiple DNA markers for evaluation of resistance against potato virus Y, potato virus S and potato leafroll virus. Czech J Genet Plant Breed 54(1):30–33

Nie X, Dickison VL, Brooks S, Nie B, Singh M, De Koeyer DL, Murphy AM (2018) High resolution DNA melting assays for detection of Rx1 and Rx2 for high-throughput marker-assisted selection for extreme resistance to potato virus X in tetraploid potato. Plant Dis 102(2):382–390

Nie X, Sutherland D, Dickison V, Singh M, Murphy A. M, De Koeyer D (2016) Development and validation of high-resolution melting markers derived from Ry sto STS markers for high-throughput marker-assisted selection of potato carrying Ry sto. Phytopathology, 106(11), 1366–1375

Nóbrega NR, Silberschmidt K (1944) Sobre uma provável variante do vírus "Y" da batatinha (Solanum vírus 2, Orton) que tem peculiaridade de provocar necroses em plantas de fumo. Arq Inst Biol 15:307–330

Nyalugwe EP, Wilson CR, Coutts BA, Jones RAC (2012) Biological properties of potato virus X in potato: effects of mixed infection with Potato virus S and resistance phenotypes in cultivars from three continents. Plant Dis 96:43–54

Ohki T, Sano M, Asano K, Nakayama T, Maoka T (2018) Effect of temperature on resistance to potato virus Y in potato cultivars carrying the resistance gene Rychc. Plant Pathol 67:1629–1635. https://doi.org/10.1111/ppa.12862

Orfanidou CG, Pappi PG, Efthimiou KE, Katis NI, Maliogka VI (2016) Transmission of Tomato chlorosis virus (ToCV) by Bemisia tabaci biotype Q and evaluation of four weed species as viral sources. Plant Dis 10:2043–2049

Palukaitis P (2012) Resistance to viruses of potato and their vectors. Plant Pathol J 28(3):248–258

Pantoja KFC, Rocha KCG, ELL B, Pavan MA, Krause-Sakate R (2014) Evaluation of the primary and secondary dispersal of tomato severe rugose virus to Capsicum spp. genotypes by *Bemisia tabaci* MEAM1. Summa Phytopathol 40:375–377. https://doi.org/10.1590/0100-5405/2001

Pérez Barrera W, Valverde Miraval M, Barreto Bravo M, Andrade-Piedra J, Forbes GA (2015) Pests and diseases affecting potato landraces and bred varieties grown in Peru under indigenous farming system. Rev Latinoam Papa 19:29–41

Petek M, Rotter A, Kogovšek P, Baebler Š, Mithöfer A, Gruden K (2014) Potato virus Y infection hinders potato defence response and renders plants more vulnerable to Colorado potato beetle attack. Mol Ecol 23(21):5378–5391

Pundhir VS, Akram M, Ansar M, Rajshekhara H (2012) Occurence of stem necrosis disease in potato caused by groundnut bud Necrosis virus in Uttarakhand. Potato J 39:81–83

Qiu CL, Zhang ZX, Li SF, Bai YJ, Liu SW, Fan GQ, Gao YL, Zhang W, Zhang S, Lu WH, Lü DQ (2016) Occurrence and molecular characterization of Potato spindle tuber viroid (PSTVd) isolates from potato plants in North China. J Integr Agric 15(2):349–363

Raigond B, Sharma M, Chauhan Y, Jeevalatha A, Singh BP, Sharma S (2013) Optimization of duplex RT-PCR for simultaneous detection of Potato virus Y and S. Potato J 40(1):22–28

Raigond B, Sharma P, Kochhar T, Roach S, Verma A, Jeevalatha A, Verma G, Sharma S, Chakrabarti SK (2017) Occurrence of groundnut bud necrosis virus on potato in north western hills of India. Ind Phytopathol. https://doi.org/10.24838/ip.2017.v70.i4.76993

Ramalho TO, Galvino-Costa SBF, Figueira AR (2012) Predominance of potato cultivar Agata in Brazil and its effect in the dissemination and variability of Potato virus Y. Phytopathology 102(S):97–97

Rowley JS, Gray SM, Karasev AV (2015) Screening potato cultivars for new sources of resistance to Potato virus Y. Am J Potato Res 92:38–48

Rotenberg D, Jacobson AL, Schneweis DJ, Whitfield AE (2015) Thrips transmission of tospoviruses. Curr Opin Virol 15:80–89. https://doi.org/10.1016/j.coviro.2015.08.003

Saha A, Saha B, Saha D (2014) Molecular detection and partial characterization of a begomovirus causing leaf curl disease of potato in sub-Himalayan West Bengal, India. J Environ Biol 35:601–606

Sahi G, Hedley PE, Morris J, Loake GJ, MacFarlane SA (2016) Molecular and biochemical examination of spraing disease in potato tuber in response to Tobacco rattle virus infection. Mol Plant-Microbe Interact 29(10):822–828

Salazar L, Muller G, Querci M, Zapata J, Owens R (2000) *Potato yellow vein virus*: its host range, distribution in South America and identification as a Crinivirus transmitted by Trialeurodes vaporariorum. Ann Appl Biol 137(1):7–19

Salvalaggio AE, Lambertini PML, Cendoya G, Huarte MA (2017) Temporal and spatial dynamics of Tomato spotted wilt virus and its vector in a potato crop in Argentina. Ann. Appl. Biol. 171:5–14. https://doi.org/10.1111/aab.12357

Santala J, Valkonen JPT (2018) Sensitivity of small RNA–based detection of plant viruses. Front Microbiol 9:939. https://doi.org/10.3389/fmicb.2018.00939

Santala J, Samuilova O, Hannukkala A, Latvala S, Kortemaa H, Beuch U, Kvarnheden A, Persson P, Topp K, Ørstad K, Spetz C, Nielsen SL, Kirk HG, Uth JG, Budziszewska M, Wieczorek P, Obrepalska-Steplowska A, Pospieszny H, Kryszczuk A, Sztangret-Wisniewska J, Yin Z, Chrzanowska M, Zimnoch-Guzowska E, Jackeviciene E, Taluntytė L, Pūpola N, Mihailova J, Lielmane I, Järvekülg L, Kotkas K, Rogozina E, Sozonov A, Tikhonovich I, Horn P, Broer I, Kuusiene S, Staniulis J, Adam G, Valkonen JPT (2010) Detection, distribution and control of Potato mop-top virus, a soil-borne virus, in northern Europe. Ann Appl Biol 157:163–178

Santillan FW, Fribourg CE, Adams IP, Gibbs AJ, Boonham N, Kehoe MA, Maina S, Jones RAC (2018) The biology and phylogenetics of potato virus S isolates from the Andean region of South America. Plant Dis 102:869–885. https://doi.org/10.1094/PDIS-09-17-1414-RE

Schulte-Geldermann E, Gildemacher PR, Struik PC (2012) Improving seed health and seed performance by positive selection in three Kenyan potato varieties. Am J Potato Res 89(6):429–437

Sharma S, Kang SS, Sharma A, Kaur S (2016) Mixed infection by cucumber mosaic virus and potato virus x in potato with yellow mosaic in India. J Plant Pathol 98:693

da Silva GO, Carvalho ADF, Ponijaleki RS, Bortoletto AC (2015) Desempenho de cultivares de batata para produtividade de tubérculos. Batata Show 42:21–23

Singh BP, Raigond B, Sridhar J, Jeevalatha A, Ravinder K, Venkateswarlu V, Sharma S (2014) Potato seed production systems in India. Conference: national seminar on emerging problems of potato, 1–2 Nov 2014. ICAR-CPRI, Shimla

Sohrab SS, Karim S, Varma A, Abuzenadah AM, Chaudhary AG, Mandal B (2013) Role of sponge gourd in apical leaf curl disease of potato in Northern India. Phytoparasitica 41:403–410

Souza-Dias JAC, Eiras M, Fernandes CRF, Beloni C, Charkowski A (2017) Occurrence of spindle tuber like malformation in seed-potato lots associated with a somatic mutation in tissue culture. In: 101st-Ann Meeting-Potato American Association, Fargo, ND. Book of Abstracts, 78

Souza-Dias JAC, Iamauti MT, Fischer IH (2016a) Doenças da Batateira, Capt 16. In: Amorin L, Rezende JAM, Bergamin Fo A, Camargo LEA (eds) Manual de Fitopatologia v.2. Editora Agron Ceres Ltda, Ouro Fino, pp 125, 772p–147

Souza-Dias JAC, Jefrries C, Rezende JAM, Lima MF (2013) A new virus threat to seed-potato certification in Brazil: the whitefly-transmitted tomato chlorosis virus (Genus: Crinivirus). In: Gaba V, Tsror L (eds) EAPR-pathology section meeting. EAPR, Jerusalem, p 45. http://www.agri.gov.il/download/files/Caram_Souza_Dias_1.pdf

Souza-Dias JAC, Menarim E, Rentz R, Kitajima EW, Sawasaki HE, Duarte M (2016b) Alerta Necessário. Cultivar HF, Junho/Julho ano XIV 98:16–19

Souza Dias, JAC, Trevisan-junior O, Fernandes CRF, Oliveira-Preto TF, Trevisan LH (2018) The sprout/seed-potato technology: establishing as an example of resource usage and alternative income to citrus nurseries, in Brazil. In: 102nd Potato American Association Meeting (PAA 2018) Boise, ID. Book of Proceedings & Abstracts, vol 120, pp 72–73

Souza Richards R, Adams IP, Kreuze JF, De Souza J, Cuellar W, Dulleman AM, RAA VDV, Glover R, Hany U, Dickinson M, Boonham N (2013) The complete genome sequence of two isolates of potato black ringspot virus and its relationship to other isolates and nepoviruses. Arch Virol 159:811–815. https://doi.org/10.1007/s00705-013-1871-8

Taylor M (2018) Routes to genetic gain in potato. Nat Plants 4:631–632

Tenorio J, Chuquillanqui C, Garcia A, Guillen M, Chavez R, Salazar LF (2003) Symptomatology and effect on potato yield of achaparramieto rugoso. Fitopatologia 38(1):32–36

Thomas-Sharma S, Abdurahman A, Ali S, Andrade-Piedra JL, Bao S, Charkowski AO, Crook D, Kadian M, Kromann P, Struik PC, Torrance L (2016) Seed degeneration in potato: the need for an integrated seed health strategy to mitigate the problem in developing countries. Plant Pathol 65(1):3–16

Thomas-Sharma S, Andrade-Piedra J, Carvajal Yepes M, Hernandez Nopsa JF, Jeger MJ, Jones RAC, Kromann P, Legg JP, Yuen J, Forbes GA, Garrett KA (2017) A risk assessment framework for seed degeneration: Informing an integrated seed health strategy for vegetatively propagated crops. Phytopathology 107(10):1123–1135

Tian YP, Valkonen JPT (2013) Genetic determinants of Potato virus Y required to overcome or trigger hypersensitive resistance to PVY strain group O controlled by the gene Ny in potato. Mol Plant-Microbe Interact 26:297–305

Treder K, Chołuj J, Zacharzewska B, Babujee L, Mielczarek M, Burzyński A, Rakotondrafara AM (2018) Optimization of a magnetic capture RT-LAMP assay for fast and real-time detection of potato virus Y and differentiation of N and O serotypes. Arch Virol 163(2):447–458

Valkonen JPT (2007) Viruses: economical losses and biotechnological potential. In: Potato biology and biotechnology. Elsevier Science BV, pp 619–641

Villela JA, Pap T, Salas FJS (2017) Levantamento de viroses presentes em batatas-semente da produção nacional e Importadas. Batata Show 49:20–24

Virmond EP, Kawakami J, Souza-Dias JAC (2017) Seed-potato production through sprouts and field multiplication and cultivar performance in organic system. Hortic Bras 35:335–342. https://doi.org/10.1590/s0102-053620170304

Wang QC, Valkonen JPT (2009) Cryotherapy of shoot tips: novel pathogen eradication method. Trends Plant Sci 14:119–122

Wilson CR (2001) Resistance to infection and translocation of Tomato spotted wilt virus in potatoes. Plant Pathol 50:402–410

Chapter 12
Potato Seed Systems

Gregory A. Forbes, Amy Charkowski, Jorge Andrade-Piedra, Monica L. Parker, and Elmar Schulte-Geldermann

Abstract Good quality seed is almost universally considered a requirement for high productivity in all potato production systems. Much of the yield gap currently constraining productivity in low-income countries is attributed to the poor quality of seed. Potato seed sector development is thus a major concern of governments, researchers, development agencies, and civil society organizations. Potato seed systems are often characterized as formal or informal, although the informal seed system is complex and particularly in low income countries there are many linkages between the two systems. Informal seed potato systems in the Andes have existed for centuries, and for a number of reasons often produce seed of relatively high quality. In other low-income countries, informal systems produce seed of variable and frequently poor quality, contributing to very large yield gaps, characteristic of those areas. In regions of high potato productivity (e.g., the USA and Europe), formal systems, with seed of certified high quality, are dominant, although some productions subsectors (e.g., organic producers) often use seed that is not certified. Efforts to implement formal seed systems in low-income countries have been largely unsuccessful; consequently the vast majority of low-resource potato farmers source their seed via the informal system. Sectors of the development community

G. A. Forbes
Independent Consultant, Servas, Gard, France

A. Charkowski
Colorado State University, Fort Collins, CO, USA
e-mail: amy.charkowski@colostate.edu

J. Andrade-Piedra (✉)
International Potato Center, Lima, Peru
e-mail: j.andrade@cgiar.org

M. L. Parker
International Potato Center, Nairobi, Kenya
e-mail: m.parker@cgiar.org

E. Schulte-Geldermann
International Potato Center, Nairobi, Kenya

TH Bingen, University of Applied Sciences, Bingen, Germany
e-mail: e.schulte-geldermann@th-bingen.de

H. Campos, O. Ortiz (eds.), *The Potato Crop*,
https://doi.org/10.1007/978-3-030-28683-5_12

are pushing for alternative solutions, which generally involve some form of integrating formal and informal seed systems or semi-formal systems such as quality declared seed, and a policy structure that preserves farmers' rights to save and trade seed. Given the role that seed quality is currently playing in the low yields of potato in low-income countries, which is not the case in wealthier parts of the world, the review focuses primarily on seed sector development in resource-poor areas.

12.1 The Seed System Context

Seed system research and lexicon In keeping with the large social and economic dimensions of potato seed systems (discussed in more depth later in the article), research on seed systems is extensive and multifaceted, and terminology used to describe seed-related issues is varied and often confusing. Thiele (1999) broadly defined a seed system as "an interrelated set of components including breeding, management, replacement and distribution of seed." This definition is generally consistent with one established earlier at a workshop held in 1995 in Indonesia, "the total of physical, organizational and institutional components, their actions and interactions, that determine seed supply and use, in quantitative and qualitative terms" (Amstel et al. 1996), and with other definitions given since then (Camargo et al. 2004; Muthoni et al. 2010; Kromann et al. 2016). In recent years, patents and plant variety protection have added the additional dimension of intellectual property and germplasm ownership, and these impact plant breeding, crop management, seed replacement, and distribution of seed.

Seed systems have also been classified by type, with the major classes being formal and informal (Thiele 1999). The concept of a formal seed system is relatively clear, being characterized by components that are regulated by the public sector, usually by an inspection process known as "certification" and including controls over variety release, to ensure that available seed is of a recognized variety and with a low incidence of disease (Louwaars 1994; Amstel et al. 1996; Thiele 1999). The informal system is complex and conceptually less clear in that it basically includes all that is not formal, including self-saved seed, seed traded among farmers, and that acquired at local markets (Thiele 1999; Almekinders and Louwaars 2002; McGuire and Sperling 2016). While all seed outside the formal system is frequently referred to as "informal," it is also referred to using other terms, such as "farmers' seed" (Almekinders and Louwaars 2002), "local seed" (Almekinders et al. 1994), or "traditional seed" (McGuire and Sperling 2016). For this chapter, we will use the term "informal."

Several authors have highlighted the importance of the informal seed system in middle-low and low-income countries (hereafter referred to as low- income, Thiele 1999; Almekinders and Louwaars 2002; Louwaars and de Boef 2012; McGuire and Sperling 2016), including a particular focus on potato (Thiele 1999; Thomas-Sharma et al. 2015). The literature is also strongly supportive of the need to integrate formal and informal sectors in countries where the former provides only a small portion of seed that is needed (Louwaars 1994; Amstel et al. 1996; Thiele 1999; Louwaars and de Boef 2012; Thomas-Sharma et al. 2015; De Jonge and Munyi 2017). The strategy of integrating different seed systems has had practical implications in the development sector with reports on specific cases (e.g., Kromann et al. 2016) and through the development of programs focusing on an integrated approach to seed system development.

In spite of broad recognition of the role of the informal seed system by the academic community, seed sector development has reflected "varied and often opposing philosophies" (McGuire and Sperling 2016). Many development projects have been designed only (or primarily) to support the formal seed sector in low-income countries and have relied on outside expertise, without significantly increasing the minimal role that certified seed plays in providing planting material to low-income farmers (McGuire and Sperling 2016); this is particularly true for potato seed systems (Thomas-Sharma et al. 2015). Even in countries with strong formal seed systems, the formal seed system can fail farmers who do not grow the crop as a commodity. For example, organic farmers in the US commonly use informal seed for specialty and heirloom varieties and this is yet more evidence that formal seed systems only work for growers who specialize in large acreages of potato.

Why are potato seed systems important? Healthy seed systems have been described as providing access to quality planting material, at the time needed, at a fair price, to all who need it (Sperling 2008). Access to quality seed is widely considered one of the main requirements for bridging large yields gaps for potato still found in most low-income countries (Hidalgo et al. 2009; Schulte-Geldermann 2013). Healthy seed systems also act as to reduce risk of disease outbreaks by keeping spread of a disease in check or even as part of a pest eradication plan. Conversely, seed systems without effective quality control can be very efficient at spreading seed-borne pathogens. Seed systems are also important for the diffusion of new varieties with beneficial traits and the maintenance of crop diversity in the landscape (Pautasso et al. 2013; Arce et al. 2018). In the case of a new or emerging pathogen in a region, the seed system acts as the conduit through which locally adapted resistant varieties can be distributed (if these are available).

Arguably, the primary impetus for development of seed systems in potato is the vegetative nature of propagation of the crop, and the phenomenon of what is now referred to as degeneration (Fig. 12.1). The importance and causes of seed degeneration, a process through which yield is lost in vegetatively propagated crops through pathogen accumulation in consecutive cycles of propagation, are of particular concern to informal seed systems. Globally, seed degeneration is among the leading limitations to potato yield (Thomas-Sharma et al. 2015; Bertschinger et al. 2017). In high-income countries, which have the highest potato yields, this problem has been effectively managed, at least for large commercial growers, through the utilization of seed certified to have high quality (low incidence of pathogens, varietal purity, and appropriate physiological age). This process has been highly successful for most producers in these countries by providing access to economically priced seed of high quality.

Fig. 12.1 Degeneration in potato. Small plants with low yield due to the accumulation of pathogens in consecutive cycles of vegetative propagation. (Photo credits: G. Forbes)

For smallholder farmers from low- to middle-low-income countries, certified seed is often not available, or the cost is prohibitive. Instead, farmers acquire seed of unknown quality via the informal system, either from the previous year's crop, or from other informal sources such as those mentioned above. In informal systems, degeneration is often a problem because seed is not tested and may be produced under conditions of high disease pressure with little or no quality control.

12.2 Traditional Potato Seed Systems in the Andes

The Andean region is the origin of the cultivated potato and represents an interesting case for studying potato seed systems (Fig. 12.2). Potato seed systems in the Andes have been informal for millennia, and even today only small amounts of formal seeds are used by Andean farmers (Hidalgo et al. 2009; Devaux et al. 2014). With even relatively low rates of disease spread, one could assume that high levels of degeneration would occur in areas where the informal system has been dominant for centuries. However, studies done in traditional Andean potato seed systems over the past 30 years often found relatively low frequencies of tubers infected with yield-limiting potato viruses (Bertschinger et al. 1990; Fankhauser 1999; Pérez et al. 2015; Navarrete et al. 2017).

Several factors have been identified that could contribute to the continued quality of seed potato in the informal systems in the Andes. Andean farmers have

Fig. 12.2 Native varieties in the Andes. Seed for producing most of these varieties come from informal systems, but major yield-limiting potato viruses, which are the most common cause of seed degeneration and are often found in relatively low frequencies. (Photo credits: J.L. Gonterre, in association with the International Potato Center)

traditionally had complex farming practices that conceivably help reduce the soil-borne phases of diseases leading to degeneration, such as sectoral (Orlove and Godoy 1986) or other types of fallowing and rotation (Thurston 1990), high hilling, or reduced cultivation methods (e.g., Cartagena et al. 2004). Other factors characteristic of Andean potato systems may contribute to reduce disease spread among plants or pathogen transmission within plants. Resistance to PVY and PLRV have been found in *S. tuberosum* subsp. *andigena,* and *Ry* genes have been found in other Andean taxa making up local potato landraces (Machida-Hirano 2015), which may partly explain low incidences of PVY and PLRV found in these varieties. Bertschinger et al. (2017) also found that virus transmission from infected mother tubers to daughter tubers was greatly suppressed at high altitudes in the Andes. This is consistent with traditional practices in which farmers moved virus-infected seed to higher altitudes to reduce infection (Thiele 1999). High levels of agrobiodiversity may also help mitigate degeneration in traditional Andean potato fields. One study found that Peruvian farmers growing native "floury" cultivars between 3500 and 4250 m altitude had on average between 8 and 20 different genotypes per field (de Haan et al. 2010). This biodiversity is "uneven" in that it is highly dependent on the type of farmer, but it is also an important component of a complex seed exchange network that represents "a strong safety net through which smallholders can respond to crop failure and seed stress" (Arce et al. 2018).

12.3 Potato Seed Systems in Europe and North America

The basic outline of the seed system used in Europe and North America for potato was developed in the late nineteenth and early twentieth centuries and has its roots in how seed potatoes are grown in the Andes (Shepard and Claftin 1975; Frost et al. 2013). Considerable advances were made in the 1980s, when both pathogen testing and potato micropropagation became widespread. Despite the use and availability of technology, seed potatoes are produced primarily in the northern agricultural regions of these two continents to avoid insect virus vectors, and a wide range of bacterial and fungal diseases common in warmer climates. In Europe and North America, commercial growers who plant large acreages of potato almost exclusively use certified seed. However, in the United States, farmers who manage mixed vegetable farms, and particularly organic farmers, generally use informal seed, demonstrating that current seed systems tend to best serve growers who produce potato as a commodity.

In these seed systems, potato varieties are maintained in tissue culture as micropropagated plantlets. These initial plantlets, often called mother plants, are tested for all pathogens of concern, including the major potato viruses, potato spindle tuber viroid, and common bacterial and fungal pathogens (Frost et al. 2013). Propagation in tissue culture is relatively inexpensive, requires little space, and the plantlets grow quickly, so hundreds of thousands of plantlets can be produced annually in a relatively small facility of tens to a few hundred square meters (Naik and Buckseth 2018). The micropropagated plantlets are then planted into greenhouses or screenhouses into either pots or into hydroponic or aeroponic systems. The potatoes harvested from these greenhouse-grown plants are called nuclear seed or minitubers. The minitubers must be stored until the subsequent season to break tuber dormancy. Minitubers are planted into seed potato fields and the progeny from these plants are generally field-multiplied another 2–5 years before being sold to farmers who grow potatoes for fresh use or processing.

At each stage, specific inspections and pathogen tests are required for the certification schemes used in each country, state, or province where potatoes are grown. Generally accepted protocols are collected and verified by entities such as the European and Mediterranean Plant Protection Organization (EPPO), the North American Plant Protection Organization (NAPPO), and the United Nations Economic Commission for Europe (UNECE) to aid in trade in certified seed potato. The efforts have had some success, with increases in yield and near-elimination of diseases such as spindle tuber and bacterial ring rot (Frost et al. 2013). Policy harmonization has also engendered debate in low-income countries (De Jonge and Munyi 2017) and in Europe (Prip and Fauchald 2016). In the United States, seed potato producing states still have relatively little similarity in their certification regulations across the different states. Currently, potato virus Y is the most important potato virus on these continents, but losses due to this virus are relatively small (Zeng et al. 2018).

These certification systems primarily focus on diseases that are only spread by seed potatoes and not on important soil-borne pathogens. As a result, diseases such as powdery scab, corky ringspot, golden cyst nematode, and other difficult-to-control diseases are now widespread and increasing in importance (e.g., Beuch et al. 2014; Contina et al. 2018). A second challenge is that plant variety protection has resulted in the proliferation of similar, but protected varieties, for which little information on disease response is available. This poses a significant challenge to certification agencies, which are tasked with insuring varietal purity and disease thresholds on an ever-increasing number of new potato varieties. Finally, growers in both Europe and North America are investing in the development of inbred and hybrid diploid varieties that can be produced through true seed rather than plant micropropagation (Lindhout et al. 2011; Jansky et al. 2016). If these varieties become popular, they will alter the current seed system.

12.4 Potato Seed Systems in Low-Income Countries

Earlier we stated that low-income countries are characterized by informal seed systems with very low use of certified seed. This is generally the case, but it is worth examining in some detail efforts that have gone into establishment of certified seed programs in some of these countries, as well as a number of recent innovations aimed at improving seed systems of resource-poor farmers.

The highly conspicuous absence of certified seed in the potato seed sector in most low-income countries has recently been documented, at least in part. In specific reference to potato, Thomas-Sharma et al. (2015) list percentages of formal and informal seed in 14 low- income countries. In China and India formal seed usage is listed at 20%, but in all other countries it is below 10% and in most it is below 5%. McGuire and Sperling (2016) provide a more extensive examination of how farmers source many kinds of seed in low-income countries and note that for potato over 95% of seed comes from own stock, friends, neighbors, relatives, or local markets, i.e., the informal system (Fig. 12.3). It is worth noting for context that

Fig. 12.3 Informal seed potato in a local market in Bangladesh. (Photo credits: J. Andrade-Piedra)

these authors also show that in low-income countries there is a similar pattern for all vegetatively propagated crops and, somewhat surprisingly, for legumes and cereals as well.

The lack of certified seed for potato and other crops in low-income countries is not easily attributed to a lack of effort on the part of governments and development agencies. McGuire and Sperling (2016) provide an impressive list of projects funded by the World Bank and by the Alliance for a Green Revolution in Africa (AGRA) as an indication of development support to seed systems in low-income countries. In the preparation of this document we were not able to find data on investments specifically in the potato seed sector in developing countries, but there is no doubt that many millions of dollars have been spent by development agencies over the last half a century to improve the potato seed sector in low-income countries.

Seed sector actors, and specifically donors, in low-income countries take different approaches to the problem, which may be generally classified into two types: those that predominantly support development of a formal seed sector and those that support a broader approach to seed sector development (McGuire and Sperling 2011; Thomas-Sharma et al. 2015; Otieno et al. 2017). While a number of donors subscribe to broad seed sector development (Lossau et al. 2000), it is the experience of the authors that the large majority of projects, and certainly the larger projects supporting potato seed sector development, tend to focus primarily on formal seed sector development.

12.5 Perspectives on Potato Seed System Development

Policy Low-income countries are struggling with numerous policy issues related to seed. Many governments and regional organizations are developing policies and laws modeled on the guidelines of the International Union for the Protection of New Varieties of Plants (UPOV). This has led to much concern in civil society and in the research community of the impacts that such policies could have on resource-poor farmers and informal seed trade (Tripp and Louwaars 1997; De Jonge and Munyi 2017; Otieno et al. 2017; Vernooy 2017). The controversies surrounding these policies have given rise to both proponents and opponents of regional harmonization laws based on UPOV standards; opposing actors apparently rarely meet to debate options (De Jonge and Munyi 2017).

It is unclear what the eventual effects of this struggle will have on resource-poor potato farmers in these countries. Because of its vegetative nature, perishability, and bulkiness, seed potato (namely tubers whereas potato seed refers to true, botanical seed) presents particular difficulties for establishing breeding programs, implementing certified seed systems and marketing seed in a way that is commercially viable in low-income countries where the infrastructure and other elements of the business ecosystem are not favorable. This could be a major reason why there has been very little activity of major seed potato companies in low-income countries (Thomas-Sharma et al. 2015).

Funding for seed sector development represents an area where seed policy and seed sector development philosophy can affect resource-poor potato farmers. As noted, most funding in potato seed sector development over the years has been in support of the formal sector, with relatively little funding to optimize and promote on-farm seed management, which has been shown to be effective in slowing down or even reversing seed degeneration important in areas without access to certified seed (see below).

Underlying the struggle over seed regulation in low-income countries is the contrast, often seen as a dichotomy, between formal and informal seed systems. As noted, integration of seed systems (Louwaars and de Boef 2012; Kromann et al. 2016; Ferrari et al. 2017) has been proposed as a way to find common ground between those promoting commercial seed industry, plant variety protection, and harmonization of seed standards, and those promoting farmers rights to save and trade their own seed.

Although it was not intended as a mechanism to integrate seed systems, the quality declared seed (QDS) approach offers a more flexible alternative for seed quality assurance than strict certification programs. Developed by FAO (2006) and later adapted for potato and other vegetatively propagated crops (Fajardo et al. 2010), the QDS approach is being used in Ethiopia (Schulz et al. 2013) and some elements of it are applied in Ecuador (Kromann et al. 2016) and Peru (MINAGRI 2018). Seed potato produced under a QDS approach was shown to be a profitable business for seed multipliers in Kenya, but at the same time it has been ineffective in limiting the dissemination of bacterial wilt (*Ralstonia solanacearum*) and potato cyst nematode (*Globodera pallida*), which points out the need of rigorous testing and validation of the QDS approach to local conditions. However, formal seed systems also are unable to effectively limit the dissemination of pathogens such as *R. solanacearum* and *G. pallida*, so the QDS approach is not deficient compared to formal seed systems in this respect.

Technology Innovations A number of the technological innovations have been or are being evaluated and promoted to improve seed potato quality in low-income countries. Many of these are relatively old but have been recently revisited, and often adapted, for their application to certain situations, particularly where resources are scarce. Some of these technologies are reviewed here under two categories: those that relate to on-farm management of seed, and those relating to rapid multiplication of early generation seed.

On-Farm Seed Management Some relatively old technologies are receiving renewed consideration by seed specialists. Positive selection is implemented by farmers and consists of identifying and marking plants that have no visible symptoms of disease or abiotic stress. Seeds for the next planting are then taken from these plants. This sounds relatively simple and the activity itself is simple, however, the efficacy of selection can depend on many factors including, the type of virus, environmental conditions, and farmer skill. Positive selection is particularly important in areas where there is no access to seed produced under a quality control system, thus the impetus is on the farmer to manage quality control. Nonetheless, a number of studies have demonstrated significant improvements in seed quality at the farm level as a result of positive selection implemented by farmers (Gildemacher et al. 2011; Schulte-Geldermann et al. 2012; Okeyo et al. 2018; Priegnitz et al. 2018), which could be due to the fact that it is easier to identify a fully healthy plant than it is to identify symptoms potentially caused by virus or that could also be due to abiotic stresses (Gildemacher et al. 2011). When positive selection is used by farmers there is an increase in yield that could be attributed to several factors, one of these being a reduction in the incidence of virus infection (Schulte-Geldermann et al. 2012). To improve the utility of this relatively old approach, CIP and its partners developed a number of training guides for positive selection aimed at both farmers and trainers (Gildemacher et al. 2007).

Another old tactic that has received some renewed attention is the seed plot technique, which consists of producing small amounts seed of relatively high quality in a confined area that is free of or has a low incidence of soil-borne pathogens (Vashisth 1979; Bryan 1983; Kakuhenzire et al. 2005; Ali et al. 2013). There are many variations on this very simple principle that can be applied to devise flexible systems that adapt to different contexts. The initial seed may be purchased or may be derived from positive selection (Bryan 1983). The best seed coming from the seed plot, i.e., that produced with positive selection in the seed plot and further post-harvest selection, can be used for a new seed plot in the next season. The remaining seed from the seed plot is used for ware production. The seed plot technique can very easily be integrated with the purchase of small amounts of high-priced certified seed (Kinyua et al. 2015; Ochieng-Obura et al. 2016).

Rapid multiplication technologies Aeroponics is a more recent technology that has made inroads into the potato sector in the last few decades. This technique consists of a soilless culture, in which the underground part of the plant is enclosed in a dark chamber and supplied with nutrients through a misting system. Plants grown in this way produce minitubers in the dark chamber, which are harvested as they reach the desired size. Within an aeroponics system, plants may produce very high numbers of minitubers, with plants on average sometimes producing over 45 tubers (Mateus-Rodriguez et al. 2013). There is no shortage of research (and opinion) expounding the benefits of aeroponics (Muthoni et al. 2011; Chiipanthenga et al. 2012; Kakuhenzire et al. 2017; Lakhiar et al. 2018), but implementation of aeroponics in low-income countries where resources are scarce and power supplies unreliable is costly, difficult and risky (Mateus-Rodriguez et al. 2013). The introduction of aeroponics in sub-Saharan Africa has resulted in a large increase minituber production (Harahagazwe et al. 2018), although total numbers are still low and minitubers are expensive.

Unfortunately, feasibility and even economic analyses of rapid multiplication technologies used in development projects generally do not consider the role of the development community's purchase of seed, which is often a market-distorting factor in fledgling seed programs in low-income countries (Bentley and Vasques 1998; Bentley et al. 2001). Aeroponics is just one form of hydroponic plant production and other forms are also used in seed potato production (Lommen 2007; Corrêa et al. 2009), many of which are simpler than aeroponics and may be more appropriate for many low-income countries (Mateus-Rodriguez et al. 2013). For example, nutrient film hydroponic systems are widespread in North America and also present in Brazil and China. These systems are relatively easy to manage and yield and cost of production information is available (Guenthner et al. 2014). The appropriateness of a complex technology like aeroponics appears to depend on the capacity of local players and local infrastructure (Mateus-Rodriguez et al. 2013), hence sand hydroponics not relying on power and highly skilled labor is an attractive alternative.

Rapidly growing young vegetative tissue of potato can be cut and rooted in a number of ways (Bryan et al. 1981). Apical rooted cuttings have long been used in SE Asia (Vander Zaag and Escobar 1990), and particularly in Vietnam (Tran et al. 1990). This technique is being introduced into sub-Saharan Africa to provide a simple but effective technique for multiplying early generation seed (Parker et al. 2019) (Fig. 12.4). In the current application used in sub-Saharan Africa, two-node apical cuttings (4–5 cm long) are harvested several times at intervals of 2–3 weeks from in vitro-derived mother plants. The cuttings are then rooted in trays with a substrate of coconut sawdust, clean subsoil, and sterilized decomposed manure. Once rooted, the cuttings can be transplanted directly to the field to produce the first generation of tubers.

Cuttings are penetrating the seed system, and opportunities they present are being validated in Kenya to scale out the technology through diversified use (Fig. 12.5).

Strategic innovations Given the contrasting approaches of seed sector development actors (McGuire and Sperling 2016), it is not surprising that many people interested in this subject have called for greater integration, both in development and research. Thomas-Sharma et al. (2015) proposed an integrated approach to managing the problem of degeneration through a strategy called integrated seed

Fig. 12.4 Cutting almost ready for transplanting (left), and soon after transplanted in nursery beds in the field (right). (Photo credits: M. Parker)

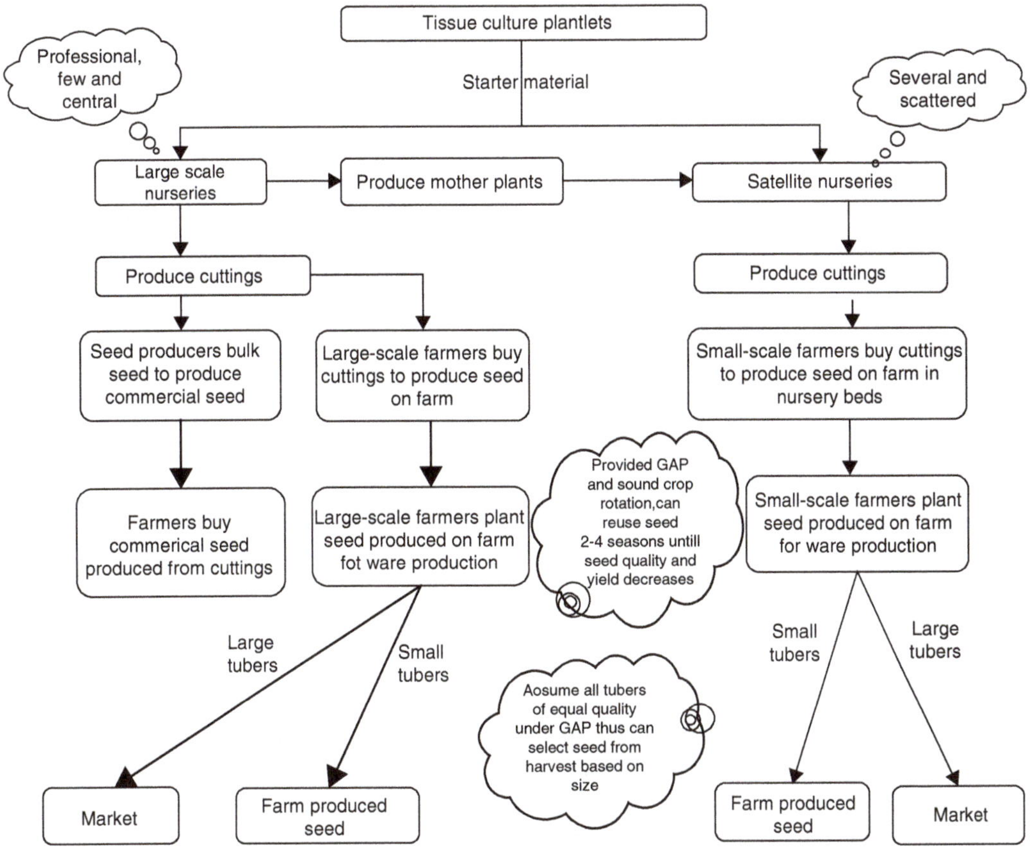

Fig. 12.5 Diversified pathways to use cuttings in seed production to scale out the technology; GAP: good agricultural practices

health. This involves the integration of three different classes of tactics farmers may employ to maintain or even improve seed quality: (1) on-farm practices such as seed plot technique or positive selection; (2) the use of varieties that degenerate slowly due to their natural resistance to degeneration-causing pathogens; and (3) a more strategic use of certified seed. The latter may involve the less frequent purchase of clean replacement seed, or purchase of small quantities that may be put in a designated seed plot (Ochieng-Obura et al. 2016).

At a higher level, researchers have also called for integration throughout the seed sector, with emphasis on the interaction between formal and informal systems (Louwaars 1994; Tripp 1996; Munyi and De Jonge 2015). The most visible incarnation of this approach is the Integrated Seed Systems Development (ISSD) program.[1] This program is managed globally by the Centre for Development Innovation (CDI), Wageningen University & Research (WUR) and the Royal Tropical Institute (KIT), but has many local partners in Africa and country programs in Uganda and Ethiopia.

The CGIAR has long been another major player in seed sector development in low-income countries, with CIP leading the potato component. For many years seed

[1] See ISSD Website http://www.issdseed.org/.

programs resided within specific centers but this has been consolidated to some extent within the CGIAR Research Programs. All potato work resides within the Roots, Tubers and Banana program (RTB), which initiated a project in 2012 to address biophysical (especially seed degeneration) and socioeconomic constraints of seed production in RTB crops. This led to fruitful collaborations with advanced research institutions in the US and Europe and has produced some novel approaches to studying seed systems, including a conceptual framework for intervening in RTB seed systems (Bentley et al. 2018), a multi-crop analysis (Almekinders et al. 2019), the integrated seed heath approach explained above (Thomas-Sharma et al. 2015), epidemiological modeling of seed potato degeneration (Thomas-Sharma et al. 2017), and the geographic analysis of seed system dynamics using network analysis (Buddenhagen et al. 2017). Hence, since seed systems are complex, more research is needed to identify the right entry points and multiple angles for innovations to enhance the systems as a whole and according to local conditions.

References

Ali S, Kadian MS, Ortiz O, Singh BP, Chandla VK, Akhtar M (2013) Degeneration of potato seed in Meghalaya and Nagaland states in north-eastern hills of India. Potato J 40:122–127

Almekinders CJ, Louwaars NP (2002) The importance of the farmers' seed systems in a functional national seed sector. J New Seeds 4:15–33

Almekinders CJM, Louwaars NP, De Bruijn GH (1994) Local seed systems and their importance for an improved seed supply in developing countries. Euphytica 78:207–216

Almekinders CJM, Walsh S, Jacobsen KS, Andrade-Piedra JL, McEwan M, De Haan S, Kumar L, Staver C (eds) (2019) Why interventions in the seed systems of roots, tubers and bananas crops do not reach their full potential. Food Security. First Online 23 Jan 2019, 20 p. ISSN 1876-4525

Amstel H van, Bottema JW, Sidik M, Santen CE van (1996) Integrating seed systems for annual food crops. Proceedings of a workshop held in Malang, Indonesia October 24-27, 1995. Regional Co-ordination Center for Research and Development of Coarse Grains, Pulses, Roots and Tuber Crops in the Humid Tropics of Asia and the Pacific, Research Institute for Legume and Tuber Crops, Agronomy Society of Indonesia

Arce A, De Haan S, Burra DD, Ccanto R (2018) Unearthing unevenness of potato seed networks in the high Andes: a comparison of distinct cultivar groups and farmer types following seasons with and without acute stress. Front Sustain Food Syst 2:43

Bentley JW, Vasques D (1998) The seed potato system in Bolivia: organisational growth and missing links. ODI, London

Bentley JW, Tripp R, De La Flor RD (2001) Liberalization of Peru's formal seed sector. Agric Hum Values 18:319–331

Bentley JW, Andrade-Piedra JL, Demo P, Dzomeku B, Jacobsen K, Kikulwe E, Kromann P, Lava Kumar P, McEwan M, Mudege N, Ogero K, Okechukwu R, Orrego R, Ospina B, Sperling L, Walsh S, Thiele G (2018) Understanding root, tuber, and banana seed systems and coordination breakdown: a multi-stakeholder framework. J Crop Improv. Published online 18 Jun 2018, 23 p. ISSN 1542-7528

Bertschinger L, Scheidegger UC, Luther K, Pinillos O, Hidalgo A (1990) La incidencia de virus de papa en cultivares nativos y mejorados en la sierra peruana. Revista Latinoamericana de la Papa 3:62–79

Bertschinger L, Bühler L, Dupuis B et al (2017) Incomplete infection of secondarily infected potato plants—an environment dependent underestimated mechanism in plant virology. Front Plant Sci 8:74

Beuch U, Persson P, Edin E, Kvarnheden A (2014) Necrotic diseases caused by viruses in Swedish potato tubers. Plant Pathol 63:667–674

Bryan JE (1983) On farm seed improvement by the potato seed plot technique. Technical information bulletin 7. International Potato Center, Lima

Bryan J, Jackson M, Meléndez N (1981) Rapid multiplication techniques for potatoes. International Potato Center (CIP), Lima

Buddenhagen CE, Hernandez Nopsa JF, Andersen KF, Andrade-Piedra J, Forbes GA, Kromann P et al (2017) Epidemic network analysis for mitigation of invasive pathogens in seed systems: potato in Ecuador. Phytopathology 107:1209–1218

Camargo CP, Bragantini C, Monares A (2004) Small-scale seed production systems: a non-conventional approach. In: Seed multiplication by resource-limited farmers. Proceedings of the Latin American workshop. Goiânia, Brazil, 7–11 April 2003. FAO plant production and protection paper, vol 180, pp 63–66

Cartagena Y, Toapanta G, Valverde F (2004) Mas papas con huacho rozado. Instituto Nacional Autónomo de Investigaciones Agropecuarias (INIAP), Porteviejo

Chiipanthenga M, Maliro M, Demo P, Njoloma J (2012) Potential of aeroponics system in the production of quality potato (Solanum tuberosum l.) seed in developing countries. Afr J Biotechnol 11:3993–3999

Contina JB, Dandurand LM, Knudsen GR (2018) A spatial analysis of the potato cyst nematode Globodera pallida in Idaho. Phytopathology 108:988–1001

Corrêa RM, Pinto J, Faquin V, Pinto C, Reis ES (2009) The production of seed potatoes by hydroponic methods in Brazil. Fruit Veg Cereal Sci Biotechnol 3:133–139

de Haan S, Núñez J, Bonierbale M, Ghislain M (2010) Multilevel agrobiodiversity and conservation of Andean potatoes in central Peru: species, morphological, genetic, and spatial diversity. Mt Res Dev 30:222–231

De Jonge B, Munyi P (2017) Creating space for 'informal'seed systems in a plant variety protection system that is based on UPOV 1991. Integrated Seed Sector Development (ISSD) Africa

Devaux A, Kromann P, Ortiz O (2014) Potatoes for sustainable global food security. Potato Res 57:185–199

Fajardo J, Lutaladio N, Larinde M, Rosell C, Barker I, Roca W, Chujoy E (2010) Quality declared planting material: protocols and standards for vegetatively propagated crops. Expert consultation, Lima, 27–29 November 2007. Food and Agriculture Organization of the United Nations, Rome

Fankhauser C (1999) Main diseases affecting seed degeneration in Ecuador: new perspectives for seed production in the Andes. European Association for Potato Research, Italy, pp 50–51

FAO (2006) Quality declared seed system. FAO Plant protection and protection paper 185. FAO, Rome, 2430 p

Ferrari L, Fromm I, Jenny K, Muhirec A, Scheidegger U (2017) Formal and informal seed potato supply systems analysis in Rwanda. In: Future agriculture: social-ecological transitions and bio-cultural shifts organised by the University of Bonn and the Centre for Development Research. University of Bonn and the Centre for Development Research, Bonn

Frost KE, Groves RL, Charkowski AO (2013) Integrated control of potato pathogens through seed potato certification and provision of clean seed potatoes. Plant Dis 97:1268–1280

Gildemacher P, Demo P, Kinyae P, Wakahiu M, Nyongesa M, Zschocke T (2007) Select the best: positive selection to improve farm saved seed potatoes: trainers manual. International Potato Center, Lima

Gildemacher PR, Schulte-Geldermann E, Borus D et al (2011) Seed potato quality improvement through positive selection by smallholder farmers in Kenya. Potato Res 54:253–266

Guenthner JF, Charkowsi A, Genger R, Greenway G (2014) Varietal differences in minituber production costs. Am J Potato Res 91:376–379

Harahagazwe D, Andrade-Piedra J, Parker M, Schulte-Geldermann E (2018) Current situation of rapid multiplication techniques for early generation seed potato production in Sub-Saharan Africa. International Potato Center, CGIAR Research Program on Roots, Tubers and Bananas (RTB). RTB Working paper no. 2018-1

Hidalgo O, Manrique K, Velasco C, Devaux A, Andrade-Piedra JL (2009) Diagnostic of seed potato systems in Bolivia, Ecuador and Peru focusing on native varieties. In: Tropical roots and tubers in a changing climate: a convenient opportunity for the world. Fifteenth triennial symposium of the International Society for Tropical Root Crops, Lima, Peru, 2–6 November 2009, pp 41–46

Jansky SH, Charkowski AO, Douches DS et al (2016) Reinventing potato as a diploid inbred line-based crop. Crop Sci 56:1412–1422

Kakuhenzire R, Musoke C, Olanya M et al (2005) Validation, adaptation and uptake of potato small seed plot technology among rural, resource-limited households in Uganda. In: African crop science conference proceedings, pp 1355–1361

Kakuhenzire R, Tibanyendera D, Kashaija IN, Lemaga B, Kimoone G, Kesiime VE et al (2017) Improving minituber production from tissue-cultured potato plantlets with aeroponic technology in Uganda. Int J Agric Environ Res 3:3948–3964

Kinyua ZM, Schulte-Geldermann E, Namugga P, Ochieng-Obura B, Tindimubona S, Bararyenya A et al (2015) Adaptation and improvement of the seed-plot technique in smallholder potato production. In: Low J, Nyongesa M, Quinn S, Parker M (eds) Potato and sweetpotato in Africa: transforming the value chains for food and nutrition security. CABI, Wallingford, pp 218–225

Kromann P, Montesdeoca F, Andrade-Piedra J (2016) Integrating formal and informal potato seed systems in Ecuador. In: J Andrade Piedra, J Bentley, C Almekinders, K Jacobsen, S Walsh, G Thiele (eds) Case studies of roots, tubers and bananas seed systems, Lima (Peru), 244 p. RTB working paper. ISSN 2309–6586. no. 2016–3, pp 14–22

Lakhiar IA, Gao J, Syed TN, Chandio FA, Buttar NA (2018) Modern plant cultivation technologies in agriculture under controlled environment: a review on aeroponics. J Plant Interact 13:338–352

Lindhout P, Meijer D, Schotte T, Hutten RC, Visser RG, van Eck HJ (2011) Towards F 1 hybrid seed potato breeding. Potato Res 54:301–312

Lommen WJ (2007) The canon of potato science: 27. Hydroponics. Potato Res 50:315

Lossau A, Weiskopf B, Kasten W (2000) Support for the informal seed sector in development co-operation-conceptual issues. GTZ, Germany and Centre for Genetic Resources, Wageningen

Louwaars NP (1994) Integrated seed supply: a flexible approach. In: Seed production by smallholder farmers: proceedings of the ILCA/ICARDA research planning workshop held in ILCA, Addis Ababa, Ethiopia, pp 39–46

Louwaars NP, de Boef WS (2012) Integrated seed sector development in Africa: a conceptual framework for creating coherence between practices, programs, and policies. J Crop Improv 26:39–59

Machida-Hirano R (2015) Diversity of potato genetic resources. Breed Sci 65:26–40

Mateus-Rodriguez JR, de Haan S, Andrade-Piedra JL et al (2013) Technical and economic analysis of aeroponics and other systems for potato mini-tuber production in Latin America. Am J Potato Res 90:357–368

McGuire S, Sperling L (2011) The links between food security and seed security: facts and fiction that guide response. Dev Pract 21:493–508

McGuire S, Sperling L (2016) Seed systems smallholder farmers use. Food Security 8:179–195

Ministerio de Agricultura y Riego del Peru (MINAGRI) (2018) Reglamento específico de semillas de papa. http://minagri.gob.pe/portal/download/pdf/normas-legales/decretossupremos/2018/ds10-2018-minagri.pdf. Accessed 26 Mar 2019

Munyi P, De Jonge B (2015) Seed systems support in Kenya: consideration for an integrated seed sector development approach. J Sustain Dev 8:161

Muthoni J, Mbiyu MW, Nyamongo DO (2010) A review of potato seed systems and germplasm conservation in Kenya. J Agric Food Inform 11:157–167

Muthoni J, Mbiyu M, Kabira JN (2011) Up-scaling production of certified potato seed tubers in Kenya: potential of aeroponics technology. J Hortic Forest 3:238–243

Naik PS, Buckseth T (2018) Recent advances in virus elimination and tissue culture for quality potato seed production, Biotechnologies of crop improvement, vol 1. Springer, New York, pp 131–158

Navarrete I, Panchi N, Andrade-Piedra JL (2017) Potato crop health quality and yield losses in Ecuador. Revista Latinoamericana de la Papa 21:51–70

Ochieng-Obura B, Parker ML, Bruns C, Finckh MR, Schulte-Geldermann E (2016) Cost benefit analysis of seed potato replacement strategies among smallholder farmers in Kenya. In: Solidarity in a competing world: fair use of resources, Vienna, Austria

Okeyo GO, Sharma K, Atieno E, Narla RD, Miano DW, Schulte-Geldermann E (2018) Effectiveness of positive selection in managing seed-borne potato viruses. J Agric Sci 10:71

Orlove BS, Godoy R (1986) Sectoral fallowing systems in the Central Andes. J Ethnobiol 6:169–204

Otieno GA, Reynolds TW, Karasapan A, Lopez Noriega I (2017) Implications of seed policies for on-farm agro-biodiversity in Ethiopia and Uganda. Sustain Agric Res 6:12–30

Parker ML, Low JW, Andrade M, Schulte-Geldermann E, Andrade-Piedra J (2019) Climate change and seed systems of roots, tubers and bananas: the cases of potato in Kenya and Sweetpotato in Mozambique. In: Rosenstock TS, Nowak A, Girvetz E (eds) The climate-smart agriculture papers. Springer, Cham, pp 99–111

Pautasso M, Aistara G, Barnaud A et al (2013) Seed exchange networks for agrobiodiversity conservation. a review. Agron Sustain Dev 33:151–175

Pérez W, Valverde M, Barreto M, Andrade-Piedra J, Forbes GA (2015) Pests and diseases affecting potato landraces and bred varieties grown in Peru under indigenous farming system. Revista Latinoamericana de la Papa 19:29–41

Priegnitz U, Lommen WJ, van der Vlugt RA, Struik PC (2018) Impact of positive selection on incidence of different viruses during multiple generations of potato seed tubers in Uganda. Potato Res 62:1–30

Prip C, Fauchald OK (2016) Securing crop genetic diversity: reconciling EU seed legislation and biodiversity treaties. Rev Euro Comp Int Environ Law 25:363–377

Schulte-Geldermann E (2013) Tackling low potato yields in Eastern Africa: an overview of constraints and potential strategies. In: G Woldegiorgis, S Schulz, B Berihun (eds) Seed potato tuber production and dissemination, experiences, challenges and prospects: proceedings. National workshop on seed potato tuber production and dissemination, 12–14 Mar 2012. Ethiopian Institute of Agricultural Research (EIAR); Amhara Regional Agricultural Research Institute (ARARI); International Potato Center, Bahir Dar. pp. 72–80 . ISBN 978–99944–53-87-x

Schulte-Geldermann E, Gildemacher PR, Struik PC (2012) Improving seed health and seed performance by positive selection in three Kenyan potato varieties. Am J Potato Res 89:429–437

Schulz S, Woldegiorgis G, Hailemariam G, Aliyi A, van de Haar J, Shiferaw W (2013) Sustainable seed potato production in Ethiopia: from farm-saved to quality declared seed. Seed potato tuber production and dissemination, experiences, challenges and prospects. Proceedings. Ethiopian Institute of Agricultural Research (EIAR); Amhara Regional Agricultural Research Institute (ARARI); International Potato Center, Bahir Dar, pp 60–71

Shepard JF, Claftin LE (1975) Critical analyses of the principles of seed potato certification. Annu Rev Phytopathol 13:271–293

Sperling L (2008) When disaster strikes: a guide to assessing seed system security. CIAT, Cali

Thiele G (1999) Informal potato seed systems in the Andes: why are they important and what should we do with them? World Dev 27:83–99

Thomas-Sharma S, Abdurahman A, Ali S et al (2015) Seed degeneration in potato: the need for an integrated seed health strategy to mitigate the problem in developing countries. Plant Pathol 65:3–16

Thomas-Sharma S, Andrade-Piedra J, Carvajal Yepes M, Hernandez Nopsa JF, Jeger MJ, Jones RAC et al (2017) A risk assessment framework for seed degeneration: informing an integrated seed health strategy for vegetatively propagated crops. Phytopathology 107:1123–1135

Thurston HD (1990) Plant disease management practices of traditional farmers. Plant Dis 74:96–102

Tran VM, Nguyen VU, Vander Zaag P (1990) Rapid multiplication of potatoes: influence of environment and management on growth of juvenile apical cuttings. Am Potato J 67:789–797

Tripp R (1996) Supporting integrated seed systems: institutions, organizations and regulations. In: Amstel H v, Bottema JW, Sidik M, Santen CE v (eds) Integrating seed systems for annual food crops, Integrating seed systems for annual food crops. Proceedings of a workshop held in Malang, Indonesia October 24–27, 1995. Regional Co-ordination Center for Research and Development of Coarse Grains, Pulses, Roots and Tuber Crops in the Humid Tropics of Asia and the Pacific, Research Institute for Legume and Tuber Crops, Agronomy Society of Indonesia, Indonesia

Tripp R, Louwaars N (1997) Seed regulation: choices on the road to reform. Food Policy 22:433–446

Vander Zaag P, Escobar V (1990) Rapid multiplication of potatoes in the warm tropics: rooting and establishment of cuttings. Potato Res 33:13–21

Vashisth KS (1979) Seed plot technique: its significance in seed potato production in India. Seeds Farms (India) 5:13–14

Vernooy R (2017) Options for national governments to support smallholder farmer seed systems: the cases of Kenya, Tanzania, and Uganda

Zeng Y, Fulladolsa AC, Houser A, Charkowski AO (2018) Colorado seed potato certification data analysis shows mosaic and blackleg are major diseases of seed potato and identifies tolerant potato varieties. Plant Dis 103:192–199

Part IV
Seed Systems, Participatory Research and Gender

Chapter 13
Participatory Research (PR) at CIP with Potato Farming Systems in the Andes: Evolution and Prospects

Oscar Ortiz, Graham Thiele, Rebecca Nelson, and Jeffery W. Bentley

Abstract Participatory Research (PR) at the International Potato Center (CIP) included seven major experiences. (1) Farmer-back-to-farmer in the 1970s pioneered the idea of working with farmers to identify their needs, propose solutions, and explain the underlying scientific concepts. The ideas were of great influence at CIP and beyond. (2) With integrated pest management (IPM) pilot areas in the early 1990s, entomologists and social scientists developed technologies with farmers in Peru and other countries to control insect pests. Households that adopted just some of the techniques enjoyed high economic returns, and this showed the importance of IPM specialists, social scientists, and farmers working together. (3) Farmer field school (FFS) was adapted for participatory research in the 2000s. Farmers learned that late blight was caused by a microorganism, while testing resistant varieties and fungicides, and researchers took into account more specifically farmer knowledge for training and PR purposes. (4) CIP used participatory varietal selection (PVS) after 2004 to form consortia of farmers, local government, NGOs, and research. Farmers' preferences were disaggregated by gender. Selection criteria of other market actors were included, and new varieties were released, showing the importance of combining farmer and researcher knowledge in this process. (5) Participatory approaches to develop native potato variety value chains. After 2000, CIP used the PMCA (participatory market chain analysis) and stakeholder platforms to improve smallholders' access to markets. PMCA brought farmers and other market actors together to form stakeholder platforms which created market innovations, including

O. Ortiz (✉)
International Potato Center, Lima, Peru
e-mail: o.ortiz@cgiar.org

G. Thiele
CGIAR Research Program on Roots, Tubers and Bananas, Lima, Peru
e-mail: g.thiele@cgiar.org

R. Nelson
Cornell University, Ithaca, NY, USA
e-mail: rjn7@cornell.edu

J. W. Bentley
Independent Consultant, Cochabamba, Bolivia

H. Campos, O. Ortiz (eds.), *The Potato Crop*,
https://doi.org/10.1007/978-3-030-28683-5_13

new potato-based products, expanding the inclusion of diverse actors in the PR processes. (6) Advocacy for PR and policy change with the Andean Change Alliance tested PR methods including PVS and PMCA from 2007 to 2010, providing evidence to influence policies to include smallholders in research and development. (7) After 2010, nutrition-related PR documented anemia among children in the high Andes, which could be mitigated by eating native potatoes rich in zinc and iron. CIP partnered with 20 organizations to improve household incomes and nutrition. Over four decades, CIP continues evolving in using PR, showing that combining social and biological scientists' input and keeping farmers' views upfront was key for PR. The experience also showed that the participation of other actors related to the value chains was needed in order to create successful agronomic, market, and social innovations. Future participatory research at CIP may be improved by using ICT to enrich diversity and richness of information sharing among PR actors.

13.1 Introduction

Spanning nearly 40 years, PR at CIP covered most major areas of agricultural research, including new seed and storage technologies, integrated management of pests and diseases, plant breeding, in situ conservation of genetic resources, and value chains. Participatory research at CIP influenced academic and applied research at many other institutions. CIP played a pioneering role in PR in the 1970s. Since then, PR has waxed and waned (Thiele et al. 2001) and moved to higher levels of scale from farm-level management to value chains, food systems, and policy. This chapter provides an update, focusing on work in the Andes, where several PR experiences took place.

The following case studies of PR at CIP are organized roughly chronologically (although some methods overlapped in time). They start with farmer-back-to-farmer, and continue through IPM pilot units, participatory variety selection, FFS, participatory approaches for native potato variety value chains using the participatory market chain analysis (PMCA), advocacy for PR and policy change (via the Alliance for Andean Change), ending with nutrition-related PR in more recent years.

13.2 Cases of Potato-Related PR in the Andes: Learning from Experience

13.2.1 The Farmer-Back-to-Farmer Model

The farmer-back-to-farmer model emerged from an interdisciplinary CIP team that included both social and biophysical researchers. CIP's first director, Richard Sawyer, hired anthropologists and expected them to work in an integrated team with the center's breeders and agronomists. The combination of different views amongst the scientist around seed potato storage led to the farmer-back-to-farmer model (Rhoades and Booth 1982). This model, with its insistence at looking at actual farmer practice in a pragmatic way, was a major influence on much of the farmer participatory research that would follow at CIP and elsewhere. In many ways farmer-back-to-farmer was ahead of its time.

In the late 1970s, a team of CIP researchers was studying post-harvest losses of potato in the Mantaro Valley of Peru. CIP started its participatory research in the Mantaro Valley because it was then one of Peru's main potato producing areas and one of the nearest to the main market in Lima. CIP anthropologist Robert Werge, and CIP biological scientist Robert Booth and other researchers were developing technology that farmers could use to avoid post-harvest losses, working within the framework of Farming Systems Research (FSR, which later would be largely replaced by farmer participatory research). FSR encouraged researchers from different disciplines to work together, and in their seminal paper, Rhoades and Booth (1982) noted that interdisciplinary research could easily become merely multidisciplinary, with researchers working alone within their disciplinary boundaries, and seldom interacting. Their paper was as much about getting social scientists to work with other researchers as it was about involving farmers.

In the 1970s interaction between anthropologists, economists, and biological scientists often provoked stressful but constructive arguments (Rhoades et al. 1986). Two teams were working from different perspectives, a production team with agronomists and an economist, and a post-harvest team with agronomists and an anthropologist (Thiele et al. 2001). An early breakthrough in the Mantaro Valley potato research came when the anthropologist told the biologists that post-harvest potato losses were really of little concern to farmers, who could use the smaller tubers for seed or for animal feed. Even damaged potatoes could be salvaged for the cooking pot and shriveled tubers could be made into chuño—the Andean method for freeze-drying at high altitudes. However, seed potatoes grow long sprouts that farmers disliked having to break off before planting.

Now it was the biologist's turn to be helpful. Booth explaining to the anthropologist (Werge) that the long sprouts were induced by the darkness where the potatoes were stored, inside the farm houses, and that while seed potato needs to be sprouted, the shorter sprouts were more vigorous than longer ones. This insight helped the anthropologist to refine his questions. By working together, the biologist and the anthropologist refined their problem topic. The issue was no longer post-harvest losses, but how best to store seed potatoes on-farm.

On the Santa Ana research station in the Mantaro Valley, the biologist showed the anthropologist how potatoes stored in diffused light (not in total darkness) developed short vigorous sprouts and a greenish skin. Such tubers were ideal for planting. The anthropologist then took some of the wooden greening trays from the station to the homes of some collaborating farmers and stacked the trays under the porch roof and tried storing seed potato there, in diffused light.

Farmers liked the seed tubers with short sprouts, but observed that the fine wooden trays would be too expensive, so a CIP technologist made some simple racks from local lumber. CIP soon began teaching diffused light storage (DLS) to farmers in the Philippines (where Booth was now working) and in Peru, teaching farmers how to make and stack wooden seed trays, but also explaining the underlying scientific principle that potato sprouts are shorter and more vigorous, increasing yield if seed is stored in diffused light.

This experience with DLS became the foundation story for farmer-back-to-farmer, a four-step model. Step 1: the anthropologist and the biologist go to the field to understand the problem from the farmer's perspective based on observation and action research, and to reach a common definition of the research problem (e.g. how to improve the storage of seed potatoes). In steps 2 and 3, the researchers develop a technology through a mix of on-farm and on-station research. Finally, the researchers present a prototype technology to farmers who act as advisors on how to adapt it to suit their own conditions (Rhoades and Booth 1982).

CIP economist Douglas Horton later wrote that the greatest benefit of the Mantaro Valley Project was not improved potato production, but institutional change: within a few years CIP was conducting research in various countries with national programs, based on farmer-back-to-farmer (Horton 1986). While Rhoades and Booth never mention the word "participation" in their 1982 paper, farmer-back-to-farmer influenced much of the farmer participatory research (FPR) that would follow (Veteto and Crane 2014). Thirty-five years after it was published, the short, engaging farmer-back-to farmer paper can still be read profitably for its practical philosophy of working with farmers.

Rhoades and Booth (1982) came up with several ideas that still guide much of participatory research. Researchers must work with farmers to identify the right problem to solve, interact creatively with them, work on-farm and on-station, and present the results back to farmers for feedback. Rhoades and Booth understood the importance of telling farmers the underlying principles (why the technology works) and not just how to use it, making it easier for farmers to adapt the technique to their own circumstances. All of these ideas are as sound now as when they were written. A decade later, farmer-back-to-farmer was one of the influences on interdisciplinary participatory research at CIP, but by the 1990s, these research methods were becoming more formal.

13.2.2 IPM Pilot Areas

In the early 1990s CIP entomologist Fausto Cisneros led research that included biological scientists and members of the Social Sciences Department, such as Oscar Ortiz an agronomist with specialization in agricultural extension. They studied the Andean potato weevil and how to control it in 15 pilot areas, mostly in Peru, and also in Ecuador, Bolivia, and Colombia. There was also research on the potato tuber moth at six sites (in Peru, Colombia, Bolivia, and the Dominican Republic), and on the sweetpotato weevil in Cuba and the Dominican Republic. Researchers needed to adapt IPM (integrated pest management) to the local conditions (farming practices) and farmers' knowledge in pilot communities, called pilot areas. In those communities, researchers would select the most appropriate technologies, and assess farmer knowledge so that training could be provided according to knowledge gaps using practical demonstrations (Cisneros et al. 1995). The technologies were based on the results of previous research conducted by CIP in the late 1980s with Peru's national agricultural research institute, INIA. Through this pioneering experience CIP started working with nongovernmental organizations (NGOs) that were oriented to agricultural extension and development in some of the pilot units (Ortiz et al. 2009).

The researchers realized that each of the IPM technologies had to be validated through participatory research, akin to step 4 of the farmer-back-to-farmer model, presenting results back to communities for advice. The consultation with farmers was held in formal pilot areas[1] and pilot units[2] (see definitions below) in five countries. An agronomist was stationed in each pilot area to mentor the farmers on the technologies, to see which technologies worked and to get an idea of how to explain the innovations to other farmers (Cisneros et al. 1995).

This integrated pest management (IPM) research aimed to lower costs, manage pests, reduce environmental damage, and minimize health risks. The key word was "integrated," using many techniques, especially natural enemies, cultural practices, and pheromones. Insecticides were to be used as a last resort, and as little as possible. However, the researchers also realized that IPM had been vaguely described; abstractions would not convince farmers who needed concrete results (Cisneros et al. 1995). Therefore, specific training techniques were developed to support farmers' learning of complex concepts related to IPM (Ortiz et al. 1997).

The pilot area work came after some celebrated IPM research by IRRI with the rice planthopper in Indonesia. However, farmers could manage the planthopper largely by avoiding insecticide and letting natural enemies control the pest (see Heong et al. 2014). In contrast, the Andean potato weevil was becoming a more serious pest largely because of increased intensity of cropping (Ortiz 2006; Parsa 2010). The Andean potato weevil was a difficult pest that needed to be managed with several proactive and preventive options; it was not enough to abandon insecticides and let nature take its course. And, the approach was called "pilot areas" to refer to the communities where the research took place, because for the IPM to work, it needed to be applied in most of or the entire the community for it to work.

The pilot area work was led by an entomologist or an agronomist with experience in pest management who collaborated with a social scientist on training and impact assessment, testing various management options in many pilot areas, for three problems (Andean potato weevil, tuber moth, and sweetpotato weevil). The Andean potato weevil technologies were based on an understanding of local knowledge and on the pest behavior. Andean farmers recognized the adult weevil, and its grub, but did not understand that there were four life stages of the same species (from egg to adult). Entomologists had learned that the Andean potato weevil ate

[1] Communities where CIP and collaborators carried out the participatory research.

[2] Portions of farmer's fields where the management options were tested.

only potato, and that the insect could be killed in different stages (egg, larva, pupa, and adult) depending on the season of the year. Technologies included hand-picking adult weevils at night (the adult insect had nocturnal habits), uprooting the volunteer plants that harbored weevils, winter plowing to destroy pupae in the soil, piling potatoes on sheets so the larvae could not pupate in the soil, and making barriers around the field to keep the flightless weevils from walking in from neighboring plots (Cisneros et al. 1995; Ortiz et al. 2009).

Farmers rejected some technologies, including early harvesting and eliminating volunteer plants. But pilot farmers adopted barriers around the fields and piling potatoes on sheets. To teach the technologies chosen by pilot farmers, extensionists designed their own materials, including field visits, dramas, games, and manuals. The most acceptable technologies were then shared with other communities as part of a USAID-funded project, MIPANDES, implemented by the international NGO CARE with technical support from CIP (Cisneros et al. 1995; Ortiz et al. 2009).

Rhoades (1987) had claimed that only 2% of farmers surveyed had adopted DLS as it was taught, but that the other 98% had adopted the underlying principles. This may have struck the entomologists as imprecise thinking, and they wanted a more objective way of measuring adoption. Yet, the pilot areas research was designed to give farmers a hand in selecting and rejecting technologies. Farmers could also choose technologies from an integrated menu. Researchers never expected farmers to adopt all of the technologies. However, because so many techniques were taught, it was hard to decide how many components had to be adopted to count as "adoption of IPM" (Ortiz et al. 2009).

Early results showed that adopting even some of the management options led to a large reduction in pest damage. For example, in the pilot area of Huatata, Peru, weevil damage had gone from 44 to 8.5% in just 4 years. Mizque, Bolivia had a favorable environment for weevils, which damaged all of the tubers before the project, but only 10% in later years. Results from other pilot areas were comparable. Using even some of the technologies managed the pest (Cisneros et al. 1995).

Therefore, the team opted to study adoption by measuring the economic impact of the technologies. It did not matter how many technologies farmers managed, but the value of the harvest saved by using them. Farmers who adopted some of the IPM practices could achieve an average benefit of about $100 per hectare (Ortiz et al. 2009). The pilot areas ended with a large project (MIPANDES) implemented with the NGO CARE in 1995 and 1996 in Cajamarca and three other provinces of Peru to teach farmers the menu of IPM technologies for managing the Andean potato weevil and the tuber moth (Chiri et al. 1997).

The pilot areas work was an important formative experience for the FFS experience which followed it (see Sect. 13.2.3 below), particularly because several of the biological and social scientists who participated in one experience continued with the FFS, and they had realized the importance of farmer learning for IPM and the need for appropriate teaching methods (Ortiz et al. 1997). The pilot areas research proposed a systematic way of validating technologies with farmers and an objective method of measuring the results, yet this method probably did not receive all of the recognition that it deserved because there were limited journal publications. The most influential participatory methods tend to be the ones that are well documented. CIP's later research with FFS was well-documented and frequently cited, eclipsing the effort with pilot areas.

13.2.3 Participatory Research and Training: The FFS Experience

Following the pilot areas experience, some of the same researchers continued with IPM, but this time with a potato disease, late blight, and a new method of participatory research: FFS (farmer field school). By 1997, plant pathologist Rebecca Nelson had recently transferred to CIP from IRRI, which had been working with FFS since 1987, first with rice planthoppers and rice stemborers in Indonesia, and later with rice blast in Vietnam (Nelson et al. 2001). For the CIP team, FFS made sense because they had already experienced the importance of facilitating farmers' learning of complex concepts involved in IPM, which were even more complex when dealing with the pathogen that causes potato late blight.

IRRI entomologists had explicitly conceived of FFS as an extension method, not for research (Kevin Gallagher, pers. com.) But FFS was based on experiential learning and the learning field so it did have an informal experimental content. FFS at CIP went several steps further (perhaps too many steps, as the research design was often too complicated for farmers). In 1997, FFS became a PR method (Thiele et al. 2001) as Nelson and colleagues redesigned FFS as a research method dubbed FPR-FFS (farmer participatory research through farmer field schools).

The IPM pilot areas had worked in 16 sites from Bolivia to Cuba (see previous section) between 1990 and 1996, and the FPR-FFS started in four communities, in a single municipality, in San Miguel, Cajamarca, Peru. Two years later field schools would expand to about 20; and a similar experience took place in Cochabamba, Bolivia. CIP was still partnering with CARE, which had already worked in San Miguel during several projects, including MIPANDES and one called ANDINO, which had taught farmers about late blight (Godtland et al. 2004). The experience later expanded, between 1999 and 2004, to other communities in Peru, and also in Bolivia, Ethiopia, Uganda, China, and Bangladesh with a similar approach through an IFAD-funded project (Ortiz et al. 2011).

In the FFS, farmers observed pest ecology in the field. Farmers studied late blight lesions under small microscopes and cultured the disease in plastic bags to watch its spread (Nelson et al. 2001). Farmers had not previously known that late blight was caused by a microorganism. But the FFS provided also the opportunity to test advanced potato clones with resistance to late blight, something that had effects easily perceived by farmers (Figs. 13.1, 13.2, and 13.3 show different stages of participatory evaluation of potato clones as part of FFS implementation), so that they could select those materials that were suitable for their conditions and preferences (Ortiz et al. 2004).

Fig. 13.1 FFS participant evaluating potato varieties during flowering in Cajamarca, Peru (2004) (Credits: O. Ortiz)

The Indonesian FFS focusing on insect pest management used the simplest possible design: an IPM plot next to a farmer's plot. The late blight FPR-FFS tested three potato varieties, ranging in resistance to late blight. Each variety was treated with fungicide at three different intervals, for a total of nine treatments. This research would show that late blight could be managed through varietal resistance combined with a fungicide regime that was adjusted based on the variety and the environment (Sherwood et al. 2000).

Some farmers seem to have felt that 1 year of field school was enough. The second season, 1998–1999, two of the four communities dropped out of FPR-FFS, but CIP and CARE replaced them with eight new communities. The second year these ten communities replicated their experiment with the potato varieties and fungicides from the previous year (Sherwood et al. 2000).

FPR-FFS was small-scale for its first 2 years, but after that the Dutch government funded a large IPM project with the Andean field schools, led by the FAO (Nelson et al. 2001). Between the second and third seasons of FPR-FFS, 35 extensionists

Fig. 13.2 FFS farmer during variety evaluation at harvest in Cajamarca, Peru (2003) (Credits: R. Orrego)

from Peru, Bolivia, and Ecuador attended a 3-month practical ToT (training-of-trainers) FFS course in Ecuador (Sherwood et al. 2000).

After the ToT course, some of the returning extensionists taught the third year of FPR-FFS in San Miguel. It was expanded to 19 communities and three new experiments were added. Farmers continued doing trials of varieties and fungicides. Second, they also tried an IPM vs. farmer's treatment (per the original FFS design), plus a third experiment to grow potatoes from true seed (Nelson et al. 2001). In the fourth experiment, the field schools tested 50 breeding lines of potatoes from CIP. Each community tested just some of the lines, but each clone was tried by two or three communities. This was in 1999–2000, and it foreshadowed the mother-and-baby trials which would be CIP's next participatory method (see following section). Based in part on these evaluations with farmers, one of the varieties (Chata Roja) would be released in 2000 (Nelson et al. 2001).

By now the work with FPR-FFS seemed to be paying off. An economic evaluation showed that late blight management, because of its direct effect on increasing yield, was worth a net benefit of some $236 per hectare and year for the farmers of San Miguel (Ortiz et al. 2004). There was also strong implementation of CIP-affiliated field schools in Ecuador, which generated more learning material than

Fig. 13.3 FFS farmer assessing culinary quality of potato varieties in Cajamarca, Peru (2005) (Credits: O. Ortiz)

could be covered in a single season; the material was compiled as a manual so that trainers could choose from it to custom-design FFS for other communities (Pumisacho and Sherwood 2000). CIP and other organizations were using FFS in other departments of Peru for bacterial wilt, insect pests, and other topics, but the method was also used in Bolivia, Ecuador, Ethiopia, and Uganda (Ortiz et al. 2008). In 2017 the IPM Association in the US gave the IPM Team Award to CIP and partners for their pioneering work with FFS research for the integrated management late blight, addressing the human and technical components of this technology (Ortiz et al. 2019).

However, CIP's research also looked at the factors that limited farmers' involvement in participatory research through FFS, and CIP observed that "the experimental designs tested sometimes were found to be overly complex for training of farmers" (Ortiz et al. 2004). There was also concern over the costs of FFS, as much as $1000 per one FFS per season (Ortiz et al. 2011). Farmers joined the field school to learn, but were discouraged by the time demands; organizations were limited by the lack of qualified trainers (Ortiz et al. 2011). CIP did not continue using FFS as a PR method after 2008, but the method has continued its evolution, and was institutionalized in several public and private agricultural organizations in Peru reaching more than 1000 up to 2008 (Orrego et al. 2010), something that has continued to evolve. More recently, the Peruvian Ministry of Education officially formalized a training and certification program for FFS facilitators (Sineace 2016).

13.2.4 Participatory Variety Selection (PVS)

Plant breeding may take 10 or 12 years or longer to produce a new crop variety. This enormous investment of time, money, and energy can be wasted if the new variety is not adopted. Plant breeders can anticipate farmers' preference for varieties by learning which traits farmers (and consumers) want. For example, there is no point in breeding for processing quality when farmers prefer varieties for table consumption.

In 1987–1988, a Peruvian government project was collaborating with CIP to test varieties with farmers. CIP anthropologist Gordon Prain and agronomist Urs Scheidegger had helped to set this up. Their method was heavily influenced by farmer-back-to-farmer; collaborating farmers were invited to evaluate eight or nine clones, as they saw fit, to find "friendly" varieties. Direct sales of seed to farmer collaborators were seen as an indicator of acceptance (Thiele et al. 2001).

Then Sieglinde Snapp, an agronomist at ICRISAT (International Crops Research Institute for the Semi-Arid Tropics), visited CIP/Lima to share her experiences with a new method of PVS called the mother-and-baby trial design (MBT; Snapp 1999). The MBT method, also known as the central/satellite design, involves a complete, replicated trial in a central location in a community, with satellite mini-trials in farmers' fields. Each farmer trial only includes a subset of the treatments of the central trial, with different farmers testing different combinations of lines such that all the material in the central trial is tested on multiple farms. This visit influenced the types of experiments implemented with the CIP version of FFS. Since plant breeders needed more replicability and numbers for statistical analysis, this MBT method was considered a useful advance. The PVS model thus gave way to an MBT approach, allowing the breeder to collect quantitative, replicated data from the mother trial, while gauging farmers' reactions to the lines or clones grown under farm conditions. The method was widely used at CIMMYT and elsewhere as well (Snapp 2002).

By 1999 CIP had 110 clones of a new breeding group called B1C5, which contained genes from native *Solanum tuberosum spp andigena* potatoes. The clones combined traits like high yield and resistance to late blight with tolerance for low inputs, early maturity (120 days instead of 180) and the taste qualities that Andean consumers preferred (Allauca 2011; Janampa 2012; Camacho-Henriquez et al. 2015). In 2004, CIP plant breeders sent 20–30 advanced clones to communities in Cusco in the high Andes. Over the subsequent decade, CIP facilitated the formation of a consortium of farm communities, municipalities, NGOs, universities, and INIA—Peru's national agricultural research agency so use PVS, and trained key members of staff to use the MBT method (Camacho-Henriquez et al. 2015).

Mother and baby trials were evaluated by the farmers who grew them, but also by other community members who met three times during the season (flowering, harvest, and post-harvest). Farmers' selection criteria were free-listed and then prioritized, giving villagers maximum flexibility to define the traits that they demanded. Women and men farmers worked in separate groups so that the results could be disaggregated by gender. At harvest, the yields were measured and farmers evaluated the potatoes by appearance, taste, and texture. Farmers evaluated the potatoes at post-harvest for weight loss, dormancy, and sprouting. The mother-and-baby trials were gender sensitive and appropriate for illiterate farmers (Camacho-Henriquez et al. 2015).

In 2007, as a result of the mother and baby trials, several potato varieties were released in Peru, including Pallay Poncho, Puca Lliclla, Altiplano (Arcos et al. 2015), and Kawsay (Camacho-Henriquez et al. 2015) followed by others later, e.g. INIA 325—Poderosa (CIP 2014). Some farmers in participating communities even set up businesses to produce and sell seed of the new varieties (Camacho-Henriquez et al. 2015).

In 2008–2009, as part of the Cambio Andino Program, CIP established further trials with four consortia (six in 2009–2010) with mother and baby trials in two or three places per consortium, with 10–20 clones at each site. In each region, each consortium had a leader (an NGO or another organization) that facilitated the mother-and-baby trials. By the third year (2010–2011) three clones were selected at each cite (Fonseca et al. 2011). CIP held annual workshops with the consortia members to improve the mother-and-baby method (Fonseca et al. 2011).

The mother-and-baby trial design allowed plant breeders to understand farmers' selection criteria (e.g. large tubers, resistance to frost and to late blight). There were some gender differences, with women preferring potatoes for boiling. MBT allowed farmers to receive the varieties they wanted. Participation with food manufacturers, wholesalers, and restaurants allowed selecting potatoes that met market demand (Fonseca et al. 2011); this foreshadowed methods CIP would pioneer later to engage with other market actors, besides farmers.

In order to improve the quality and standardization of the mother-and-baby method CIP published a manual to describe it in detail for trainers and facilitators (de Haan et al. 2017). MBT was originally designed to be used by plant breeders and farmers. By adding government, private sector, NGOs, vocational schools, and other actors, CIP was able to involve representative of the whole value chain in selecting varieties. This helped not just to select desired traits, but to facilitate registration of new varieties and to disseminate them with farmers, buyers, processors, and consumers. PVS at CIP went beyond the mother-and-baby approach and also combined participatory approaches with online data collection and analysis tools, which was another innovation in participatory research.

13.2.5 Participatory Development of Native Variety Value Chains

An analysis of different market opportunities revealed that innovation in selling native potatoes had the greatest potential to improve the incomes of smallholders in the high Andes (Devaux et al. 2009). During the 1990s, the road network expanded dramatically in Peru, allowing trucks to reach remote areas previously unreachable, and lowering transportation costs for potatoes and other produce. With long-term support from the Swiss Development Cooperation, CIP began a decade of addressing market opportunities for the native potato, developing and using two tools: the PMCA (participatory market chain analysis) and stakeholder platforms (Devaux et al. 2009). CIP researchers Thomas Bernet, André Devaux, Graham Thiele, and others created the PMCA in 2000 (Bernet et al. 2006), based on RAAKS (rapid appraisal of agricultural knowledge systems—see Engel and Salomon 2003). In the early 2000s, PMCAs conducted with support from CIP started to address the question of how to improve smallholders' access to markets, especially for farmers cultivating potato landraces. One key bottleneck was that native potatoes were unfamiliar to consumers of Lima, Peru's capital and largest city. Through its work with gastronomy schools, the PMCA helped Peruvian chefs to appreciate the culinary potential of native potatoes. This gave native potatoes a new image; and they were now acceptable to middle-class consumers (Horton and Samanamud 2013).

A single PMCA can last for 2 years and comprises three phases. Phase one includes a market survey to identify the stakeholders and includes a large meeting that brings stakeholders together in one room (Devaux et al. 2009). While previous participatory research engaged with researchers, extensionists and farmers, the PMCA links with many more actors (including ministry officials, market specialists, food technologists, transporters and wholesalers, food processors, supermarkets, and chefs—Ordinola et al. 2014). The event must be expertly facilitated to form a community from actors who all work in the same value chain, but many of whom have never met for co-innovation (Devaux et al. 2009).

At this first event, the facilitator helps processors (e.g. supermarkets and food manufacturers) to present a diagnosis and set up groups on promising possible innovations. By the end of Phase 3, innovations move to the promotion phase. In Peru, the PMCA stimulated the following innovations: colored potato chips, made from native varieties, tunta and chuño, clean and well sorted in small bags, and fresh potatoes of native varieties in net bags for supermarkets (Devaux et al. 2009).

At the start of the PMCA in Peru, and as part of the Papa Andina project, the team used what they called a "pro-poor filter," to identify market segments where resource-poor farmers would be especially likely to benefit. This led to the selection of native potatoes produced in the high Andes mostly by resource-poor farmers, as an attractive market segment for innovation. In Peru, multinational companies dominated much of the market for conventional potato chips from improved potatoes supplied by larger producers.

During the second phase of the PMCA, different actors begin to work together, researching and developing different innovations (which could be commercial products or institutional innovations, such as multi-stakeholder platforms). The third phase sees these innovations finished and ends with a large, final event where companies, farmers, and others come together to launch the new products and other innovations. It is important that the companies own the new brands. For example, Wong Supermarket created a brand, T'ikapapa, of mixed, colorful native potato varieties in an attractive net bag, as an outcome of the PMCA (Thiele et al. 2011; Devaux et al. 2009).

As the actors of the value chain begin to work together, buying and selling their products, they learn to trust each other which can lead them to take collective actions, such as Peru's National Potato Day, celebrated on 30 May every year since 2005 with displays of native varieties and gourmet food (Horton and Samanamud 2013; Ordinola et al. 2014). This achievement led the stakeholders, via the Ministry of Agriculture, to successfully petition the UN to name 2008 as the International Year of the Potato (Horton and Samanamud 2013). A new stakeholders' platform emerged from the PMCA to advocate for native potatoes and to consolidate their market position (Devaux et al. 2009; Thiele et al. 2011).

Creating products from native potatoes allowed smaller processors to open a niche market for smallholders, at a price 10–30% higher than before (Devaux et al. 2009). For example, Jalca Chips, which was directly influenced by PMCA, innovated by selling colored potato chips made from native varieties, at the Lima International Airport. This triggered other innovation processes as PMCA intended with the chips have since been copied and improved upon by other food manufacturers and are now widely sold in Peru under various brand names.

An important outcome of the PMCA in Peru was the inclusion of native varieties in the National Registry of Commercial Cultivars (Horton and Samanamud 2013). This allows native varieties to be legally sold as seed and is an important step in their continued survival.

The PMCA stimulated a broader innovation process. In Peru, the wholesale price of native potatoes is increasing as is the volume sold (Horton and Samanamud 2013). The smallholders who supply these native potatoes benefit from higher prices, so that the PMCA has many more indirect beneficiaries than direct ones (Ordinola et al. 2014).

CIP replicated the PMCA with partners in Bolivia (Proinpa) and Ecuador (INIAP). Bolivia also experienced a revived interest in native crops. The PMCA fostered the creation of bagged and clean chuño, native potato chips and net bags of fresh native potatoes sold in supermarkets (Thiele et al. 2011; Oros et al. 2011).

The PMCA stimulated the creation of stakeholder platforms as an institutional innovation. In Ecuador, the potato stakeholder platform triggered the creation of an organization, the Consortium of Small Potato Producers (CONPAPA) which linked highland farmers to others who produce quality seed and bulked potatoes for sale so smallholders can consistently supply the market for modern varieties with a high commercial demand (Thiele et al. 2011; Devaux et al. 2009; Kromann et al. 2016).

Participatory research started at CIP triggered by the interaction of social scientists and agronomists as described earlier around seed systems. Over the years, researchers learned to collaborate with farmers across a broader range of useful technologies including varieties. But PMCA and stakeholder platforms went one step further, involving the other actors of the value chain to explicitly demand new research and launch new commercial products. Farmers responded eagerly to the new market demands for native potatoes (Thiele et al. 2011).

13.2.6 Advocacy for PR and Policy Change (Experience of Cambio Andino)

From 2007 to 2010 a project managed by Graham Thiele at CIP and Carlos Arturo Quirós at CIAT, called Andean Change Alliance (Alianza Cambio Andino, in Spanish), partnered with a broad group of about 20 public and civil organizations, including Proinpa (Promotion and Research for Andean Products) in Bolivia, INIAP (National Agricultural and Livestock Research Institute in Ecuador), INCOPA (a CIP-led project for Innovation and Competitiveness of the Potato) in Peru, the PBA Foundation in Colombia, and many others to advocate for participatory research by showing that participatory methods enhanced research outcomes and helped give the poor a voice to ensure their demands were attended to (Thiele et al. 2012). In Colombia, Ecuador, Peru, and Bolivia, the Alliance tested various participatory methods including: Participatory Monitoring and Evaluation (SEP), Local Committees for Agricultural Research (CIALs), Participatory Variety Selection (see Sect. 13.2.4), and PMCA (Thiele et al. 2012).

To build an evidence base for advocacy for participatory research, the Andean Change Alliance developed a knowledge bank, including an online catalog of participatory methods (www.cambioandino.org) and a book of case studies (Thiele et al. 2012). The Andean Change Alliance was broad ranging including NGOs, universities, farmer associations, local and national governments of Peru, Bolivia, Ecuador, and Colombia. Advocacy was systematically and sequentially organized identifying both "pain points" around participation or a lack of it and a constituency to leverage change (Flores 2010):

- 2007—Capture demands for participatory methods within the innovation systems and organize the methods to meet that demand.
- 2008—Promote the use of participatory methods according to the demand for them and evaluate their outcomes and impacts.
- 2009—Improve the participatory methods.
- 2010—Use the evidences of the outcomes and impacts for policy advocacy to enable participatory research and improve its quality and relevance.

Advocacy was driven by evidence that the inclusion of smallholders in development research-&-development improved the interventions (Flores 2010).

From 2007 to 2009 the Andean Change Alliance facilitated the use of and evaluated participatory approaches, including eight cases of the PMCA in Peru, Bolivia, Ecuador, and Colombia (Table 13.1) (Devaux et al. this volume). Andean Change provided training, backstopping, and coaching.

Local teams consistently sought to adapt PMCA to fit local circumstances; respecting to different degrees the PMCA's fundamental principles. However, in Santa Cruz, Bolivia during a fruit PMCA and a vegetable PMCA, the facilitators tried to rush the model by skipping phase 2, and farmers rejected the contract offered to them because they had not built up enough trust with the buyer.

Table 13.1 Eight cases of PMCA in the Andes, facilitated by the Andean Change Alliance

Case	New product	Completed PMCA	Analyzed[a]	Deviations/problems
1. Coffee/ Peru	Roasted coffee, packaged and labeled	Yes	Yes	
2. Cheese/ Bolivia	Mozzarella packed for retail market	Yes	Yes	NGO had low interest in creating links with other market actors
3. Potatoes/ Bolivia	Native varieties in net bags for supermarket	Yes	Yes	NGO had low interest in developing new market products and links
4. Yams/ Colombia	No	Yes	Yes	Outside knowledge had less to offer
5. Potatoes/ Ecuador	Semi-formal seed	Yes	No	Tried to work at national scale immediately. Consumers had less interest in native potatoes
6. Cheese/ Peru	No	No	No	Facilitators stressed a new production technology, not market innovation
7. Fruit/ Bolivia	No	No	No	Facilitators skipped phase 2 of PMCA
8. Veg/ Bolivia	No	No	No	Same as above

[a]I.e. discussed by Horton et al. (2013)

Horton et al. (2013) focused on four completed cases. Results were greatest where PMCA was implemented with the greatest fidelity. Most important is the engagement of market actors, not just farmers. Adapting a method (lack of fidelity) may be innovative, but if local teams are encouraged to change any and all parts of the model they may skip key components (Horton et al. 2013).

The analysis and documentation of experience with the implementation of SEP and CIALs showed positive outcomes related to the agency of farmers and farmer organizations with enhanced self-respect of farmers, mutual respect among different actors, enhanced capacity for negotiation and stronger organizations. The enhanced agency of farmers promoted structural changes that affected local traditional organizations and development institutions. Institutional innovation, changes in rules, roles, and practices involving power sharing occurred initially at the local organizational level and gradually permeated into the institutional level reaching practices and processes in the institutions that delivered the services (Polar et al. 2012). There were positive changes in the quality of agricultural advisory services as well as in coverage and inclusion. However, farmers' agency was still unable to generate institutional changes in public sector actors (Polar 2014).

Perhaps the furthest reaching, and longest lasting advocacy of the Andean Change Alliance was with the many national organizations, public and civil, which participated with farmers in research, commissioned and mentored by the Alliance (Thiele et al. 2012; Flores 2010).

13.2.7 Nutrition-Related Participatory Research Through Potato

In a cruel irony, people who spend much of their day producing food are often unable to properly feed their own kids. The children of many Andean farm families are chronically undernourished. The causes and contexts of poverty and a poor diet are complex. Nonetheless, in about 2010, CIP crop scientist André Devaux and colleagues, in line with government diagnosis, recognized that anemia is a major problem which research could tackle. In the Andes over 20% of children under three suffer from chronic undernourishment, mainly due to deficiencies of micro elements such has iron and zinc; 39% of children under two are anemic (Devaux et al. 2015).

Anemia can result from a diet poor in zinc and iron. Fortunately, some native potato varieties are rich in zinc and iron, and the vitamin C in potatoes helps to body to absorb these essential minerals. Potatoes are also well endowed with vitamin B, antioxidants, and have as much protein as grains, so a diet with generous portions of native potatoes can be nutritious (Devaux et al. 2012; Ordinola 2015; Creed-Kanashiro et al. 2015).

From 2011 to 2015, Devaux and colleagues led a project, IssAndes (Innovation for food security and sovereignty in the Andes) to address undernutrition with research. IssAndes started with a baseline study in rural communities in Apurímac and Huancavelica, Peru. The results confirmed that 42% of children under 2 years of age where chronically undernourished. Fifty-four percent of the children were not getting enough iron in their diets and zinc was deficient in 48% (Creed-Kanashiro et al. 2014). The baseline study also found a positive relationship between children's iron and zinc intake with native potato production, raising small animals, and area of production (Creed-Kanashiro et al. 2015). Area of production would correlate with household income, suggesting that better-off families could feed their children a more diverse diet.

Building on the work of plant breeders, Devaux and colleagues identified 200 native potato varieties in the CIP genebank with high levels of zinc and iron. The next step was to see which of those varieties would appeal to farm families. This was somewhat like the previous work with PMCA (see Sect. 13.2.5), but then the emphasis had been on finding potatoes of the right size, shape, color, and taste to appeal to the gourmet food market in Lima. Now, the goal was to find nutritious potato varieties for the poor (Devaux and Kromann 2016). Farmers in the highland communities participated in assessing CIP's improved clones from a breeding population that resembled native varieties using PVS (see Sect. 13.2.4) and selected one particular variety. This appealing variety was named "Kawsay" (Quechua for "to live"). The Peruvian government released Kawsay, as mentioned in Sect. 13.2.4, as a resistant variety with higher content of iron and zinc.

Since the baseline study had suggested that higher incomes led to healthier children (Creed-Kanashiro et al. 2015), IssAndes worked to improve livelihoods by helping farm families to create something to sell. The project helped to launch

Kiwa® nutritious: colored potato chips, made from native potatoes and sold around the world (Devaux and Kromann 2016).

Through IssAndes, over 5000 rural farm households in Peru, Bolivia, Ecuador, and Colombia, were encouraged to grow, eat, and sell native potatoes rich in zinc and iron, and to experiment with vegetable home gardens and with raising guinea pigs and other small animals.

13.3 The Prospects of PR in the Andes and Globally

The previous sections of the chapter have described different PR stages and approaches used by CIP over the last three decades, which has shown to be a valuable mechanism to bring together different stakeholders to develop technologies, methodological approaches, and business opportunities. It has been observed that at the beginning of CIP's PR experience, the types of actors involved in agricultural research for development were relatively few (farmers, extension workers, and researchers, mostly from government organizations), but the range of new stakeholders started to increase over the years, including NGOs, farmer organizations, civil society, and private sector. New PR methods such as the PMCA evolved to include the views of these diverse actors; but within a single value chain and for particular business opportunities. However, agri-food systems are becoming more complex with interconnectedness established from rural to urban areas and vice versa, and future PR approaches will need to take this into account. Therefore, methods will need to continue evolving to deal with increasingly complex agricultural innovation systems (Hall et al. 2004; Ortiz et al. 2013; Hellin 2012) to face emerging challenges, such as climate change. In addition, the ways in which farmers are connected to the world is also changing.

The penetration of information and communication technologies (ICTs) to rural areas has increased significantly in recent years, opening new opportunities for capturing farmers' views in ways not attempted before. In the developed world, farmers are already immersed in extensive and complex networks of information sharing and decision support tools facilitated by ICTs; and although in developing countries, this is not yet the case, ICTs can facilitate gathering the opinions of more farmers, and also diverse actors in the value chains, and to share those views with actors located in other regions and countries. In addition, farmers in developing countries are slowly moving towards being part of the big data movement, which will connect information in new and faster ways. For example, farmer preferences differentiated by gender could be linked with the genetics that explain those traits, making it possible to establish a connection between trait preferences and trait genetics in a way that was not possible before.

Adaptation and resilience to climate change requires farmers to adapt faster than even before, for which the flow of information from and to them should also be faster. ICTs offer the possibility of connecting a larger number of farmers with sources of information and advice that could help them to make decisions to tackle climate change—CIP already experimented with this (Sperling et al. 2008), but in the future, there will be the need to connect even more people with more information sources, and, particularly, to support making sense of the available big data for farmers' decision making.

The essence of PR approaches has been maintained over the years as a mechanism to facilitate the dialogue between scientists from different disciplines, and particularly between them and farmers or other actors of the value chain. This will continue evolving and in the future, PR methods will need to include a larger number of viewpoints, from the disciplinary view point, but also from the innovation and agri-food systems viewpoints, and new ICTs offer the possibility of improving communication, analysis, and decision making. We can envision a larger number of PVS using mother and baby trials, all interconnected in real time, so that farmer groups not only analyze their own results, but access results of many more, and also breeders can provide the genetic explanation of user preferences. The future is full of opportunities for PR.

13.4 Concluding Remarks

The early experience with participatory research at CIP, with farmer-back-to-farmer in the early 1980s, showed the importance of identifying research topics with farmers, working with communities creatively, and presenting results back to farmers. These ideas informed much of the later participatory research at CIP and elsewhere (Thiele et al. 2001).

By the early 1990s, social scientists were accepted as full team members of research teams. Research methods were more formal, unlike farmer-back-to-farmer with its "loose guidelines rather than polished and elaborate research methods" (Thiele et al. 2001). The Pilot Areas approach now measured results with economic impact (Ortiz et al. 2009), which allowed evaluating several technologies at once.

The IPM research (Pilot Areas and FFS) showed the importance of understanding gaps in farmers' knowledge and sharing ideas with them to further collaboration. By the late 1990s, FFS became a formal research method at CIP. FFS and PVS (mother-and-baby) started at about the same time and were influenced by mother-and-baby trials at CIMMYT and work by CIP with "friendly" varieties (Prain et al. 1992).

The Pilot Areas had started working with NGOs. FFS and PVS continued to collaborate with NGOs, but with PVS the NGOs were incorporated into larger consortia, with more forming training and workshops for staff. Food manufacturers, wholesalers, restaurants, and the rest of the value chain were brought into research.

In the early 2000s, the PMCA involved other market chain actors, besides farmers. With PMCA it was no longer enough to select a new variety or a technology: PMCA stimulates innovations by actors. PMCA contributed to a progressive shift in consumers' perceptions of native potato varieties to reposition them as prestigious, even gourmet food, and improving livelihoods.

In the 2010s, the Alliance for Andean Change explicitly intended to promote broader uptake of PMCA, PVS, Participatory Monitoring and Evaluation, and other participatory methods, validating them in four Andean countries and using the evidence of success to advocate for the mainstreaming of participatory methods with national research systems. It was not possible to document the influence of Andean Change on national systems, because the project ended prematurely, but there may have been a lasting influence on many partner organizations in Peru, Bolivia, Ecuador, and Colombia.

By 2012, multidisciplinary research linked with CIP plant breeding which provided native varieties with high zinc and iron content. These varieties were then evaluated with farmers, using techniques refined by previous work with PVS. The nutrition research followed on the successful work with PMCA, but now the emphases was not on finding a more marketable potato, but on choosing a more nutritious one. Like the Andean Change Alliance, the nutrition work with the IssAndes project collaborated with 20 organizations in four Andean countries (actually, many of the same organizations that had been part of Andean Change). INIAP, the government agricultural research institute in Ecuador, built a large aeroponics unit devoted in part to producing seed of native potato varieties. This can be seen as an example of successful advocacy from Andean Change and previous CIP work (e.g. with PVS).

While participatory research at CIP may seem fragmented and at times uncertain (Thiele et al. 2001), CIP never stopped using participatory research. CIP social scientists, breeders, IPM specialists and agronomists have continued working together, innovating, creating influential participatory methods that integrate disciplines, bring farmers on board, and widen participation to include all the actors of the value chain, which in the future will continue evolving to face more complex challenges and link more diverse stakeholders taking advantage of emerging technologies (ITCs) and the possibility of using the big data revolution.

References

Allauca Saguano SM (2011) Análisis de los criterios de preferencias en la selección de nuevas variedades de papa, con enfoque en la influencia del mercado. Thesis, Universidad Nacional Agraria, La Molina, Lima, Peru, 239 p

Arcos J, Gastelo M, Holguín V (2015) INIA 317 – Altiplano, variedad de papa con buena adaptación en la región altiplánica del Perú. Revista Latinoamericana de la Papa 19(2):68–75

Bernet T, Thiele G, Zschocke T (2006) Participatory market chain approach (PMCA)—user guide. International Potato Center (CIP) – Papa Andina, Lima

Camacho-Henriquez A, Kraemer F, Galluzzi G, de Haan S, Jäger M, Christinck A (2015) Decentralized collaborative plant breeding for utilization and conservation of neglected and underutilized crop genetic resources. In: Al-Khayri JM, Jain SM, Johnson DV (eds) Advances in plant breeding strategies: breeding, biotechnology and molecular tools, vol 1. Springer, Heidelberg, pp 25–61

Chiri A, Ccama F, Fano H, Dale W (1997) Final evaluation of the integrated pest management for Andean Communities (MIPANDES) project (no. 527-0372). Report submitted to CARE/Peru

CIP (2014) Liberan nueva variedad de papa 'INIA 325 – PODEROSA' en Perú. https://cipotato.org/es/press-room/blog/liberan-nueva-variedad-papa-inia-325-poderosa-peru/. Accessed 9 Aug 2017

Cisneros F, Alcázar J, Palacios M, Ortiz O (1995) A strategy for developing and implementing integrated pest management. CIP Circular 21(3):2–7

Creed-Kanashiro HM, Astete-Robilliard L, Abad Arrue M, Marin M, Bartolini R (2014) Línea de Base Nutricional Perú. Proyecto IssAndes. CIP, Lima, p 65

Creed-Kanashiro H, Hareau G, Devaux A, Maldonado L, Ordinola M, Fonseca C, Suarez V, Astete L, Marin M, Penny M (2015) Agriculture-nutrition linkages: nutritional outcomes in potato based production systems of Peru. Poster presented at the conference on global food security

De Haan S, Salas E, Fonseca C, Gastelo M, Amaya N, Bastos C, Hualla V, Bonierbale M (2017) Selección participativa de variedades de papa (SPV) usando el diseño mamá y bebé: una guía para capacitadores con perspectiva de género. CIP, Lima, 82 p

Devaux A, Kromann P (2016) Strengthening innovation in agri-food systems for food security in the Andes. CIP, Lima, 2 p

Devaux A, Horton D, Velasco C, Thiele G, López G, Bernet T, Reinoso I, Ordinola M (2009) Collective action for market chain innovation in the Andes. Food Policy 34(1):31–38

Devaux A, Andrade-Piedra J, Ordinola M, Velasco C, Hareau G, López G, Rojas A, Flores P, Fonseca C, Kromann P (2012) Agricultura, seguridad alimentaria y nutrición en los Andes: potenciales aportes de la innovación en papa. Paper read at the Congreso de la Asociación Latinoamericana de la Papa, 17–20 Sep 2012, Uberlândia, Brazil

Devaux A, Flores P, Velasco C, Babini C, Ordinola M (2015) Innovation to enhance agriculture, nutrition, and health linkages. IssAndes project brief. International Potato Center, Lima, 6 p

Engel P, Salomon M (2003) Facilitating innovation for development: a RAAKS resource box, Amsterdam

Flores R (2010) Avances, oportunidades y temas estratégicos de incidencia política en la región. Paper read at the Taller de Evidencias y Argumentos, Metodologías de la Alianza Cambio Andino y Metodologías Invitadas. Cambio Andino, Lima, 8–10 Nov 2010

Fonseca C, De Haan S, Miethbauer T, Maldonado L, Ruiz R (2011) Selección participativa de variedades de papa en Perú. In: Thiele G, Quirós CA, Ashby J, Hareau G, Rotondo E, López G, Paz Ybarnegaray R, Oros R, Arévalo D, Bentley J (eds) Métodos participativos para la inclusión de los pequeños productores rurales en la innovación agropecuaria: experiencias y alcances en la región andina 2007–2010. Programa Alianza Cambio Andino, Lima, pp 169–184

Godtland E, Sadoulet E, de Janvry A, Murgai R, Ortiz O (2004) The impact of farmer-field-schools on knowledge and productivity: a study of potato farmers in the Peruvian Andes. Econ Dev Cult Change 53:63–92

Hall A, Mytelka L, Oyeyinka B (2004) Innovation systems: what's involved for agricultural research policy and practice? ILAC brief 2

Hellin J (2012) Agricultural extension, collective action and innovation systems: lessons on network brokering from Peru and Mexico. J Agric Educ Exten 18(2):141–159

Heong KL, Escalada MM, Chien HV, Cuong LQ (2014) Restoration of rice landscape biodiversity by farmers in Vietnam through education and motivation using media. Surv Perspect Integr Environ Soc 7(2)

Horton D (1986) Farming systems research: twelve lessons from the Mantaro Valley Project. Agric Admin 23:93–107

Horton D, Samanamud K (2013) Peru's native potato revolution. Papa Andina innovation brief 2. International Potato Center, Lima, 6 p

Horton D, Rotondo E, Paz Ybarnegaray R, Hareau G, Devaux A, Thiele T (2013) Lapses, infidelities, and creative adaptations: lessons from evaluation of a participatory market development approach in the Andes. Eval Program Plann 39:28–41

Janampa Martínez AM (2012) Selección participativa bajo el diseño mamá &bebé de 20 clones de papa *Solanum tuberosum* spp. *andígena* (población B1C5), con resistencia horizontal a la rancha (*Phytophthora infestans*). Thesis, Universidad para el Desarrollo Andino, Lircay, Huancavelica, Peru, 127 p

Kromann P, Montesdeoca F, Andrade-Piedra J (2016) Integrating formal and informal potato seed systems in Ecuador. In: Andrade-Piedra J, Bentley J, Almekinders C, Jacobsen K, Walsh S, Thiele G (eds) Case studies of root, tuber and banana seed systems. RTB working paper no. 2016-3. RTB, Lima, pp 13–32

Nelson R, Orrego R, Ortiz O, Tenorio J, Mundt C, Fredrix M, Vien NV (2001) Working with resource-poor farmers to manage plant diseases. Plant Dis 85(7):684–695

Ordinola M (2015) Agricultura, nutrición y seguridad alimentaria: un enfoque diferente. Agro Enfoque 199:65–67

Ordinola M, Devaux A, Bernet T, Manrique K, Lopez G, Fonseca C, Horton D (2014) The PMCA and potato market chain innovation in Peru. Papa Andina innovation brief 3. International Potato Center, Lima, 8 p

Oros R, Rodríguez F, Gonzales F, Thiele G (2011) Caso de implementación EPCP 2: la papa nativa en Norte Potosí, Bolivia. In: Thiele G, Quirós CA, Ashby J, Hareau G, Rotondo E, López G, Paz Ybarnegaray R, Oros R, Arévalo D, Bentley J (eds) Métodos participativos para la inclusión de los pequeños productores rurales en la innovación agropecuaria: experiencias y alcances en la región andina 2007–2010. Programa Alianza Cambio Andino, Lima, pp 99–109

Orrego R, Ortiz I, Tenorio J (2010) Interactuando para aprender: el caso de las escuelas de campo de agricultores (ECAs) en el Perú. LEISA 26(4):33–35

Ortiz O (2006) Evolution of agricultural extension and information dissemination in Peru: an historical perspective focusing on potato-related pest control. Agric Human Values 23(4):477–489

Ortiz O, Alcazar J, Palacios M (1997) La enseñanza del manejo integrado de plagas en el cultivo de papa: la experiencia del CIP en la Zona Andina del Perú. 9/10:1–22

Ortiz O, Garret KA, Heath JJ, Orrego R, Nelson RJ (2004) Management of potato late blight in the Peruvian highlands: evaluating the benefits of farmer field schools and farmer participatory research. Plant Dis 88(5):565–571

Ortiz O, Frias G, Ho R, Cisneros H, Nelson R, Castillo R, Orrego R, Pradel W, Alcázar J, Bazán M (2008) Organizational learning through participatory research: CIP and CARE in Peru. Agric Human Values 25(3):419–431

Ortiz O, Kroschel J, Alcázar J, Orrego R, Pradel W (2009) Evaluating dissemination and impact of IPM: lessons from case studies of potato and sweet potato IPM in Peru and other Latin American countries. In: Peshin R, Dhawan AK (eds) Integrated pest management: dissemination and impact. Springer, New York, pp 419–434

Ortiz O, Orrego R, Pradel W, Gildemacher P, Castillo R, Otiniano R, Gabriel J, Vallejo J, Torres O, Woldegiorgis G, Damene B, Kakuhenzire R (2011) Incentives and disincentives for stakeholder involvement in participatory research (PR): lessons from potato-related PR from Bolivia, Ethiopia, Peru and Uganda. Int J Agric Sustain 9(4):522–536

Ortiz O, Orrego R, Pradel W, Gildemacher P, Castillo R, Otiniano R, Gabriel J, Vallejo J, Torres O, Woldegiorgis G, Damene B, Kakuhenzire R, Kasahija I, Kahiu I (2013) Insights into potato innovation systems in Bolivia, Ethiopia, Peru and Uganda. Agric Syst 114(2013):73–83

Ortiz O, Nelson R, Olanya M, Thiele G, Orrego, R, Pradel W, Kakuhenzire R, Woldegiorgis G, Gabriel J, Vallejo J, Xie, K (2019) Human and Technical Dimensions of Potato Integrated Pest Management Using Farmer Field Schools: International Potato Center and Partners' Experience With Potato Late Blight Management. Journal of Integrated Pest Management 10(1):1–8

Parsa S (2010) Native herbivore becomes key pest after dismantlement of a traditional farming system. Am Entomol 56(4):242–251

Polar V (2014) Participation for empowerment: an analysis of agricultural innovation in two contrasting settings of Bolivia. PhD thesis. School of Oriental and African Studies, University of London, 331 p, La Paz – SOAS. http://eprints.soas.ac.uk/20311/1/Polar_Funez_3625.pdf

Polar V, Fernández J, Ashby J, Quiros CA, Roa JI (2012) Participatory methods and the co-production of agricultural advisory services. Results from four case studies in Bolivia and Colombia. International Potato Center (CIP), Lima. Working paper 2012-1, 107 p. https://cgspace.cgiar.org/bitstream/handle/10568/90952/CIP-Participatory-methods-and-the-co-production-of-agricultural-advisory.pdf?sequence=1

Prain G, Uribe F, Scheidegger U (1992) "The friendly potato": farmer selection of potato varieties for multiple uses. In: Moock JL, Rhoades RE (eds) Diversity, farmer knowledge and sustainability. Cornell University Press, Ithaca, pp 52–68

Pumisacho M, Sherwood S (2000) Herramientas de aprendizaje para facilitadores: manejo integrado del cultivo de papa. INIAP, CIP, FAO, Quito

Rhoades R (1987) Farmers and experimentation. ODI AgREN discussion paper 21

Rhoades RE, Booth RH (1982) Farmer-back-to-farmer: a model for generating acceptable agricultural technology. Agric Admin 11(2):127–137

Rhoades RE, Horton DE, Booth RH (1986) Anthropologist, biological scientist and economist: the three musketeers or three stooges of farming systems research. In: Jones JR, Wallace BJ (eds) Social sciences and farming systems research. Methodological perspectives on agricultural development. Westview, Boulder, pp 21–40

Sherwood S, Nelson R, Thiele G, Ortiz O (2000) Farmer field schools for ecological potato production in the Andes. LEISA-Leusden 16:24–25

Sineace (National System for Evaluation, Accreditation and Certification of Educational Quality—Ministry of Education, Peru) (2016) Facilitadores de Escuelas de Campo de Agricultores podrán certificar sus competencias. https://www.sineace.gob.pe/facilitadores-de-escuelas-de-campo-de-agricultores-podran-certificar-sus-competencias/. Accessed 11 Jan 2018

Snapp S (1999) Mother and baby trials: a novel trial design being tried out in Malawi. TARGET. The newsletter of the Soil Fertility Research Network for Maize-Based Cropping Systems in Malawi and Zimbabwe. January 1999 issue. CIMMYT, Harare

Snapp S (2002) Quantifying farmer evaluation of technologies: the mother and baby trial design. In: Bellon MR, Reeves J (eds) Quantitative analysis of data from participatory methods in plant breeding. CIMMYT, Mexico City, pp 9–17

Sperling F, Validivia C, Quiroz R, Valdivia R, Angulo L, Seimon A, Noble I (2008) Transitioning to climate resilient development: perspectives from communities in Peru, Environment department papers, no.115. Climate change series. Washington, DC, World Bank, 103p

Thiele G, van de Fliert E, Campilan D (2001) What happened to participatory research at the International Potato Center? Agric Human Values 18:429–446

Thiele G, Devaux A, Reinoso I, Pico H, Montesdeoca F, Pumisacho M, Andrade-Piedra J, Velasco C, Flores F, Esprella R, Thomann A, Manrique K, Horton D (2011) Multi-stakeholder platforms for linking small farmers to value chains: evidence from the Andes. Int J Agric Sustain 9(3):423–433

Thiele G, Quirós CA, Ashby J, Hareau G, Rotondo E, López G, Paz Ybarnegaray R, Oros R, Arévalo D, Bentley J (eds) (2012) Métodos participativos para la inclusión de los pequeños productores rurales en la innovación agropecuaria: experiencias y alcances en la región andina 2007-2010. Programa Alianza Cambio Andino, Lima. https://cgspace.cgiar.org/handle/10568/65712

Veteto JR, Crane TA (2014) Tending the field: special issue on agricultural anthropology and Robert E. Rhoades. Cult Agric Food Environ 36(1). https://doi.org/10.1111/cuag.12023

Chapter 14
Gender Topics on Potato Research and Development

Netsayi Noris Mudege, Silvia Sarapura Escobar, and Vivian Polar

Abstract Sustainable Development Goals 5 calls for addressing gender equality and women empowerment by, among other things, eliminating all forms of discrimination against women. At CIP we interpret this to mean strengthening the use of gender approaches in research and ensuring that research products are responsive to the needs of men and women. This chapter reviews lessons learnt over the years on integrating gender into potato research and development. The chapter discusses how gender has been approached in five key themes in potato research, namely (1) conserving and accessing genetic resources, (2) genetics and crop improvement, (3) managing priority pests and disease, (4) access to seed (seed flows and networks), and (5) marketing, postharvest processing and utilization. This chapter discusses how gender relations that favor men influence women's participation in and their ability to benefit from potato production, marketing, and research for development. The review shows that potato research has been increasingly focusing on social determinants of potato farming because of the realization that purely technical solutions will not solve inefficiencies in potato production. Using a gender relations approach, the chapter attempts to draw out lessons that can contribute to the design of future potato interventions including research aimed at reducing the gender gap in agriculture in general and potato farming in particular.

N. N. Mudege (✉)
International Potato Center, Nairobi, Kenya
e-mail: n.mudege@cgiar.org

S. Sarapura Escobar
Royal Tropical Institute, Amsterdam, The Netherlands
e-mail: sarapura@uoguelph.ca

V. Polar
CGIAR Research Program on Roots, Tubers and Bananas (RTB), Lima, Peru
e-mail: v.polar@cgiar.org

H. Campos, O. Ortiz (eds.), *The Potato Crop*,
https://doi.org/10.1007/978-3-030-28683-5_14

14.1 Introduction

It has already been established that gender differences matter in agricultural production in various farming systems all over the world, where the ownership and management of farms and natural resources by men and women are often defined by culturally specific gender roles (Meinzen-Dick et al. 2010). Evidence indicates that agriculture and development projects should be gender responsive and take into consideration the needs, aspirations, knowledge, opportunities, constrains, and challenges faced by men and women farmers, young and old, if hunger and poverty are to be alleviated (Njenga and Gurung 2011). Additionally, it is also clear in research that gender intersects with other structures of social hierarchy such as class, race, caste, and age (German and Taye 2008). Certainly, farmers are not a uniform group. Men and women play different roles within agricultural systems occupying different socioeconomic positions linked to these roles, and may suffer from different vulnerabilities (Carr 2008). These differences and vulnerabilities should be considered when new technologies are being developed. Kingiri (2010) noted that generally in farming systems research and innovation, unequal relationships between men and women in households are taken for granted. As a result, development of new technologies may end up increasing the gender gap and benefiting men more than women because social relations of gender are not properly understood. In some cases, as noted by Quisumbing and Pandolfelli (2010), new technologies may even harm women if they are not properly thought through.

The interest of gender in research in potato-related research organizations such as CIP and partners started more than two decades ago. During this period, several studies have looked at the key technical potato production constraints. However, a limited number of studies have discussed the key technical constraints of potato production from gender perspective (see for example, Tapia and de la Torre 2000; Mera-Orcés 2001; Polar et al. 2017; Mudege et al. 2016). In a study promoting participatory technology development in potato farming and production in Ethiopia, Jibat et al. (2007) suggest that it is important to understand gender roles in agriculture production and decision-making to ensure that research address men and women's needs making the results of research more demand oriented. A key limitation of these studies is that they often look at gender roles, and pay little attention to gender relations (see for example, Tapia and de la Torre 2000; Mera-Orcés 2001).

Most of the research has focused on gender division of labor along the potato value chain. For example, a study by Muhinyuza et al. (2012) in Rwanda showed that both men and women are involved in main "potato production activities" as well as decision-making on production and marketing. "However, some activities such as weeding, cooking, and storage protection are exclusively done by women while predominantly men are totally concerned with pest management". Similar work has also been conducted in the Andes in Latin America on gender division of roles in potato production (Tapia and De La Torre 1998; Laub and Muir 2008). Such research often conducts sex disaggregated analysis of labor distribution but does not disaggregate when it comes to constraints because of the focus on technical constraints such as pests and disease (see Muhinyuza et al. 2012; Sah et al. 2007).

However, while early research focused on division of labor and social demographic variables, more recent research has started focusing more and more on gender relations, the normative environment including institutional factors along the potato value chain (Mudege and Demo 2016). Institutional factors such as access to credit and markets by men and women, and their influence on the adoption of improved potato varieties as well as productivity. As part of gender relations, gender norms and how they shape the opportunity structure for men and women in agriculture have been studied (Petesch et al. 2018). Some research also focuses on the gender gap in access to resources, (such as capital), assets (such as land and other productive assets), and knowledge (for example assess to extension services and credible information sources) which if not addressed lead to a gender gap in productivity between men and women (Quisumbing et al. 2014).

In discussing gender topics in potato research, this chapter will go beyond the usual focus on gender roles and asset gaps to look at how the normative environment creates and shape the opportunity structure for men and women. This chapter will use the gender relations approach promoted by Kabeer and Subrahmanian (1996), who defines gender relations as social interactions that embody both the material and the ideological aspects that are revealed not only in the division of labor and allocation of resource between women and men but also in how "value is given, and power is mobilized."

Thus, the gender relations approach focuses on unequal relations between men and women (Little and Panelli 2003). This is an important aspect of the approach because while understanding gender division of labor is good, it falls short in explaining the social reasons that limit women's access to resources and information, which in turn influence adoption of new technologies and accessing the benefits that they can generate. Gender inequalities which in many cases favor men often limit the resources that women have access to and what they are able or not able to do. For example, gender relations influence access to resources such as land and water sources which are essential for agriculture production and productivity.

This chapter explores how gender matters in several key topics including: (a) conservation and access to genetic resources; (b) breeding and crop improvement; (c) access to seed; (d) managing priority pests and diseases; and (e) marketing, postharvest management, processing and utilization. The chapter will look at how gender relations can influence research processes to draw lessons that can contribute in the design of future interventions that can help reduce the gender gap in agriculture and to ensure agricultural research benefits both men and women. This chapter will examine how gender issues have been considered in key potato research topics.

14.2 Conserving and Accessing Genetic Resources

The assumption that guides all conservation efforts is that genetic resources are under threat and need to be safeguarded (Brush 2004) as essential source of foods to sustain healthy diets and a source of genes to supply resistance and functional traits in breeding programs (Jones et al. 2018). Maintaining or conserving diversity in situ is an active and purposeful part of farm management (Brush et al. 1981) where men and women have differentiated roles and responsibilities. The role of women in conserving genetic biodiversity for potato has been well documented in different crops and agroecologies.

Regarding potatoes, studies in this area have been mostly conducted in the Andean region in Latin America. It has been noted that women play a key role in maintaining genetic diversity for potato particularly in selection of seed and varieties, storage of seed and utilization (GRAIN 2000; Sarapura 2013). Women in the Andes Mountains have been significantly involved in conserving genetic diversity particularly through their direct involvement in selecting and preserving different native potato varieties to meet different needs both in terms of culinary characteristics as well as to meet different social rituals and obligations. Sarapura (2013) notes that potato genetic conservation is done with care and tuber seed is stored in special places as the tubers are the nexus of relationships and commitments between the community, nature, cosmos, and deities (Sarapura 2013). Selection and conservation of native potato varieties are mostly done by women who know specific characteristics of potato varieties and have knowledge on how to select them according to different purposes the different potato varieties meet.

Tapia and de la Torre (1998), for example, note that in Peru and Bolivia women act as conservationists preserving native potatoes such as the bitter potato species (*Solanum juzepczukii* and *S. curtilobum)* which can survive temperatures as low as −3 °C and can be freeze-dried into traditional products. Female producers in the highlands and especially in peasant communities do not only play a decisive role in food security (Tapia and de la Torre 2000), but also perform a significant role in seed management and food provision (Tapia and de la Torre 2000; Aguayo and Hinrichs 2015). Women select the seed of native varieties based on the crops' in situ morphological and yield interpretation, culinary quality and crop yield, processing quality, and resistance to diseases, drought, or floods (Tapia and de la Torre 2000). Management of genetic diversity through careful management of combination of varieties enables communities to manage risks, particularly where climate stress is more frequent and intense (De Haan 2009).

Tapia and de la Torre (1998) noted that although some women adopt new improved varieties, they also keep and conserve native potatoes, because genetic diversity increases food security in the Andean highlands. In addition, potato research has been increasingly focusing on how women can be recognized and benefit from their role of conserving biodiversity. For example, Sarapura et al. (2016) illustrate that women involved in the Management Consortium of Native Potato Producers of Junin and Huancavelica in Peru (COGEPAN) took part in a project called Papa Andina, coordinated by the International Potato Center (CIP), which helped linking farmers to markets and created formal market chains for native potatoes. This project empowered women who conserved native potatoes, since they were able to use proceeds from selling native potato to buy land under their own name or be able to rent or sharecrop land to increase agricultural production. Men and women involved in the project were also able to access credit, and open bank accounts compared to counterparts who were not part of this project.

Research on potato genetic biodiversity in the Andes has illustrated that in many instances farmer knowledge on genetic biodiversity cannot be separated from the natural and cultural contexts from which it has emerged, including resources, relationships (kinship), and community relations. The traditional knowledge that women possess has been verbally transmitted from one generation to another, from person to person (mothers to daughters). It is based on the *saber campesino* (spiritual, ecological, geographical knowledge), sense and wisdom. It is entrenched in the Andean cosmovision in which the differences between elusive knowledge and physical things are frequently imprecise and vague. Research initiatives and interventions like the Papa Andina project have illustrated that the role that women potato farmers play in the Andes in safeguarding the traditional information, knowledge, traditions, and practices of producing and reproducing the Andean potatoes cannot be overestimated.

Based on our experience in potato research and review of existing literature, several opportunities and challenges related to conservation of genetic biodiversity and the role of gender relations are illustrated below.

As illustrated in Table 14.1, how men and women's work is valued and the resources they control may favor or limit the ability of women to not only participate in conserving potato genetic resources but also benefit from their efforts

Table 14.1 Gender relations in conservation and access to potato genetic resources

Gender relations aspects	Aspects that favor women's engagement	Aspects that limit women's engagement
Valuing of men and women's work	• Women conserve and control genetic resources • Women's work in conservation is recognized and highly valued	• Women's role in conserving genetic resources is not highly valued by actors in the potato sector
Access to resources, assets and control of benefits	• Women have highly significant local knowledge on genetic conservation of landraces • Women are able to financially benefit from their genetic conservation efforts and be able to decide on how to use their benefits.	• Men control information means and channels • Men control access to land
Gender norms and opportunity structure	• It is women's work to save seed of different varieties and store it • Women have access to local knowledge on genetic conservation	• Men control marketing channels and land which makes it hard for women to benefit from their conservation efforts. • Men have more access to new technical information on potato production since they are targeted by extension.

Projects such as the Papa Andina Initiative integrated this traditional or "emic knowledge" with scientific and technical knowledge to give place to three different areas or spheres of innovation interconnected with the utmost essential and most original opening points for recognizing change in peasant women—(1) technology use, (2) social norm change or social innovation, and (3) economic resilience. For example, while it is acknowledged that women in peasant communities possess an intrinsic adaptive capacity to maintain, manage, and preserve the native potatoes and know how to adapt native potatoes to different climatic conditions, pathogens, and plagues (Sarapura 2013), through the Papa Andina initiative, women were able to strengthen their innovation capabilities as well as their ability to conserve genetic diversity in situ. For example, they were able to access new technical information and knowledge on potato production as well as receive market information using Information and Communication Technology (ICT) tools. Gender relations of power that restrict access to information for women may affect their ability to benefit from their genetic conservation efforts.

The analysis presented above highlights the importance of understanding gender relations in potato genetic resources conservation and use. Furthermore, it illustrates that the involvement of both men and women in the design and implementation of interventions and policies to support conservation should be prioritized.

14.3 Crop Improvement

It has been suggested that conventional breeding has not been able to benefit farmers in marginal areas such as in Rwanda because farmer traits are not considered in the breeding process, which leads to relatively low adoption (Muhinyuza et al. 2012). Hence, failing to consider the trait preferences of farmers, especially those in marginal areas, can lead to the promotion of varieties ill-suited to the needs of vulnerable groups such as women. For instance, one of the persistent gender gaps in agriculture is lower adoption of modern varieties among women producers (Ashby and Polar 2019). Overlooking traits important to women farmers and consumers may lead to women's disempowerment and aggravated household food insecurity and poverty (Tufan et al. 2018).

More broadly, the need for the involvement and participation of farmers in the development of new crop varieties for smallholder farmers was explicitly explained by DeVries and Toenniessen (2001). As noted by these authors, farmers should be involved in all aspects of variety development that include priority setting, early generation breeding, variety testing, and selection so that breeders obtain regular input from farmers that enables them to structure their selection indices accurately. Thus, farmers should not be just technology recipients and beneficiaries but actors who influence and provide key inputs to the technology development process (Machida et al. 2014).

The literature on the gender dimension of agricultural production in Africa and elsewhere points to the connection between gender and crop preferences as well as gender-related dynamics and constraints in technology adoption. Indeed, given that women and men have different roles in providing for household food security, it is not surprising that research generally portray that they have different preferences as well (Meinzen-Dick et al. 2010; Tufan et al. 2018). However, until recently the gender dimensions of trait preferences have gone largely unrecognized and unappreciated as a distinct area of research on potato breeding. To address this weakness, potato research in increasingly integrating men and women farmers' views in crop improvement initiatives by collecting sex disaggregated data to identify differences and similarities between men and women's trait preferences. A recent study in Ethiopia, for example, suggested that there were only a few significant differences between men and women's desired traits (Kolech et al. 2015, 2017). They suggested that women in one of the study sites were more concerned with long stolons than men since this was an indicator that the potato could be harvested sequentially, according to needs, and not all at once addressing women's food security concerns, while on the other hand men were more concerned with low soil fertility which lowered yields, and limited market access. In their analysis the authors concluded that men were more concerned with market demand while women concerned more with food security. Gender relations that promote engagement of men in high value market chains while relegating women to the domestic economy with concerns mainly for family food may disadvantage women in the long run. In Kabale, Uganda, one of the potatoes growing districts, it was noted that both men and women traders prefer large sized potato suitable for making French fries (Bonabana-Wabbi et al. 2013). Therefore, understanding what men, women farmers and traders prefer is important for a breeding program.

14.3.1 Breeding Objectives

In line with the growing evidence of the need to integrate gender considerations into crop improvement, a key objective of the potato breeding program at CIP is to characterize gender differentiated preferences for traits, in different agri-food systems, and what the consequences would be of having those traits available to help breeding strategies accelerate varietal development. Research has been conducted in Latin America and Africa to understand gender differences in trait preferences. Research in Peru for example highlighted that men often preferred improved potatoes that are high in yield, resistant to hail and frost and to diseases (particularly late blight), while women often focused on culinary quality for fresh consumption, they prefer potato varieties with shallow "eyes," and as well as yellow/cream flesh color. A study in Ethiopia (Mudege et al. forthcoming) also noted some differences in men and women's trait preferences (see Box 14.1).

14.3.2 *Gender Relations and Adoption of Improved Potato Varieties*

Due to the different roles that men and women play in families and communities, traits related to consumption exhibits some sharp gender differences. Based on our experience in potato research and review of existing literature, below we summarize examples of how gender relations may shape the opportunity space for adoption of potato varieties.

As Table 14.2 illustrates, when women's preferred traits are valued and integrated into the breeding program, this may improve women's willingness to adopt new improved varieties. Thus, breeding programs need to go beyond profits to ensure that key important traits that may not have an immediate economic value but are important to women are not neglected.

In line with their gender roles, women prefer traits that lessen their burden and time in food preparation. For example, women in Ethiopia preferred potato that did not have deep eyes because it is easy to peel and prepare local dishes (Mudege et al. forthcoming). In Peru it was noted that women preferred some varieties because they could make soup or different traditional dishes or because they could be used for traditional rituals such as testing the patience of a new bride by asking her to peel a particularly difficult-to-peel potato (Tapia and De la Torre 1998). Some potato varieties are regarded as more nutritious, particularly for pregnant women (Tapia and De la Torre 1998). Preferences of certain varieties for their perceived maternal

Table 14.2 How gender relations shape the likelihood for adoption and use of improved potato varieties

Gender relations aspects	Aspects that favor women's adoption and use of improved varieties	Aspects that limit women's adoption and use of improved varieties
Valuing of men and women's preferred traits	• Men and women are consulted on preferred traits and these traits are considered in a breeding program	• Gender relations that limit women's access to markets and their ability to benefit from crop marketing may limit their need and ability to adopt improved potato varieties • Traits that women value may be neglected by a breeding program if they are deemed not to have any economic value
Access to resources, assets and control of benefits		• Men control access to land and finances, thus women may not be able to adopt or benefit from improved varieties that demand high inputs such as fertilizers.
Gender norms and opportunity structure	• Women are interested in food security so are likely to adopt improved potato varieties that are high yielding	• Domestic use by women of traditional varieties for important social rituals.

health effects, i.e. more nutritious, is a fundamental revelation that can help breeding programs to better target women by developing varieties with traits that directly benefit them. Thus, to be able to breed potato that farmers can adopt more easily, research is increasingly realizing that we need to understand the socioeconomic and institutional contexts in which farmers operate.

Furthermore, farmers' trait preferences should also be understood in a holistic manner, that not only looks at gender roles but also at the sociocultural environment in which variety and trait decisions are made. When farmers consider agronomic traits such as yield, for instance, they do not do that in a discrete manner. The yield trait is important for commercial and food security purposes, as illustrated by the following example from Ethiopia (Box 14.1):

Box 14.1 : Men and Women's Differences in Potato Trait Preferences in Ethiopia

Gender mainstreamed Participatory Varietal Selection Activities (PVS) in Ethiopia showed that man and woman farmer perspectives need to be integrated to ensure that released varieties meet their needs. Men and women had different preferences in their selection of potato clones. For example, out of five important clones, men and women's preferences matched in the top two selected clones. However, in the top three clones that men and women selected, they preferred resistance to disease and pest attack, high yield as well as tuber sizes that are preferred by the market as criteria. However, while men only selected clones which they perceived as free from pests and diseases, for their second and third best clone, women selected some potato clones that had some insect and pest damage because these clones had a size which was preferred by markets, had a good shape and superficial eyes (shallow eyes) which made processing and peeling easier. For the clone which they selected as second, women stated that the clone had a disadvantage in that it had cracks which increased loss upon processing and did not make good potato stew as this type of cracked potato would disintegrate upon boiling. This study shows that although men and women are interested in marketable traits, women had additional requirements particularly related to processing that men did not have (Mudege et al. forthcoming).

In some instances, the introduction of new improved potato varieties does not require only the promotion of new potato varieties as alternatives to traditional/native or local potatoes, but also involves the construction of different social configurations, it requires new patterns of farming and the enrolment of many different social actors and different ways of interaction. For example, introduction of improved varieties in the Andes also entails training of farmers on new production technologies. New production technologies may have a gender implication. Breeding programs, on their own, cannot change social configurations to be more empowering to women. To achieve this, interventions should be linked to other programs focusing on gender transformation; for example, in terms of access to and control of resources, finance, access to markets, all of which may determine farmer's ability to adopt new improved varieties. New social configurations could mean that breeding programs are linked more to other interventions focusing more on gender transformation in communities. Therefore, gaining support for improved varieties is not a simple process, especially if the new improved varieties are being promoted in contexts where there are already established potato regimes and landscapes. As others have already argued, although end users may not have all information, they are often the experts when it comes to knowledge of the local context (in which new varieties are being introduced), hence it is imperative that their needs and aspirations are taken into considerations in all initiatives that touch on them right from idea conception (Njenga and Gurung 2011). It therefore becomes important to consider farmers' preferences, values, practices, and behaviors because these matter in understanding the complexities surrounding adoption or rejection of new improved potato, and puts farmers' agency and knowledge at the center of analysis.

Additionally, gender differentiated access and control over assets and resources can influence the crop and/or variety selected for production (Njenga and Gurung 2011). In the Andean potato-based production systems, women have less access to labor, have difficult or limited access to farm equipment for land preparation, and face restrictions to access and use inputs such as fertilizers and pest control products. These constraints shape women's preferences for lower yielding native potatoes that have lower market value but also require less inputs and labor (Polar et al. 2017). Studies on gender-differentiated crop trait preferences show evidently that varietal choice is related to access to and control of resources, rights, and responsibilities differentially shared by men and women engaged in production, processing, and marketing (Christinck et al. 2017; Bentley et al. 2018; Ashby and Polar 2019).

14.3.3 How to Integrate Gender Concerns into Breeding and Crop Improvement

To integrate gender concerns into the potato breeding program, CIP is using a two-pronged approach:

The first prong is to ensure that the interest of men and women are taken into account in the setting of breeding objectives. This can be done if in depth research is conducted through analysis of secondary data and/or collection of new data to identify the needs and interests of men and women who are end users and target groups. The second prong is to ensure that men and women groups are engaged in the evaluation of new potato clones either through participatory varietal selection (PVS) or participatory plant breeding (PPB) (Box 14.2). It is felt that these approaches will increase likelihood of adoption of varieties and technologies while at the same time addressing issues of gender equity since the needs of men and women will be addressed by the breeding program. In line with this, CIP has developed and tested a manual for gender mainstreamed participatory varietal selection to collect information on trait preferences as well as to allow men and women farmers to evaluate potato clones before release.

Box 14.2: PVS in the Peruvian Andes

In the mid-1980s, potato breeders in the Instituto Nacional de Innovación Agraria of Peru (INIA) and the International Potato Center (CIP) jointly evaluated advanced potato clones from a diverse late blight-resistant population. These evaluations were conducted in farmers' fields. Three hot spots for late blight were selected in the Department of Huanuco in central Peru. In return for their support, farmers received one-half of the output of the trials. Lastly, the retained seed provided farmers the opportunity to start multiplying and using any clone that fits their circumstances. In the final evaluation, six of the most promising clones from 6 years of on-station selection and 3 years of testing in farmers' fields were selected. By the time one of the selected clones, Canchan-INIAA, was released, dozens of farmers were growing the variety, and a considerable amount of seed had been distributed via the informal seed system. Both men and women were involved in the evaluation of clones. The early adoption of Canchan INIAA before its release was a result of positive evaluation by men in terms of yield, earliness, and resistance to late blight, and by women in terms of the skin color, storability, and consumption quality. Women played a significant role in conserving and managing Canchan INIAA potato variety seeds, while men had a strong role during weeding, hilling, and harvest. It was actually woman farmer- managed Canchan INIAA seed that was eventually used to release the variety (C. Fonseca, personal communication), and the inclusion of both men and women in the selection of Canchan-INIAA, not only because of resistance to late blight, but the quality attributes, may be the reason why it is still a popular variety more than 20 years after its release, and after having lost the resistance attribute.

In addition, it is important that potato gender research contributes to current work on crop ontologies. In this way, end user priorities can be integrated into breeding programs by standardizing farmer priorities while ensuring that the breeding programs are not overwhelmed. The CGIAR Research Program on Roots Tubers and Bananas (RTB) for instance has developed next-generation breeding systems based on the collection and application of genetic, metabolite, and phenotypic data together with participatory, gender-responsive research on farmers' trait preferences, aiming at establishing a connection between preferred traits and the genetics that explain them. Thus, efforts in potato research should continue to contribute to the selection of traits that can be used in genomic prediction, and use of weighted selection indices that aim to ensure that new varieties have wide and gender-equitable impact (RTB 2016). By building up and contributing to sex disaggregated crop ontology database, gender mainstreamed potato research will ensure that potato breeding objectives continue to evolve in ways that are responsive to men and women farmers in different agrifood systems. This effort is being integrated in the definition of more specific breeding product profiles that can take user and gender-differentiated preferences into account.

14.4 Managing Priority Pests and Diseases

The adoption of pest and disease management practices is an important topic to reduce production losses in potato. In some regions of the Andes, potato production is associated with heavy use of chemical inputs to manage pests and optimize profits (Mera-Orcés 2001). Similarly, in highland regions of Africa diseases and insect pests have the greatest potential for potato yield reduction, thus farmers rely heavily on pesticide use (Okonya and Kroschel 2015). However, most studies on the topic adopt a limited approach of only looking at gender roles (see Malena 1994) and how differentiated access to land, labor, finance, and education, shape women's technological needs differently (Malena 1994). In addition, writing on gender in the Andes, Paulson (2003) suggests that women are not homogenous as we have widows, married women, young, and single who have different experiences and different needs. Thus instead of focusing on a dichotomy of static gender roles, we should instead focus on the "possibilities for a more dynamic conceptualization of social roles in relation to changing social, economic, and environmental condition" (Paulson 2003). Since women differ by age, socioeconomic status, and other variables, we instead look at gender relations, which help to explain the roles of men and women and the decisions they make in relation to pests and disease management.

14.4.1 Gender Relations and Pest and Disease Management in Potato

Gender relations play a critical role in the management of pests and diseases in potato. Recent research in Malawi (Mudege, unpublished results) illustrates that social relations that privileged men's potato crop over women's crop for spraying meant that women's potato fields were more likely to be affected by late blight compared to men's crops. This in turn affected the availability of quality planting material for women if their crop was diseased, since farmers selected planting material from their ware potato crop for the next cropping season. Additionally, since men in men-headed households were expected to grow potato for the market, while women's potato plots were mostly for family consumption, men controlled the family budget. This had gender implications in pest and disease management. While lack of money to buy chemicals for spraying was mentioned as a key obstacle for both men and women, women were disproportionately more affected as they were only able to buy small quantities of chemicals and also because as mentioned, men's potato crop was privileged when it came to application of chemicals (see Box 14.3). While some women could afford the chemicals, many often mentioned they may not have access to knapsack sprayers that were even more expensive to purchase and difficult to rent from other farmers. Where men and women from the same household cultivated and managed different plots, the men's plot were prioritized when it came to spraying for diseases. Women often mentioned that they could not afford to buy the knapsack sprayer because in families where men and women had separate plots often there was no cooperation, leaving women with a higher burden of taking care of family consumption needs and little to invest in agriculture. Additionally, women often mentioned that because they had no money to buy quality seed, they sometimes purchase diseased seed because it is cheaper.

Box 14.3: Using Pest and Disease Control Methods in Malawi

Women focus group participants in a study conducted by CIP in Malawi debated whether it was easy for women to purchase and use chemicals if their potato crop was affected by disease:

Participant: …some [women] don't have the money we have seen the whole field being infected by disease but people failing to get money to buy the chemical. (*People arguing, many people speaking at the same time*)

Participant: The chemical is 100 kwacha it is not expensive anyone who want can buy the chemical.

Participant: Let me speak for myself, there was a year when my potato was destroyed because I had no money for chemicals. …. My husband was working but he was not being paid so I thought that if I go and borrow money from somewhere how will I repay the money so the whole field for potatoes was destroyed.

Participant: You could have sold 2kgs of maize and got 100 kwacha.

Facilitator: Your friends are surprised.

Participant: They are surprised because they can afford but I am talking about what happened at my home I am not talking about someone else (Dedza, women farmer group members)

Even if you have money to buy the chemical the problem is we have one [knapsack sprayer] in the house and since it was bought on the husband's budget he takes it with him to the field everyday and you will have nothing to use to spray with until all your potato is destroyed (Ntcheu, women nongroup members).

Source data (Unpublished data Malawi- Integrating gender into RTB research to improve development outcomes project)

Table 14.3 How gender relations shape engagement of women in pest and disease management

Gender relations aspects	Aspects that favor women's adoption and use of improved pest and disease management options	Aspects that limit women's adoption and use of improved pest and disease management options
Valuing of men and women's crops	• Cheap farmer-based pest and disease management practices • Pest control methods which may be labor-intensive but require very little use of outside inputs	• Men's potato crop is valued more than potato cultivated on women-controlled plots because men's crop is for commercial purposes and women's crop is for domestic use. Men's crops get preferential treatment when it comes to pest and disease management (this is context specific in some African countries but not the same in the Andes)
Access to knowledge and other resources	• Women have access to information about pest and disease control which often is a preserve of men	• Information on pest and disease management is packaged in ways that favor/promote men's access
Gender norms and opportunity structure	• Women are consulted and engaged in decision about crop protection • Cooperative household decision-making on agricultural investments • High outmigration of male labor which leads to feminization of most practices related to pest and disease management	• Men dominate decision-making about adopting and using pest and disease management technologies

Box 14.3 shows that power relations in the household may determine distribution of resources and women's ability to use household resources in controlling pests and diseases in women managed plots. In a different context in the Andes, there is gender differentiated specialization in terms of pest management, which is usually perceived as men's job. Therefore, men have access to information and knowledge on pesticide use and pest control in general, including IPM in the field, while women are more interested in controlling pests in the store, which is more under their control.

Sharma et al. (2017) identified the use of disease-free tubers as seed as one of the key ways of controlling bacterial wilt. Since farmers normally use saved seed from their crop, small holder-friendly seed technologies will benefit women immensely in terms of increased productivity even if the land under production does not expand because they will have lower pests and disease burden.

Based on our experience with potato research and gleaning from some of the literature we reviewed, Table 14.3 shows how gender relations shape the opportunity space for women potato farmers.

If women are able to gain access to pest control information and methods are responsive to the resources available to them, women will benefit.

In many communities, such as in the Andean region, men-biased sources of information and knowledge on pest and disease management are used. For instance, information and knowledge on pest and disease management accessed through training events or the fact that men have higher literacy and can read labels of commercially available products more easily. Women on the other hand had lower literacy levels and lower command of Spanish, which limited their access to information provided in written form or through capacity building events (Polar et al. 2017). In Uganda, women and men's sources of information also differed. For example, in Eastern Uganda men obtained their information from vendors at local markets or from labels in pesticide packages, while women indicated that extension agents were their most important source of pesticide information (Erbaugh et al. 2003). However, in Uganda like in many other Africa countries extension is often underfunded. The gap in terms of access to information may need to be reduced, for example, by packaging information and presenting it in ways which makes it equally available to men and women, including use of local languages and bringing information and training closer to villages since women may not be as highly mobile as men.

While some studies in Latin America (see Polar et al. 2017; Mera-Orcés 2001) have focused on access to information by men to promote the use of pest and disease management technologies, approaches to gender and pest and disease management in potato need to go beyond just access to knowledge, information, and pest management technologies. Access to training and information in many cases intersect with control over assets and resources and power relations within households which affect use of improved pest and disease management technologies by men and women within households and communities. For example, although both men and women in Malawi had the same knowledge related to spraying of chemicals, women often mentioned that they did not have money to buy chemicals nor access to other resources that were essentially controlled by men, as illustrated in Box 14.3. This case clearly describes how access to assets and resources interplays with gender relations of power to determine the use of pest and disease management technologies. Even if economic resources are available to women, priorities are established based on whose crop is valued, who has access to and control of resources such as spraying equipment, and who makes decisions about use of available resources.

As relations of power, gender relations shape men and women's access to assets and resources. Access to and control of resources in turn may determine the type of potato pest and disease management technologies that men and women can adopt. A cross country study in Bolivia, Ecuador, and Perú on potato-based agricultural systems found that men and women have different perceptions and usage of chemical or organic inputs for pest and disease management. Women often preferred organic inputs and/or management practices because of their low cost, even if they were time-consuming; men on the other hand preferred chemical control because, they regarded chemicals as more effective against potato late blight (Polar et al. 2017). However, in adverse environments, both men and women were inclined to use chemical inputs to reduce the risk of economic losses in potatoes that were produced for markets (Polar et al. 2017; Mera-Orcés 2001). In this case the use of chemical inputs is conditioned by the women's lower access to economic resources, as well as by the purpose why the crop is cultivated: domestic versus market.

The role of women as caregivers influences the distribution of activities linked to potato production. Although pest and disease management in general is perceived as both men and women's responsibility, the actual application of pesticides is more associated with men, while food preparation for field workers is a woman's responsibility (Mera-Orcés 2001 [Ecuador]; Erbaugh et al. 2003 [Uganda]). The examples above show that it is important to analyze both division of labor and decision-making processes related to use and adoption of pest and disease management options.

Hierarchical gender relations that denote men as decision makers at community level may result in women's needs not addressed or considered in programs. Sarapura (2013) reported that in peasant communities in the Andes, decisions on applying a new technique or a new method of cultivating native potatoes had to be approved by the community council. These councils are dominated by men as leaders and number of members. Even though, women are in charge of most of the production processes related to potato, in the case of a disease outbreak, community councils decide on the way forward. For example, in the community of Racracalla in Peru, peasant producers who were concerned about a new disease, late blight, caused by *Phytophthora infestans* and did not have access to formal extension systems, tried to resolve this issue by using their community organizations. For example, they tried to deal with the disease using different methods such as high tilling, construction of canals inside the plots, use of ashes in order to repel the disease. All the solutions were agreed upon by the community councils and tried. Sarapura (2013) identified gender implications in that if extension systems are just directed to the community councils which are men dominated, women may be left behind. In this specific case, decision-making is also controlled at the collective level, where women's perceptions, needs, and limitations may not be adequately considered, since all decision makers are men.

Pest and disease management in the potato crop is a more complex issue and depends on the type of pests or disease, the alternative control methods available, the inputs to be used, the information and knowledge available and the access to external sources of information. There is the need to also differentiate the perceptions of men and women regarding alternatives, and also understand the factors that influence access to and control of resources, so that suitable control methods can be identified during participatory research and used by farmers. This aspect becomes even more critical since climate change is likely to influence the increase in pests and diseases in the potato crop in different parts of the world.

14.5 Access to Potato Seed (Seed Flows and Networks)

The lack of disease-free, quality planting material has been mentioned in several studies as one of the key reasons for low potato productivity (Hoque and Sultana 2012; Lutaladio et al. 2009; Gildemacher et al. 2009). For example, in Malawi,

Demo et al. (2008) suggests that lack of quality planting material is a key barrier to improved productivity. In many developed countries, formal systems are the source of quality planting seed for many crops including vegetatively propagated crops such as potato. Although the formal sector dominates the seed systems in developed countries, in developing countries, in spite of huge investments in the sector, "90–95% of the world's small holder farmers still obtain seed from informal sources, largely from other farmers" (Reddy et al. 2007). The situation is particularly dire for vegetatively propagated crops in many Sub-Saharan Africa countries. There has been less focus on tuber crops, legumes, and horticultural crops among the formal seed systems of SSA (Biemond et al. 2012). It had been noted that in many Southern African countries many commercial seed companies are not interested in producing seed for vegetatively propagated crops because of added complicating problems such as low multiplication rate, bulkiness, short shelf life, and difficult maintenance during the dry season.

There has been debate regarding whether formal seed systems or the integration of formal and informal elements are good for potato seed systems as and to understand which are better for women and men farmers in marginal areas. Potato projects are increasingly encouraged to collect sex disaggregated data as well as to conduct studies to understand how gender relations affect the dissemination of new seed technologies. Based on our experience in potato research and the literature we reviewed, below is a table outlining some of the ways gender relations may impact on the efficacy of seed systems for women.

Table 14.4 gives examples of how gender relations permeate this sector. Approaches to seed certification clearly show how gender relations have an impact on the seed system. For example, some approaches regard formal seed certification

Table 14.4 How gender relations shape engagement of women in potato seed systems

Gender relations aspects	Aspects that favor women's access and use of quality seed	Aspects that limit women's access to and use of quality seed
Which channels are valued for seed dissemination?	• Improved technologies that promote availability of affordable quality seed • Availability of quality seed in local seed networks and local markets friendly to women	• Unaffordable quality seed • Methods of availing seed that value masculine channels • Formalization of potato seed systems (certification) that dispossess women of their role as seed producers and conservers
Access to knowledge on quality seed and other resources	• Women access not only knowledge but also other resources, mainly credit, so they can run own seed businesses or gain access to quality seed for purchase	• Men dominate decision-making on seed at both household and community level • Men control the resources needed to purchase quality seed
Gender norms and opportunity structure	• Farmer-based quality seed management technics • Gender-responsive farmer-based seed producer groups at community level	

as essential to make available good quality seed to farmers. However, although certification could guarantee good quality seed, evidence from other crops have shown that in some cases women are less able to benefit from certification schemes than men. For example, women may lack the resources needed to have them certified as seed producers, thus are dispossessed from their role as seed producers and keepers. In addition, currently in many countries where CIP is intervening, new improved varieties of potato are available, but taking longer to disseminate, which means technologies that include public–private partnerships for rapid multiplication would benefit both men and women potato farmers especially if they lead to the availability of cheaper, good quality seed. Development of an innovative seed systems model called the "3G approach," which combines rapid multiplication of tuber seed (through aeroponics) with technologies for good farmer seed management in about three generations rather than in the conventional seven have the potential to increase the availability of good quality seed and may also lower the cost of seed thereby improving seed security (Demo et al. 2015).

However, a CIP study in Malawi (Mudege and Demo 2016) shows that commercialization of seed could ensure the availability of seed but may not ensure accessibility of seed. Although both men and women prefer quality seed, women often lack the income to afford clean potato seed. Both men and women expressed a willingness to pay more for good quality, but men could afford to pay much more than women. On one hand, women preferred noncash transactions—such as paying for seed with labor or asking their friends for loans of seed to be returned after harvest. On the other hand, most men said they had opportunities for odd jobs in the community and surrounding areas which they could use to raise money to purchase seed. Women said that they could rely on their friends to get seed to repay after harvest, but it was usually difficult for men to give another man seed to plant (Mudege and Demo 2016). Likewise, in the Andes it has been suggested that because women have lower access to economic resources, they prefer low cost technologies and use local seed which they can access through barter, in kind payments, rituals in festivals and farmer to farmer exchange (Thiele 1999; Tapia and de la Torre 1998; Zimmerer 2003; Sperling and McGuire 2012). As a result of limited access to cash, private sector markets may not increase women's access to quality seed, but local markets because of mechanisms where women can provide labor in exchange for seed. Thus, it is not only important to target multiplication of seed to increase availability of quality seed, but to focus on the mechanisms and pathways that can ensure women's access to seed despite the limitations they face in terms of access and control over other resources.

Furthermore, FAO (2008) acknowledges that commercialization of agriculture, including seed trade, tends to exclude women. Women are usually excluded because they lack the resources which are needed to participate in commercialized systems. Zimmerer (2003) regards women as important in seed flows but men are often engaged in seed dissemination beyond the community in the Andes. Women are key players at procuring seed within the community. Both men and women are engaged in seed flows "at the extra community level"; however, when it comes to procurement of seed from the market and development institutions women's role reduces.

Therefore, suggestions for commercial seed systems need to be evaluated from a gender perspective—in order not to do harm to women (Gibson et al. 2009). In addition, formal markets may not be able to sell fertilizers and other inputs in the smaller quantities that women can afford. Thus, vibrant local markets which are not only cash reliant but based on other forms of reciprocity and relationships within the community would be able to meet women's needs. However, when commercially produced seed is available, the effects have the potential to trickle down when those farmers who can afford to buy clean seed may have relatively clean planting material to sell to other farmers at harvest time.

To accommodate women and other poor marginalized farmers, research is also moving increasingly towards developing farmer friendly seed management technics such as negative and positive selection, which have been promoted to improve the quality of seed in the informal sector. Positive and negative selection are two methods introduced in Malawi to help farmers to access healthier planting material. Positive selection requires marking potato plants as parent stock. Plants chosen must display good growth and most importantly should show no signs of bacterial wilt and/or viruses. Negative selection is selecting plants that will not be used as parent stock. Plants marked are those infected with bacterial wilt and/or viruses (Tantowijoyo and van de Fliert 2006 see also Salazar 1996; Njukeng et al. 2007). Farmers who belong to farmers groups are taught how to identify health plants and select them as parent stock for seed. Farmers are taught how to identify the health plants which are supposed to be "big"; have many and thick stems; have dark green leaves without malformations; have many, large and well-shaped tubers; do not show obvious disease symptoms (Gildemacher et al. 2007, 2011). A study by Gildemacher et al. (2011) in Kenya shows that positive selection provided small holder farmers with better quality seed and led to high yields. However, potato research in Malawi has for example illustrated that women are often left out on training on agronomic practices and thereby lack the knowledge they need to improve their productivity and accessing clean seed (Mudege et al., 2015a, b).

Since positive and negative selection relies on visual inspection, it is not entirely reliable as it needs a farmer to be experienced in spotting the diseased plants (Chiipanthenga et al. 2012). In addition, even when farmers have knowledge, cultural practices and norms can militate against dealing with pests and diseases. For example, in the Andes it was noted that farmers were afraid to rogue diseased plants for fear that if they tampered with food crops they could be punished by the ancestors (Thiele 1999). Very few studies have also discussed what men and women know regarding seed quality and how to maintain it. Given the critical importance of quality seed it is important to know what farmers know about seed quality and how this can affect the fight against potato diseases and pests in the informal seed system. Thus, work has been conducted in different countries, including Malawi and Uganda, to understand what men and women's knowledge and perceptions are regarding seed quality.

Potato research has also looked for innovative ways to link the formal and informal seed systems to ensure accessibility of seed. For example, the Consortium of Potato Producers from Ecuador CONPAPAA initiative in Ecuador sought to produce good quality seed by providing producers (farmers) seed grown through aeroponics and training them to multiply it as quality declared seed (Kromann et al. 2016). This approach managed to make available seed which was less expensive than certified seed. Using this approach allowed women and indigenous farmers to access quality planting material through the merging of formal and informal seed systems than would not have been possible with conventional approaches. In Peru, the informal system satisfies the seed needs of 99% of the potato growers. Seed in the informal sector in Peru is locally available and cheaper than certified seed. This shows that an integrated seed system under certain conditions can outperform standalone formal and informal seed systems. Orrego and Andrade-Piedra (2016) present a case in Peru where quality seed from the formal sector was distributed through both formal and informal channels for further multiplication and dissemination. Having locally available clean and cheap seed is particularly important for women, since they have low access to monetary income. Research on potato in the Andes has already shown that women are often more engaged in sourcing seed within the Andes using family and community networks (Tapia and De la Torre 1998; Zimmerer 2003). Some women's movements in Latin America are building on traditional roles of indigenous women in seed management to positioning themselves as privileged custodians of seeds and biodiversity (Aguayo and Hinrichs 2015).

Given the current challenges to design seed systems that contribute to resilience of farming systems to threats, such as climate change, it is important to understand and take into account both women and men perceptions for seed-related alternatives, particularly, if seed businesses could become an important source for income for women and youth.

14.6 Marketing, Postharvest Processing and Utilization

This section will look at how female farmers face gender-specific challenges in relation to potato markets, and the discussion of key findings describes how gender matters in marketing. Urbanization has provided potential markets for potato and potato products due to demand from urban consumers (Bonabana-Wabbi et al. 2013). However, farmers have very limited market information and cannot take advantage of emerging opportunities to meet demand. Bonabana-Wabbi et al. (2013) suggest that because of lack of market information farmers are prone to exploitation by middleman. To address this, farmer collective marketing has often been used to improve the bargaining position of farmers in marketing (Bonabana-Wabbi et al. 2013). However, CIP research on gender and collective marketing in Malawi, illustrated that gender inequalities were often reinforced in marketing groups (Mudege et al. 2015b). For example, in groups, payment was often given to male household heads even when women had submitted the potato for sale in cases where their husbands were also group members. This was a different experience from the COGEPAN in Peru (Sarapura et al. 2016) were women directly received their money and invested it for their own benefit. While in COGEPAN, the approach

Table 14.5 How gender relations shape engagement of women in potato markets

Gender relations aspects	Aspects that favor women's participation in markets and their ability to benefit	Aspects that limit women's participation in markets and their ability to benefit
Power relationship between male household heads and women	• Approaches that allow women to have direct access to markets and benefits	• Ideologies that regard male household heads as the official representatives of the family and therefore in charge of markets and marketing proceeds • Men control all decisions related to marketing.
Access to markets	• Women participate in case of low potato quantities	• In men-headed households, high volumes of potato sold are usually decided by men
Gender norms and opportunity structure	• Training on business management and market access targeting women	• Men make decision on how much to sell, where to sell and whom to sell to • Lack of mobility for women restricts access to high value markets

challenged gender relations of power, in the Malawi case, collective marketing reinforced existing gender inequalities.

In Kabale, Uganda, it was noted that men and women were equally engaged in potato trading (Bonabana-Wabbi et al. 2013). However, studies often rank types of trade in order of how many people are engaged without looking at the division of participation by sex to ensure that targeted gender responsive interventions are made. For example, Bonabana-Wabbi et al. (2013) reported that in some districts in Uganda the majority of those engaged in potato trade where in retail compared to wholesale followed by collectors, agents, and transporters. However, her study and other studies do not segment these value chain actors from a gender perspective. For example, it is not clear how many men and women are engaged at these levels and what profit margins accrue to men and women at these different nodes (Box 14.4). Research on potato marketing for example could collect this information since it is critical to know who benefits from potato technologies including improved varieties (Table 14.5).

Potato research has also looked at the gender-related constraints to marketing. For example, it was noted that in Uganda men sold more potato than women because they had contact with buyers and brokers whom they meet at local markets (Sebatta et al. 2014). In order to gain insight on the lack of participation by women in potato commercialization, the structures influencing men and women's participation need to be understood as well. For example, Jibat et al. (2007) notes that in Ethiopia women are engaged in potato markets when selling at low quantities and in local markets while men dominated bulk sales as well as sales at markets far from the village. However, the potato story is not all doom and gloom since it is possible to conduct research and implement interventions that are gender responsive as the example below shows (Box 14.4):

Box 14.4: Markets and Inclusive Value Chains

Research in potato marketing in Uganda revealed that both men and women mentioned that they had limited access to markets and to timely information on potato market prices. Women also suggested that because on unequal relationships within households, men often decided on who to sell to, how much potato to sell and where to sell as well as deciding on use of potato income without necessarily consulting women. In addition, it was clear that the market itself was structured in ways that did not favor women's participation. For example, gender norms that designated potato as a men's crop meant that women who tried to sell potato on their own without their husbands were viewed with suspicion while husbands could sell crops on their own without their wives. Lack of mobility for women, poor transportation system and infrastructure, selling potato in large bags that women could not handle, and distance to markets were all mentioned by women as barriers to participating in potato marketing. To address some of these issues within the remit of the project, both men and women farmers and traders were targeted with training of trainers on business skills including marketing and net profit and loss calculation. Sixty nine men and 34 women were trained in marketing to ensure that they understood what marketing was, demand and supply forces, customer analysis and customer feedback mechanism, strategies, segmentation, product differentiation, and marketing information in the context of potato business. Effort was made to ensure involvement of women in this training even if they were not involved in management committees of associations. Since both men and women mentioned lack of market information as limiting their ability to negotiate with buyers, the project through its partner Self Help Africa (SHA) developed mechanisms for disseminating market-related information such as potato prices in Kampala through text messaging to group leaders who would then disseminate this information to other group members to ensure that farmers negotiate from an informed position. (Summary of our experiences with the ENDURE project in Uganda).

In Uganda (Mudege et al. 2016) notes that men mentioned lack of knowledge on what the market wants and also on price intelligence. Although men participated more in markets than women, both men and women did not have enough market skills or adequate market information. For example, women in Wanale mentioned not even knowing where exactly their husbands sell the potato, whilst men mentioned that they were often told that their potato was poor quality and also that they were not sure about the prices their potato would reach. Lack of engagement by women in markets was also noted elsewhere by Mudege et al. (2015b), who showed that in Malawi wherever husband and wife made joint decisions over use of income from potato, money was used to buy seed and fertilizer and other equipment; whereas when male household heads made decisions on their own, women mentioned that they often did not benefit from potato income. Thus, potato research and interventions should not just focus on technical aspect of potato production but also include social aspects such as promoting joint decision-making to ensure that both men and women benefit. In addition, methods to improve access to markets and bargaining power, for example, the use of Information and Communication Technologies (ICT) to inform women and men about market prices or weather forecasts could increase the ability of women and men to benefit from potato markets. In reference to de facto female- headed cotton farming households, Horrell and Krishnan (2007) state that "even without additional resources, greater profitability could be achieved from their existing agricultural output through access to better selling networks and buying consortia for inputs." The same can also be true for women potato farmers, if social norms that prevent them from marketing potato or deciding on household expenditure are challenged, women may be able to invest more into potato production resulting in better quality crop and better storage. Based on our experience and literature reviewed, Table 14.5 shows examples on how gender relations may influence the ability of women to engage and benefit from potato marketing.

Power relations between men and women heads of households and the position of women in the community relative to the position of men may shape the opportunity space for women to not only engage in marketing but also benefit from it.

In other geographical and social setups, such as in the Andes, research has shown that women dominate potato markets including price negotiation and control of income (Amaya and Alwang 2012), because men regard women as better negotiators. Additionally, Amaya notes that traders in the market are mostly women, and men regard it as undignified to argue with women when bargaining for a better price, thus let their wives sell potato. However, it is noted that women increasingly need information to participate in regional markets although marketing decisions are made jointly by husbands and wives. It was also noted that men monopolize the use of cellular phones to get information on markets (Amaya and Alwang 2012). Therefore, the cellular phone has not fundamentally changed gender roles: market decisions continue to be jointly made, and men continue to control access to market information.

14.6.1 Postharvest Utilization

Gender relations also influence women's access to postharvest technologies. In many parts of the world women are responsible for postharvest activities at household level. Knowledge and skills are passed on dynamically from generation to generation and these actions have long subsisted outside public and private sectors, R&D and agricultural extension systems (Tapia and de la Torre 2000), and oftentimes when technologies are being designed, the innate skills that men and women farmers have and the roles they play are not taken into account. However, since postharvest technologies have been developed and disseminated, there are concerns that men take over the latest technology and women can be left behind. A study in Bolivia (Polar et al. 2017) illustrates that women are the ones who are mostly engaged in grading potato by size. However, when technologies to mechanise potato selection were tested, only men were engaged in validation meetings. These technologies were later introduced but were not adopted by the women who were supposed to benefit. Women were not involved in evaluating this technology and found it difficult to use because the machine was tall and needed substantial physical strength to manually load heavy bags of potato. The machines were later adapted to meet women's needs, and their use helped reduce the amount of time women devoted to selection of tubers for the market. Ogunlana (2004) also suggests that women farmers can easily adopt innovations that can enhance their economic status if constraints pertaining to access to the technology (e.g. information and ease of utilization) are taken into consideration. Box 14.5 highlights a case where gender-related concerns were integrated during the introduction of ambient stores for storing ware potato in Uganda:

Box 14.5: Introducing Ambient Stores for Storing Ware Potato

A potato postharvest project implemented by the International Potato Center (CIP) in collaboration with the CGIAR Research Program on Roots, Tubers and Bananas (RTB) in Eastern Uganda to introduce ambient stores for storing ware potato under ambient conditions was rolled out through farmer and trader associations. Evidence from the associations demonstrated that women were underrepresented in leadership positions and almost nonexistent in storage management committees of the four associations.

Women expressed concern that if only men hold leadership positions in the management of the store, women may not be able to benefit from and to fully utilize the stores. For example, they noted that training targeted group leaders who were often men. Women insisted that for them to benefit from the stores, there should be gender balance in the people selected to manage the store. And although women had expressed the will to be active in store management, personnel from an NGO who facilitated the process of selecting store management committees revealed that most women refused to occupy these positions when elected. Some of the reasons women used to explain why they refused to occupy leadership positions included lack of time to commit for such duties, fear that their husbands would not allow them, meetings times may not be conducive for women, and some women regarded their illiteracy as a limiting factor.

While recognizing the need to have women represented in the group management committees, women also mentioned the risk of women being given token positions where they would not be involved in decision-making, stating for instance that women could be designated to deputy or committee member positions which did not have much influence in terms of decision-making. Some associations noted that since men had been the original members of the associations they dominated positions and it was difficult for women to break into top leadership.

The potato team worked with partners such as Self Help Africa (SHA) as well as farmer and traders' associations to review association management rules for gender inclusiveness. Where it was not possible, the team made a concerted effort to ensure that even if they were not in leadership positions, women representatives were also trained on store management to ensure that they are knowledgeable and actively involved. Under this project, 119 farmers and traders were trained on structure and governance (41 women and 78 men), 102 farmers (32 women and 70 men) were trained on enterprise analysis focusing on cost benefit analysis. Additionally a total of 106 farmers (69 male and 37 female) were trained on business planning in order to equip participants with skills to develop and use business plans to maximize profitability. A total of 111 (82 male and 29 female) were also trained on record keeping and store management. The aim of the module was to equip participants, in particular the store management committees, with skills to manage the ambient stores effectively and ultimately develop store management guidelines and records (Mudege and Mayanja 2016).

The case described in Box 14.5 illustrates that in some cases it is important to be gender intentional when designing potato interventions and research, otherwise researchers run the risk of not including women. It is important to identify the constraints that women face, so that these can be taken into consideration and addressed during the design of research and interventions.

14.7 Access to Extension and Training

Research has shown that women in most potato-growing areas have very limited access to training (Dersseh et al. 2016; Mudege et al. 2015a; Polar et al. 2017). For example, when women and other poor households are not targeted with training on new improved potato varieties, they did not regard lack of training as a key constraint to production (Dersseh et al. 2016). While the cause and effect relationship is unclear, it may be because oftentimes, women lack information about the importance of access to training and of improved varieties, and therefore they do not engage in training to the extent they should. Based on our experience and literature review, Table 14.6 gives some examples on how gender relations shape women's engagement and access to training and information.

If training is organized where it is physically accessible to women, women are directly invited to participate, and training methodologies are taking cognizance of women and men's capabilities, women will be able to access the information they

Table 14.6 How gender relations shape engagement of women in access to and benefit from potato-related extension services

Gender relations aspects	Aspects that favor women's access to and benefit from extension services	Aspects that limit women's access to and benefit from extension services
Valuing of men and women's access to extension	• Training venue and time selected according to women's physical mobility and time availability • Use of female extension workers • Using appropriate language and training methods	• Women and poor households not targeted on invited for training • Training targets men as heads of households and decision makers • Training targets group leaders. Men usually lead farmer groups
Gender norms and opportunity structure	• Access to information, technology, and resources to implement knowledge and skills acquired from training	• Men may make household decisions on who will attend training

need. However, whether women can use the knowledge and information they gain may depend on gender and decision-making power within households.

Research has shown that in East Wollega and West Shewa Zones in Ethiopia, even though the gender division of labor regards potato as a woman's crop, it was mostly men who participated in training on potato production, regardless the roles they played in its cultivation (Jibat et al. 2007). There are similar findings in the Andes where women play a key role in potato farming and management, yet they have low participation in training events compared to men (Polar et al. 2017). On the other hand, where potato cultivation is regarded as a men's activity women's contributions are overlooked (Mera-Orcés 2001) and women are denied access to the resources and training they need. In Malawi, potato research has also illustrated that women are often left out of training which further reinforces gender stereotypes that women know nothing (Mudege et al. 2015a).

The potato seed system intervention in Malawi relies on farmer participation both as group members and as lead farmers in the knowledge cascade approach. Men and women were targeted with training as part of groups. However, Mudege et al. (2015a) found that unless properly managed and monitored, delivery mechanisms that depend on participatory activities can be gender-blind. Interventions based on participation can become gender-blind if they do not recognize the differences between men and women. Gender-blind approaches "make assumptions, which lead to a bias in favor of existing gender relations … gender blind policies tend to exclude women" (March et al. 1999). For example, Mudege et al. (2015a) notes that farmer group training mostly targeted group leaders who were frequently men. If women do not have access to same extension services as men, their productivity and incomes may be limited. A study of factors influencing potential adoption of technology in potato-based systems in the Andes found that women had limited exposure to innovations through capacity building because they lacked skills in the official languages, had lower literacy levels, and the spaces where technical information was provided were dominated by men (Polar et al. 2017). Thus, to enhance women's access to and benefit from extension services, services need to be accessible to women in terms of language used, methodological tools, schedule times, and delivery spaces.

14.8 Conclusions

The analysis presented in this chapter discusses how gender relations shape the opportunity space for men and women potato farmers along the potato value chain. Power to make decisions and act on them is important and influences the ability of men and women to participate along the potato value chain and benefit from it. The analysis shows that lack of access to and control over assets and resources and decision-making can restrict women's engagement in potato production, marketing, and utilization. Institutional innovations and more gender responsive programming that consider the opportunities and constraints of men and women can contribute to equitable development of the sector. How men and women efforts are valued and the resources they control do not only affect women's ability to participate in the potato sector but also shapes their participation and ability to reap benefits from the sector at the same level as men.

Gender research related to potato production, use, and commercialization has in many cases attempted to show the role and importance of women in different potato-related activities from production to market, for example, as laborers or as custodians of genetic diversity and local knowledge. While most of this research has been concerned for example in bringing to the forefront hitherto hidden women's contributions to potato farming, it has also succeeded in developing initiatives to address both men and women's needs depending on the roles they play and the needs they have. However, research has also shown the folly of limiting gender analysis just to gender roles and access to resources and assets. This is so because even where women are the ones engaged in certain roles, they may fail to benefit from their labor because they defer to male households' heads to make decisions. Women may also not access training even for tasks they are engaged in because recruitment may favor male household heads to attend training even if they are not engaged in potato activities. As a result, it is important for potato research to also investigate gender relations and how the valuing of men and women's labor and contributions determine their ability to engage in potato production, management, and marketing. This chapter has illustrated the importance of going beyond just understanding gender roles and access to resources and to also understand gender power relations within households and communities since these determine whether women and men are able to equitably benefit from the products of research.

Non-pecuniary benefits of cultivating certain varieties of potato may influence adoption. Non-pecuniary benefits are those nonmonetary benefits related to social and ritual needs in communities where potato is part of the culture such as in the Andes. These types of benefits need to be understood as part of the value proposition to ensure that released improved varieties meet man and woman farmer's needs. Breeding programs should ensure that key important traits that may not have an immediate economic value but are important to women are not neglected. Farmers may continue to cultivate varieties that have lost some of their key attributes because these varieties may meet some needs which are not readily quantifiable. This means gender work in potato breeding should continue to contribute to knowledge on men and women user preferred traits that can be standardized and integrated into breeding programs to ensure that they continue to be responsive to farmers' needs. Failure to do so may jeopardize farmers' ability to benefit from genetic gains resulting from new improved varieties.

A supportive social and institutional environment is needed to promote adoption of improved varieties and other potato-related technologies by men and women farmers. For example, access to resources such as land, cash, and decision-making power influences whether men and women can use and adopt new technologies and methods. Having knowledge only may not result in adoption of technologies in the absence of a supportive institutional and social environment. For instance, if extension services are gender biased women will not be able to benefit from them. However, having access to information is itself not enough condition of benefitting from potato research. Unequal gender relations and other relations of inequality may prevent men and women from adopting technologies that can help them.

Commercialization and commoditization of seed runs the risk of dispossessing women of their control of seed and may also jeopardize their ability to access seed. Seed is often part of the social fabric where women may gain status or capital by distributing seed freely in their communities. Although noncash transactions such as women working for seed are often not valued by policy makers who promote commercialization and certification, they often ensure redistributive potential within communities where women and other poor and vulnerable groups are able to access seed. More research and investment may be needed to improve farmer-based seed to ensure circulation of better-quality planting material in communities. This also will be able to address the needs of women and other vulnerable groups from poor communities who prefer culturally recognized noncash transactions for accessing seed.

Certainly, more research on gender and potato still needs to be conducted because most of the research that has been conducted so far is at a relatively small scale. More large-scale research including surveys need to be conducted on a variety of topics to ensure that results are more generalizable and lead to the development of gender integration approaches suited to various regions and countries.

References

Amaya ARU, Alwang J (2012) Women rule: potato markets, cellular phones and access to information in the Bolivian highlands. Agric Econ 43:403–413

Ashby J, Polar V (2019) The implications of gender for modern approaches to crop improvement and plant breeding. In: Sachs C (ed) Gender, agriculture and agrarian transformation. Routledge, London

Aguayo EC, Hinrichs JS (2015) Curadoras de semillas: entre empoderamiento y esencialismo estratégico. Revista Estudos Feministas 23(2):347–370

Bentley JW, Andrade-Piedra JL, Demo P, Dzomeku B, Jacobsen K, Kikulwe E, Kromann P et al (2018) Understanding root, tuber, and banana seed systems and coordination breakdown: a multi-stakeholder framework. J Crop Improve. https://doi.org/10.1080/15427528.2018.1476998

Biemond CP, Stomph TJ, Kamaraa A, Abdoulaye T, Hearne S, Struik PC (2012) Are investments in an informal seed system for cowpea a worthwhile endeavour? Int J Plant Product 6(3):1735–80430

Bonabana-Wabbi J, Ayo S, Mugonola B, Taylor DB, Kirinya J, Tenywa M (2013) The performance of potato markets in South Western Uganda. J Dev Agric Econ 5(6):225–235

Brush SB (2004) Farmers' bounty: locating crop diversity in the contemporary world. Yale University Press, New Haven. https://www.jstor.org/stable/j.ctt1np9rd

Brush SB, Carney HJ, Humán Z (1981) Dynamics of Andean potato agriculture. Econ Bot 35(1):70–88. https://doi.org/10.1007/BF02859217

Carr ER (2008) Men's crops and women's crops: the importance of gender to the understanding of agricultural and development outcomes in Ghana's Central Region. World Dev 36(5):900–915

Chiipanthenga M, Maliro M, Demo P, Njoloma J (2012) Potential of aeroponics system in the production of quality potato (Solanum tuberosum l.) seed in developing countries. Afr J Biotechnol 11(17):3993–3999. http://www.academicjournals.org/AJB

Christinck A, Eva Q, Fred R, Ashby JA (2017) Gender Differentiation of Farmer Preferences for Varietal Traits in Crop Improvement: Evidence and Issues. Working Paper. https://cgspace.cgiar.org/handle/10947/4660

De Haan, S. (2009). Potato diversity at height. Multiple dimensions of farmer-driven in situ conservation in the Andes. Thesis. Wageningen University, Wageningen.

De Vries J, Toenniessen G (2001) Securing the harvest: biotechnology, breeding and seed systems for African crops. The Rockefeller Foundation, CABI Publishing, New York

Demo P, Mwenye OJ, Pankomera P, Chiipanthega M (2008) Investigation of Appropriate Fertilizer Doses for Potato Production Using Different Planting Spacing in Major Growing Areas of Malawi. http://www.cabi.org/gara/FullTextPDF/2008/20083323853.pdf

Demo P, Lemaga B, Kakuhenzire R, Schulz S, Borus D, Barker I, Woldegiorgis G, Parker M, Schulte-Geldermann E (2015) 1 Strategies to Improve Seed Potato Quality and Supply in Sub-Saharan Africa: Experience from Interventions in Five Countries

Dersseh WM, Gebresilase YT, Schulte RPO, Struik PC (2016) The analysis of potato farming systems in Chencha, Ethiopia: input, output and constraints. Am J Potato Res 93:436–447

Erbaugh JM, Donnermeyer J, Amujal M, Kyamanywa S (2003) The Role of Women in Pest Management Decision Making in Eastern Uganda 10(3) Journal of International Agricultural and Extension Education. http://citeseerx.ist.psu.edu/viewdoc/download?doi=10.1.1.577.5951&rep=rep1&type=pdf

FAO (2008) The state of food and agricultures: biofuels, prospects risks and opportunities. FAO, Rome. ftp://ftp.fao.org/docrep/fao/011/i0100e/i0100e.pdf

German L, Taye H (2008) A framework for evaluating effectiveness and inclusiveness of collective action in watershed management. J Int Dev 20(1):99–116

Gibson RW, Mwanga ROM, Namanda S, Jeremiah SC, Barker I (2009) Review of sweetpotato seed systems in East and Southern Africa. Integrated crop management working paper 2009-1. CIP, Lima, p 48. http://cipotato.org/wp-content/uploads/2014/08/004730.pdf

Gildemacher P, Demo P, Kinyae P, Nyongesa M, Mundia P (2007) Selecting the best plants to improve seed potato. LEISA Mag 23(2):10–11

Gildemacher PR, Demo P, Barker I, Kaguongo W, Woldegiorgis G, Wagoire WW, Wakahiu M, Leeuwis C, Struik PC (2009) A description of seed potato systems in Kenya, Uganda and Ethiopia. Am J Potato Res 86:373–382

Gildemacher P, Schulte-Geldermann E, Borus D, Demo P, Kinyae P, Mundia P, Struik P (2011) Seed potato quality improvement through positive selection by smallholder farmers in Kenya. Potato Res 54:253–266

GRAIN (2000) Potato: a fragile gift from the Andes GRAIN|15 September 2000|Seedling-September 2000

Horrell S, Krishnan P (2007) Poverty and productivity in female headed households in Zimbabwe, Journal of Development Studies, Taylor & Francis Journals, 43(8):1351–1380

Hoque AM, Sultana MS (2012) Disease free seed potato production through seed plot technique at farmers' level in Bangladesh. J Plant Protect Sci 4(2):51–56

Jibat GA, Belisa M, Gudeta H (2007) Promotion of participatory technology in potato farming–Ethiopia. Chapter 2. In: Flintan F, Tedla S (eds) Natural resource management: the impact of gender and social issues. OSSEREA & IDRC, Addis Ababa, pp 19–54

Jones AD, Creed-Kanashiro H, Zimmerer KS, de Haan S, Carrasco M, Meza K, Cruz-Garcia GS et al (2018) Farm-level agricultural biodiversity in the Peruvian Andes is associated with greater odds of women achieving a minimally diverse and micronutrient adequate diet. J Nutr 148(10):1625–1637. https://doi.org/10.1093/jn/nxy166

Kabeer N, Subrahmanian R (1996) Institutions, relations and outcomes. Framework and tools for gender aware planning. IDS, Brighton. https://www.ids.ac.uk/files/Dp357.pdf

Kingiri A (2010) Gender and agricultural innovation revisiting the debate through an innovation systems perspective. Discussion Paper 06. Research Into Use (RIU), Department for International Development (DFID), UK

Kolech SA, Halseth D, Perry K, De Jong W, Tiruneh MF, Wolfe D (2015) Identification of farmer priorities in potato production through participatory variety selection. Am J Potato Res 92:648–661

Kolech SA, De Jong W, Perry K, Halseth D, Tirineh MF (2017) Participatory variety selection: a tool to understand farmers' potato variety selection criteria. Open Agric 2:453–463

Kromann P, Montesdeoca F, Andrade-Piedra J (2016) Integrating formal and informal. In: Andrade-Piedra J, Bentley J, Almekinders C, Jacobsen K, Walsh S, Thiele G (eds) Case studies of roots, tubers and bananas seed systems. CGIAR Research Program on Roots Tubers and Bananas (RTB), Lima, pp 14–32. RTB working paper no. 2016-3. ISSN 2309-6586

Laub R, Muir G (2008) Potato and gender-international year of the potato 2008. FAO factsheets_Gender. Hidden Treasure (blog). http://www.fao.org/potato-2008/en/potato/gender.html

Little J, Panelli R (2003) Gender research in rural geography. Gend Place Cult 10(3):281–289. https://doi.org/10.1080/0966369032000114046

Lutaladio N, Ortiz O, Haverkort A, Caldiz D (2009) Sustainable potato production; guidelines for developing countries. FAO, Rome

Machida L, Derera J, Tongoona P, Langyintuo A, Mac Robert J (2014) Exploration of farmers' preferences and perceptions of maize varieties: implications on development and adoption of Quality Protein Maize (QPM) varieties in Zimbabwe. J Sustain Dev 7(2):194–207

Malena C (1994) Gender issues in integrated pest management in African agriculture. NRI Socio-economic Series 5. Chatham, United Kingdom: Natural Resources Institute. http://www.nzdl.org/gsdlmod?e=d-00000-00---off-0hdl--00-0----0-10-0---0---0direct-10---4-------0-1l--11-en-50---20-about---00-0-1-00-0--4----0-0-11-10-0utfZz-8-00&a=d&cl=CL1.7&d=HASH01c963ec65781a15133c31bf.2

March C, Smyth I, Mukhopadhyay M (1999) a guide to gender analysis framework. Oxfam, GB, Oxford

Meinzen-Dick R, Quisumbing A, Behrman J, Biermayr-Jenzano P, Wilde V, Noordeloos M, Ragasa C, Beintema N (2010) Engendering Agricultural Research, IFPRI discussion paper 00973

Mera-Orcés V (2001) Paying for survival with health: potato production practices, pesticide use and gender concerns in the Ecuadorian highlands. J Agric Educ Ext 8(1):31–40. https://doi.org/10.1080/13892240185300061

Mudege NN, Demo P (2016) Seed potato in Malawi: not enough to go around. Chapter 10. In: Andrade-Piedra J, Bentley J, Almekinders C, Jacobsen K, Walsh S, Thiele G (eds) Case studies of roots, tubers and bananas seed systems. CGIAR Research Program on Roots, Tubers and Bananas (RTB), Lima, pp 146–163. RTB working paper no. 2016-3. ISSN 2309-6586

Mudege NN, Tafadzwa C, Ted N, Eliya K, Paul D (2015a) Gender norms and access to extension services and training among potato farmers in Dedza and Ntcheu in Malawi. J Agric Educ Ext

Compet Rural Innov Transform 22(3):291–305. http://www.tandfonline.com/doi/abs/10.1080/1389224X.2015.1038282

Mudege NN, Kapalasa E, Chevo T, Nyekanyeka T, Demo P (2015b) Gender norms and the marketing of seeds and ware potatoes in Malawi. J Gender Agric Food Secur 1(2):18–41. http://www.agrigender.net/views/marketing-of-seeds-and-ware-potatoes-in-Malawi-JGAFS-122015-2.php

Mudege NN, Mayanja S, Naziri D (2016) Technical Report: Gender situational analysis of the potato value chain in Eastern Uganda and strategies for gender equity in postharvest innovations. CRP RTB. 53 p

Mudege NN, Biazin BT, Brouwer R (forthcoming) Gender situational analysis of the sweetpotato value chain in selected districts in Sidama and Gedeo Zones, Southern Ethiopia. International Potato Center, Lima, Peru

Muhinyuza JB, Shimelis H, Melis R, Sibiya J, Nzaramba MN (2012) Participatory assessment of potato production constraints and trait preferences in potato cultivar development in Rwanda. Int J Dev Sustain 1(2):358–380

Njenga M, Gurung J (2011) Enhancing gender responsiveness in putting nitrogen to work for smallholder farmers in Africa (n2africa). Women Organising for Change in Agriculture and NRM

Njukeng AP, Chewaching MG, Chofong G, Demo P, Sakwe P, Njualem D (2007) Determination of virus-free potato planting materials by positive selection and screening of tuners from seed stores in the Western Highlands of Cameroon. Afr Crop Sci Conf Proc 8:809–815

Ogunlana EA (2004) The technology adoption behavior of women farmers: The case of alley farming in Nigeria. Renewable Agriculture and Food Systems 19(1):57–65

Okonya JS, Kroschel J (2015) A Cross-Sectional Study of Pesticide Use and Knowledge of Smallholder Potato Farmers in Uganda Biomed Res Int

Orrego R, Andrade-Pedra J (2016) Aeroponic seed and native potatoes in Peru. In: Andrade-Piedra J, Bentley J, Almekinders C, Jacobsen K, Walsh S, Thiele G (eds) Case studies of roots, tubers and bananas seed systems. CGIAR research program on Roots, Tubers and Bananas (RTB), Lima, pp 33–46. RTB working paper no. 2016-3. ISSN 2309-6586

Paulson S (2003) Gendered practices and landscapes in the Andes: the shape of asymmetrical exchanges. Hum Organ 62(3):242–254

Petesch P, Bullock R, Feldman S, Badstue L, Rietveld A, Kamanzi A, Tegbaru A, Yila J (2018) Local normative climate shaping agency and agricultural livelihoods in Sub-Saharan Africa. J Gender Agric Food Secur 3(1):23

Polar V, Babini C, Velasco P, Fonseca C (2017) La technología no es neutral: Factores que influyen en la potencial adopción de technología agrícola por hombres y mujeres. International Potato Center, Lima

Quisumbing AR, Pandolfelli L (2010) Promising approaches to address the needs of poor female farmers: resources. Constrain Interv World Dev 38(4):581–592

Quisumbing AR, Meinzen-Dick R, Raney TL, Croppenstedt A, Behrman JA, Peterman A (2014) Gender in agriculture: closing the knowledge gap. Springer, New York

Reddy RC, Tonapi VA, Bezkorowajnyj PG, Navi SS, Seetharama N (2007) Seed system innovations in the semi-arid tropics of Andhra Pradesh. Research report. International Crop Research Institute for Semi Arid Tropics, International Livestock Research Institute (ILRI), ICRISAT, Patancheru, 224 pp. isbn:978-92-9066-502-1

RTB (2016) RTB proposal 2017-2022, vol 1. RTB, Volucella

Sah U, Kumar S, Pandey NK (2007) Gender analysis of potato cultivation in Meghalaya. Potato J 34(3–4):235–238

Salazar L (1996) Potato viruses and their control. International Potato Center (CIP), Lima

Sarapura S (2013). Gender and agricultural innovation in peasant production of native potatoes in the Central Andes of Peru. PhD Thesis, University of Guelph, Canada. p. 351.

Sarapura ES, Hambly-Odame H, Thiele G (2016) Gender and innovation in Peru's native potato market chains. In: Transforming gender and food systems in the Global South. IDRC, Taylor and Francis, Canada

Sebatta C, Mugisha J, Katungi E, Kashaaru A, Kyomugisha H (2014) Smallholder farmers' decision and level of participation in the potato market in Uganda. Mod Econ 5:895–906

Sharma K, Shawkat B, Miethbauer T, Schulte-Geldermann E (2017) Strategies for bacterial wilt (Ralstonia solanacearum) management in potato field: Farmers' guide. International Potato Center. Lima (Peru). 2 p

Sperling L, McGuire S (2012) Fatal gaps in seed security strategy. Food Security 4(4):569–579

Tantowijoyo W, van de Fliert E (2006) In: Widagdo H, Ketelaar JW (eds) All about potatoes. An ecological guide to potato integrated crop management. FAO Regional Office for Asia and the Pacific, Bangkok. https://research.cip.cgiar.org/typo3/web/fileadmin/icmtoolbox/ICM_Toolbox/Integrated_crop_management/All_about_potatoes_-_complete_EN_0602.pdf

Tapia ME, de la Torre A (1998) Women farmers and Andean seeds. Gender and genetic resource management, FAO, Lima. http://www.bioversityinternational.org/uploads/tx_news/Women_farmers_and_Andean_seeds_308.pdf

Tapia ME, de la Torre A (2000) La mujer campesina y las semillas andinas. FAO and IPGRI, Lima

Thiele G (1999) Informal potato seed systems in the Andes: why are they important and what should we do with them? World Dev 21(1):83–99

Tufan HA, Stefania G, Catherine M (2018) State of the Knowledge for Gender in Breeding: Case Studies for Practitioners. Working Paper, https://cgspace.cgiar.org/handle/10568/92819

Zimmerer KS (2003) Geographies of seed networks for food plants (potato, ulluco) and approaches to agrobiodiversity conservation in the Andean countries. Soc Nat Resour 16(7):583–601. https://doi.org/10.1080/08941920309185

Index

H. Campos, O. Ortiz (eds.), *The Potato Crop*,
https://doi.org/10.1007/978-3-030-28683-5

J

K

L

T

www.ingramcontent.com/pod-product-compliance
Lightning Source LLC
LaVergne TN
LVHW062156200926
848484LV00007B/1921